Direct and Large-Eddy Simulation VIII

ERCOFTAC SERIES

VOLUME 15

Aims and Scope of the Series

ERCOFTAC (European Research Community on Flow, Turbulence and Combustion) was founded as an international association with scientific objectives in 1988. ERCOFTAC strongly promotes joint efforts of European research institutes and industries that are active in the field of flow, turbulence and combustion, in order to enhance the exchange of technical and scientific information on fundamental and applied research and design. Each year, ERCOFTAC organizes several meetings in the form of workshops, conferences and summerschools, where ERCOFTAC members and other researchers meet and exchange information.

The ERCOFTAC Series will publish the proceedings of ERCOFTAC meetings, which cover all aspects of fluid mechanics. The series will comprise proceedings of conferences and workshops, and of textbooks presenting the material taught at summerschools.

The series covers the entire domain of fluid mechanics, which includes physical modelling, computational fluid dynamics including grid generation and turbulence modelling, measuring-techniques, flow visualization as applied to industrial flows, aerodynamics, combustion, geophysical and environmental flows, hydraulics, multiphase flows, non-Newtonian flows, astrophysical flows, laminar, turbulent and transitional flows.

For further volumes:
www.springer.com/series/5934

Hans Kuerten · Bernard Geurts ·
Vincenzo Armenio · Jochen Fröhlich
Editors

Direct and Large-Eddy Simulation VIII

Editors
Prof. Hans Kuerten
Eindhoven University of Technology
Department of Mechanical Engineering
PO Box 513
5600 MB Eindhoven
The Netherlands
j.g.m.kuerten@tue.nl

Prof. Vincenzo Armenio
Università di Trieste
Dipto. Ingegneria Civile e Ambientale
Piazzale Europa 1
34127 Trieste
Italy
armenio@dica.units.it

Prof. Bernard Geurts
University of Twente
Faculty EEMCS
PO Box 217
7500 AE Enschede
The Netherlands
b.j.geurts@utwente.nl

Prof. Dr.-Ing. Jochen Fröhlich
Technical University of Dresden
Institute of Fluid Mechanics
George-Bähr-Str. 3c
01062 Dresden
Germany
Jochen.Froehlich@tu-dresden.de

ISSN 1382-4309 ERCOFTAC Series
ISBN 978-94-007-2481-5 e-ISBN 978-94-007-2482-2
DOI 10.1007/978-94-007-2482-2
Springer Dordrecht Heidelberg London New York

Library of Congress Control Number: 2011939572

Cover design: VTeX UAB, Lithuania

Printed on acid-free paper

Springer is part of Springer Science+Business Media (www.springer.com)

Preface

Simulation of turbulent flow by means of direct numerical simulation (DNS) and large-eddy simulation (LES) started almost fifty years ago. Probably the earliest paper on the application of LES was by Smagorinksy in the March 1963 issue of Monthly Weather Review. Although Smagorinsky did not mention the term large-eddy simulation explicitly, he proposed a model to represent the effects of small-scale eddies on the large-scale dynamics of the flow, which was treated explicitly. Smagorinsky applied his now famous model to the simulation and study of the dynamics of the general circulation in the earth's atmosphere.

Direct numerical simulation of wall-bounded flows started some twenty years later with the well-known 1987 paper in the Journal of Fluid Mechanics by Kim, Moin and Moser on DNS of turbulent channel flow at a bulk Reynolds number of 3300. Although DNS had been applied before on homogeneous, isotropic turbulence and some preliminary studies on under-resolved channel flow had been performed, this paper presented the first fully resolved DNS of a wall-bounded turbulent flow. The large number of citations reported on Web of Science proves the tremendous impact both papers had and still have on research in turbulence.

The continuing growth of computational power has increasingly stimulated the usage of DNS and LES, since LES and in some applications even DNS can now be used as a design tool for several practical and industrial problems. This is reflected by the possibility of CFD software packages to perform LES, although this should still be treated with care. On the other hand, for flow in simple geometries, such as channel flow, DNS has been extended to higher and higher Reynolds numbers, which brings the study of fundamental properties of turbulent flow at large Reynolds numbers within reach. These examples highlight the two major reasons for usage of DNS and LES: application-driven research and fundamental research into the nature of turbulence and into turbulence models.

The history of this research over the past two decades can well be grasped from the contents of the ERCOFTAC series of Workshops on Direct and Large-Eddy Simulation. This series started in 1994 and, with intervals of approximately two years, has led to the eighth DLES workshop organized at Eindhoven University of Technology in July 2010. Like the previous editions, this workshop has been formatted around approximately ten invited contributions in different areas of DNS and LES, ranging from fundamental properties to industrial applications and treating various application areas, such as two-phase flow, environmental flow and combustion. Around 70 of the submitted abstracts have been selected for oral presentation during the workshop.

Most of the invited and contributed papers have been submitted to be included in the Proceedings of DLES8 and after a careful review procedure most of them can be found in this volume. The papers are grouped in themes, with slight reordering compared to the program of the workshop. The contributions provide a broad overview of the most important current issues and application areas ofDNS and LES. Fundamental issues related to the usage of LES and the development of subgrid models are still an important research topic. Contributions to this topic can be found in the first two parts of the Proceedings on fundamentals and on methodologies and modeling techniques. These two parts also contain contributions on fundamental studies of turbulent flow and on numerical issues, such as novel numerical techniques.

During the workshop two special sessions were held. One centered around Lagrangian turbulence and had been planned long before the start of the workshop. The other was a result of the submitted abstracts. It appeared that the number of abstracts on Rayleigh-Bénard flow justified a special session devoted to this topic. The contributions in the session on Lagrangian turbulence were regrouped in Part III on multiphase flow, together with more general contributions on two-phase flow. The contributions on Rayleigh-Bénard flow were allocated a separate part in the Proceedings. The remaining three parts of the Proceedings are devoted to the application areas environmental flows, compressible and reactive flows and industrial applications. Each of these application areas was discussed by an invited presentation as well.

The organization of the ERCOFTAC DLES VIII Workshop and the preparation of this Proceedings would not have been possible without the support of many. We gratefully acknowledge financial support from Eindhoven University of Technology, TU/e, the Royal Netherlands Academy of Sciences, KNAW, Universiteitsfonds Eindhoven, UFe, the Netherlands Research School on Fluid Mechanics, J.M. Burgerscentrum and the Netherlands Organisation for Scientific Research, NWO. The European Research Community on Flow, Turbulence and Combustion, ERCOFTAC, enabled the attendance and contribution of many young scientists to DLES8 by making available scholarships to 39 PhD students. We also thank the members of the Scientific Committee of DLES8 for their contribution to the reviewing process of the papers.

Eindhoven,
March, 2011

Hans Kuerten
Bernard Geurts
Vincenzo Armenio
Jochen Fröhlich

Contents

Part I
Fundamentals

Direct simulations of wall-bounded turbulence

Javier Jiménez and Ricardo García-Mayoral

1 Introduction

Direct simulations have become indispensable tools in turbulence research, and, in the last two decades, especially so for the study of wall-bounded turbulence. Early simulations dealt mostly with the viscous and buffer layers near the wall, because their Reynolds numbers had to be necessarily limited [14]. They led very soon to a fairly complete description of the kinematics of this part of the flow [17], and, later, to the qualitative understanding of their dynamics [12]. While most of those studies were carried out in turbulent channels, the results are generally believed to apply to all attached wall-bounded turbulent flows, because the time scales of the near-wall region are too fast to interact strongly with the slower processes of the non-universal outer layers.

In the last few years, simulations have moved beyond those 'universal' aspects of wall-bounded turbulence into the properties of specific flows, because we are now able to compute Reynolds numbers that are high enough for the outer layers, which are typically the seat of the differences among different flows, to be relatively uncontaminated from low-Reynolds-number effects. Those newer simulations include boundary layers [18, 19, 11] and pipes [21], but they are beginning to extend to more complicated, although still canonical, flows.

Another trend that has been stimulated by the availability of faster and larger computers is the introduction of more complex physics in simpler flows, such as roughness, rotation, or MHD.

Javier Jiménez
School of Aeronautics, Universidad Politécnica, 28040 Madrid, Spain, e-mail: jimenez@torroja.dmt.upm.es

Ricardo García-Mayoral
School of Aeronautics, Universidad Politécnica, 28040 Madrid, Spain, e-mail: ricardo@torroja.dmt.upm.es

H. Kuerten et al. (eds.), *Direct and Large-Eddy Simulation VIII*,
ERCOFTAC Series 15, DOI 10.1007/978-94-007-2482-2_1,

We briefly discuss below examples of each of those two kinds of simulations, one canonical and the other one complex. In the first one, we will show how the differences between zero-pressure-gradient boundary layers and equilibrium channels can be traced to the effect of intermittency in the outer layers. In the second one, we will see that the breakdown of the viscous regime of drag reduction over riblets is due to a hydrodynamic instability that develops just above the wall when the Reynolds number of the groove section is large enough.

2 The turbulent boundary layer

Turbulent boundary layers are harder to compute than channels, mainly because the inflow, outflow, and free-stream boundary conditions make the numerics more involved, and complicate the use of standard spectral methods. Another important reason is that they are not equilibrium flows, and depend on the details of the inflow perturbations. This is roughly equivalent to the 'tripping' used in laboratory experiments, and makes boundary layers less universal than channels [19]. While channel flow depends essentially only on the Reynolds number, boundary layers have more parameters, ranging from their receptivity to external perturbations to the details of the trip [4]. Even so, with due care, and if the Reynolds number is large enough, boundary layers can be correctly simulated and compared with channels, and some differences are found. Basically, the velocity fluctuations are stronger in boundary layers than in channels (figure 1), and the spanwise spectra of the large-scale streamwise velocity are somewhat narrower. While those effects were known from laboratory experiments [10, 2], it was difficult to ascertain their causes from the available information. On the other hand, the more detailed data sets that can be obtained from simulations can be used, for example, to show that the differences in the intensities are due to the intermittency between irrotational and turbulent flow in the outer edge of the boundary layer. Essentially, when fast irrotational fluid is engulfed by turbulence, it can only be decelerated by pressure, which is less effective than the Reynolds stresses of the turbulent regions. The first consequence is the appearance of a 'wake' component in the mean velocity profile, which is stronger in boundary layers because of the presence of that faster fluid. The stronger velocity gradient in the wake creates extra energy production, and strong pressure pulses that redistribute the energy to the transverse components, and tend to isotropize turbulence in the intermittent region [11].

The analysis of the simulation results allows us to trace in detail the different conditional means of the irrotational and rotational fluids, the effect of the pressure, and the nearly-isotropic nature of the additional fluctuations (figure 2).

A similar analysis can be used to explain the reasons for the different sizes of the largest structures in the two cases. What really happens is not that the channel spectra are wider that those of boundary layers, but that they are longer [11]. When observed at similar streamwise wavelengths, the spanwise scales of boundary layers and channels are identical, but channels extend to longer wavelengths, and those

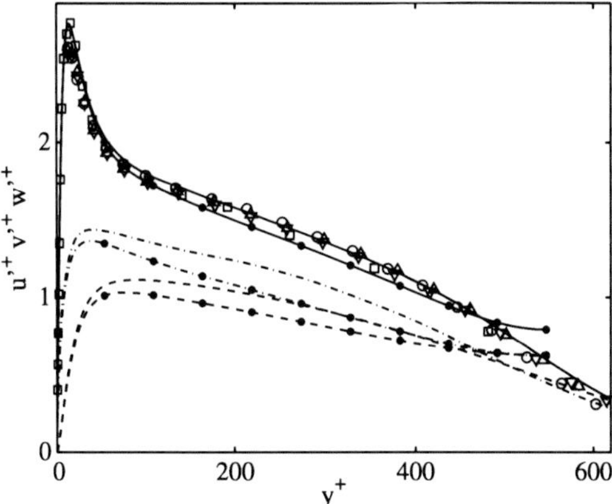

Fig. 1 Intensities of the velocity fluctuations. The lines with heavy dots are a channel, and those without are a boundary layer; solid: streamwise; dashed: wall-normal; chain-dot: spanwise. The friction Reynolds number is 550. Open symbols are experimental boundary layers at approximately the same Reynolds number [11].

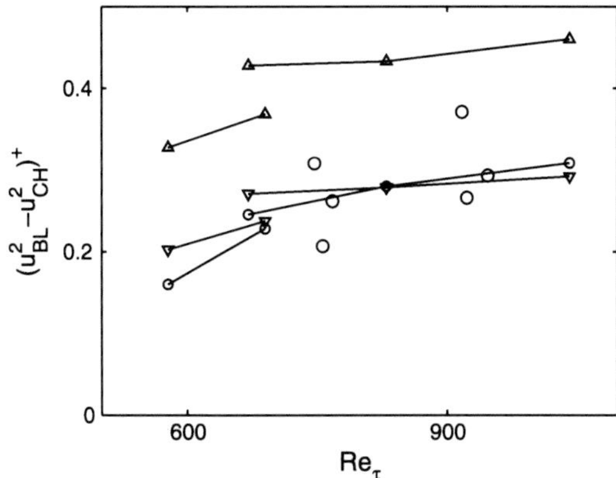

Fig. 2 Maximum difference between the fluctuation energy of boundary layers with respect to channels, as a function of the friction Reynolds number. Lines are simulations; isolated symbols are experiments; circles: streamwise; △: wall-normal; ▽: spanwise.

long structures, which are missing from the boundary layers, are also wider because they branch and meander [9, 8]. For example, it can be shown that, while the longer streaks of the streamwise velocity in the buffer layer look wider than the shorter ones when measured in terms of their bounding boxes, the widths implied by their surface areas are the same in both cases [9]. Similar results were obtained in the logarithmic layer by [8], although using different methods. The reason for why the structures in the boundary layer are shorter than those in channels is unclear, and requires further analysis of the data, but the scales involved are consistent with the length scales of the growth of the boundary layer thickness. Essentially, the structures stop getting longer when they run into parts of the boundary layer which are appreciable thicker than the locations at which they have originated [11].

3 The breakdown of viscous flow over riblets

The second example deals with the dynamics of skin-friction reduction by riblets. It is well known that longitudinal riblets decrease the friction drag of turbulent boundary layers over flat walls, and there is a reasonable consensus that the mechanisms involved in the limit of very small riblets is that the anisotropic surface inhibits the transverse fluctuations with respect to the longitudinal ones, thus interfering with the natural friction-generation processes of the buffer layer. The result is an offset between the virtual origins of the spanwise and streamwise flows. The drag reduction is proportional to that offset, and therefore to the riblet size. This would seem to imply that larger riblets should be better, but their efficiency degrades for spanwise spacings larger than approximately 10–15 wall units, beyond which the drag reduction becomes an increase. The reason for this break-down has been under discussion for a long time, and limits the maximum drag reduction that can be obtained from a given riblet design.

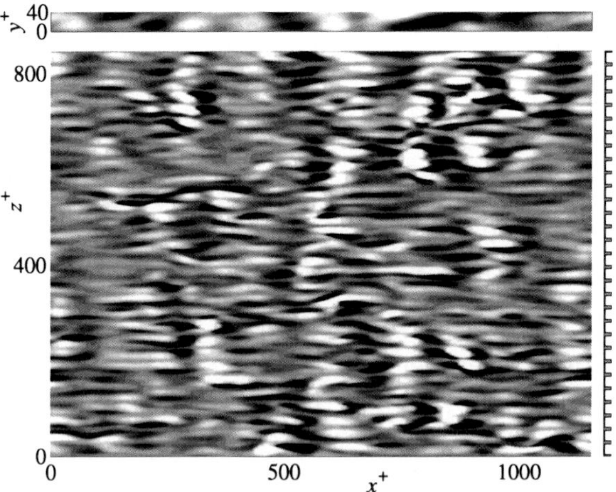

Fig. 3 Top view of the wall-normal velocity in the neighborhood of the wall, for a ribbed surface. The riblets cross section is shown on the right, with a spacing $s^+ = 29$. The flow is from left to right, and light colours are positive. In the top of the figure there is a side view of the spanwise-averaged stream function below $y^+ = 20$, with light colours denoting clockwise rotation. Note that, since the side view has been averaged over the whole span, the vortices visible in it are coherent across the box. Dimensions are in wall units [6].

Several theoretical proposals have been made to explain that performance degradation, mostly centring on the changes of the cross-flow above the larger riblets.

They fall into two broad groups. The first one is that the riblets lose effectiveness once they become so large that the Stokes approximation ceases to apply to the cross-flow. For example, [7] suggested that the deterioration was due to the generation of secondary streamwise vorticity over the riblets, because the unsteady cross-flow separates and sheds small-scale vortices that create extra dissipation. The second group of theories assumes that the observed optimum spacing, $s^+ \approx 10-20$, is related to the scale of the turbulent structures in the wall region, such as in the observations by [20, 15, 3] that the increase in drag coincides with the lodging of the quasi-streamwise vortices within the riblet grooves. However, all those observations referred to spacing of 30–40 wall units, well past the optimum size.

Numerical simulations in [6] suggests a third alternative that, somewhat surprisingly, involves the longitudinal flow, rather than the transverse one. It is observed that the onset of the drag increase coincides with the appearance of a hydrodynamic instability that takes the form of spanwise rollers (figure 3), whose extra Reynolds stresses account quantitatively for the drag increase. Those rollers are essentially the same ones that form over vegetable canopies [16, 5], and over permeable [13] and porous [1] surfaces, and can be related to the Kelvin-Helmholtz instability of the approximate inflection point that the mean velocity profile has at the wall, once the impermeability condition is relaxed by the possibility of flow along the riblet grooves [6].

4 Conclusions

We have discussed two examples in which numerical simulations were used to clarify aspects of the flow physics that had been known for some time, but whose dynamics could not be easily untangled from the previously available data.

Beyond the interest of those two particular cases, a wider point can be made that the most profound influence of direct simulations on the study of turbulence has been to make large amounts of data freely available for analysis. The two effects discussed above had been well known for a long time, and theoretical models had been proposed for them. The difference is now that the data available from simulations allows us to test any proposed model in detail, and that free speculation is much harder than it used to be. Because of their essentially complete information about the flow field, simulations also provide the possibility of serendipitous discovery, as was the case for the riblet instability. The most important transformation induced by simulations on turbulence research has been to change the emphasis from the accumulation of new data to the analysis of existing, or easily obtained, ones.

This work has been supported in part by the CICYT grant TRA2009-11498, and by the sixth framework AVERT program of the European Commission, AST5-CT-2006-030914. Ricardo García-Mayoral was supported by an FPI fellowship from the Spanish Ministry of Education and Science.

References

1. Breugem, W.P., Boersma, B.J., Uittenbogaard, R.E.: The influence of wall permeability on turbulent channel flow. J. Fluid Mech. **562**, 35–72 (2006)
2. Buschmann, M.H., Gad-el-Hakl, M.: Normal and cross-flow Reynolds stresses: differences between confined and semi-confined flows. Exp. in Fluids **49**, 213–223 (2010)
3. Choi, H., Moin, P., Kim, J.: Direct numerical simulation of turbulent flow over riblets. J. Fluid Mech. **255**, 503–539 (1993)
4. Erm, L.P., Joubert, P.N.: Low-Reynolds-number turbulent boundary layers. J. Fluid Mech. **230**, 1–44 (1991)
5. Finnigan, J.: Turbulence in plant canopies. Ann. Rev. Fluid Mech. **32**, 519–571 (2000)
6. García-Mayoral, R., Jiménez, J.: Hydrodynamic stability and the breakdown of the viscous regime over riblets. J. Fluid Mech. (Submitted)
7. Goldstein, D.B., Tuan, T.C.: Secondary flow induced by riblets. J. Fluid Mech. **363**, 115–151 (1998)
8. Hutchins, N., Marusic, I.: Evidence of very long meandering features in the logarithmic region of turbulent boundary layers. J. Fluid Mech. **579**, 467–477 (2007)
9. Jiménez, J., del Álamo, J.C., Flores, O.: The large-scale dynamics of near-wall turbulence. J. Fluid Mech. **505**, 179–199 (2004)
10. Jiménez, J., Hoyas, S.: Turbulent fluctuations above the buffer layer of wall-bounded flows. J. Fluid Mech. **611**, 215–236 (2008)
11. Jiménez, J., Hoyas, S., Simens, M.P., Mizuno, Y.: Turbulent boundary layers and channels at moderate Reynolds numbers. J. Fluid Mech. (2010). DOI 10.1017/S0022112010001370
12. Jiménez, J., Kawahara, G., Simens, M.P., Nagata, M., Shiba, M.: Characterization of near-wall turbulence in terms of equilibrium and 'bursting' solutions. Phys. Fluids **17**, 015105 (2005)
13. Jiménez, J., Uhlman, M., Pinelli, A., Kawahara, G.: Turbulent shear flow over active and passive porous surfaces. J. Fluid Mech. **442**, 89–117 (2001)
14. Kim, J., Moin, P., Moser, R.D.: Turbulence statistics in fully developed channel flow at low Reynolds number. J. Fluid Mech. **177**, 133–166 (1987)
15. Lee, S.J., Lee, S.H.: Flow field analysis of a turbulent boundary layer over a riblet surface. Experiments in Fluids **30**, 153–166 (2001)
16. Raupach, M.R., Finnigan, J., Brunet, Y.: Coherent eddies and turbulence in vegetation canopies: the mixing-layer analogy. Boundary-Layer Meteorology **78**, 351–382 (1996)
17. Robinson, S.K.: Coherent motions in the turbulent boundary layer. Ann. Rev. Fluid Mech. **23**, 601–639 (1991)
18. Schlatter, P., Örlü, R., Li, Q., Fransson, J., Johansson, A., Alfredsson, P.H., Henningson, D.S.: Turbulent boundary layers up to $Re_\theta = 2500$ through simulation and experiments. Phys. Fluids **21**, 051702 (2009)
19. Simens, M., Jiménez, J., Hoyas, S., Mizuno, Y.: A high-resolution code for turbulent boundary layers. J. Comput. Phys. **228**, 4218–4231 (2009)
20. Suzuki, Y., Kasagi, N.: Turbulent drag reduction mechanism above a riblet surface. AIAA J. **32**(9), 1781–1790 (1994)
21. Wu, X., Moin, P.: A direct numerical simulation study on the mean velocity characteristics in turbulent pipe flow. J. Fluid Mech. **608**, 81–112 (2008)

Structure of a turbulent boundary layer studied by DNS

Philipp Schlatter, Qiang Li, Geert Brethouwer,
Arne V. Johansson, Dan S. Henningson

1 Introduction and numerical method

Turbulent boundary layers constitute one of the basic building blocks for understanding turbulence, particularly relevant for industrial applications. Although the geometries in technical but also geophysical applications are complicated and usually feature curved surfaces, the flow case of a canonical boundary layer developing on a flat surface has emerged as an important setup for studying wall turbulence, both via experimental and numerical studies. However, only recently spatially developing turbulent boundary layers have become accessible via direct numerical simulations (DNS). The difficulties of such setups are mainly related to the specification of proper inflow conditions, the triggering of turbulence and a careful control of the free-stream pressure gradient. In addition, the numerical cost of such spatial simulations is high due to the long, wide and high domains necessary for the full development of all relevant turbulent scales. We consider a canonical turbulent boundary layer under zero-pressure-gradient via large-scale DNS. The boundary layer is allowed to develop and grow in space. The inflow is a laminar Blasius boundary layer, in which laminar-turbulent transition is triggered by a random volume force shortly downstream of the inflow. This trip force, similar to a disturbance strip in an experiment [7, 8], is located at a low Reynolds number to allow the flow to develop over a long distance. The simulation covers thus a long, wide and high domain starting at $Re_\theta = 180$ extending up to the (numerically high) value of $Re_\theta = 4300$, based on momentum thickness θ and free-stream velocity U_∞. Fully turbulent flow is obtained from $Re_\theta \approx 500$. The numerical resolution for the fully spectral numerical method [2] is in the wall-parallel directions $\Delta x^+ = 9$ and $\Delta z^+ = 4$, resolving the relevant scales of motion. The simulation domain requires a total of $8 \cdot 10^9$ grid points in physical space, and was thus run massively parallel with 4096 processors.

Philipp Schlatter *et al.*,
Linné FLOW Centre, KTH Mechanics, Stockholm, Sweden, and
Swedish e-Science Reserach Centre (SeRC), e-mail: `pschlatt@mech.kth.se`

H. Kuerten et al. (eds.), *Direct and Large-Eddy Simulation VIII*,
ERCOFTAC Series 15, DOI 10.1007/978-94-007-2482-2_2,

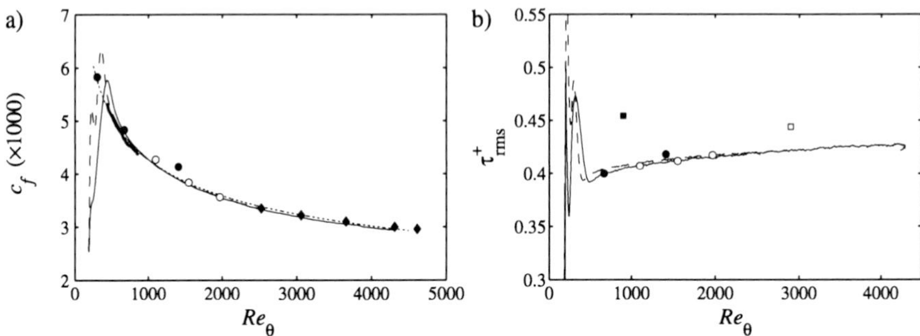

Fig. 1 Skin-friction coefficient c_f and wall-shear-stress fluctuations τ^+_{rms}; solid: present DNS [7], dashed: DNS Schlatter *et al.* [8], dotted: $c_f = 0.024 Re_\theta^{-0.25}$, thick solid: DNS Li *et al.* [5]. • DNS Spalart [10], ∘: DNS Simens *et al.* [9], □: DNS Ferrante and Elghobashi [3], ■: DNS Wu and Moin [11], and ♦: experiments Österlund [6].

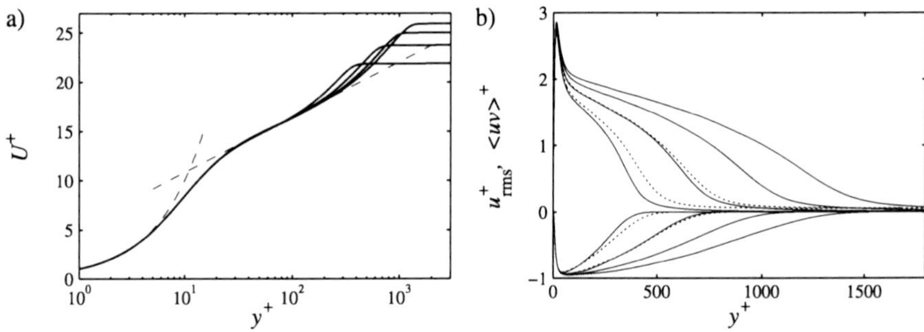

Fig. 2 Mean velocity profile U^+, streamwise fluctuations u^+_{rms} and shear stress $\langle uv \rangle^+$ at four $Re_\theta = 1100, 2000, 3000$ and 4100; solid: present DNS, dotted: DNS by Simens *et al.* [9]. dashed: log law with $\kappa = 0.41$ and $B = 5.2$.

2 Results

2.1 Turbulent statistics and streamwise development

Turbulence statistics obtained in the boundary layer such as mean profiles, fluctuations, two-point correlations *etc.* of the flow are in good agreement with other simulations, see also Refs. [7, 8], and experimental studies at similar Reynolds numbers [6]. To illustrate, Fig. 1 shows the skin-friction coefficient c_f and the fluctuating wall-shear stress τ_{rms}, compared to literature data. It turns out that the skin friction can be well described using the simple relation $c_f = 0.024 Re_\theta^{-0.25}$ for the present Re_θ-range. In particular $\tau^+_{\mathrm{rms}} = u'/U|_{y \to 0}$ appears to be a sensitive measure for the development of the near-wall turbulence [7]. Mean velocity and stress profiles are shown in Fig. 2, compared to the DNS by Simens *et al.* [9]; the comparison to experiments at the higher *Re* is not shown here due to space restrictions (see however the comparisons provided *e.g.* in Ref. [8]).

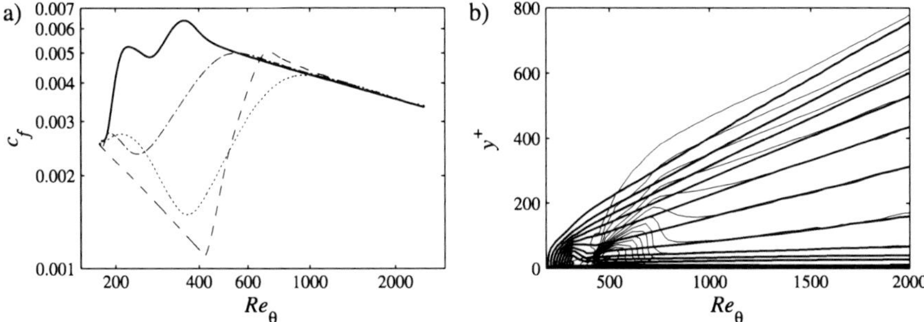

Fig. 3 a) Skin-friction coefficient c_f for various trippings: solid: "baseline" case, dot-dashed: lower amplitude and dotted: lower frequency trippings, dashed: classical transition via exponential growth of TS waves. b) Contours of $u^+_{\rm rms}$ (spacing 0.25) for baseline case (thick) and TS-wave case (thin lines).

It is striking to see differences in the fluctuations profiles at $Re_\theta = 1100$ between our DNS and Ref. [9], which seem to have greatly reduced at $Re_\theta = 2000$. To further study the dependence of the boundary layer on initial conditions (*i.e.* the way how turbulence is induced), a series of DNS with varying tripping parameters has been performed. Fig. 3a) shows c_f for these DNS: Depending on tripping, laminar-turbulent transition is induced at different Re_θ, but the curves quickly settle on a common $c_f(Re_\theta)$ distribution, indicating a rather quick adaptation of the near-wall turbulence. Note also the typical overshoots of c_f as a result of transition. On the other hand, as shown in Fig. 3b), the outer-layer convergence, illustrated by contours of $u_{\rm rms}$ takes considerably longer than near the wall. Precisely this effect causes the discrepancies shown in Fig. 2b). It is worth noting that the profiles of the TS-wave-tripped DNS closely match the rms-profiles of Ref. [9] in Fig. 2. It remains an open question what conditions to apply for a "fully-developed" state of the boundary layer, and, consequently, whether a generic boundary layer at such low Re_θ exists.

When it comes to spectral information recorded in the boundary layer (not shown), spatial structures of two very distinct spatial (and temporal) scales are detected, *i.e.* the well-known turbulent streaks close to the wall scaling in inner (viscous) units, and long and wide structures scaling in outer units. These large-scale structures persist throughout the boundary layer from the free stream down to the wall and act as a modulation of the near-wall streaks. It is thus clear that the interplay of these two structures, scaling in different way, is strongly dependent on the Reynolds number, *i.e.* the downstream distance.

2.2 Vortical coherent structures

However, the characteristics of these large-scale structures, and their relation to actual coherent structures present in the flow are not entirely clear, in particular for higher Reynolds numbers. Recent studies, summarized by Adrian [1] and also observed in a DNS of low-*Re* boundary layers [11] suggest a dominance of hairpin-shaped vortices of various sizes throughout the boundary layer. Hairpins are well-known structures in transitional flows, *e.g.* appearing as the results of shear-layer roll-up. It is now interesting to see whether hairpin-like structures persist in fully turbulent flow (either in the outer region or close to the wall), or whether they are mainly restricted to low-*Re* or transitional flows. Thus, we would like to study the structural characteristics of a turbulent boundary layer as a function of Reynolds number, starting from the transitional phase up to fully turbulent flow at higher *Re* than available in previous studies.

Figures 4 to 7 included here show small sections of the large simulation domain, each of them only visualising about 5% of the length of the full simulation domain. Each figure highlights different development stages of the boundary layer. Isocontours of negative λ_2 [4] identifying vortical structures are colored by the wall distance. Laminar-turbulent transition (Fig. 4) is induced by a trip forcing as described above. The subsequent breakdown to turbulent flow is characterized by the appearance of velocity streaks and unambiguous hairpin vortices, which are seen to dominate the whole span of the flow, see *e.g.* Ref. [1]. The hairpin vortices increase in number, and individual distinctive heads of such vortices are clearly visible for some distance downstream (Fig. 5 at $Re_\theta = 600$). This feature of low-*Re* turbulent flow is also put forward in Ref. [11], there denoted as "forest of hairpins". We can thus confirm that, at least at low-*Re*, hairpins are indeed the dominant structure in a turbulent boundary layer. However, as the Reynolds number is further increased above about 1000, the scale separation between inner and outer units is getting larger, and the flow is less and less dominated by these transitional flow structures. At $Re_\theta = 2500$, as shown in Fig. 6, isolated instances of arches belonging to hairpin vortices can still be observed riding on top of the emerging outer-layer streaky structures. But the dominance of hairpin-like structures is clearly lower than in the previous figures closer to transition. This effect is even increased by considering the highest present Reynolds number, $Re_\theta = 4300$ as shown in Fig. 7. Then, individual hairpins or arches cannot be seen any longer. The boundary layer now is truly turbulent, and the outer layer is dominated by large-scale streaky organization of the turbulent vortices. Ongoing analysis of the flow structures close to the wall (*i.e.* the near-wall cycle) reveals characteristic oscillations of the near-wall streak, mainly in a sinuous manner leading to a staggered appearance of the vortex cores.

3 Summary and conclusions

A large DNS of a turbulent boundary layer has been performed reaching up to $Re_\theta = 4300$. The statistics are in very good agreement with available experimental measurement. The focus of the present contribution is twofold: First, an assessment of the importance of initial conditions at low Re_θ, and secondly, an analysis of coherent structure at higher Re_θ. It is for instance shown that various turbulence statistics can be significantly influenced up to $Re_\theta \approx 2000$ by the choice of tripping

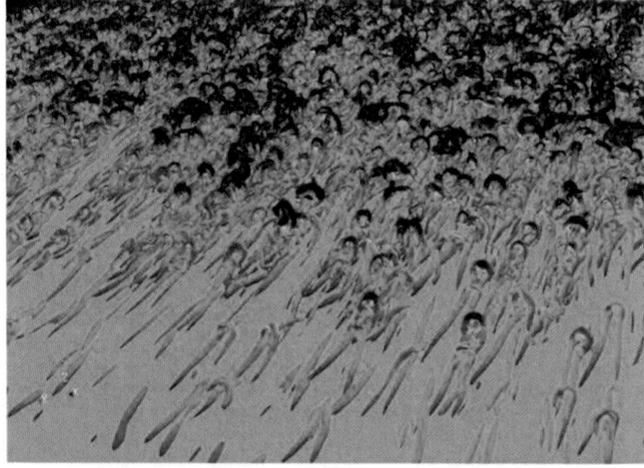

Fig. 4 Visualization of the structures in a turbulent boundary layer by isocontours of negative λ_2 [4]; color code represents wall distance. The middle of the visualized domain is located at about $Re_\theta = 400$. Transition to turbulence can be observed by the appearance of Λ and hairpin vortices. No individual turbulent spots can be seen as the breakdown is happening simultaneously along the span.

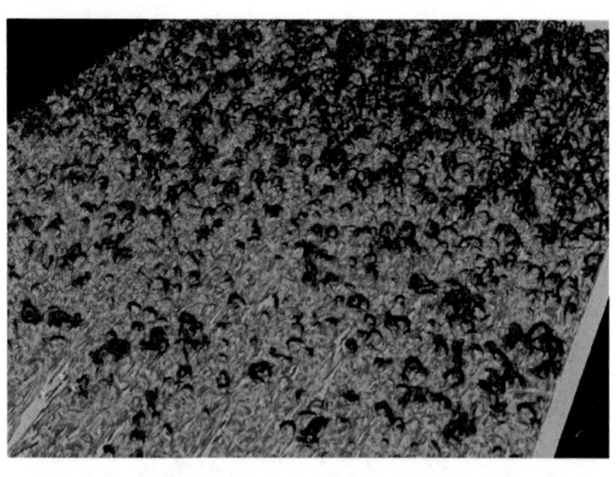

Fig. 5 Visualization of the structures in a turbulent boundary layer at about $Re_\theta = 600$. The whole span of the flow is clearly turbulent, *i.e.* dominated by random, unsteady vortical motion. However, the dominance of hairpin vortices as a remainder of transition are still the dominating coherent structures. Note that there is no clear scale separation yet given the low Re_θ.

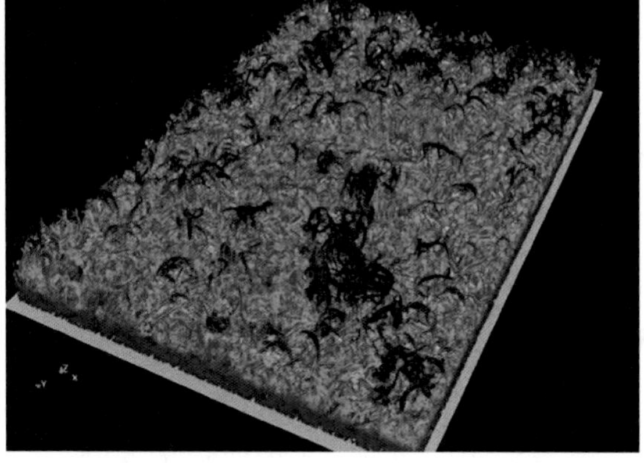

Fig. 6 Visualization of the structures in a turbulent boundary layer at about $Re_\theta = 2500$. The visualized domain has a length of about 7000 and a width of about 4500 viscous units. The intermittent structures close to the boundary-layer edge are clearly visible; they even arrange in large-scale bulges. The vortices do not span the whole boundary-layer height; the near-wall region is characterized by its own dynamics.

position and type; these differences relate to the behavior of the flow in the outer region of the boundary layer. Whether a generic boundary layer exists at such low Reynolds numbers clearly deserves more in-depth analysis. In a second part, the

Fig. 7 Visualization of the structures in a turbulent boundary layer at about $Re_\theta = 4300$. As in the previous frame, large-scale corrugation of the boundary-layer edge can be seen. The regular hairpin vortices seen close to transition have completely disappeared and are replaced by much more chaotic structures. Note also that the typical eddy diameter is practically unchanged compared to the previous figure.

coherent structures in the outer layer are studied. It is clearly shown that hairpin vortices, being characteristic transitional remainders at lower Reynolds numbers, are not visible any longer for higher *Re*.

Simulation data: `www.mech.kth.se/~pschlatt/DATA`

Acknowledgements Computer time was provided by the Swedish National Infrastructure for Computing (SNIC) with a generous grant by the Knut and Alice Wallenberg Foundation (KAW).

References

1. R. J. Adrian. Hairpin vortex organization in wall turbulence. *Phys. Fluids*, 19(041301):1–16, 2007.
2. M. Chevalier, P. Schlatter, A. Lundbladh, and D. S. Henningson. SIMSON - A Pseudo-Spectral Solver for Incompressible Boundary Layer Flows. Technical Report TRITA-MEK 2007:07, KTH Mechanics, Stockholm, Sweden, 2007.
3. A. Ferrante and S. Elghobashi. Reynolds number effect on drag reduction in a microbubble-laden spatially developing turbulent boundary layer. *J. Fluid Mech.*, 543:93–106, 2005.
4. J. Jeong and F. Hussain. On the identification of a vortex. *J. Fluid Mech.*, 285:69–94, 1995.
5. Q. Li, P. Schlatter, L. Brandt, and D. S. Henningson. DNS of a spatially developing turbulent boundary layer with passive scalar transport. *Int. J. Heat Fluid Flow*, 30:916–929, 2009.
6. J. M. Österlund. *Experimental studies of zero pressure-gradient turbulent boundary-layer flow*. PhD thesis, Department of Mechanics, KTH Stockholm, Sweden, 1999.
7. P. Schlatter and R. Örlü. Assessment of direct numerical simulation data of turbulent boundary layers. *J. Fluid Mech.*, 659:116–126, 2010.
8. P. Schlatter, R. Örlü, Q. Li, G. Brethouwer, J. H. M. Fransson, A. V. Johansson, P. H. Alfredsson, and D. S. Henningson. Turbulent boundary layers up to $Re_\theta = 2500$ studied through numerical simulation and experiments. *Phys. Fluids*, 21(051702):1–4, 2009.
9. M. P. Simens, J. Jiménez, S. Hoyas, and Y. Mizuno. A high-resolution code for turbulent boundary layers. *J. Comput. Phys.*, 228:4218–4231, 2009.
10. P. R. Spalart. Direct simulation of a turbulent boundary layer up to $R_\theta = 1410$. *J. Fluid Mech.*, 187:61–98, 1988.
11. X. Wu and P. Moin. Direct numerical simulation of turbulence in a nominally zero-pressure-gradient flat-plate boundary layer. *J. Fluid Mech.*, 630:5–41, 2009.

A physical length-scale for LES of turbulent flow

Ugo Piomelli, Bernard J. Geurts

1 Introduction

The fundamental assumption underlying large-eddy simulations (LES) is that the large, energy-carrying, eddies are resolved, while only the smaller eddies are modeled. An implication of this assumption is that the filter-width Δ, the length scale that separates the resolved from the unresolved eddies, should be a fraction of the integral scale, which is characteristic of the large eddies. In practice, however, the filter width is taken to be proportional to the grid size, h. This approach is generally legitimate, since the grid is usually refined where the important turbulence scales are smaller; it presents, however, two problems. First, rapid variations of the mesh (especially in methods that use local mesh refinement) may cause commutation and aliasing errors, and unphysical results [9]. Second, it requires knowledge, on the part of the user, on the characteristics of turbulence; in complex flows it may not be possible to predict the turbulence behavior *a priori*.

In this work we report initial results of an investigation aimed at achieving a definition of Δ directly based on local turbulence quantities that are characteristic of the flow that is considered. We are aiming to define a length-scale distribution that is grid-independent, with the view that flow physics are decoupled from any numerical element such as spatial discretization method and computational grid. This approach to large-eddy simulation follows the classical PDE interpretation of the closed LES equations [3]. The mesh size h can in our PDE approach be chosen in

Ugo Piomelli
Department of Mechanical and Materials Engineering,
Queen's University, Kingston (Ontario) K7L 3N6, Canada
e-mail: ugo@me.queensu.ca

Bernard J. Geurts
Multiscale Modeling and Simulation, Faculty EEMCS, J.M. Burgers Center,
University of Twente, P.O. Box 217, 7500 AE Enschede, The Netherlands
e-mail: b.j.geurts@utwente.nl

H. Kuerten et al. (eds.), *Direct and Large-Eddy Simulation VIII*,
ERCOFTAC Series 15, DOI 10.1007/978-94-007-2482-2_3,

such a way that the smallest resolved eddy, of size Δ, can be reproduced accurately by the numerical method.

We examine a definition of the integral length scale that can be applied to shear flows, with several potential advantages: grid refinement studies are straightforward and a clear measure of the adequacy of the resolution is provided by the ratio Δ/h achieved. While the proposed definition presents some shortcomings for the application to high-Reynolds numbers flows with little or no shear, its use allows us to evaluate the conceptual applicability of the decoupling between filter-width and grid size.

In the following, after introducing the governing equations and describing the numerical scheme and boundary conditions, we will present the model for the unresolved eddies and specify the definition of the filter width. There will then be the application of the proposed model to a turbulent plane channel flow at Reynolds numbers up to $Re_\tau = 2,000$ (based on channel half-width and friction velocity u_τ). Some concluding remarks and recommendations for future work will be presented.

2 Governing equations and numerical model

In the classical approach to large-eddy simulations, the velocity field is separated into a resolved (large-scale) and a small-scale, unresolved field, by a spatial filtering operation. The continuity and the Navier-Stokes equations for the resolved field are:

$$\frac{\partial \overline{u}_i}{\partial x_i} = 0, \tag{1}$$

$$\frac{\partial \overline{u}_i}{\partial t} + \frac{\partial (\overline{u}_j \overline{u}_i)}{\partial x_j} = \nu \frac{\partial^2 \overline{u}_i}{\partial x_j \partial x_j} - \frac{\partial \tau_{ij}}{\partial x_j} - \frac{1}{\rho}\frac{\partial \overline{p}}{\partial x_i} + \delta_{i1} f. \tag{2}$$

Here, x_1, x_2 and x_3 are the streamwise, wall-normal and spanwise directions, also referred to as x, y and z. The resolved velocity components in these directions are, respectively, $\overline{u}_1$, $\overline{u}_2$ and $\overline{u}_3$ (or $\overline{u}$, $\overline{v}$ and $\overline{w}$). In these equations, f is the driving pressure gradient for the plane channel flow calculations and $\tau_{ij} = \overline{u_i u_j} - \overline{u}_i \overline{u}_j$ are the unresolved stresses that constitute the well-known central closure problem in large-eddy simulation.

Periodic boundary conditions are used in the streamwise and spanwise directions, while no-slip conditions are imposed at the wall. The Navier-Stokes equations are discretized using second-order central differences on a staggered grid, and integrated using a fractional step method [1]. The spatial discretization of the convective terms conserves momentum and energy discretely [8]. The code has been validated extensively for a variety of turbulent flows [7, 6].

We investigate large-eddy simulations based on a filter width Δ that is a *fixed* fraction α of a given definition of the integral scale L. To explore this idea, we concentrate on a definition of L based on turbulent kinetic energy (TKE), $\mathscr{K}$, and resolved vorticity $\omega = \langle \omega_i \omega_i \rangle^{1/2}$, where $\omega_i = \varepsilon_{ijk} \overline{u}_{j,k}$ is the resolved vorticity, and

ε_{ijk} is the permutation symbol, to give:

$$L = C\mathscr{K}^{1/2}/\omega \Rightarrow \Delta = \alpha C\mathscr{K}^{1/2}/\omega, \tag{3}$$

where C is a constant. We use the total vorticity (instead of the fluctuating one) to preserve the property that in a laminar boundary layer the length-scale will go to zero. The resulting model for the unresolved stresses is:

$$\tau_{ij} - \frac{\delta_{ij}}{3}\tau_{kk} = -2C_k^2\frac{\mathscr{K}}{\omega^2}|\overline{S}|\overline{S}_{ij}, \tag{4}$$

where we have absorbed all the coefficients into C_k. The definition of Δ proposed in (3) is expected to work well in shear-dominated flows, in which ω is largely due to the mean shear; in high-Reynolds-number flows at zero or low shear the effect of the small scales becomes significant, and the length-scale proposed here would not be representative of the integral scale of the flow. It is important to test the filter-scale definition in shear-dominated flows and to quantify under what flow conditions the required shear-dominance is actually attained to a sufficient degree. For the channel this is expected to hold near the wall, while the assumption will be more challenged in the core of the channel. Generalizations of this concept that are applicable in the high-*Re* limit are currently being investigated.

The new model has some desirable characteristics: first, the eddy viscosity vanishes in laminar flow (in laminar irrotational regions the ratio $\mathscr{K}^{1/2}/\omega$ becomes undefined, but can be set to zero). Secondly, the eddy viscosity (and hence the stresses) vanish near a wall; the asymptotic behavior is not exact (y^2 instead of y^3) but the decay is rapid enough that no additional damping is required.

The coefficient C_k is determined by error minimization on coarse grids, using the Successive Inverse Polynomial Interpolation (SIPI) method [4]. This procedure consists in performing several coarse LES systematically varying the value of the coefficient C_k with the objective of minimizing a user-defined error function. This procedure has been shown to converge rapidly (requiring only 5–6 evaluations); since it is carried out on a grid that is much coarser than that used for production, the coefficient evaluation does not add significantly to the overall cost of the simulation.

3 Results for channel flow

We applied the new model (4) to the simulation of turbulent channel flow at Reynolds numbers Re_τ between 400 and 2,000 (based on channel half-height δ and friction velocity u_τ). The error measure chosen for the SIPI optimization is the skin-friction coefficient, $C_f = 2\tau_w/\rho U_b^2$ where U_b is the average velocity in the channel. The error is defined as $\varepsilon = |C_f - C_{f,\mathrm{tgt}}|/C_{f,\mathrm{tgt}}$ and the desired value $C_{f,\mathrm{tgt}}$ is obtained from the experimental correlation [2] $C_{f,\mathrm{tgt}} = 0.073(2Re_b)^{-1/4}$ where $Re_b = U_b\delta/\nu$.

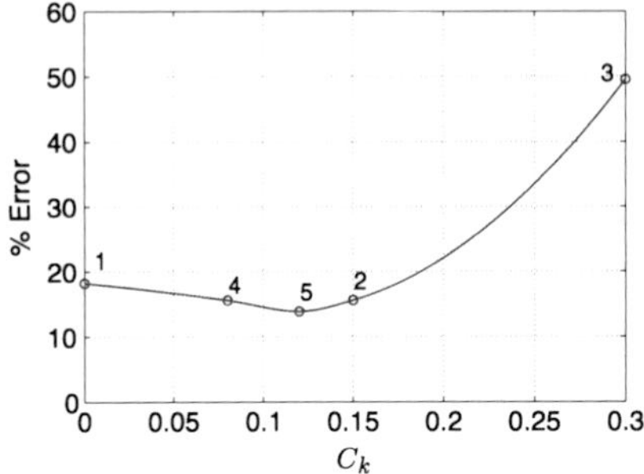

Fig. 1 Error landscape for the $Re_\tau = 1,000$ calculations. 48×65×48 grid points. The iterates are marked by circles.

For the case $Re_\tau \simeq 1,000$ case ($Re_b = 19,000$) the error is minimum at $C_k = 0.12$ as seen in Fig. 1. The turbulence statistics were found relatively insensitive to variations in the range $0.1 < C_k < 0.15$. Once the coefficient is determined on a coarse grid, refining the grid results in grid-converged LES; the formulation was found to be very robust. For example, decreasing the grid size by a factor of 16 results in a 40% decrease of the eddy viscosity; the Dynamic or Smagorinsky models would predict a decrease of the eddy viscosity by a factor of approximately 16^2. The results of three calculations with progressively refined grids, Figure 2, shows very good convergence towards the DNS results of [5]: in this instance agreement is achieved with 128×129×128 points (corresponding to $\Delta x^+ \simeq 46$ and $\Delta z^+ \simeq 22$).

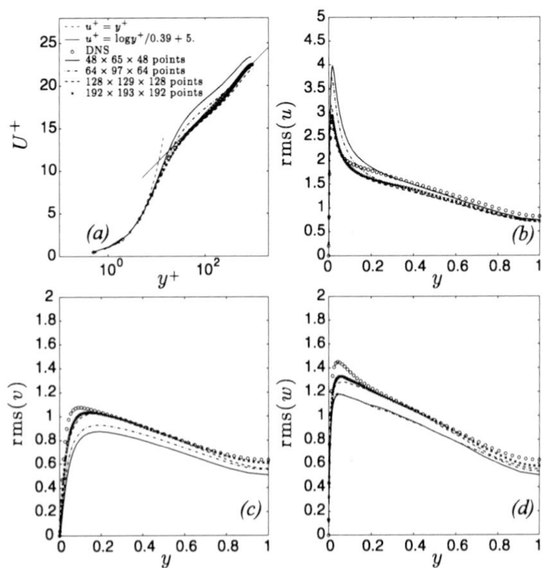

Fig. 2 Turbulence statistics for the $Re_\tau \simeq 1,000$ calculations. $C_k = 0.12$. *(a)* Mean velocity; *(b)* rms of u; *(c)* rms of v; *(d)* rms of w.

The LES should result in a DNS as *both* grid size *and* filter-width are refined to allow accurate numerical representation of a more and more complete fraction of the range of turbulence scales, respectively. The model for Δ proposed here preserves this property. We performed simulations at $Re_\tau = 400$ and found that the optimum value of C_k reached a value of 0 for 256^3 grid points: this is the resolution at which a model for the unresolved eddies does not improve the results, and a coarse DNS is more accurate than any LES with an eddy viscosity model.

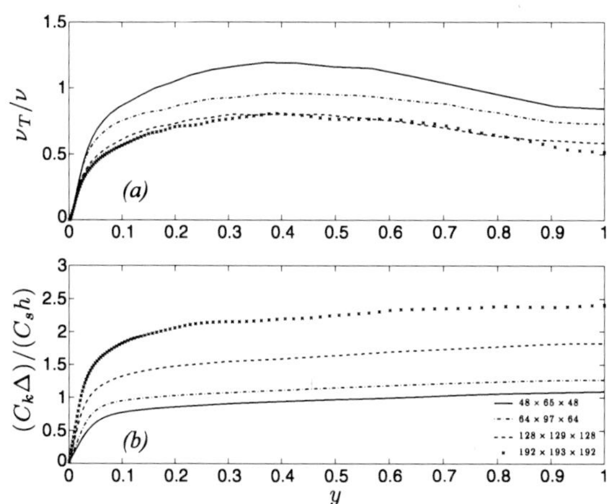

Fig. 3 Profiles of filter-width and eddy viscosity; $Re_\tau = 1{,}000$. *(a)* ν_T/ν. *(b)* Ratio $C_k\Delta/C_s h$.

In Figure 3*(a)* we show ν_T/ν for four simulations, with varying grid spacing, at $Re_\tau = 1{,}000$. The finer mesh results in better prediction of the vorticity; the eddy viscosity is robust and only decreases by a small amount. Figure 3*(b)* shows the ratio between $C_k\Delta$ and the length-scale used in the Smagorinsky model, $C_s h$ (with $C_s = 0.1$). Near the walls this ratio is significantly less than one, reflecting the better wall behavior obtained with the new definition of Δ.

We also conducted simulations at $Re_\tau \simeq 2{,}000$. The SIPI procedure was carried out on a $64\times97\times64$ grid using the error based on the skin-friction coefficient yielding $C_k = 0.13$. The turbulence statistics demonstrate improved agreement with the DNS data as the mesh is refined while keeping C_k fixed. This indicates that with the grid-independent filter-width definition, as proposed here, the convergence to a grid-independent LES can be pursued efficiently.

An important issue to be considered when developing SFS models is their cost. The model proposed here, in terms of computational cost, is comparable to the Smagorinsky model, and significantly cheaper than dynamic models that, due to the repeated filtering operations, require 60% more CPU time (for a given grid). The proposed model requires, of course, 5 or 6 extra simulations to determine the optimum value of the coefficient; since those are performed on a grid that is much coarser than the one used for production runs, however, this overhead is not excessive. Moreover, the SIPI test runs not only indicate a 'near-optimum' value of C_k but also give an impression of the sensitivity of predictions to numerical and modeling parameters, thereby increasing the fidelity of the simulations.

4 Conclusions

We have proposed a new definition of the filter width Δ based on turbulence quantities instead of the computational grid. The filter width is taken a fraction of the local integral length scale L, which is approximated as the ratio between the square root of the turbulent kinetic energy (TKE) and the resolved-vorticity magnitude. The model coefficient, C_k, represents a turbulence resolution parameter, which determines what fraction of the integral scales is resolved. The proposed definition was found to work well in a wall-bounded flow at moderate Reynolds numbers of up to $Re_\tau = 2000$.

The coefficient C_k can be determined by assigning it in such a way as to minimize the error in the prediction of some desired quantity. Here, we chose the friction coefficient for the turbulent channel flow. Good results were also obtained by requiring for example that the sub-filter scales do not contribute more than a certain percentage to the total stress or the total energy dissipation. In all cases the determination of C_k can be carried out on a very coarse grid, using any error minimization procedure. The one we chose, the Successive Inverse Polynomial Interpolation (SIPI), proposed for a similar problem by [4] converged to the minimum in 5 or 6 function evaluations. The proposed definition of Δ is, in principle, grid-independent. The model has been tested in a plane channel flow. It allows accurate prediction of the turbulent statistics, at a cost lower than that of the dynamic eddy viscosity model.

References

1. Chorin, A. J. 1968 Numerical solution of Navier-Stokes equations. *Math. Comput.* **22** (104), 745–762.
2. Dean, R. B. 1978 Reynolds number dependence of skin friction and other bulk flow variables in two-dimensional rectangular duct flow. *ASME J. Fluids Eng.* **100**, 215–223.
3. Geurts, B. J. 2003 *Elements of direct and large-eddy simulation*. Philadelphia: Edwards.
4. Geurts, B. J. & Meyers, J. 2006 Successive inverse polynomial interpolation to optimize Smagorinsky's model for large-eddy simulation of homogeneous turbulence. *Phys. Fluids* **18**, 118102.
5. Hoyas, S. & Jiménez, J. 2006 Scaling of the velocity fluctuations in turbulent channels up to $Re_\tau = 2003$. *Phys. Fluids* **18**, 011702.
6. Keating, A. & Piomelli, U. 2006 A dynamic stochastic forcing method as a wall-layer model for large-eddy simulation. *J. Turbul.* **7** (12), 1–24.
7. Keating, A., Piomelli, U., Balaras, E. & Kaltenbach, H.-J. 2004*a* A priori and a posteriori tests of inflow conditions for large-eddy simulation. *Phys. Fluids* **16** (12), 4696–4712.
8. Morinishi, Y., Lund, T. S., Vasilyev, O. V. & Moin, P. 1998 Fully conservative higher order finite difference schemes for incompressible flows. *J. Comput. Phys.* **143**, 90–124.
9. Vanella, M., Piomelli, U. & Balaras, E. 2008 Effect of grid discontinuities in large-eddy simulation statistics and flow fields. *J. Turbul.* **9** (32), 1–23.

Regularization modeling of buoyancy-driven flows

F.X. Trias, A. Gorobets, A. Oliva, R.W.C.P. Verstappen

1 Introduction

Since direct numerical simulations (DNS) of buoyancy-driven flows cannot be computed at high Rayleigh numbers, a dynamically less complex mathematical formulation is sought. In the quest for such a formulation, we consider regularizations (smooth approximations) of the nonlinear convective term. The first outstanding approach in this direction goes back to Leray [1]; the Navier-Stokes-α model also forms an example of regularization modeling (see [2], for instance). The regularization methods basically alter the convective terms to reduce the production of small scales of motion. In doing so, we propose to preserve the symmetry and conservation properties of the convective terms exactly. This requirement yields a family of *symmetry-preserving regularization* models [3, 4]: a novel class of regularizations that restrain the convective production of smaller and smaller scales of motion in an unconditional stable manner, meaning that the velocity cannot blow up in the energy-norm (in 2D also: enstrophy-norm). In this work, a method to dynamically determine the regularization parameter (local filter length) from the requirement that the vortex-stretching must be stopped at the scale set by the grid is also proposed and tested. The numerical algorithm used to solve the governing equations preserves the conservation properties too [5] and is therefore well-suited to test the proposed simulation model. Here, the performance of the method is tested for a turbulent differentially heated cavity (DHC). Due to the complex behavior exhibit [6, 7] an accurate turbulence modeling of this configuration is as a great challenge.

F.X. Trias, A. Gorobets, A. Oliva
Heat and Mass Transfer Technological Center, Technical University of Catalonia, e-mail: cttc@cttc.upc.edu

F.X. Trias, R.W.C.P. Verstappen
Institute of Mathematics and Computing Science, University of Groningen, e-mail: R.W.C.P.Verstappen@rug.nl

H. Kuerten et al. (eds.), *Direct and Large-Eddy Simulation VIII*,
ERCOFTAC Series 15, DOI 10.1007/978-94-007-2482-2_4,

2 $\mathscr{C}_4$-regularization modeling

We restrict ourselves to the $\mathscr{C}_4$ approximation (see [3, 4], for details): the convective term in the Navier-Stokes (NS) equations is then replaced by the following $\mathscr{O}(\varepsilon^4)$-accurate smooth approximation $\mathscr{C}_4(u,v)$ given by

$$\mathscr{C}_4(u,v) = \mathscr{C}(\overline{u},\overline{v}) + \overline{\mathscr{C}(\overline{u},v')} + \overline{\mathscr{C}(u',\overline{v})}, \tag{1}$$

where $\mathscr{C}(u,v) = (u \cdot \nabla)v$ represents the convective operator. Note that here a prime indicates the residual of the filter, *e.g.* $u' = u - \overline{u}$, which can be explicitly evaluated, and $\overline{(\cdot)}$ represents a normalized self-adjoint linear filter with filter length ε. Therefore, the dimensionless governing equations result to

$$\begin{aligned} \partial_t u + \mathscr{C}_4(u,u) &= PrRa^{-1/2}\Delta u - \nabla p + f; \qquad \nabla \cdot u = 0, \\ \partial_t T + \mathscr{C}_4(u,T) &= \quad Ra^{-1/2}\Delta T, \end{aligned} \tag{2}$$

where Ra and Pr are the Rayleigh and the Prandtl numbers, respectively. To account for the density variations, the Boussinesq approximation is used. Note that the $\mathscr{C}_4$ approximation is also skew-symmetric like the original convective operator, *i.e.* $(\mathscr{C}_4(u,v),w) = -(\mathscr{C}_4(u,w),v)$ where $(f,g) = \int_V (f \cdot g) dV$ denotes the usual scalar product and $\nabla \cdot u = 0$. Hence, unlike other regularization methods [1, 2], the same inviscid invariants -kinetic energy, enstrophy in 2D and helicity- that the original NS equations are kept for the new set of equations (2). The performance of the $\mathscr{C}_4$ approximation has been successfully tested for a turbulent channel flow [3] and an air-filled ($Pr = 0.71$) DHC of height aspect ratio $A = 4$ at Ra-numbers (based on the cavity height) 10^{10} [8] and 10^{11} [9] by means of direct comparison with the DNS results [6, 7]. In the latter cases, the filter width, ε, was treated as a parameter and, therefore, it needed to be prescribed in advanced. Figure 1 displays illustrative results for the aforesaid DHC problem at $Ra = 10^{10}$ as a function of ε/h. These results show that the numerical solution is weakly dependent for sufficiently large values of ε. Similar behavior has also been observed for the other configurations.

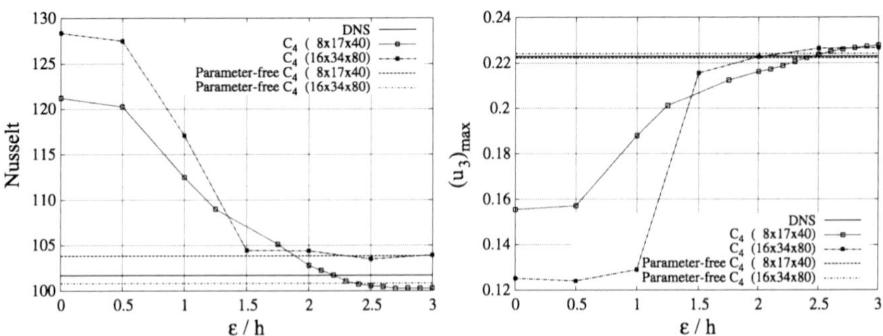

Fig. 1 The overall averaged Nusselt number (left) and the maximum of the averaged vertical velocity at the horizontal mid-height plane (right) as a function of the ratio of the filter length ε to the average grid width h at $Ra = 10^{10}$.

2.1 Stopping the vortex-stretching mechanism

In the present work, we propose to determine ε dynamically from the requirement that the vortex-stretching must be stopped at the scale set by the grid. The idea behind is to modify convective operator sufficiently to guarantee that the following inequality is locally hold

$$\lambda_k(\varepsilon) \leq \left(PrRa^{-1/2}\right) k^2, \tag{3}$$

where $\lambda_k(\varepsilon) = \omega_k \cdot \mathscr{C}_{4,k}(\omega, u)/(\omega_k \cdot \omega_k)$ is the Rayleigh quotient of the vortex-stretching at the grid scale, $k = \pi/h$. In practice, the value of λ_k has been bounded by the largest (positive) eigenvalue of the straintensor S, $\lambda_k(\varepsilon) \leq f_{4,k}(\varepsilon)\ \lambda_{max}(S)$. For the $\mathscr{C}_4$-approximation, the damping function, $0 < f_{4,k} \leq 1$, at the highest frequency is given by $3\hat{g}_k^2(\varepsilon) - 2\hat{g}_k^3(\varepsilon)$ (see [3, 4], for details), where $0 < \hat{g}_k(\varepsilon) \leq 1$ is the transfer function of the linear filter. Therefore, it suffices that the following inequality be locally hold

$$3\hat{g}_k^2(\varepsilon) - 2\hat{g}_k^3(\varepsilon) \leq \frac{Pr}{Ra^{1/2}} \frac{k^2}{\lambda_{max}(S)} \quad \longrightarrow \quad \hat{g}_k(\varepsilon) \quad \longrightarrow \quad \varepsilon \tag{4}$$

to guarantee that the vortex-stretching stops at the smallest grid scale.

	DNS	RM1	RM2
$Ra = 10^{10}$	$128 \times 190 \times 462$	$16 \times 34 \times 80$	$8 \times 17 \times 40$
$Ra = 10^{11}$	$128 \times 682 \times 1278$	$12 \times 45 \times 85$	$8 \times 30 \times 56$

Table 1 Description of meshes: the spanwise (N_x), the wall-normal horizontal (N_y), and the vertical (N_z) resolutions for tested cases. Further details about the meshing can be found in [6].

3 Results for a turbulent differentially heated cavity

The performance of the proposed method to dynamically determine the regularization parameter ε of the $\mathscr{C}_4$ approximation has also been tested for the aforesaid DHC problem. Again two very coarse meshes (see Table 1) have been solved. In this work, a fourth-order accurate Gaussian filter [10] has been chosen. The boundary conditions that supplement the NS equations are also applied to the filter. Details about the filtering procedure can be found in [4, 11]. First results displayed in Figure 1 exhibit the great potential of this method. Note that in the parameter-free $\mathscr{C}_4$ cases, the results do not depend on ε/h. At least as good results as the optimal ε/h ratio determined by trial-and-error procedure have been obtained.

3.1 Grid (in)dependence analysis

A reliable modeling of turbulence at (very) coarse grids is a great challenge. The coarser the grid, more convincing model quality is perceived. However, it might happen that solution be strongly dependent on meshing parameters and thus some particular combinations could 'accidentally' provide good results. An example of this behavior can be found in [12] for a turbulent channel flow. In order to elucidate this point, the same DHC problem has been solved on a series of 50 randomly generated meshes: with (N_x, N_y, N_z)-values limited by those given by meshes RM1 and RM2 (see Table 1), *i.e.* $8 \leq N_x \leq 16$, $17 \leq N_y \leq 34$, and $40 \leq N_z \leq 80$. Results for the overall Nusselt (Nu), the centerline stratification, the maximum vertical velocity and the wall shear stress are displayed in Figure 2. We can observe that the parameter-free $\mathscr{C}_4$ modeling predicts good results irrespective of the meshing whereas very poor and dispersed results are obtained when the model is switched off. Especially significant is the fairly good prediction of the stratification (note the large dispersion obtained without model!).

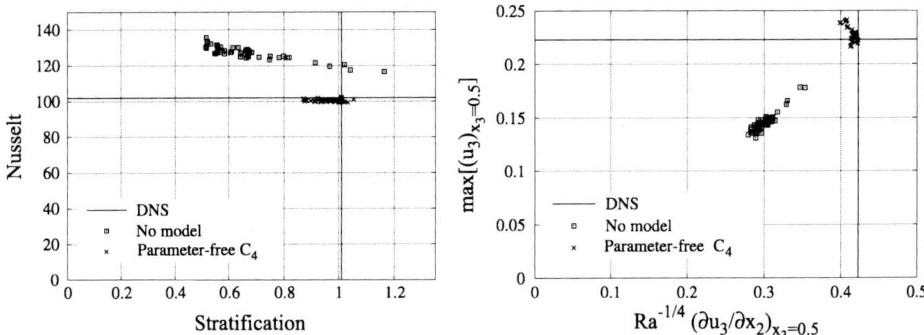

Fig. 2 Left: the overall Nusselt number and the centerline stratification. Right: the maximum vertical velocity and the wall shear stress scaled by $Ra^{-1/4}$ at the horizontal mid-height plane. Results have been obtained for 50 randomly generated grids at $Ra = 10^{10}$.

3.2 Performance at higher (and lower) Ra-numbers

The performance of the parameter-free $\mathscr{C}_4$-regularization has also been tested at higher (and lower) Ra by means of direct comparison with DNS data [6, 7]. This study covers a relatively wide range, $6.4 \times 10^8 \leq Ra \leq 10^{11}$, from weak to fully developed turbulence. The meshes used to carry out these simulations have been generated keeping the same number of points in the boundary layer as in the coarse mesh RM1 for $Ra = 10^{10}$. Results for the overall Nusselt number corresponding to 56 simulations within the whole range of Rayleigh numbers studied by DNS, *i.e.*, $6.4 \times 10^8 \leq Ra \leq 10^{11}$, are displayed in Figure 3. At first sight, we observe again a fairly good agreement with the DNS results (solid dots) and the power-law correlation obtained from the DNS data [7]. It must be noted that the *Nu-Ra* dependence obtained with the parameter-free $\mathscr{C}_4$ is smooth suggesting again that the proposed model is performing well 'independently' of *Ra* and meshing parameters that may suddenly change for two consecutive points in the graph.

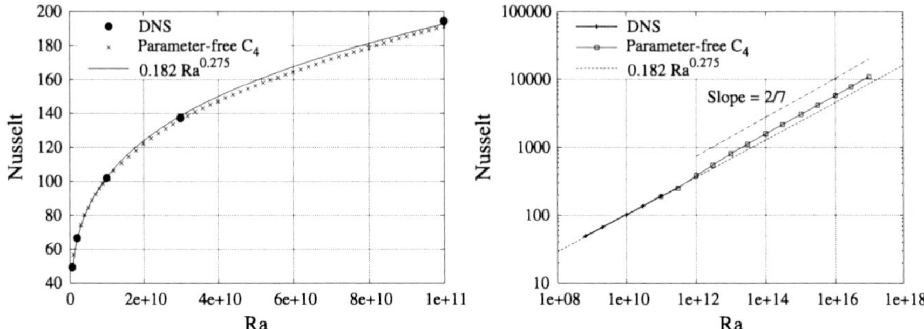

Fig. 3 Left: overall Nusselt number for $6.4 \times 10^8 \leq Ra \leq 10^{11}$. Right: overall Nusselt number for Rayleigh numbers up to 10^{17}.

Since performing computations with the parameter-free $\mathscr{C}_4$ approximation is rather cheap we were tempted to perform simulations at very high *Ra*. Of course, for the range $10^{11} < Ra \leq 10^{17}$ there is no DNS (or experimental) data to compare with. Anyhow, it is interesting to see that results displayed in Figure 3 show a good agreement with a $2/7$ power-law scaling of Nusselt (*Nu* increases approximately from 10^2 to 10^4, that is 2 orders of magnitude, when *Ra* is increased 7 orders, from 10^{10} to 10^{17}). This scaling law, predicted by alternative theories [13] of turbulent natural convection, has also been experimentally measured for Rayleigh-Bénard configurations [14].

4 Concluding remarks

The performance of the proposed method to dynamically determine the regularization parameter ε of the $\mathscr{C}_4$ approximation has been tested for an air-filled DHC problem of aspect ratio 4. Results presented here exhibit the great potential of the proposed method as new turbulence model: the parameter-free $\mathscr{C}_4$ approach is able to predict at least as good results as the optimal ε/h ratio determined by trial-and-error procedure. For further details about the results presented here the reader is referred to [4].

In conclusion, the proposed method constitutes a parameter-free turbulence model. Moreover, since no *ad hoc* phenomenological arguments that cannot be formally derived for the NS equations are used it suggests that this method should be valid for other configurations. Nevertheless, more simulations for a wide variety of cases and meshes will be necessary to confirm these conclusions.

Acknowledgements This work has been financially supported by the *Ministerio de Educación y Ciencia*, Spain, (ENE2007-67185) and a postdoctoral fellowship *Beatriu de Pinós* (2006 BP-A 10075) by the *Generalitat de Catalunya*. Reference DNS results have been performed on the IBM MareNostrum supercomputer at the Barcelona Supercomputing Center. The authors thankfully acknowledge this institution.

References

1. Leray, J. (1934) Sur le movement d'un liquide visqueaux emplissant l'espace. *Acta Mathematica*, **63**, 193–248.
2. Geurts, B. J. and Holm, D. D. (2003) Regularization modeling for large-eddy simulation. *Physics of Fluids*, **15**, L13–L16.
3. Verstappen, R. (2008) On restraining the production of small scales of motion in a turbulent channel flow. *Computers & Fluids*, **37**, 887–897.
4. Trias, F. X., Verstappen, R. W. C. P., Gorobets, A., Soria, M., and Oliva, A. (2010) Parameter-free symmetry-preserving regularization modeling of a turbulent differentially heated cavity. *Computers & Fluids*, **39**, 1815–1831.
5. Verstappen, R. W. C. P. and Veldman, A. E. P. (2003) Symmetry-Preserving Discretization of Turbulent Flow. *Journal of Computational Physics*, **187**, 343–368.
6. Trias, F. X., Gorobets, A., Soria, M., and Oliva, A. (2010) Direct numerical simulation of a differentially heated cavity of aspect ratio 4 with Ra-number up to 10^{11} - Part I: Numerical methods and time-averaged flow. *International Journal of Heat and Mass Transfer*, **53**, 665–673.
7. Trias, F. X., Gorobets, A., Soria, M., and Oliva, A. (2010) Direct numerical simulation of a differentially heated cavity of aspect ratio 4 with Ra-number up to 10^{11} - Part II: Heat transfer and flow dynamics. *International Journal of Heat and Mass Transfer*, **53**, 674–683.
8. Trias, F. X., Soria, M., Oliva, A., and Verstappen, R. W. C. P. (2006) Regularization models for the simulation of turbulence in a differentially heated cavity. *Proceedings of the European Computational Fluid Dynamics Conference (ECCOMAS CFD 2006)*, Egmond aan Zee, The Netherlands, September.
9. Trias, F. X., Verstappen, R. W. C. P., Soria, M., Gorobets, A., and Oliva, A. (2008) Regularization modelling of a turbulent differentially heated cavity at $Ra = 10^{11}$. *5th European Thermal-Sciences Conference, EUROTHERM 2008*, Eindhoven, The Netherlands, May.
10. Sagaut, P. and Grohens, R. (1999) Discrete Filters for Large Eddy Simulations. *International Journal for Numerical Methods in Fluids*, **31**, 1195–1220.
11. Trias, F. X. and Verstappen, R. W. C. P. (2011) On the construction of discrete filters for symmetry-preserving regularization models. *Computers & Fluids*, **40**, 139–148.
12. Meyers, J. and Sagaut, P. (2007) Is plane-channel flow a friendly case for the testing of large-eddy simulation subgrid-scale models? *Physics of Fluids*, **19**, 048105.
13. Shraiman, B. I. and Siggia, E. D. (1990) Heat transport in high-Rayleigh-number convection. *Physical Review A*, **42**, 3650–3653.
14. Sommeria, J. (1999) The elusive 'ultimate state' of thermal convection. *Nature*, **398**, 294–295.

On the relevance of discrete test-filtering in the integral-based dynamic modelling

Filippo Maria Denaro

1 Introduction

Large Eddy Simulation (LES) is based on a separation between large resolved, and small unresolved Sub-Grid Scales (SGS), obtained by applying a low-pass filtering operator on the governing equations. Actually, such filtering operation is often nothing but a formalism while writing the LES equations in continuous form. Indeed, in the so-called *implicit filtering approach* the discretization of both domain and differential operators is practically used as built-in filtering, e.g. see [1]. The characteristic filter length Δ is implicitly linked to the computational grid size step h (the only user-defined parameter) by the so-called *sub-filter resolution parameter* $Q = \Delta/h$. Hence, numerical representation of the filtered variables is associated with a finite number of resolved scales and marginal resolution.

The importance of recognizing the dependence of LES results on the numerically induced filter width was highlighted, for example [2, 3], the statistics being demonstrated to be affected by the numerical resolution for Q lower than 4. Reducing the grid size h, that is the filter length, alters the statistics, unless the energy containing eddies are completely resolved by the numerical grid. Stolz and co-authors [4] discussed the behaviour of the mean velocity profile in a channel flow LES simulation. Only a no-model LES surprisingly appears as the best solution. The tendency to overshoot the law of the wall and the sensitivity of the mean velocity to the test-filtering parameters is observed also for spectral LES with the dynamic Smagorinsky model, as demonstrated by several tests on channel flow reported by Sarghini et al. [5]. Turbulent channel flow simulations are also presented in Ref. [6] where a systematic comparison of the LES results for different grid resolutions, finite difference (FD) schemes and turbulence closure models is performed. Very recently,

F.M. Denaro
Department of Aerospace and Mechanical Engineering, Via Roma 29, 81031, Aversa (CE), Italy, e-mail: denaro@unina.it, also affiliated at ICAR-CNR, Via P. Castellino, Napoli, Italy

H. Kuerten et al. (eds.), *Direct and Large-Eddy Simulation VIII*,
ERCOFTAC Series 15, DOI 10.1007/978-94-007-2482-2_5,

Brasseur and Wei [7] discussed the problem of the overshoot in the law of wall profile, addressing three key-parameters that can affect the LES result.

These studies clearly confirm the existence in LES of a complex "mixing" between numerical, filtering and modeling effects. Thus, the aim of the present work is to investigate the role of the built-in filter parameters on the LES solution obtained by using the new integral-based eddy viscosity dynamic procedure [8, 9] discretized in a Finite Volume (FV) manner. Indeed, while using the integral form instead of the differential divergence form, many new aspects emerge that are not fully explored and assessed in the LES framework [9, 10]. Depending on the accuracy order of the discretization of the divergence form, the implicit shape of the filter is not constrained to be the top-hat filter, e.g. see [11]. Conversely, the integral form drives to solve univocally for an approximated top-hat filtered variable [9]. Provided that a particular decomposition between resolved and unresolved fluxes is introduced in the integral-based filtered equation, no commutation property must be invoked to write the filtered equations [8, 9, 12, 13]. The proper choice of the test-filtering length is a consequence of the resulting value Q in effect for the chosen FV method and the closure modeling contains important contribution that is sensitive to the adopted spatial discretization. In the following, a section depicts the framework of the integral-based filtered equations and the consequent new dynamic procedure. Then, the discretization of both equations and test-filtering is introduced and an assessment of the role of the discrete filter parameters is given. The conclusion paragraph addresses the main achievement of this study.

2 The framework of the analysis

The continuous integral-based filtered momentum equation writes in compact way [8, 9] as

$$\frac{\partial \overline{\mathbf{v}}}{\partial t} + \overline{\underline{\nabla} \cdot \mathbf{F}_R(\overline{\mathbf{v}})} = \overline{\underline{\nabla} \cdot \mathbf{T}(\overline{\mathbf{v}}, \mathbf{v})}, \tag{1}$$

governing the volume-averaged velocity $\overline{\mathbf{v}}(\mathbf{x},t) = \frac{1}{|\Omega(\mathbf{x})|} \int\limits_{\Omega(\mathbf{x})} \mathbf{v}(\mathbf{x}',t)\, d\mathbf{x}'$, equivalent to filtering with the top-hat filter kernel of uniform characteristic length Δ. A proper decomposition of the total flux $\mathbf{F}$ in the resolvable $\mathbf{F}_R(\overline{\mathbf{v}})$ and irresolvable $\mathbf{T}(\overline{\mathbf{v}}, \mathbf{v}) = \mathbf{F}_R(\overline{\mathbf{v}}) - \mathbf{F}(\mathbf{v}) = (\overline{\mathbf{v}}\,\overline{\mathbf{v}} - \mathbf{v}\mathbf{v}) - (2\nu \underline{\nabla}^s \overline{\mathbf{v}} - 2\nu \underline{\nabla}^s \mathbf{v})$ fluxes, was introduced. This way, the commutation property between filter and derivatives is never invoked. Note that the symbol bar over the divergence operators in (1) is used only for a compact notation since in the FV implementation one considers the surface integral of the fluxes [8, 9]. By applying the top-hat test-filter of width Δ_t on the Eq.(1) one defines the sub-test tensor $\mathbf{M} = (\overline{\overline{\mathbf{v}}}^{\Delta_t}\, \overline{\overline{\mathbf{v}}}^{\Delta_t} - \mathbf{v}\mathbf{v}) - (2\nu \underline{\nabla}^s \overline{\overline{\mathbf{v}}}^{\Delta_t} - 2\nu \underline{\nabla}^s \mathbf{v})$, resulting the identity

$$\mathbf{M} - \mathbf{T} = \left(\overline{\overline{\mathbf{v}}}^{\Delta_t}\, \overline{\overline{\mathbf{v}}}^{\Delta_t} - \overline{\mathbf{v}}\,\overline{\mathbf{v}} \right) - \left(2\nu \underline{\nabla}^s \overline{\overline{\mathbf{v}}}^{\Delta_t} - 2\nu \underline{\nabla}^s \overline{\mathbf{v}} \right) \equiv \mathbf{L}, \tag{2}$$

the integral-based resolved tensor stress being denoted by $\mathbf{L}$ in analogy with the classic Leonard tensor. The (2), written for the deviatoric parts (denoted by the subscript *dev*), represents the counterpart of the Germano identity here originally extended to the integral form of the equation [8, 9]. After introducing the eddy viscosity assumption for modelling the deviatoric part of the tensor, that is $\mathbf{T}_{dev} \cong 2\nu_{LES}\underline{\nabla}^s\overline{\mathbf{v}}$ and $\mathbf{M}_{dev} \cong 2\nu'_{LES}\underline{\nabla}^s\overline{\overline{\mathbf{v}}}^{\Delta_t}$, by assuming the same coefficient and the same rate of dissipation at primary filter and test-level (scale-invariance hypothesis) the eddy viscosity can be determined by means of the scaling law model [8, 9]. That is, one writes $\nu_{LES} = C_\varepsilon \Delta^{4/3}$ and $\nu'_{LES} = C_\varepsilon \Delta_t^{4/3}$, being $C_\varepsilon \equiv C^{2/3}\varepsilon^{1/3}$ with ε the dissipation rate. Thus, after projecting along the test-direction tensor $\underline{\nabla}^s\overline{\overline{\mathbf{v}}}^{\Delta_t}$, one gets

$$2\nu_{LES}\left(\underline{\nabla}^s\overline{\mathbf{v}} - \alpha^{4/3}\underline{\nabla}^s\overline{\overline{\mathbf{v}}}^{\Delta_t}\right) : \underline{\nabla}^s\overline{\overline{\mathbf{v}}}^{\Delta_t} = \mathbf{L}_{dev} : \underline{\nabla}^s\overline{\overline{\mathbf{v}}}^{\Delta_t}, \tag{3}$$

being $\alpha = \Delta_t/\Delta$ the ratio of the *test to FV-based implicit filter width* which is an input parameter to be prescribed by the user. The relevance of (3) appears as, owing to the integral-based formulation, the practical determination of $\nu_{LES}(\mathbf{x},t)$ does not require to arbitrarily extract the model viscosity out of the filtering operation, as conversely happens for the differential-based LES dynamic formulations [1]. Congruently, no additional averaging is necessary during the computation and the (3) can be used to determine an effective three-dimensional eddy viscosity field [8, 9].

3 Computational assessment. Results

As a matter of fact, any LES is always a numerical procedure thus, the filtered velocity turns to be only approximated by the computed finite-dimensional vector field, say $\overline{\mathbf{v}}^d$. The problem in practically resolving (3) is that the effective filter length Δ is not prescribed but its value Qh is implicitly defined by the grid-size and the chosen type of discretization, e.g. see [1, 9, 11–13]. Moreover, one has to compute the discrete test-filtered velocity, say it $\overline{\overline{\mathbf{v}}^d}^{d,\Delta_t}$, its characteristic length Δ_t depending on the adopted computational formula. Thus, fixing a-priori a correct value for α is not straightforward, in a LES computation we need to prescribe some suitable value α_d, prescribed while considering the actual discrete test filter employed. Hence, the effective eddy viscosity, computed by discretizing the (3), is

$$\nu^d_{LES} = \frac{\mathbf{L}^d_{dev} : \underline{\nabla}^d\overline{\overline{\mathbf{v}}^d}^{d,\Delta_t}}{2\left(\underline{\nabla}^d\overline{\mathbf{v}}^d - \alpha_d^{4/3}\underline{\nabla}^d\overline{\overline{\mathbf{v}}^d}^{d,\Delta_t}\right) : \underline{\nabla}^d\overline{\overline{\mathbf{v}}^d}^{d,\Delta_t}}. \tag{4}$$

For instance, Sarghini *et al.* [5] have demonstrated that better results for channel turbulent flow can be obtained by properly tuning the value of Δ_t/h according to the effective filter discretization.

In order to properly assess the congruent filter parameters, one needs to estimate the ratio $\Delta_t/h = (\Delta_t/\Delta)(\Delta/h) = \alpha Q$. Substantially, the three computational filter parameters $\Delta_t/h, \alpha, Q$, linked each other as $\alpha = \alpha_d(Q, \Delta_t/h)$, have to be taken into account for a correct dynamic modeling.

Here, the discrete test-filtered velocity is computed by means of the fourth order accurate 2D extension of the Simpson rule, according to two different local stencils, which are the $3\Delta x \times 3\Delta z$ (case 1) and $5\Delta x \times 5\Delta z$ FVs (case 2), respectively. An approach to estimate the test-filtering width of such formulas, consists in interpreting the wave number k_{eff}, for which the transfer function gets the value 0.5, as the effective cut-off frequency, e.g. [1]. By computing the transfer function expressions $\widehat{G}(\xi,\zeta)$ for cases 1, 2, assuming $h = (\Delta x^2 + \Delta z^2)^{1/2}$ and fixing $\xi = k_x\Delta x = \zeta = k_z\Delta z$, one has $\Delta_t/h = \pi/(k_{eff}h) = \pi/\xi_{eff} = 2.168$ and 4.51, respectively. A different evaluation of the test-filter width is obtained according to the standard deviation of the corresponding discrete filter function, that is $\Delta_t/h = \sqrt{12 \sum\limits_{I,K}(I^2+K^2)W_{IK}}$ being W_{IK} the coefficients of the discrete filtering expression centered around the $I,K = (0,0)$ local grid point. By evaluating for the coefficients of the case 1 and 2, one gets $\Delta_t/h = 2\sqrt{2} = 2.828$ and $4\sqrt{2} = 5.657$, respectively.

An assessment of the effective role of the filter parameters $\alpha_d, Q, \Delta_t/h$ is here obtained exploiting the classical problem of the plane turbulent channel flow at Re_τ =590, e.g. see [13]. The LES is performed by adopting a second order accurate projection-based FV method, see [8, 9, 15]. The dependence of the results under two parameters are analyzed: *a*) the computational length h (and the induced filter length Δ=Qh) and *b*) the chosen ratio α_d. To this aim, three computational grids were used. All of them have 64 cells, non-uniformly distributed in vertical direction such that the computational cell-center near the walls is located at $y^+ \approx 0.33$. Along the stream-wise and span-wise directions three uniform grids of 32×42 (Coarse grid - CG), 48×64 (Mid grid - MG) and 64×80 (Fine grid - FG) FVs, are used. The computational mesh sizes in wall-units are $\Delta x^+ \cong 116, \Delta z^+ \cong 44$, $\Delta x^+ \cong 77, \Delta z^+ \cong 29$ and $\Delta x^+ \cong 58, \Delta z^+ \cong 23$, for CG, MG and FG respectively. Then, for each grid, the solutions are obtained for a set of three arbitrarily chosen values $\alpha_d = 3,4,5$ and $\alpha_d = 2,3,4$, tested for case 2 and 1, respectively. The computational time step is fixed to $\Delta t = 5 \times 10^{-5}$. For the sake of completeness, also a no-model simulation is performed on each one of the three grids. The resulting statistically averaged streamwise velocity profiles are organized in Fig. 1, from top to bottom, for CG, MG and FG, respectively along with the no-model and DNS [14] solutions. The case 1 is shown on the right column, the case 2 on the left one. It is worthwhile remarking for vanishing h, the effective filter width is accordingly reduced as Δ=Qh, hence one gets no convergence towards a filtered solution. Therefore, it is not possible to clearly distinguish between the effect due to the reduction of the local truncation error magnitude and the one due to the reduction of the filtering length. The only proper behavior would be that the LES solution tends to the DNS for vanishing computational size, as is shown in Fig. 1. One can see that the results are strongly affected by both the ratio α_d and the formula of the test filtering. In case 2, one sees a systematic deviation from the logarithmic slope as well as a greater values in the inertial region than case 1. For this latter, it appears that $\alpha_d = 2$, perhaps a commonly adopted value, produces the worst solution, the best one appearing for $\alpha_d = 4$. The no-model simulation on the fine grid appears quite in accordance with the DNS solution. Actually, in terms of LES solutions there is also no theoretical reasons to expect an exact concordance of the statistically averaged velocity with the logarithmic law produced by DNS [4].

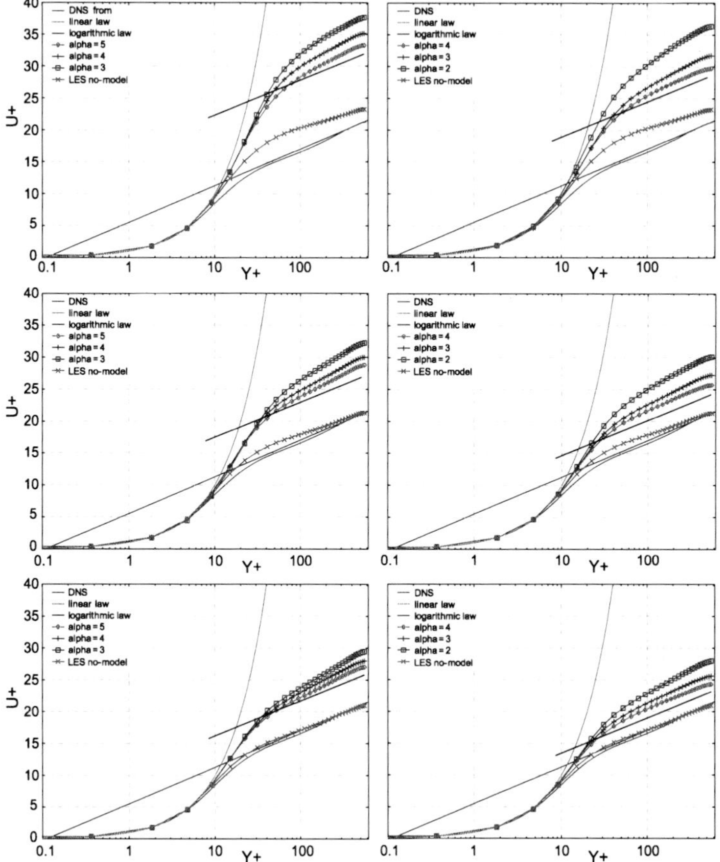

Fig. 1 Statistically averaged stream-wise velocity profiles for (from top to bottom) Coarse Grid, Mid Grid, Fine Grid. Right column case 1, left column case 2. DNS data are taken from Ref. [14].

4 Conclusions

This paper focused onto the relevance of the (primary and test) filter parameters in the LES of the turbulent channel flow. A new integral-based eddy viscosity dynamic SGS procedure was addressed, allowing for a mathematical consistent three-dimensional eddy viscosity to be deduced. The scaling law model is tested for several values of the computational parameters. When considering the discretization framework, the relevance of using a value of the ratio of the *test to FV-based implicit filter width* greater than the theoretical (and commonly used in LES codes) one is discovered. This is in accord to the use of an implicit primary filter. However, the scaling law model reveals somehow excessively dissipative, probably due to the lack of a real scaling in the non-homogenous part of the flow. Some simulations using the classic Smagorinsky model are in progress, preliminary results revealing an improved velocity profile since from the coarse grid.

References

1. P. Sagaut, Large Eddy Simulation for Incompressible Flows. An Introduction. 3 ed., Springer, 2006.
2. B.J. Geurts, J. Frohlich, A framework for predicting accuracy limitations in large-eddy simulation, *Phys. Fluids*, 14, L41, 2002.
3. S.B. Pope, Ten questions concerning the large-eddy simulation of turbulent flows, *New Journal of Physics*, 6, 35, 2004. Online at `http://www.njp.org/`.
4. S. Stolz, N.A. Adams, L. Kleiser, An approximate deconvolution model for large-eddy simulation with application to incompressible wall-bounded flows, *Phys. Fluids*, 13, 4, 2001.
5. F. Sarghini. U. Piomelli, E. Balaras, Scale-similar model for large-eddy simulations, *Phys. Fluids*, 11, 6, 1596–1607, 1999.
6. J. Gullbrand, F.K. Chow, The effect of numerical errors and turbulence models in large-eddy simulations of channel flow, with and without explicit filtering, *J. Fluid Mech.*, 495, 323–341, 2003.
7. J.G. Brasseur, T. Wei, Designing large-eddy simulation of the turbulent boundary layer to capture law-of-the-wall scaling, *Phys. Fluids*, 22, 021303, 2010.
8. F.M. Denaro, G. De Stefano, D. Iudicone, V. Botte, A finite volume dynamic large-eddy simulation method for buoyancy driven turbulent geophysical flows, *Ocean Modelling*, 17, 3, 199–218, 2007.
9. F.M. Denaro, G. De Stefano, A new development of the dynamic procedure in large-eddy simulation based on a Finite Volume integral approach. Application to stratified turbulence, *Theor. Comp. Fluid Dyn.*, online DOI10.1007/s00162-010-0202-x, 2010.
10. J.H. Ferziger, M. Peric, Computational Methods for Fluid Dynamics, Springer, 2001.
11. B.J. Geurts, F. van der Bos, Numerically induced high-pass dynamics in large-eddy simulation, *Phys. Fluids*, 17, 125103, 2005.
12. G. De Stefano, F.M. Denaro, G. Riccardi, High order filtering for control volume flow simulations, *Int. J. Num. Methods in Fluids*, 37, 7, 2001.
13. P. Iannelli, F.M. Denaro, G. De Stefano, A deconvolution-based fourth order finite volume method for incompressible flows on non-uniform grids, *Int. J. Num. Methods in Fluids*, 43, 4, 431–462, 2003.
14. R.D. Moser, J. Kim, N.N. Mansour, Direct numerical simulation of turbulent channel flow up to $Re_\tau = 590$, *Phys. Fluids*, 11, 4, 1999.
15. F.M. Denaro, A 3D second-order accurate projection-based finite volume code on non-staggered, non-uniform structured grids with continuity preserving properties: application to buoyancy-driven flows, *Int. J. Num. Methods in Fluids*, 52, 393–432, 2006.

Hybrid assessment method for LES models

Cassart Benjamin, Teaca Bogdan, Carati Daniele

1 Introduction

LES is the study of turbulent flows obtained by resolving only the largest scales of Navier-Stokes equations. The restriction to the largest scales is equivalent to the application of a low pass filter, denoted by the overbar symbol $\overline{\cdots}$, on the equations. We solve the equations for the filtered incompressible velocity field $\overline{u}_i$:

$$\partial_t \overline{u}_i = -\partial_j(\overline{\overline{u}_j \overline{u}_i}) + \nu \partial_j \partial_j \overline{u}_i - \partial_i \overline{p} + \partial_j \tau_{ij}, \tag{1}$$

$$\partial_j \overline{u}_j = 0, \tag{2}$$

where ν is the kinematic viscosity of the fluid. The difference between the original and the LES Navier-Stokes equations comes from the so-called subgrid-scale stress tensor: $\tau_{ij} = \overline{u_i u_j} - \overline{\overline{u}_i \overline{u}_j}$ acting only via the gradient of its traceless tensor $h_i = \partial_j \tau^\star_{ij}$. Its exact value cannot be computed, it is needed to approximate the exact tensor $\tau^\star_{ij}$ by a subgrid-scale model tensor $\tau^\star_{ij} \approx \tau^{\mathrm{M}}_{ij}(\overline{u}_l)$. The present study proposes a new framework to assess the LES models. So far, the assessment of subgrid-scale model relies either on a priori or a posteriori methods.

A priori testing consists in computing the exact tensor $\tau^\star_{ij}$ from an accurately simulated (direct numerical simulation; DNS) flow [1] or from an experimentally measured flow [2]. This exact tensor is then compared with the model tensor, typically by computing the correlation between the exact and model subgrid-scale ten-

Cassart Benjamin
Université Libre de Bruxelles, Campus Plaine, CP 231, B-1050 Brussels, Belgium, e-mail: bcassart@ulb.ac.be

Teaca Bogdan
Université Libre de Bruxelles, Campus Plaine, CP 231, B-1050 Brussels, Belgium, e-mail: bteaca@ulb.ac.be

Carati Daniele
Université Libre de Bruxelles, Campus Plaine, CP 231, B-1050 Brussels, Belgium, e-mail: dcarati@ulb.ac.be

H. Kuerten et al. (eds.), *Direct and Large-Eddy Simulation VIII*,
ERCOFTAC Series 15, DOI 10.1007/978-94-007-2482-2_6,

sors $C = \langle \tau^{\star}_{ij} \tau^{\text{M}}_{ij} \rangle / \langle \tau^{\star}_{kl} \tau^{\star}_{kl} \rangle^{1/2} \langle \tau^{\text{M}}_{st} \tau^{\text{M}}_{st} \rangle^{1/2}$ where $\langle \ldots \rangle$ denotes the volume average. The comparison is even sometimes limited to the effect of the subgrid-scale terms on the energy balance by comparing only the correlation between $\overline{u}_i h_i$ and $\overline{u}_i h^{\text{M}}_i$ where h^{M}_i is given by $-\partial_j \tau^{\text{M}}_{ij}$. The drawback of a priori approach is that high correlations do not guarantee accurate predictions of the evolution of the flow field.

A posteriori assessment methods consist in comparing not the causes, *i.e.* τ_{ij} or h_i as in a priori methods, but the effects, *i.e.* the velocity field computed from an actual LES. This is achieved by running a LES with the subgrid-scale model for a few eddy turn over times and comparing its statistics with the statistics measured from the filtering of either DNS run or experimental data. In a posteriori methods, point-wise comparison is not appropriate anymore, since the two turbulent fields (LES and DNS) are expected to evolve differently in the turbulent regime as soon as the subgrid-scale model is not exact. This allows to gain a better insight on the effect of the model but only through global or locally averaged comparisons.

This article presents a new method between a priori and a posteriori approaches gathering the advantages of both. First the force created to reach this goal is presented. Then different first diagnostics are discussed. Asymptotic limits show that this method can behave as the two usual assessment methods. Finally a few models are tested to compare this method with a posteriori one. We will see that this method gives at least as much information as an a posteriori model, but with the point wise accuracy of an a priori model.

2 Framework

The methodology presented here is an attempt to combine the advantages of both methods. As in an a priori method, DNS and LES are run together to compare h_i on a point-wise basis. As in a posteriori method, knowledge about the dynamic role of the model can be assessed. A time evolving a priori method should be the best method but the calculations to obtain $\tau^{\star}_{ij}$ are resource consuming. Therefore a need to shift toward a more a posteriori analysis. To avoid the correlation loss of the fields seen in a posteriori method, a numerical force term is used to bring back the LES velocity field when it deviates from the value of the filtered DNS one. In practice, DNS and LES are run. The LES is having both a model for the subgrid-scale term and the force q_i: $\partial_t \overline{u}^{\text{LES}}_i = -\partial_j (\overline{\overline{u}_j \overline{u}_i}) + \nu \partial_j \partial_j \overline{u}_i - \partial_i \overline{p} + \partial_j \tau_{ij} + q_i$. This "restoring" force q_i, can be written as

$$q_i = \alpha (\overline{u}^{\text{DNS}}_i - \overline{u}^{\text{LES}}_i) \,, \tag{3}$$

where α has the dimension of the inverse of a time. Using this force allows to keep the precision of a priori method during the whole run. Restoring force will not only serve as a way to enforce the LES field to stay in the vicinity of the DNS field, but the importance of this force field gives a point-wise information on the amount of deviation suffered by the LES field at every time step. The computation cost of this force is lower than the calculation of $\tau^{\star}_{ij}$.

The force amplitude, $A = q_i q_i$, indicates the local intensity of the restoring force and gives a point wise information on how far the fields end up. Monitoring the energy injection of this force, $P = \overline{u}_i^{\rm LES} q_i$, allows to see if energy is injected ($P > 0$) or removed ($P < 0$) when $\overline{u}_i^{\rm LES}$ deviates from $\overline{u}_i^{\rm DNS}$. The quantity $I = \nu \partial_j \partial_j \overline{u}_i^{\rm LES}\, q_i$ is the term arising in the dissipation evolution equation due to the restoring force. Two quantities that do not involve the restoring force are considered: the relative errors in the kinetic energy and in the energy dissipation: $\Delta e = (E(\overline{u}^{\rm DNS}) - E(\overline{u}^{\rm LES}))/E(\overline{u}^{\rm DNS})$ and $\Delta d_\nu = (D_\nu(\overline{u}^{\rm DNS}) - D_\nu(\overline{u}^{\rm LES}))/D_\nu(\overline{u}^{\rm DNS})$ both computed from the value of the exact field on the filtered grid, where $E = \langle \overline{u}_i \overline{u}_i \rangle/2$ is the kinetic energy and $D_\nu = \nu \langle \overline{u}_i \partial_j \partial_j \overline{u}_i \rangle = -2\nu \langle \overline{S}_{ij} \overline{S}_{ij} \rangle$ is the viscous dissipation and where $\overline{S}_{ij} = \frac{1}{2}(\partial_i \overline{u}_j + \partial_j \overline{u}_i)$ is the strain tensor.

The value given to the parameter α determines the time-scale at which the restoring force will react. In the limit $\alpha \to 0$, the restoring force vanishes and the LES and the filtered DNS fields move away from each other rapidly. In that case, the comparison between $\overline{u}_i^{\rm LES}$ and $\overline{u}_i^{\rm DNS}$ simply reduces to an a posteriori assessment of the model used in $\tau_{ij}^{\rm M}$. In the limit $\alpha \to \infty$ the restoring force seems to tend to infinity. In this limit, $\overline{u}_i^{\rm LES}$ and $\overline{u}_i^{\rm DNS}$ are also expected to remain extremely close to each other which tends to reduce the forcing amplitude. In order to determine the dominant effect of the forcing for very large value of α, the following formal solution of the LES equations is needed:

$$\overline{u}_i^{\rm LES}(t) = \int_{-\infty}^{t} e^{-\alpha(t-t')} \left(\alpha\, \overline{u}_i^{\rm DNS}(t') + rhs[\overline{u}_\ell^{\rm LES}(t')] - \partial_j \tau_{ij}^{\rm M}(t') \right) dt' . \qquad (4)$$

Since α is large, the integrand is very much dominated by the values $t' \approx t$ and can be expanded in time. Keeping only the lowest order terms in $\tau = \alpha^{-1}$, the following approximation can easily be derived:

$$\overline{u}_i^{\rm LES} = \overline{u}_i^{\rm DNS} - \tau \left(\partial_t \overline{u}_i^{\rm DNS} - rhs[\overline{u}_\ell^{\rm LES}] + \partial_j \tau_{ij}^{\rm M} \right) + \mathcal{O}(\tau^2) , \qquad (5)$$

where all the quantities appearing in this expression have to be evaluated at time t. The time derivative of the DNS field can be expressed using: $\partial_t \overline{u}_i^{\rm DNS} = \overline{rhs[u_\ell^{\rm DNS}]} = rhs[\overline{u}_\ell^{\rm DNS}] - \partial_j \tau_{ij}^\star$. Moreover, the expansion (5) shows that $\overline{u}_i^{\rm LES}$ and $\overline{u}_i^{\rm DNS}$ differ at most by a term of order τ. As a consequence $rhs[\overline{u}_\ell^{\rm DNS}] - rhs[\overline{u}_\ell^{\rm LES}]$ should also be of order τ. In that case, the expansion (5) reduces to $\overline{u}_i^{\rm LES} = \overline{u}_i^{\rm DNS} - \tau \partial_j \left(\tau_{ij}^{\rm M} - \tau_{ij}^\star \right) + \mathcal{O}(\tau^2)$ and the force reduces to $h_i^\star - h_i^{\rm M}$ with an amplitude independent of α. The analysis of the force for very large α leads to results that are independent of α. It is equivalent to a time-dependent a priori analysis performed in the evolution of the simulation. A valid choice of value for α would be to have it large enough so that its time-scale is smaller than any other time-scale present in the LES.

Many models could be assessed by this method, and will be. The first models studied here are the classical Smagorinsky eddy viscosity model [3]: $\tau_{ij}^{\rm M} = -2\,(C_S \overline{\Delta})^2\, |\overline{S}|\, \overline{S}_{ij}$ where $\overline{\Delta}$ is the LES grid spacing, $|\overline{S}| = (2\overline{S}_{ij}\overline{S}_{ij})^{1/2}$ and C_S is the Smagorinsky constant that has to be determined. Two values will be tested ($C_S = 0.1$ & $C_S = 0.2$). The dynamic Smagorinsky model in which the constant C_S is self-calibrated during the LES run using the dynamic procedure is also assessed with a volume average. The optimal value that can be taken by this type of model is the one obtained by taking $\tau_{ij}^{\text{M-1}} = \overline{S}_{ij}(\tau_{kl}^{\star}\overline{S}_{kl})/(\overline{S}_{st}\overline{S}_{st})$, *i.e.* the dissipative part of $\tau_{ij}^{\star}$.

In the tests presented here, we investigate a decaying isotropic turbulent flow. The DNS is performed with 512^3 grid points and the filtered grid for the assessed models has 32^3 grid points. The simulations are run on the TURBO code, using a 3D pseudo-spectral approach. Time advancement uses a 3rd order Runge-Kutta scheme with a time step calculated based on CFL criteria. Nonlinear products are dealiased using a two-third method. Due to the box size, its grid, the DNS resolved flow gives an initial Taylor scale Reynolds number $Re_{\lambda} = 160$. The flow is evolving in a cubic box with a length $L = 2\pi$ and periodic boundary conditions in the three directions.The DNS flow is ran simultaneously with a number of LES realizations that receive informations from the DNS to compute the q_i force or $\tau_{ij}^{\star}$ and its projections. All the simulations are run with the same time step by taking the minimum of all the CFL conditions, which always corresponds to the DNS. The Smagorinsky models are compared to the best model, $\tau_{ij}^{\star}$, the best dissipative model, $\tau_{ij}^{\text{M-1}}$ and the worst model, *i.e.* no model used.

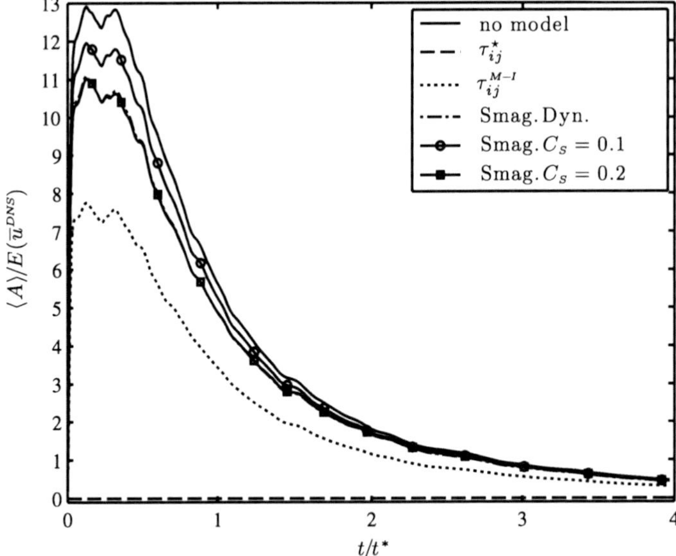

Fig. 1 The evolution of the restoring force amplitude, $\langle A \rangle$, normalized by the DNS energy on the filtered grid.

A first diagnostic is the amplitude of the restoring force applied, averaged over the volume in figure 1. This may serve as a benchmarking tool. The figure shows that $\tau_{ij}^{\star}$ behaves the best, with a force nearly at 0. On the opposite side, the LES with no model has the need for a greater force. In between, it seems that dynamical Smagorinsky model behaves as well as the constant one using $C_S = 0.2$.

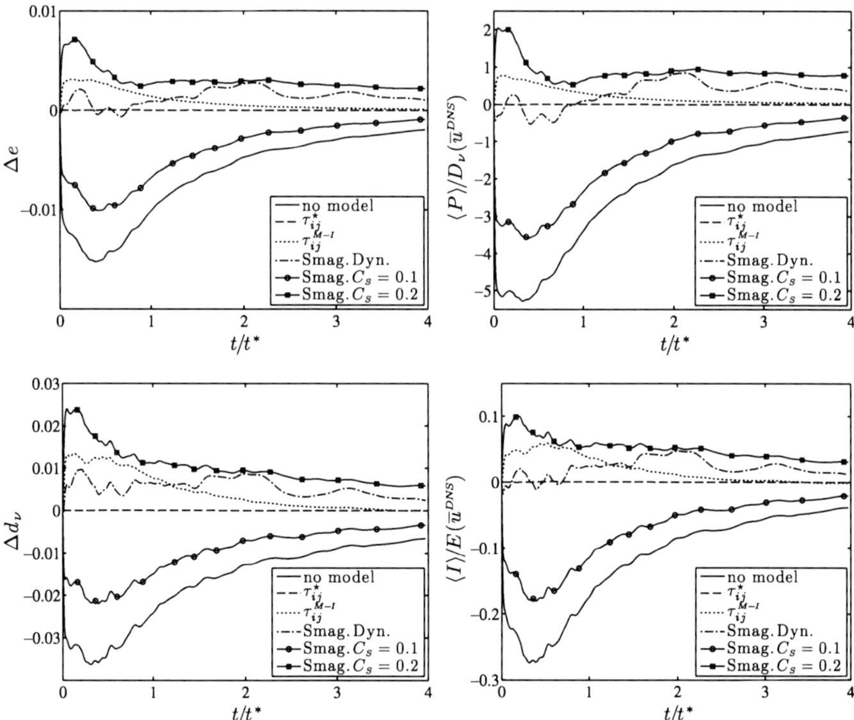

Fig. 2 The evolution of the relative energy difference for the different models (top left), of the global power injected by the restoring force, $\langle P \rangle$, normalized by the DNS energy dissipation on the filtered grid (top right), of the relative dissipation difference for the different models (bottom left) and of the global influence on dissipation from the restoring force, $\langle I \rangle$, normalized by the DNS energy on the filtered grid (bottom right).

Comparing the a posteriori diagnostic Δe to the volume average of the power injected by the restoring force shows a similar behavior, as can be seen on figure 2, top. It is expected that a LES flow with an excess of energy will see the restoring force remove the excess and vice versa. The new diagnostic P_i possesses, contrarily to Δe, a full volume information, from which can be plotted more point wise informations (still to be developed). Similar observations can be made about ΔD_ν and the volume averaged influence on dissipation of the restoring force on figure 2, bottom. We are able here, as well, to extract more than a global analysis from the LES flow and the behavior of his model.

The diagnostics shown here display only global information, it is normal to observe that a good model for a posteriori method will be good with hybrid method as well. Deeper investigations will give access to finer spectral and spatial information up to point wise accuracy to evaluate the efficiency of a model for the different scales, to see if the model is more able to take care of dissipation or helicity, etc.

More details about those results will be published soon [4].

3 Future work

This method shows a compromise between a priori and a posteriori assessments for subgrid-scale models. This requires to access during the whole run to a DNS database or parallel run for the computation of the restoring force. This article shows the first steps taken to explore this hybrid method, with a few classical cases. These promising results open new perspectives for the future.

Diagnostics can be improved by outputting more than the volume averaged value. The spectrum from A_i, P_i and I_i could grant significant information about the model efficiency at different scales. Outputting the full volume restoring force field value and displaying it next to the velocity field could help assessing the role of the tested model, giving a full 3D view with respect to turbulent structures.

More LES models are implemented in TURBO and will be tested together once a satisfactory number of them will be reached. Other codes could benefit from adapting to such an assessment method to evaluate models for near wall region.

Acknowledgements B.C. is research fellow of the FRIA (Belgium) and D.C. is research director of the FRS-FNRS (Belgium). This work has been supported by the contract of association EURATOM-Belgian State. The content of the publication is the sole responsibility of the authors and it does not necessarily represent the views of the Commission or its services.

References

1. Lund T.S. and Novikov E.A.: Parametrization of subgrid-scale stress by the velocity gradient tensor. Center for Turbulence Research - Annual Briefs, pp. 27–43 (1992).
2. Chamecki M., Meneveau C., and Parlange M.B.: The local structure of atmospheric turbulence and its effect on the Smagorinsky model for large eddy simulation. American Meteorological Society **64**, 1941 (2007).
3. Smagorinsky J.: General circulation experiments with primitive equations: 1. The basic experiment. Mon. Weather Rev. **91**, 99 (1963).
4. B. Cassart, B. Teaca, and D. Carati, *A general assessment method for subgrid-scale models in LES*, accepted by Physics of Fluids, to be published.

Time-reversibility of Navier-Stokes turbulence and its implication for subgrid-scale models

L. Fang, W.J.T. Bos, L. Shao and J.-P. Bertoglio

1 Introduction

The last two decades have seen the emergence of a large number of new subgrid models, and it is the authors' opinion that the turbulence community must devote more efforts in developing consensual criteria than in increasing the number of models. The purpose of the present letter is to investigate one possible criterion, which is the time-reversibility of a subgrid model, when the sign of the velocity is inverse, *i.e.*, under the transformation $\boldsymbol{u} \rightarrow -\boldsymbol{u}$. We report the results of Eddy-Damping Quasi-Normal (EDQN) simulations in which the velocity is reversed in all, or part, of the scales of the flow. These results are then compared to results from LES in which the velocity is reversed, to assess the quality of the predictions of the models and to check whether time-reversibility is a valid criterion to assess subgrid models.

In the absence of viscosity, the dynamics of the Navier-Stokes equations (which reduce to the Euler equations), are time-reversible under the transformation $\boldsymbol{u} \rightarrow -\boldsymbol{u}$. This means that the flow will evolve backwards in time until the initial condition is reached. This can be understood since the nonlinear interactions, which govern the cascade of energy between scales, are associated with triple velocity correlations. The sign of these triple products is changed when the velocity is reversed, so that the nonlinear energy transfer proceeds in the opposite direction. This symmetry is broken as soon as viscous dissipation is introduced. Indeed, the conversion of kinetic energy to heat through the action of viscous stresses is an irreversible process within the macroscopic (continuum) description of turbulence.

If we consider LES and we reverse the resolved velocity, it is not known what the subgrid-model is supposed to do. For convenience we will limit our discussion to the most widely used class of models, based on the concept of eddy-viscosity [1]. The eddy-viscosity model expresses the subgrid stress as a function of the resolved scales $\tau_{ij}^{<} - \frac{1}{3}\tau_{mm}^{<}\delta_{ij} = -2\nu_t S_{ij}^{<}$, where $\bullet^{<}$ denoting a filtered quantity, ν_t the

LMFA-CNRS, Université de Lyon, Ecole Centrale de Lyon, Université Lyon 1, INSA Lyon, 69134 Ecully, France, e-mail: le.fang@ec-lyon.fr

H. Kuerten et al. (eds.), *Direct and Large-Eddy Simulation VIII*, ERCOFTAC Series 15, DOI 10.1007/978-94-007-2482-2_7,

eddy-viscosity, $\tau_{ij}^{<} = (u_i u_j)^{<} - u_i^{<} u_j^{<}$ and $S_{ij}^{<} = \frac{1}{2}\left(\frac{\partial u_i^{<}}{\partial x_j} + \frac{\partial u_j^{<}}{\partial x_i}\right)$. Note that here we only consider the effect of physical filter but not the discretization. Considering the work of Carati *et al.* [2], we may also call $\tau_{ij}^{<}$ "filtered-scale stress". Following most existed works, we consider that subgrid models aim at simulating this part rather than the error of discretization. For certain models, the reversal of the velocity leads to a reversal of the subgrid stress tensor, for others not. Indeed, the dynamic model [3] is time-reversible, in the sense that the direction of the energy flux at the filter-size reverses when the resolved velocity is reversed. Another reversible model is the CZZS model [4] by Cui *et al.* as well as its recently proposed extension [5]. In this letter the simplified formulation of the CZZS model is used as an example of a time-reversible model. In this model the eddy-viscosity is given by

$$\nu_t = \frac{1}{8}\frac{D_{lll}^{<}}{D_{ll}^{<}}\Delta, \tag{1}$$

where $D_{ll}^{<} = \langle (u_1^{<}(x_1+\Delta) - u_1^{<}(x_1))^2 \rangle$ is the second-order structure function of filtered velocity, $D_{lll}^{<}$ is the third-order structure function and Δ is the filter size. This model is time-reversible since the third-order structure function changes sign when $\boldsymbol{u}^{<} \rightarrow -\boldsymbol{u}^{<}$. For the Smagorinsky [1] model this is not the case. For this model, the eddy-viscosity is given by

$$\nu_t = (C_s\Delta)^2 \sqrt{\left\langle S_{ij}^{<} S_{ij}^{<} \right\rangle}, \tag{2}$$

where $\langle\ \rangle$ indicates an ensemble average which is in practice often treated as an average in the homogeneous directions. The eddy-viscosity in Eq. (2) can not become negative, so that the net flux of energy to the subgrid scales is always greater than or equal to zero. The time-reversibility property of models, such as for example the dynamic model, is sometimes seen as a weakness [6], since the models becomes more easily (numerically) unstable. However it is well known that the backscatter of energy, to the resolved scales is a physical property, which should be taken into account in a correct model of the subgrid dynamics [7].

In the present letter we will consider by EDQN the dynamics of subgrid and resolved scale energy after transformation $\boldsymbol{u} \rightarrow -\boldsymbol{u}$. We will define a resolved velocity field, $\boldsymbol{u}^{<}$ and a subgrid velocity $\boldsymbol{u}^{>}$. In LES the small scales are not known. It is therefore interesting to know how the resolved scales can be affected by different subgrid-scale dynamics. We therefore investigate two additional cases. In the first one the subgrid scales are not reversed, in the second one these scales are set to zero. The four different cases considered are resumed in Table 1.

Table 1 Overview of the EDQN cases. The letter N denotes normal, R reversed and Z zero.

	$\boldsymbol{u}^{<}+\boldsymbol{u}^{>}$	$-\boldsymbol{u}^{<}-\boldsymbol{u}^{>}$	$-\boldsymbol{u}^{<}+\boldsymbol{u}^{>}$	$-\boldsymbol{u}^{<}+0$
EDQN cases	NN	RR	RN	RZ

2 Numerical results

Before calculating the inverse cases, a transfer spectrum $T(k)$ of full-developed turbulence is obtained from Eddy-Damping Quasi-Normal Markovian (EDQNM) theory [8], *i.e.*

$$T(k) = \left(\frac{\partial}{\partial t} + 2\nu k^2\right) E(k) = \iint dp\, dq\, \theta_{kpq} \frac{k}{pq} b(k,p,q) E(q) \left[k^2 E(p) - p^2 E(k)\right]. \tag{3}$$

In LES, considering a cutoff wave number k_c, it can be decomposed into six parts, *i.e.* $T^{<<<}$, $T^{<<>}$, $T^{<>>}$, $T^{><<}$, $T^{><>}$ and $T^{>>>}$, where the superscripts indicate k, p, q respectively, $<$ resolved scale and $>$ subgrid scale. For instance

$$\begin{aligned} T^{<<>}(k|k<k_c) = & \int_0^{k_c} dp \int_{k_c}^{\infty} dq \theta_{kpq} \frac{k}{pq} b(k,p,q) E(q) \left[k^2 E(p) - p^2 E(k)\right] \\ & + \int_{k_c}^{\infty} dp \int_0^{k_c} dq \theta_{kpq} \frac{k}{pq} b(k,p,q) E(q) \left[k^2 E(p) - p^2 E(k)\right]. \end{aligned} \tag{4}$$

Then the initial transfer spectra of the four cases can be regrouped from them:

$$\begin{aligned} T(k)^{NN} &= T^{<<<}(k) + T^{<<>}(k) + T^{<>>}(k) + T^{><<}(k) + T^{><>}(k) + T^{>>>}(k), \\ T(k)^{RR} &= -T^{<<<}(k) - T^{<<>}(k) - T^{<>>}(k) - T^{><<}(k) - T^{><>}(k) - T^{>>>}(k), \\ T(k)^{RN} &= -T^{<<<}(k) + T^{<<>}(k) - T^{<>>}(k) + T^{><<}(k) - T^{><>}(k) + T^{>>>}(k), \\ T(k)^{RZ} &= -T^{<<<}(k). \end{aligned} \tag{5}$$

These regrouped transfer spectra are shown in Fig. 1, with very high Reynolds number.

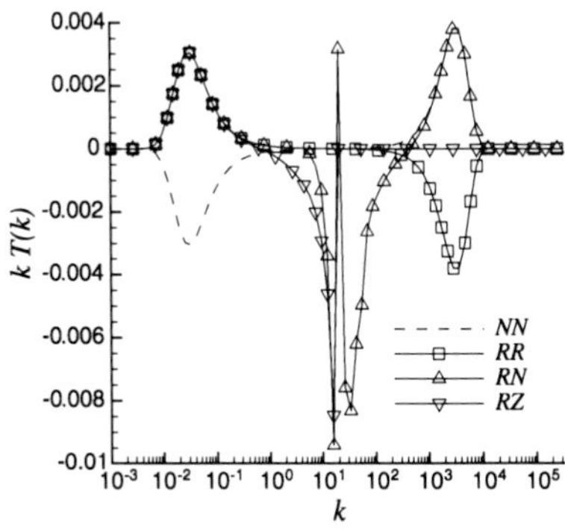

Fig. 1 Regrouping the energy transfer spectra $T(k)$ by using Eq. (5). The legends are explained in Table 1.

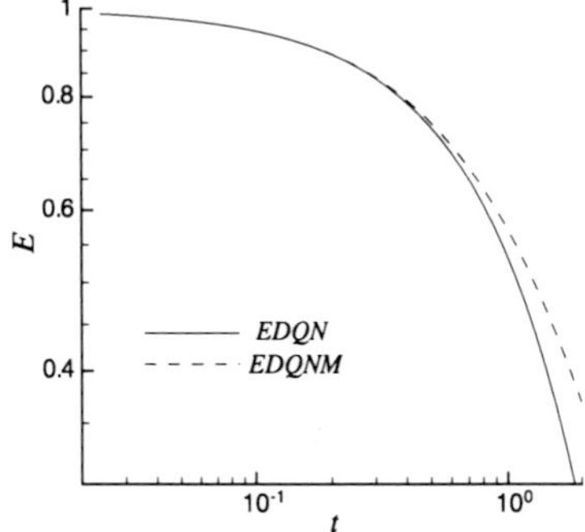

Fig. 2 Comparison of the energy decay (NN case) between EDQN and EDQNM cases.

The Markovianization of this closure does not allow to study the influence of a velocity. We therefore use the non-Markovian version of this closure (EDQN) in this research. We then compare EDQN with EDQNM result to verify its correctness. A case of free-decaying turbulence is calculated and shown in Fig. 2. The time t is normalized as $t \to (t - \tau_R)/\mathscr{T}$, with τ_R the time of reversal and $\mathscr{T}$ the turnover-time at the time of reversal, defined as $\sqrt{3/2}(\kappa/\varepsilon)$, with κ the kinetic energy and ε the viscous dissipation rate. At short time ($t \lesssim 1$) they are in acceptable agreement but at long time they are not (in fact at long time EDQN can cause negative energy spectrum [9]). It means that EDQN can not be used to predict long term turbulence dynamics, but at short times it might give a hint of the actual dynamics. Fortunately as will be shown later, the time-reversal behavior is a short-term process.

Therefore, we use EDQNM to produce an initial state, and then EDQN calculation after velocity reversal. The development of GS energy after reversal is shown in Fig. 3. All the reversed cases show the same long-term behavior in Fig. 3(a) since the GS part determines the long time behavior and this part is reversed for all three cases; however in Fig. 3(b), we find at short-time the RR case causes an increasing GS energy (backscatter). The RN case shows a constant energy for a short time and the RZ case decays like the normal case for short time but the GS energy decay slows down for longer times.

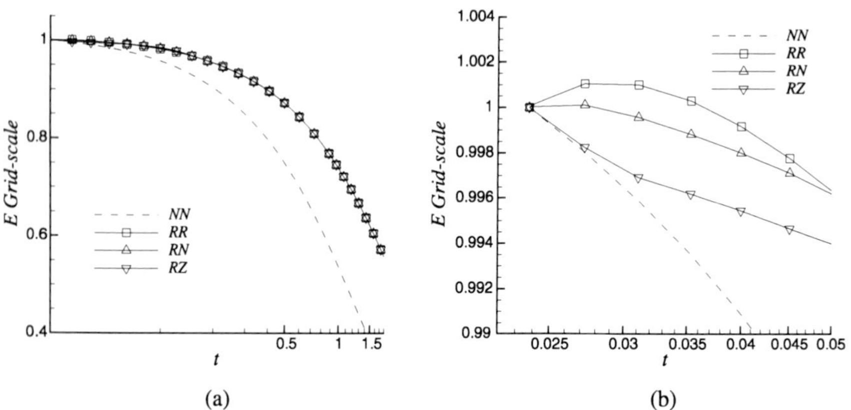

Fig. 3 Development of GS energy by using the EDQN method. (a) Long-term development. (b) Short-term development. The legends are explained in Table 1.

In order to illustrate the importance of the foregoing results for LES, we performed the same test, reversing the large scales using two distinct subgrid-models, first the simplified CZZS model (1), second the Smagorinsky model, Eq. (2) with C_s fixed at 0.14. The first model is, as mentioned before, time-reversible, the second is not. The computational mesh has 48^3 grid-points. Molecular viscosity ν is the same magnitude as eddy viscosity ν_t. As was shown by Kraichnan [10], a constant (non-scale dependent) value for the eddy-viscosity is only a good approximation in

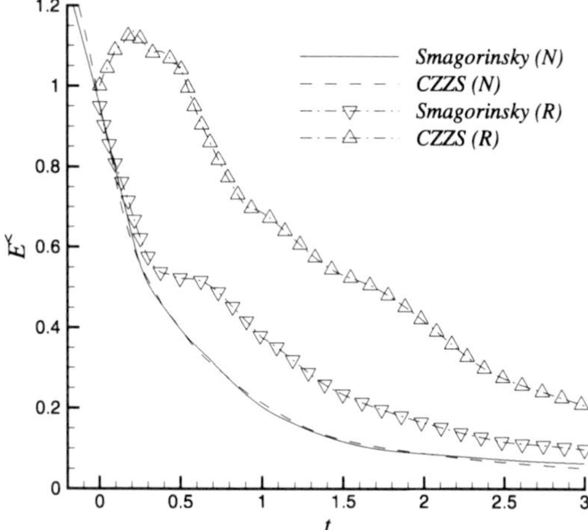

Fig. 4 Evolution of grid-scale energy in the LES cases using different subgrid models. N denotes normally decaying, R means reversal of the velocity.

the inertial range, far from the cut-off frequency. Close to the cut-off, where the role of the model is most important, the value of the eddy-viscosity strongly increases. This effect can be corrected for by adding a scale dependent cusp to the model, as was applied by Chollet and Lesieur [11]. This should be done in principle for *all* eddy-viscosity models. In the present work this cusp is introduced by modifying the eddy-viscosity to $\nu_t^*(k) = \nu_t(1+34.6\exp(-3k_c/k))$. The comparison of grid-scale energy is shown in Fig. 4. We can observe that after reversal, the simplified CZZS model yields an increase of energy, which is similar as what was observed in the RR case of EDQN calculations, before the irreversible influence of viscosity set in. The Smagorinsky model remains decaying at the same rate as the normally decaying case for some time-steps after reversal. This phenomenon is a little similar as the RZ case but still not the same. For longer times, both LES cases show decreasing GS energy, we can not simply say which is better. Briefly, each model captures a part of physics.

3 Conclusion

The *gedanken*-experiment in which the velocity at each point in a turbulent flow is reversed can be carried out in numerical simulations, and this is what we performed in the study presented in this contribution. The general response of a turbulent flow on such a reversal is that the energy decay rate decreases temporarily, and at short time the energy even increases, as would be also the case for a flow governed by the Euler equations. For LES of large Reynolds number turbulent flows, a time-reversible model is similar as reversing both GS and SGS scales, whereas a time-irreversible model is more similar to a case with reversing GS part but removing SGS part. Both these results can represent a part of physics, as shown in the EDQN results of this contribution. We should choose the correct one which we need in a practical LES calculation depending on the flow property. For example, in the atmosphere of the Earth there could be a spectral gap [12], which is similar to the RN case in this contribution, and a time-irreversible model may be appropriate; the rotating turbulence usually contains strong backscatter, and a time-reversible model may be better.

Acknowledgements The authors are grateful to Prof. G.S. Winckelmans for the many useful suggestions.

References

1. J. Smagorinsky. General circulation experiments with primitive equation. *Monthly Weather Review*, 91:99, 1963.
2. D. Carati, G.S. Winckelmans, and H. Jeanmart. On the modelling of subgrid-scale and filtered-scale stress tensors in large-eddy simulation. *Journal of Fluid Mechanics*, 441:119–138, 2001.
3. M. Germano, U. Piomelli, P. Moin, and W.H. Cabot. A dynamic subgrid-scale eddy viscosity model. *Physics of Fluids A*, 3(7):1760–1765, 1991.
4. G.X. Cui, H.B. Zhou, Z.S. Zhang, and L. Shao. A new dynamic subgrid eddy viscosity model with application to turbulent channel flow. *Physics of Fluids*, 16(8):2835–2842, 2004.
5. L. Fang, L. Shao, J.P. Bertoglio, G. Cui, C. Xu, and Z. Zhang. An improved velocity increment model based on Kolmogorov equation of filtered velocity. *Physics of Fluids*, 21(6):065108, 2009.
6. A. Pumir and B.I. Shraiman. Lagrangian particle approach to large eddy simulations of hydrodynamic turbulence. *Journal of Statistical Physics*, 113:693–700, 2003.
7. U. Schumann. Stochastic backscatter of turbulence energy and scalar variance by random subgrid-scale fluxes. *Proc. R. Soc. Lond. A.*, 451:293, 1995.
8. M. Lesieur. *Turbulence in Fluids*. Kluwer Academic, Dordrecht, 1997.
9. S.A. Orszag. *Lectures on the statistical theory of turbulence*. 1974.
10. R.H. Kraichnan. Eddy viscosity in two and three dimensions. *Journal of the Atmospheric Sciences*, 33:1521–1536, 1976.
11. J.P. Chollet and M. Lesieur. Parametrization of small scales of three-dimensional isotropic turbulence utilizing spectral closures. *Journal of the Atmospheric Sciences*, 38:2747–2757, 1981.
12. A. Pouquet, U. Frisch, and J.P. Chollet. Turbulence with a spectral gap. *Physics of Fluids*, 26:877, 1983.

Shearless turbulence - wall interaction: a DNS database for second-order closure modeling

J. Bodart, L. Joly and J.-B. Cazalbou

1 Introduction

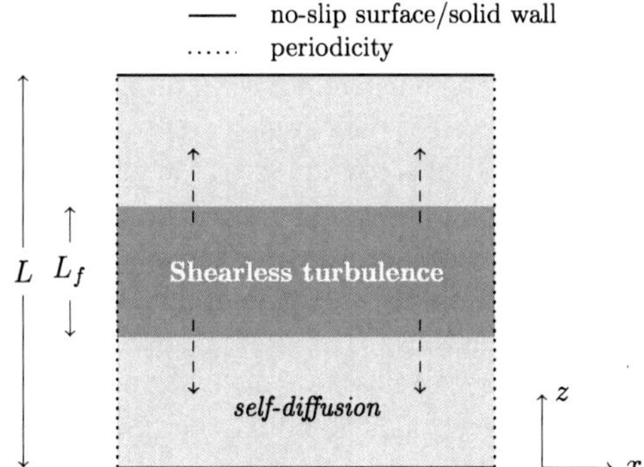

Fig. 1 Vertical cut of the computational domain.

With the aim to get a better insight of wall-turbulence interaction, we designed an original configuration where a random forcing replace usual shear-related mechanisms of turbulence production. The synthesized turbulence diffuses by itself from a central plane layer, and continuously feeds on both sides a *solid-wall or* free-slip-surface with a shearless agitation. This specific situation allows to understand the blocking effect induced by the normal velocity cancelation at the surface, and is particularly relevant to quantify the intercomponent energy transfer occurring in this region. The configuration, extensively described in [1], can be seen as the numerical

J. Bodart, e-mail: bodart@isae.fr · L. Joly, e-mail: joly@isae.fr · J.-B. Cazalbou, e-mail: cazalbou@isae.fr

Université de Toulouse, Institut Supérieur de l'Aëronautique et de l'Espace (ISAE)

H. Kuerten et al. (eds.), *Direct and Large-Eddy Simulation VIII*,
ERCOFTAC Series 15, DOI 10.1007/978-94-007-2482-2_8,

counterpart of the oscillating grid experiment (see *e.g.* [4]). The cubic domain described on the figure 1 has a side length of L. The turbulence production is confined in a central layer of thickness L_f using a random and solenoidal vector field as a source term in the Navier-Stokes equations [2]. The flow is statistically axisymmetric and the Reynolds-stresses budgets reduces to:

$$\begin{aligned} 0 &= -\frac{\partial \overline{u^2 w}}{\partial z} + \nu \frac{\partial^2 \overline{u^2}}{\partial z^2} + 2\overline{\frac{p}{\rho}\frac{\partial u}{\partial x}} - 2\nu \overline{\frac{\partial u}{\partial x_k}\frac{\partial u}{\partial x_k}} \\ &= \mathscr{D}^u_{11} + \mathscr{D}^\nu_{11} + \Pi_{11} - \varepsilon_{11} \end{aligned} \tag{1}$$

$$\begin{aligned} 0 &= -\frac{\partial \overline{w^3}}{\partial z} + \nu \frac{\partial^2 \overline{w^2}}{\partial z^2} - \frac{2}{\rho}\frac{\partial \overline{pw}}{\partial z} + 2\overline{\frac{p}{\rho}\frac{\partial w}{\partial z}} - 2\nu \overline{\frac{\partial w}{\partial x_k}\frac{\partial w}{\partial x_k}} \\ &= \mathscr{D}^u_{33} + \mathscr{D}^\nu_{33} + \mathscr{D}^p_{33} + \Pi_{33} - \varepsilon_{33} \end{aligned} \tag{2}$$

A study of the above budgets [1] reveals the main features of the flow field: Outside from the production layer, Π and $\mathscr{D}^\nu$ vanish, leading to a "pure-diffusion" region. In the vicinity of the wall, the blocking effect induces an anisotropy level between the normal component $\overline{w^2}$ and the tangential component $\overline{u^2}$ of the Reynolds-stress tensor, being in favor of $\overline{u^2}$. In this region, the intercomponent energy transfer induced by the traceless tensor Π_{ii} strengthen the anisotropy induced by the blocking effect. We recover here an interesting feature of the turbulent channel flow, in which a "slow"-"rapid" splitting of the pressure demonstrates such a kinetic-energy redistribution for the "slow" part of the pressure-strain correlation [5]. The absence of mean-shear in this configuration cancels the "rapid" part of the pressure and isolates all or part of the physical mechanisms which build up this transfer. This phenomenon is shown to rely mainly on the inherent unbalance observed between the so-called "splat" and "antisplat" events, as already observed in similar but decaying turbulent flow fields [6]. In the present case, the steadiness of the flow field as well as the remote location of the turbulence production uncover the primary origin of the disequilibrium between these kind of events: In the interacting flow field, the dissymmetry of the normal-velocity field, quantified by its skewness value, leads to a predominance of splats events in the near-wall region [1].

2 Parametric study

A first comparative study involving the same input parameters while prescribing either a *free-slip-surface (FS) or a* solid-wall (SW) demonstrates the striking independence of the anisotropy level of the velocity components with the viscous effects [1], although u and w are one order of magnitude lower in the (SW) case. Thus, this set of boundary conditions is attractive to isolate the kinematic blocking effect from the wall friction. Scaling laws or generalizations can be established by increasing the Reynolds number, which is performed in our case by keeping the same power input for the forcing while decreasing the viscosity of the fluid. Going back to a more generic parallelepiped domain of size $L_{xy} \times L_{xy} \times L_z$, we identify the other available degrees of freedom of the configuration. The distance from the source $\frac{(L_z - L_f)}{2}$ and the size l_f of the most energetic eddies synthesized in the production layer have to be set in an optimal way to characterize the wall-turbulence interaction at a reasonable computational cost. The forcing technique induces a size $l_f = \frac{4}{3}L_f$, leading to two characteristic ratio, $\frac{L_z}{L_f}$ and $\frac{L_{xy}}{L_f}$, sufficient to fully describe the parametric approach. The following constraints have to be taken into account:

1. The boundary surface should be located at least one eddy size l_f further from the edge of the production layer ($\frac{(L-L_f)}{2} > l_f$). In this way the largest structures of the flow field looses the memory of the forcing across the diffusive layer.
2. The turbulent source has to be located sufficiently close to the blocking surface to keep a non-negligible amount of energy in the near-wall region.
3. The growing of every length/time scales in the "pure-diffusion" layer requires to increase the $\frac{L_{xy}}{L_f}$ ratio when increasing $\frac{L_z}{L_f}$ to ensure the validity of the periodicity conditions in the x and y directions.

Contrary to condition (2), conditions (1) and (3) increase the computational cost. The choice adopted hereafter, $\frac{L_{xy}}{L_f} = \frac{L_z}{L_f} = \frac{1}{3}$, overcome the above constraints and keeps affordable the computational cost.

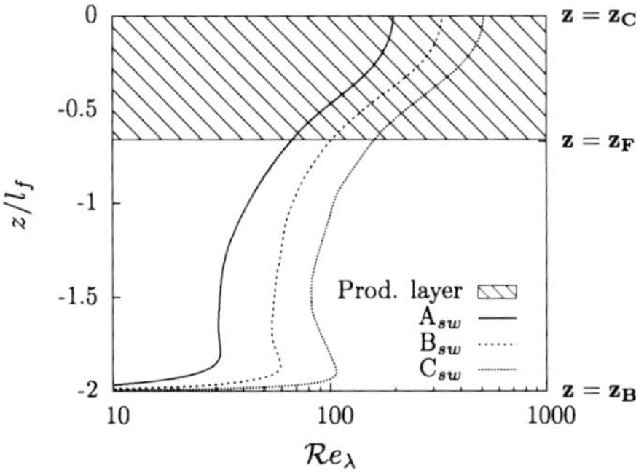

Fig. 2 Reynolds number evolution in the z−direction (SW cases).

3 DNS database

Simulations have been performed using a recently developed Navier-Stokes solver, especially designed for massively parallel architectures. The time advancement is realized through an hybrid Runge-Kutta 3/Crank-Nicolson scheme. The spatial resolution uses a mixed spectral/finite-difference discretization, with a sixth-order compact scheme in the non-homogeneous direction. The database (which will be available on-line) includes five test cases and is described in table 1. The A, B or C cases are associated with different grid resolutions, while the FS/SW subscripts stand for the *free-slip-surface* or solid-wall boundary condition respectively. Using the Taylor microscale defined in the homogeneous directions, we gives the evolution of the resulting Reynolds number value, at the midplane (C), at the edge of the forced layer (F) and at the boundary (B). The C_{FS} counterpart of the C_{SW} case will be added to the database in the future. The figure 2 shows the Reynolds number evolution along the normal direction in each (SW) case.

Case	A_{SW}	A_{FS}	B_{SW}	B_{FS}	C_{SW}
$\mathscr{R}\rceil_\lambda(z_C)$	195	195	327	329	508
$\mathscr{R}\rceil_\lambda(z_F)$	66	66	100	103	158
$\mathscr{R}\rceil_\lambda(z_B)$	0	45	0	79	0
$N_{xy}^2 \times N_z$	$224^2 \times 288$		$512^2 \times 512$		$896^2 \times 1024$

Table 1 Database description: Reynolds numbers based on the Taylor microscale λ and grid resolutions.

An estimation of the statistics reliability is made thanks to a coefficient N_e, which evaluates the number of uncorrelated samples in time, and space in the two homogeneous directions. We defined it through the classical turbulent time and length scales associated with the energy-containing motions:

$$N_e(z) = \frac{T}{\frac{k}{\varepsilon}} \left(\frac{L}{\frac{k^{\frac{3}{2}}}{\varepsilon}} \right)^2 , \quad (3)$$

where T is the simulation duration, k the turbulent kinetic energy and ε the dissipation level. Although the effects of the diffusion region lead to a growth of the different scales when approaching the boundary, we ensure $N_e(z) > 10$ all over the domain in the presented simulations.

4 Scaling of the Reynolds-stress budgets

In the near-wall region, the asymptotic behavior of the velocity fluctuations yields that the normal component w' vanishes more rapidly towards the wall than u', the tangential one. Hence the anisotropy ratio $I_u = \frac{w'}{u'}$ collapses to zero at the surface whatever the dynamic condition at the surface, let it be a free-slip surface or a solid-wall [1]. From a Reynolds-Stress-Modeling perspective, the prediction of the near-wall normal evolution of the anisotropy ratio stands both as a major issue and as a potential advantage of RSM over first-order closure-schemes that are isotropic by nature. On figure 3, we display the budget of the $\overline{uu}$ component of the Reynolds-stress tensor for both types of boundary conditions. The traceless pressure-strain

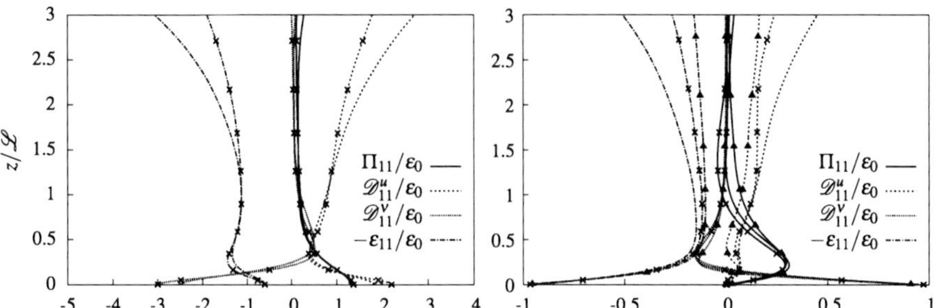

Fig. 3 Normalized $\overline{uu}$ budgets. Left: *Free-slip*, Right: *Solid-wall*. A cases (No symbols), B cases (×), C case (▲).

correlation Π_{ij} bears the intercomponent energy transfer between the normal components of the Reynolds stress tensor. Due to the rotational symmetry between horizontal velocities, Π_{11} is simply related here to its counterpart Π_{33} in the budget for $\overline{ww}$ by $\Pi_{11} = -1/2\Pi_{33}$. Within RSM closures, the pressure-strain correlation stands as the key component of the turbulence model as it sets the anisotropy level. The slow part of the pressure-strain correlation, *i.e.* the one left in shearless turbulence, is usually modeled as a "return-to-isotropy" term to comply with the physics of anisotropic homogeneous turbulence decaying remote from walls. However, the pressure-strain correlation in the vicinity of the wall is found here to enhance the anisotropy by driving energy from the less energetic normal fluctuation w' towards the tangential fluctuation u' already in energy excess. In order for a specific model to represent this non-standard behavior, we first need to scale the pressure-strain correlation with some turbulence quantity currently handled by a closure scheme, and second, to set the distance from the wall at which the non-standard anisotropy production should be switched on.

We found that the energy transfer, as well as other terms of the Reynolds-stress budget, scales nicely with the dissipation-rate at the surface, say ε_0. To estimate the thickness of the layer in which anisotropy is promoted, we propose to use the length scale $\mathscr{L}$, defined from the slope of the anisotropy ratio I_u itself. This set of scales (ε_0, $\mathscr{L}$) remarkably normalizes both budgets near the boundary as demonstrated in figure 3. The scaling is even better when focusing only on the pressure-strain correlation (black curves) in the region of interest. With a FS boundary condition, the pressure-strain correlation profiles for runs A and B collapse on the same curve meaning that the Reynolds-number effect is nicely represented in this normalized axes. With a SW boundary condition, the profile of Π_{11} in run A lies between the Π_{11} profiles of runs B and C, suggesting a residual non-monotonous Reynolds-number dependency of the peak value while the normal extent is properly scaled by $\mathscr{L}$. Beyond the relevancy of the scales set to elaborate on a turbulence model, the $\mathscr{L}$ length scale is closely related with the characteristic size of the elementary events mostly contributing to the anisotropy production, *i.e.* the characteristic size of the large scale splat events. In the present simulations, the largest eddies impacting the surface result from the length scale characteristic of the force field and was fixed through runs A to C to one fourth of the horizontal extent of the domain. The Reynolds-number increase is obtained by decreasing the viscosity thus allowing smaller and smaller eddies to contribute to large-scale splat events at the surface. The small variation of $\mathscr{L}$ with the turbulent Reynolds-number still reflects a slight sensitivity to the range of length-scales involved in the splat events. However we conjecture here that due to the non-locality of the pressure field, the off-surface turbulence structure fully determines the thickness of the blocking layer as fluid packets are approaching the wall and the energy transfer onsets. We thus expect some locking of the length scale $\mathscr{L}l$ on the largest scale of the impinging turbulence. This should be checked in our configuration by a further increase of the Reynolds number.

5 Conclusion

The proposed on-line database, includes five test cases mixing three different Reynolds numbers (from $\approx 190 < \mathscr{R}\rceil_\lambda < 500$) and, using up to one billion of grid-points for the highest Reynolds number. Two different boundary conditions have been tested *i.e.* a *solid-wall or a* free-slip-surface. In both cases, we retrieve the specific behavior of the intercomponent energy transfer observed near the boundary in wall-bounded flows (*i.e.* enhancing the anisotropy level), including situations where a mean-shear is present. Full Reynolds-stress budgets are provided, which allows straightforward tests of new second-order closure models, and especially the energy-transfer related term Π_{ij} and the dissipation term ε_{ij} . The proposed scaling of the Reynolds-stress budget and the Π_{ij} term isolates the relevant scales of the problem and gives the key-start to create an adapted correcting term of the pressure-strain correlation near the wall in second-order closure models.

Acknowledgements This work was granted access to the HPC resources of IDRIS under the allocation 2009- 92283 made by GENCI (Grand Equipement National de Calcul Intensif).

References

1. J. Bodart, J.-B. Cazalbou, and L. Joly. Direct numerical simulation of unsheared turbulence diffusing towards a free-slip or no-slip surface. *Journal of Turbulence*, (2010, accepted).
2. J. Bodart, L. Joly, and J.-B. Cazalbou. Local large scale forcing of unsheared turbulence. In *Direct and Large Eddy Simulation 7*, September 2008.
3. G. Campagne, J.-B. Cazalbou, L. Joly, and P. Chassaing. The structure of a statistically steady turbulent boundary layer near a free-slip surface. *Physics of Fluids*, 21(6), June 2009.
4. I. P. D. De Silva and H. J. S. Fernando. Oscillating grids as a source of nearly isotropic turbulence. *Physics of Fluids*, 6(7):2455–2464, 1994.
5. N. N. Mansour, J. Kim, and P. Moin. Reynolds-stress and dissipation-rate budgets in a turbulent channel flow. *Journal of Fluid Mechanics*, 194:15–44, September 1988.
6. B. Perot and P. Moin. Shear-free turbulent boundary layers. part 1: Physical insights into near wall turbulence. *Journal of Fluid Mechanics*, 295:199–227, February 1995.

Anisotropic dynamics in filtered wall-turbulent flows

A. Cimarelli and E. De Angelis

1 Introduction: anisotropic effects on the energy cascade

The most important contribution to the description of the energy transfer mechanism in the space of scales of turbulent flows is Kolmogorov theory. Under the assumption of isotropy, this theory asserts that the energy cascade in the inertial range is from large to small scales and proportional to the rate of energy dissipation. This picture is claimed to be universal since it is commonly assumed that isotropy recovery takes place at small scales of any flow for sufficiently high Reynolds number. This assumption fails to hold in wall-turbulence, where the interaction between anisotropic production and inhomogeneous spatial fluxes strongly modifies the classical energy cascade up to a reverse cascade [1].

A tool for the study of these phenomena is the balance equation for the second order structure function, $\langle \delta u^2 \rangle$, where $\delta u^2 = \delta u_i \delta u_i$ and $\delta u_i = u_i(x_s + r_s) - u_i(x_s)$. This generalized Kolmogorov equation [2], that in a turbulent channel flow reads

$$\frac{\partial \langle \delta u^2 \delta u_i \rangle}{\partial r_i} + \frac{\partial \langle \delta u^2 \delta U \rangle}{\partial r_x} + 2\langle \delta u \delta v \rangle \left(\frac{dU}{dy} \right)^* + \frac{\partial \langle v^* \delta u^2 \rangle}{\partial Y_c} =$$

$$-4\langle \varepsilon^* \rangle + 2\nu \frac{\partial^2 \langle \delta u^2 \rangle}{\partial r_i \partial r_i} - \frac{2}{\rho} \frac{\partial \langle \delta p \delta v \rangle}{\partial Y_c} + \frac{\nu}{2} \frac{\partial^2 \langle \delta u^2 \rangle}{\partial {Y_c}^2}, \qquad (1)$$

where $*$ denotes a mid-point average, i.e. $u_i^* = (u_i(x_s') + u_i(x_s))/2$, allows to quantify how the picture of the classical energy cascade is modified by anisotropy and inhomogeneity. Equation (1) can also be written in an r-averaged form in wall-parallel planes, see [1] for the details. In this form it can be written $T_r + \Pi + T_c = E + D_r + P + D_c$, where T_r, D_r are the contributions to the scale transfer due to the inertial fluctuations and viscous diffusion. T_c, D_c and P are the inhomogeneous contributions to the spatial flux related to the inertial fluctuations, viscous diffusion and

DIEM, II Facoltà di Ingegneria, Università di Bologna, Viale Risorgimento, 40136 Bologna, Italy, e-mail: andrea.cimarelli2@unibo.it, e.deangelis@unibo.it

H. Kuerten et al. (eds.), *Direct and Large-Eddy Simulation VIII*,
ERCOFTAC Series 15, DOI 10.1007/978-94-007-2482-2_9,

pressure. Π is the energy production by mean shear and E is the dissipation. Grouping together some terms in an effective production, $\Pi_e = \Pi + T_c - P$ and modified dissipation rate, $E_e = E + D_r + D_c$, the r-averaged balance reads

$$\Pi_e(r, Y_c) + T_r(r, Y_c) = E_e(r, Y_c) \tag{2}$$

whose analysis permits to characterize the different regions of the flow in term of scale-by-scale dynamics. More precisely, equation (2) allows for the description of the energy processes as function of the single scale parameter r and of the wall-distance Y_c. The analyzed data have been obtained with a pseudo-spectral simulation of a channel with a friction Reynolds number $Re_\tau = 300$. The domain is $2\pi h \times 2h \times \pi h$ with a resolution in the homogeneous directions of $\Delta x^+ = \Delta z^+ = 3.64$.

In the log-layer a large-scale production range is followed by a range dominated by the energy cascade which is closed by diffusion, see left plot of 2. Instead, in the buffer layer a large scale-energy production occurs also at small scales. A direct energy cascade at small scales and an inverse cascade at large scales exists. More details about the reverse energy cascade can be found in [1]. Such process should be related to the dynamics of the coherent structures involved in the near-wall cycle [3]. Since these structures, namely the velocity streaks and the quasi-streamwise vortices, see figure 1, are very elongated and thin, it can be expected that their action leads to a very different behavior in the streamwise and spanwise scales.

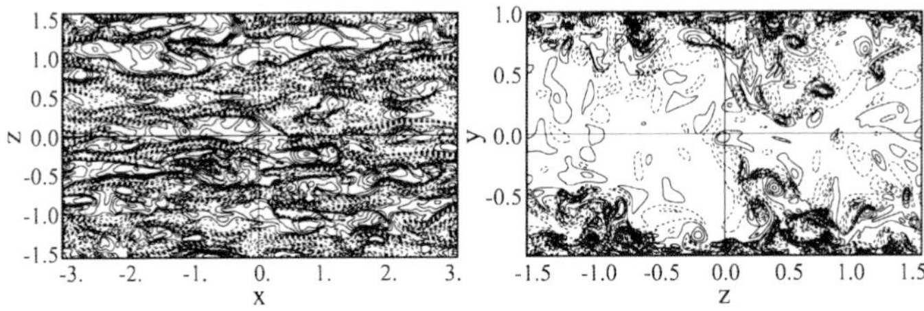

Fig. 1 Left: isocontour of the streamwise velocity fluctuations in a xz-plane at $y^+ = 20$. Right: isocontour of the streamwise vorticity, $\omega_x = (\partial w/\partial y - \partial v/\partial z)$ in a yz-plane.

In order to highlight these anisotropic dynamics, the scale-energy balance (1) can be performed in the $(r_x, 0, 0, Y_c)$ and $(0, 0, r_z, Y_c)$-space separately. In particular we identify the various regions of the (r_x^+, Y_c^+) and (r_z^+, Y_c^+)-plane where the relevant processes take place. Firstly the curve $l_{c_i}(Y_c)$ splits the space of scales into an inertial range at small r_i from a production dominated range at large r_i. This scale, identified as $\partial \langle \delta u^2 \delta u_i \rangle / \partial r_i (l_{c_i}, Y_c) = -2 \langle \delta u \delta v \rangle (dU/dy)^* (l_{c_i}, Y_c)$, is dimensionally related to the shear scale $L_s = \sqrt{\varepsilon / |S^3|}$ which is found crucial for the small scale isotropy recovery [4]. Second, the region Ω_{E_i} of the (r_i, Y_c)-plane where the energy cascade term changes sign, $-\partial \langle \delta u^2 \delta u_i \rangle / \partial r_i < 0$. This region, once integrated in the space of scales, leads to a reverse energy cascade and therefore will be hereafter referred to this phenomena. The edge of this region is the scale $l_{E_i}(Y_c)$ which splits the space of scales into a forward cascade at smaller r_i from a reverse cascade at larger r_i.

As shown in the right of figure 2, both the cross-over scales l_{c_x} and l_{c_z} show a decrease moving towards the wall. The smaller value of l_{c_z} with respect to l_{c_x} suggest that production in the spanwise scales is effective even at very small scales. This leads to a concentration of the mechanism responsible for the reverse cascade in the spanwise scales. Indeed, $-\partial\langle\delta u^2\delta u_i\rangle/\partial r_i < 0$ occurs in a wider region up to the smallest scales in the spanwise scales while in the streamwise scales take place in a smaller region for large scales, see Ω_{E_x} and Ω_{E_z} in the right plot of 2. These different features appear closely related to the anisotropic action of the coherent structures of the near-wall cycle. In particular, the scales l_{E_i} equal $r_z^+ \approx 25$ and $r_x^+ \approx 225$ at the wall-distance corresponding to the peak of production, suggesting that these phenomena are presumably the imprint of the streamwise vortices dynamics.

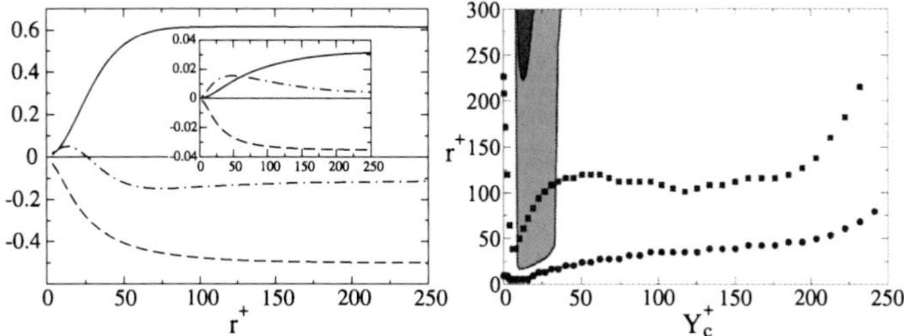

Fig. 2 Left: scale-energy balance (2) at $y^+ = 20$ and $y^+ = 160$ (inset). $-\Pi_e$ (solid line), E_e (dashed line) and $-T_r$ (dashed-dotted line). Right: the cross-over scale l_{cz} (circle), l_{cx} (square) and the reverse cascade plane Ω_{Ez} (grey region), Ω_{Ex} (dark grey region).

2 Filtered wall-turbulent dynamics

As shown in the previous section, the assessment of the anisotropic dynamics in wall-flows is very important for the comprehension of the energy cascade. This is important also in a context of turbulence modeling. Indeed, in LES, the most important feature of such models should be their ability to properly account for the energy transfer between resolved and unresolved scales and, therefore, for the energy cascade. Most of the LES models assume that the main role of the subgrid scales is to remove energy from the large scales and dissipate it accordingly to the presence of a Kolmogorov inertial range in the spectrum. For this reason this kind of approach has given good results in homogeneous flows but less in wall-turbulence where the framework of the Kolmogorov theory is modified by anisotropy and inhomogeneity.

It is therefore important to analyze how the filtered wall-turbulent physics and the subgrid stresses change in different regions and scale range as a function of the position of the filter scale l_F compared to the characteristic scales l_{c_i} and l_{E_i}. To this aim, in analogy with (1), an evolution equation for the filtered second order structure function, $\langle\delta\bar{u}^2\rangle = \langle\delta\bar{u}_i\delta\bar{u}_i\rangle$, can be used. The r-averaged form of this generalized Kolmogorov equation specialized for filtered velocity field in a channel reads

$$(\bar{\Pi}_e + T_c^{sgs}) + (\bar{T}_r + T_r^{sgs}) = (\bar{E}_e + E_{sgs}) \quad (3)$$

where together with the energy processes of equation (2) the effects of subgrid stresses appear, namely the redistribution of resolved energy in the space of scales, T_r^{sgs}, and physical space, T_c^{sgs} and the draining or sourcing of resolved energy, E_{sgs}.

The numerical analysis of equation (3) was performed in [5] using a DNS data set filtered with an equal sharp spectral cutoff filter in x and z. This approach highlighted the prominent role of the filter scale compared to l_c and l_B. These scales have the same physical meaning of l_{c_i} and l_{E_i} but are evaluated using the r-averaged equation (2) and, therefore, are scalar quantities. In the logarithmic region, the dominant energy processes can be reproduced with the resolved scales when $l_F < l_c$. In this case the main role of the subgrid stresses is to drain resolved energy, $E_{sgs} < 0$, without redistribution effects, $T_c^{sgs} \approx T_r^{sgs} \approx 0$. In the buffer layer, the scale l_B marks an intermediate range of scales where a large energy-excess occurs which is released to feed both larger (reverse energy cascade) and smaller (forward energy cascade) scales in the same physical location and other range of scales of the adjacent regions of the flow via spatial flux. When $l_F > l_B$, a large fraction of this range belongs to the subgrid scales and therefore both T_r^{sgs} and T_c^{sgs} account for a significant part of the energy cascade and spatial fluxes in the resolved scales. Furthermore, in these conditions, the large energy-excess due to turbulent production occurring in the subgrid scales feeds the resolved scales leading to a backward energy transfer, $E_{sgs} > 0$.

In the present work we extend this analysis accounting for the anisotropic dynamics exhibited by l_{c_i} and l_{E_i}. A scale-by-scale analysis will be performed using a spectral cutoff filter with different lengths in the streamwise and spanwise scales in order to separately verify the conditions $l_{F_x} < l_{c_x}$, $l_{F_x} < l_{E_x}$ and $l_{F_z} < l_{c_z}$, $l_{F_z} < l_{E_z}$.

3 Correct resolution of the wall-turbulent physics

As shown in the right plot of figure 2 the bounding condition for the filter length in the streamwise scales is the value of l_{c_x} in the logarithmic region. While in the spanwise scales is the value of l_{E_z} in the buffer layer. For that reason we consider a filter with $l_{F_x}^+ = 60$ and $l_{F_z}^+ = 15$ which will be denoted as l_{Fmix}. The resulting statistics will be compared to that obtained with a filtered field with $l_F^+ = 30$ equals in the homogeneous directions which will be denoted as l_F. This filtered field has the same amount of small scales removed by the filtering operation as l_{Fmix}.

As shown in figure 4(a) and (b), the scale-by-scale budget in the log-layer obtained with the filter l_{Fmix} is practically the same for l_F. Indeed, in this region the conditions $l_{F_x} < l_{c_x}$ and $l_{F_z} < l_{c_z}$ are verified for both the filters considered. Hence, in both cases the resolved field is able to reproduce the physics of this region with a missing fraction of energy dissipation which is recovered with E_{sgs}. The role of the subgrid stresses is to drain resolved energy, $E_{sgs} < 0$, without resolved energy redistribution effects in physical and scale-space, $T_r^{sgs} \approx T_c^{sgs} \approx 0$.

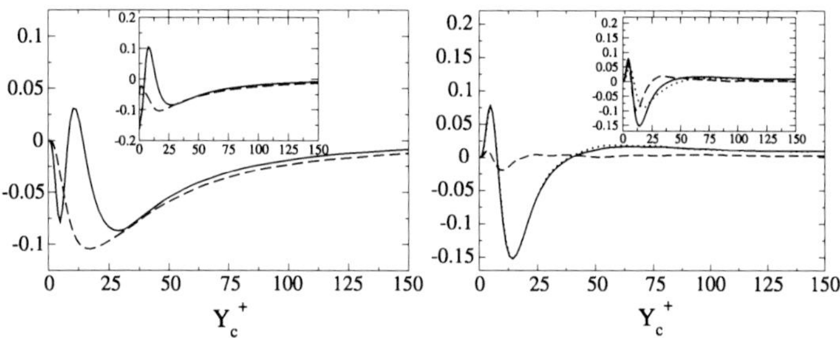

Fig. 3 Left: subgrid dissipation, E_{sgs}, and excess of scale-energy in the subgrid scales, $(\langle uv\rangle dU/dy - \langle\bar{u}\bar{v}\rangle dU/dy) - (\varepsilon - \bar{\varepsilon})$ (inset). l_F solid line and l_{Fmix} dashed line. Right: T_r (solid line), $\bar{T}_r$ (dotted line) and $\bar{T}_r^{sgs}$ (dashed line) for l_{Fmix} and l_F (inset).

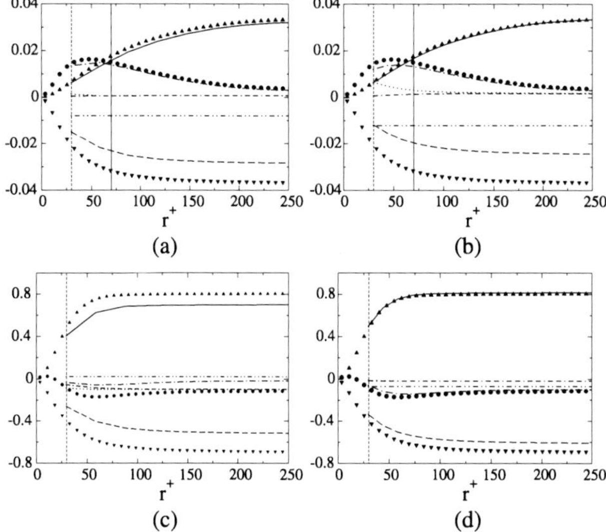

Fig. 4 Scale-by-scale budget in the log-layer, $Y_c^+ = 160$, (a)(b) and buffer layer, $Y_c^+ = 12$, (c)(d) with l_F (a)(c) and l_{Fmix} (b)(d). Filtered field: $\bar{E}_e$ dashed line, $-\bar{\Pi}_e$ solid line, $-\bar{T}_r$ dashed dotted line, $-\bar{T}_r^{sgs}$ dotted line, $-\bar{T}_c^{sgs}$ dashed-dashed-dotted line and $\bar{E}_{sgs}$ dashed-dotted-dotted line. Unfiltered field: E_e triangle down, $-\Pi_e$ triangle up and $-T_r$ circle.

In the buffer layer, the use of l_{Fmix} leads to an energy exchange between resolved and subgrid scales, E_{sgs}, which drains resolved energy, see left plot 3. The backward energy transfer observed for l_F is absent. When $l_{F_x} < l_{E_x}$ and $l_{F_z} < l_{E_z}$ the process of viscous dissipation dominates the subgrid scales also in the buffer layer where otherwise an energy-excess is observed, see the inset of the left plot 3. Therefore, in the subgrid scales prevail the viscous effects of dissipation and then of absorption of resolved scale-energy. Indeed, considering the physical meaning of the scales l_{E_i} reported in section 2, it is clear that when $l_{F_x} < l_{E_x}$ and $l_{F_z} < l_{E_z}$ the larger energy production and larger scale-energy fluxes divergence occur in the resolved motion and therefore $E_{sgs} < 0$ and the subgrid stresses effects of redistribution of resolved energy in physical and scale-space, T_r^{sgs} and T_c^{sgs} are negligible. As shown in the right of 3, the energy transfer across scales is determined by nonlinear interaction in the resolved scales also in the buffer layer for the filter l_{Fmix}, i.e. $\bar{T}_r$ is the dominant nonlinear processes while T_r^{sgs} is negligible. Otherwise, the opposite occurs as for l_F. The same conclusions apply also for the spatial flux $\bar{T}_c$ and T_c^{sgs} (not reported).

When $l_{F_i} < l_{E_i}$ the main physical processes belong to the resolved motion allowing to recover the unfiltered dynamics also in the near-wall region. As shown in figure 4(c) and (d), for filter l_{Fmix}, $\bar{\Pi}_e \approx \Pi_e$, $\bar{T}_r \approx T_r$ and $\bar{E}_e + E_{sgs} \approx E_e$. The subgrid stresses drain resolved energy, $E_{sgs} < 0$, recovering the missing fraction of dissipation and $T_r^{sgs} \approx T_c^{sgs} \approx 0$. Otherwise, for l_F, the resolved physics is very poor $\bar{\Pi}_e < \Pi_e$, $\bar{T}_r < T_r$, $\bar{E}_e + E_{sgs} < E_e$ and the subgrid scales strongly affect the resolved motion with a backward energy transfer, $E_{sgs} > 0$, and with large T_r^{sgs} and T_c^{sgs}.

4 Conclusions

How the classical picture of the energy cascade is modified by anisotropy in wall-turbulence has been analyzed in a context of filtered wall-turbulent physics and subgrid dissipation modeling. The results highlight the presence of a reverse energy cascade region in the buffer layer which strongly acts in the spanwise scales up to the smallest scales. This phenomena has been associated to the anisotropic action of the coherent structures in the near-wall cycle. In a context of filtered dynamics, the present results single out the need to adopt a filter which satisfies the constraint $l_{F_x} < l_{c_x}$ and $l_{F_z} < l_{E_z}$ leading to $l_{F_x} < 100$ and $l_{F_z} < 20$, in order to correctly capture the main physical processes of wall-turbulence. With these limits, LES of very high Reynolds number wall-bounded flows may not be possible. A runtime estimation of the relevant scales, l_{c_i} and l_{E_i}, should be used to predict from the resolved field an optimal filter scale for different wall-distances and also Reynolds number to save computational resources. Indeed, while the spanwise filtering constraint, l_{E_z}, should scale in viscous units, the streamwise should have a Reynolds number dependence since l_{c_x} is expected to increase for larger Reynolds numbers.

References

1. Marati, N., Casciola, C.M., Piva, R.: Energy cascade and spatial fluxes in wall turbulence. J. Fluid Mech. **521**, (2004).
2. Hill, R.J.: Exact second-order structure-function relationships. J. Fluid Mech. **468**, (2002).
3. Jiménez, J., Pinelli, A.: The autonomous cycle of near-wall turbulence. J. Fluid Mech. **389**, (1999).
4. Casciola, C.M., Gualtieri, P., Jacob, B., Piva, R.: Scaling properties in the production range of shear dominated flows. Phys. Rev. Lett. **95**, 2 (2005).
5. Cimarelli, A., De Angelis, E.: Analysis of the Kolmogorov equation for filtered wall-turbulent flows. Submitted to J. Fluid Mech. (2010)

Large Eddy Simulations of high Reynolds number cavity flows

Xavier Gloerfelt

1 Introduction

High Reynolds numbers cavity flows are characterized by both broadband small-scales typical of turbulent shear layers, and discrete self-sustained oscillations due to a complex feedback phenomenon between the two corners of the cavity. Intermittency of the shear-layer turbulence may lead to multiple modes which apparently coexist. The self-sustained oscillations may then exist in more than one stable state, jumping between the different modes. This complex phenomenon is often greatly simplified to build lumped models such as the Rossiter formula.

The free shear layer is viewed as two-dimensional, and the recirculating flow is neglected. At high Mach numbers, this simple formula succeeds in predicting the admissible Strouhal numbers, although it provides no information on the amplitude of the self-sustained oscillations nor it indicates which of the multiple modes will be predominant. A more comprehensive review of the numerous phenomena leading to cavity noise are available in Ref. [1].

High-Reynolds number cavity flows thus constitute a challenging test-case for Large Eddy Simulation. Several configurations are confronted to the measurements of Kegerise *et al.* [2]. The aim is also to study the influence of secondary parameters, such as the ratio L/W, the boundary layer thickness, the shear layer dynamics, and the recirculation dynamics.

X. Gloerfelt
DynFluid Laboratory, Arts et Metiers ParisTech, France, e-mail: xavier.gloerfelt@paris.ensam.fr

H. Kuerten et al. (eds.), *Direct and Large-Eddy Simulation VIII*,
ERCOFTAC Series 15, DOI 10.1007/978-94-007-2482-2_10,

2 Numerical method and configurations

In order to obtain both the vortical flow around the cavity and the radiated acoustic field, the 3-D compressible Navier-Stokes equations in conservative form are solved using high-order algorithms [3] to preserve the features of sound waves. The governing equations are integrated in time using an explicit low-storage six-step Runge-Kutta scheme, optimized in the frequency space [4]. The derivatives are calculated by using optimized finite differences with an eleven-point stencil [4]. By applying a spatial Fourier transform, the error is obtained in Fig. 1(a) and (b). The dispersion is maintained at a very low level up to the limit of resolvability $k\Delta x = \pi/2$, corresponding to four points per wavelength. As part of the algorithm, a selective filtering is incorporated in each direction to eliminate high-wavenumber oscillations. We use a centered filter built on an eleven-point stencil [4]. For the direction x, the filtered quantity f^f is computed as:

$$f^f(x_0) = f(x_0) - \alpha D_f(x_0) \quad \text{with} \quad D_f(x_0) = \sum_{j=-5}^{5} d_j^{11} f(x_0 + j\Delta x),$$

with the coefficient α chosen between 0 and 1. The filter has symmetric coefficients d_j^{11}, so that it is non-dispersive, and the dissipation error is minimized in the wavenumber space up to $k\Delta x = \pi/2$. Its transfer function is superimposed in Fig. 1(a) and (b).

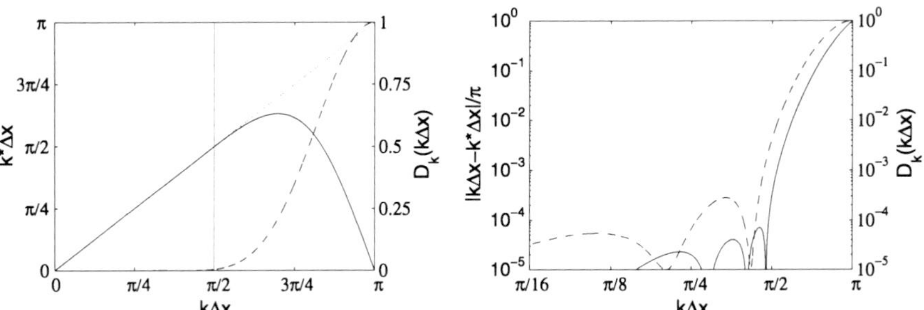

Fig. 1 On the left, the effective wavenumber $k^*\Delta x$ of the eleven-point stencil finite-difference scheme (——) is plotted versus the wavenumber $k\Delta x$ together with the exact relationship represented by the dotted line. The filter transfer function $D_k(k\Delta x)$, obtained by applying the spatial Fourier transform to D_f, is superimposed with the scale on the right (— — —). The vertical gray line indicates the limit of resolvability $k\Delta x = \pi/2$. On the right, logarithmic scale.

The LES strategy, combining a finite-difference scheme with good spectral properties with the use of a selective filtering without an additional eddy-viscosity model, bears some similarities with the new trends of the LES approach. The explicit filtering provides a smooth defiltering by removing the fluctuations at wavenumbers greater than the finite-difference scheme resolvability. As demonstrated by Mathew *et al.* [5], the effect of the Approximate Deconvolution Model (ADM) is globally similar to the convolution with an explicit filter. Moreover, the selective filtering induces a regularization similar to that used in the ADM procedure, even if the coefficient is taken constant. Since it does not affect the resolved scales, the exact value of the coefficient is not crucial [6]. Another advantage is the versatility of this LES strategy. No additional effort is required, whereas more elaborate models can induce 20% to one order of additional cost. The strategy is thus well-suited for high-Reynolds number complex applications.

Table 1 Flow conditions used in the different LES.

L/D	M_∞	p_∞, kPa	T_∞, K	Lateral b.c	W, mm
2	0.4	108.9	287.8	periodic	11.2
2	0.4	108.9	287.8	periodic	50.8
2	0.4	108.9	287.8	walls	50.8
2	0.25	308.6	307.2	walls	50.8
4	0.6	69.	277.1	walls	50.8

The chosen configuration corresponds to a canonical rectangular cavity, studied experimentally by Kegerise [7, 2]. Its length is $L = 152.4$ mm. The Reynolds number is 10^7/m, and the incoming turbulent boundary layer thickness is $\delta \simeq 6$ mm. The flow conditions are summarized in table 1. The length-to-depth ratio for the first simulated configuration is $L/D = 2$, with a free stream Mach number $M_\infty = 0.4$. The experimental cavity model spanned the width of the test section ($W = 50.8$ mm). Since the resulting shear layer is dominated by two-dimensional structures [7], we first consider only a slice of the cavity flow ($w = 11.2$ mm) by implementing periodic boundary conditions in the third direction. Then the same simulation is conducted for the full span. The third case implements no-slip side walls mimicking the lateral walls of the wind tunnel. Two others case are considered with $L/D = 2$ and $M_\infty = 0.25$, and $L/D = 4$ and $M_\infty = 0.6$. The results are obtained with a grid of 366×173 points inside the cavity and 628×200 outside in the $(x-y)$-plane, yielding an overall streamwise length $L_{x_1} = 4.28D$, and a height $L_{x_2} = 2.16D$ above the cavity. The Cartesian non-uniform grid is clustered near the cavity walls. The minimum grid sizes in each direction are $\Delta x_{1_{\min}} = 0.7$ mm, $\Delta x_{2_{\min}} = 0.4$ mm, and $\Delta x_3 = 0.8$ mm, corresponding to a grid spacing $\Delta x_2^+ = 75$ in viscous unit. 51 points are used to discretize w in the periodic case, and 128 points are used for the full span. One simulation represents roughly 400 h CPU on a NEC SX-8 machine, distributed on 8 processors using MPI communications.

3 Influence of the incoming boundary layer

The influence of the incoming turbulent boundary layer is tested for the case $L/D = 2$, $M = 0.4$, and $w = 11.2$ mm. All the calculations are initiated by imposing a turbulent mean velocity profile. In the nominal case, the initial boundary layer thickness at the inlet of the computational domain is $\delta = 3.5$ mm, and no perturbation is superimposed. Two modifications are tested. The initial boundary layer thickness at the inlet is first increased at 6.1 mm, and random Fourier modes (described in Ref. [3]) are added just after the inlet plane to feed the boundary layer turbulence. The results with one and/or the other modification are presented in the vorticity views of Fig. 2. The effect of a simple increase of the mean boundary layer thickness shows no great influence. A closer look at the vorticity pictures indicates that the size of the first structure is not related to the Kelvin-Helmholtz instability of the separated boundary layer, but is rather the results of the fusion of numerous Kelvin-Helmholtz vortices, reminiscent of the collective interaction. The adjunction of perturbations in the boundary layer may weaken the intensity of the vortices, but the same dynamics is still observed with either a thin or a thick boundary layer.

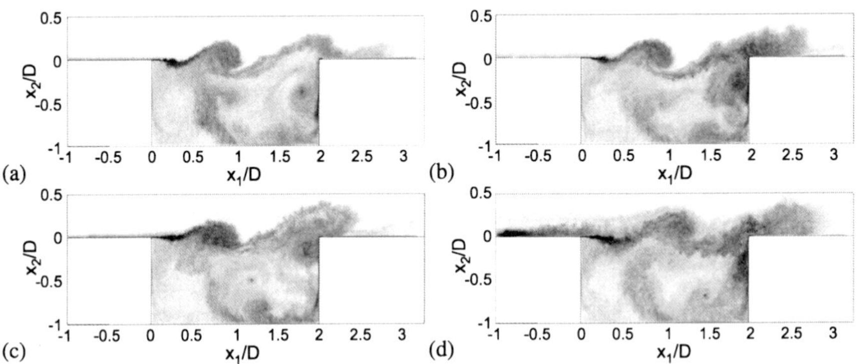

Fig. 2 Snapshots of instantaneous vorticity modulus averaged over x_3 (L/D=2, M=0.4, periodic) with: (a) $\delta_{\mathrm{ini}} = 3.5$ mm and no perturbation; (b) $\delta_{\mathrm{ini}} = 6.1$ mm and no perturbation; (c) $\delta_{\mathrm{ini}} = 3.5$ mm with turbulent inflow; (d) $\delta_{\mathrm{ini}} = 6.1$ mm with turbulent inflow.

4 Influence of lateral walls

The pressure spectra obtained in the same conditions with periodic conditions, or with no-slip side-walls are compared in Fig. 3. Three peaks at $f_1 = 460$ Hz, $f_2 = 820$ Hz, $f_3 = 1280$ Hz correspond to the first three Rossiter modes. All the components observed in the spectra may be related to the sum and difference of the primary cavity modes, and eventual harmonics, through nonlinear interactions [2]. In the experiments, mode I and III are dominating, and the peak at mode II is weaker and appears split. These features are fairly well reproduced by the LES with non slipping lateral walls. When periodic conditions are used, the hierarchy of the peaks is inversed: mode II dominates whereas modes I and III are weaker and split.

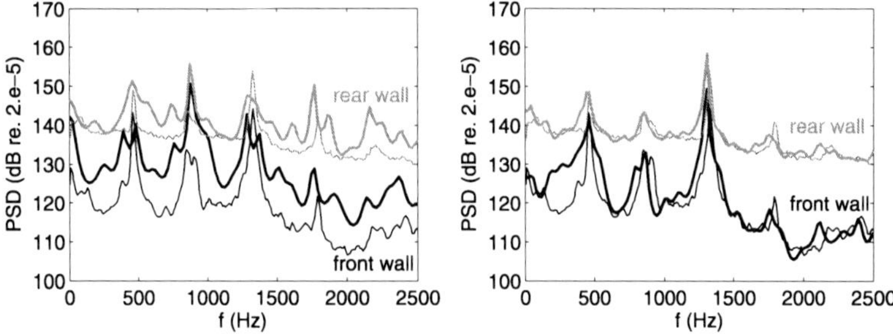

Fig. 3 Fluctuating pressure spectra at the rear wall $x_1 = 0$, $x_2 = -12.7$ mm (exp. ——; LES ▬▬), and at the front wall$x_1 = L$, $x_2 = -4.4$ mm (exp. ------; LES ▬▬). (left) Periodic lateral conditions; (right) Non slipping lateral walls.

The effect of sidewall boundary conditions on the mean recirculating flow inside the cavity is depicted in Fig. 4. The Ekman pumping due do the end walls enhances the mixing inside the cavity, leading to a single bubble, as in the experiments. The peak splitting phenomenon is also observed in feedback control experiments [8], meaning that the recirculation acts as a time delay controller for the shear layer.

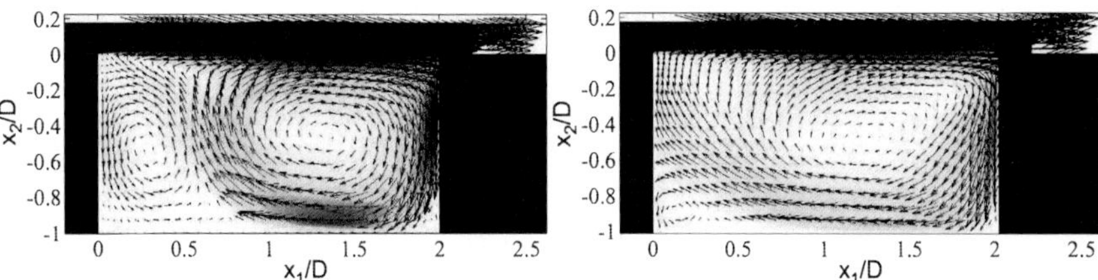

Fig. 4 Distribution of averaged velocity inside the cavity in the median plane with periodic lateral conditions (left), with lateral walls (right).

An interesting finding of Rowley *et al.* [8] is that the cavity may be a passive amplifier, very lightly damped and excited by the turbulent motions. The oscillations are then not really self-excited. By examining time-delay phase portraits and probability density functions for the case L/D=2, M=0.4, in Fig. 5, the system appears to be in a limit cycle. For shallower cavities (L/D=4), the phase portraits indicate a more stable system forced by noise.

5 Conclusions

The unsteadiness of the upward-oriented jet inside the cavity contributes to the modulated character of the separated shear layer, and thus to the appearance of multiple tones. Another striking observation is the strong unsteadiness of the recirculating

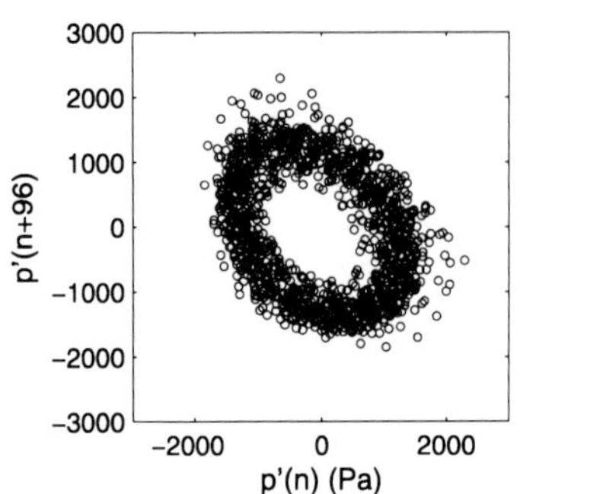

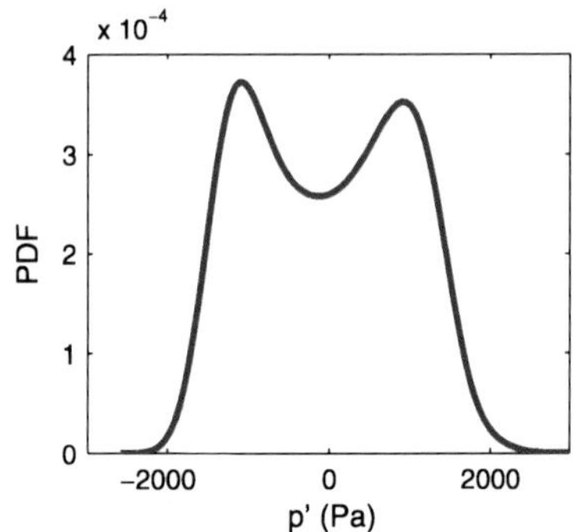

Fig. 5 Phase portrait (left) and PDF (right) of the pressure signal at the rear wall (L/D=2, M=0.4).

flow within the cavity. This suggests an important coupling of the recirculation with the separated shear layer, which form a single complex, highly non-parallel flow.

At high Reynolds numbers, the thickness of the incoming turbulent boundary layer has a limited effect since the shedding of Kelvin-Helmholtz-type vortices implies the roll-up of a sublayer, and these small vortices rapidly form large scale structure by a collective interaction. The jump from one mode to the other can proceed from a reorganization of the clusters of small scales, and can explain the coexistence of multiple Rossiter modes or the switching between two modes.

Acknowledgements This work was granted access to the HPC resources of IDRIS and CCRT under the allocation 2009-41736 made by GENCI (Grand Equipement National de Calcul Intensif).

References

1. Gloerfelt, X., Cavity noise, *VKI Lectures: Aeroacoustics of wall-bounded flow*, Von Kármán Institute, 2009, www.sin-web.paris.ensam.fr
2. Kegerise, M., Spina, E., Garg, S., and Cattafesta, L., Mode-switching and nonlinear effects in compressible flow over a cavity, *Phys. Fluids*, **16**(3), 2004, p. 678–687.
3. Gloerfelt, X., Bogey, C., and Bailly, C., Numerical evidence of mode switching in the flow-induced oscillations by a cavity, *Int. J. of Aeroacoustics*, **2**(2), 2003, p. 99–124.
4. Bogey, C. and Bailly, C., A family of low dispersive and low dissipative explicit schemes for noise computation, *J. Comput. Phys.*, **194**, 2004, p. 194–214.
5. Mathew, J., Lechner, R., Foysi, H., Sesterhenn, J., and Friedrich, R., An explicit filtering method for large eddy simulation of compressible flows, *Phys. Fluids*, **15**(8), 2003, p. 2279–2289.
6. Bogey, C. and Bailly, C., LES of a high Reynolds, high subsonic jet: Effects of the subgrid modellings on flow and noise, *AIAA Paper 2003-3557*, 2003.
7. Kegerise, M., *An experimental investigation of flow-induced cavity oscillations*, Ph.D. thesis, Syracuse University, 1999.
8. Rowley, C., Williams, D., Colonius, T., Murray, R., and MacMynowski, D., Linear models for control of cavity oscillations, *J. Fluid Mech.*, **547**, 2006, p. 317–330.

Part II
Methodologies and Modeling Techniques

On the development of a 6th order accurate compact finite difference scheme for incompressible flow

Bendiks Jan Boersma

1 Introduction

Compact finite difference methods are nowadays very popular for the simulation of compressible turbulent flows, see for instance [2] and [1]. Due to the low dissipation and dispersion errors of the compact finite difference schemes, they can be used for various type of problems including large eddy and direct numerical simulation of turbulent flow and laminar turbulent transition. However due to the low numerical dissipation compact finite difference have the tendency to be numerically quite unstable. In practice this instability issue is solved by applying a spatial filter to the calculated solution or by using a staggered layout of the flow variables. The latter is of course more appealing. In this paper we will extend the staggered formulation we have developed for compressible flow, see [1] to the incompressible flow case.

2 Governing equations and numerical method

In this section we will give the governing equations and discuss the numerical method which has been used to solve these equations. We will assume that flow is governed by the three dimensional Navier-Stokes equations in Cartesian coordinates. The equation for conservation of momentum reads:

$$\frac{\partial u_i}{\partial t} = -\frac{\partial u_i u_j}{\partial u_j} - \frac{1}{\rho}\frac{\partial p}{\partial x_i} + \nu\frac{\partial^2 u_i}{\partial x_j^2} \tag{1}$$

B.J. Boersma
Department of Process and Energy, Delft University of Technology, The Netherlands, e-mail: b.j.boersma@tudelft.nl

H. Kuerten et al. (eds.), *Direct and Large-Eddy Simulation VIII*, ERCOFTAC Series 15, DOI 10.1007/978-94-007-2482-2_11,

In which u_i is the velocity vector, ρ is the fluids density (assumed to be constant), p the pressure and ν the kinematic viscosity (also assumed to be constant). The velocity vector u_i has to satisfy the in-compressibility constrained:

$$\frac{\partial u_i}{\partial x_i} = 0 \tag{2}$$

2.1 Spatial derivatives

In this section we will discuss the computation of the spatial derivatives. We will only consider relations for uniform grids. Relations for non uniform grids can be derived easily with help of an analytic mapping, see for instance [1].

Furthermore, we will restrict the discussion to the 2D case, although the actual implementation is performed in 3D. The discretization is based on the well known marker and cell layout, as introduced by Harlow and Welch [3]. The scalar quantities, like the pressure p are stored at the cell centers and the velocity components at the cell faces, see figure 1. The spatial derivatives in equation (1) are calculated by a

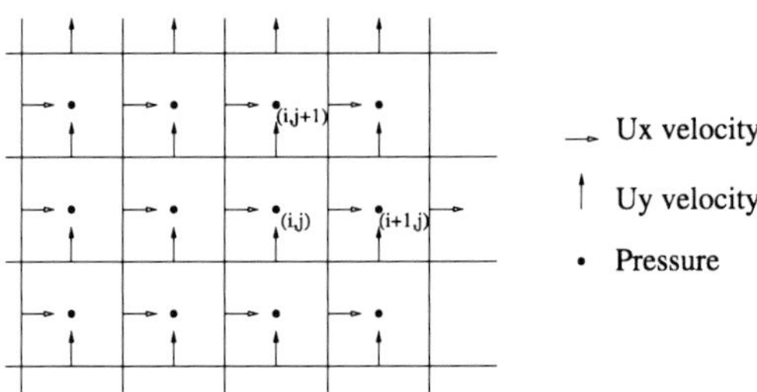

Fig. 1 The computational grid. The velocity components are stored on the cell faces and the scalar quantities at the cell center.

combination of interpolations and differentiations, similar to the approach we have used in [1]. If we, for instance, want to calculate $\partial u_1 u_1 / \partial x_1$. First the product $u_1 \cdot u_1$ is calculated at the cell face. The product is then interpolated to the cell center and subsequently differentiated with a staggered formulation which returns the value of $\partial u_1 u_1 / \partial x_1$ at the required grid location. A term like $\partial u_1 u_2 / \partial x_1$ is calculated by first interpolating the u_1 and u_2 velocity to the corner of a cell followed by a staggered differentiation.

The first derivative

In this paragraph we will consider the evaluation of the first derivative of a smooth function $f = f(x)$. If we assume that the variable f is known at points $i = 1/2, ..i + 1/2, ..n - 1/2$ and we want to obtain the values for the derivative f' at point $i =$

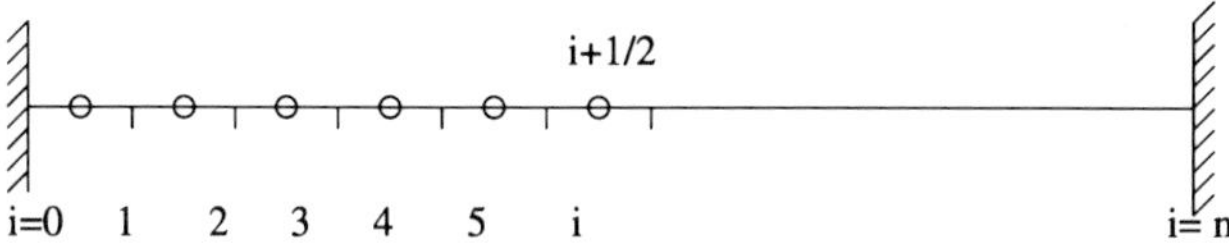

Fig. 2 The location of the grid points, the points $i = 0$ and $i = n$ are the boundary points.

$0,..i,..n$ we can use the following generic formula:

$$a(f'_{i+1} + f'_{i-1}) + f'_i = \frac{b}{\Delta X}(f_{i+1/2} - f_{i-1/2}) + \frac{c}{\Delta X}(f_{i+3/2} - f_{i-3/2}), \text{ for } 2 \le i \le n-2 \tag{3}$$

In which f'_i is derivative of f with respect to x in point i, ΔX is the (uniform) grid spacing and a, b, and c are yet unspecified coefficients. The coefficients a, b and c can be obtained from a Taylor expansions around grid point i. With the three coefficients a, b, and c in equation (3) we can obtain an 6th order accurate formulation. The values for a, b, and c for this 6th order scheme are (obtained with the Maple Software package):

$$a = 9/62,\ b = 63/62,\ c = 17/186,\ \ 2 \le i \le n-2$$

Close to the boundary at points $i = 1$ and $i = n-1$ this sixth order formulation can not be used because information from outside the domain would be required. Therefore we use a smaller stencil for these points:

$$a = 1/22,\ b = 12/11, c = 0,\ \ O(\Delta X)^4$$

Which is formally fourth order accurate in ΔX. At the boundary i.e., the point $i = 0$ we use a one sided 3rd order accurate formulation

$$f'_0 + 23 f'_1 = \frac{1}{\Delta X}\left(-25 f_{1/2} + 26 f_{3/2} - f_{5/2}\right) + O(\Delta x)^3 \tag{4}$$

The approximations above leads to a tridiagonal linear system, which can be solved easily and efficiently by means of a tridiagonal solver.

For the analysis of the compact difference scheme and also for the solution of the Poisson equation it is useful to rewrite the compact finite difference schemes in matrix form, see for instance [2], [1]. An alternative form to write equation (3)-(4) is

$$[A] f'_{n+1} = [B] f_n$$

where $[A] = [A]_{n+1,n+1}$ and $[B] = [B]_{n+1,n}]$ are matrices which contain the elements give in equation (3)-(4) and f'_{n+1} is a vector with $n+1$ elements containing the derivatives, and f a vector with n elements. The relation for f' can be written as $f' = [A]^{-1}[B] f$. Where $A_{n+1,n+1}$ is a tridiagonal matrix, the inverse of A will be a full matrix and therefore the product $[A]^{-1}[B]$ will also be a full matrix, with dimensions $n+1, n$.

2.2 Application to incompressible Navier-Stokes

The equations (1) and (2) will be solved with a standard pressure correction scheme. In such a scheme the velocity is first advanced to an intermediate velocity level, denoted by u_i^*. This can be done with various numerical schemes, amongst which the Runga-Kutta and Adams-Bashforth schemes are the most popular ones. Here we have chosen for third order Adams-Bashforth scheme, which has good stability properties for both advection and diffusion operators [8]:

$$u_i^* - u_i^n = \Delta t[\frac{23}{12}f(R^n) - 1612f(R^{n-1}) + \frac{5}{12}f(R^{n-2})] + O(\Delta t)^3 \tag{5}$$

Where Δt is the time step, $f(R^n)$ denotes the right hand side of equation (1) at time $t = n\Delta t$ and $f(R^{n-j})$ the right hand side of eq. (1) at time $t = (n-j)\Delta t)$. The choice of the time integration scheme is quite arbitrary and not very relevant for the present study. Subsequently in the pressure correction scheme, the pressure at time level $n+1/2$ is used to calculate the velocity at time level $n+1$:

$$u_i^{n+1} = u_i^* - \Delta t \frac{1}{\rho}\frac{\partial p}{\partial x_i}^{n+1/2} \tag{6}$$

So far the pressure at the time level $n+1/2$ is unknown but it can be be computed from a Poisson equation which can be derived by taking the divergence of equation (6), and enforcing the divergence to zero at time level $n+1$

$$\frac{\partial}{\partial x_i} u_i^* = \frac{\Delta t}{\rho}\frac{\partial}{\partial x_i}\left(\frac{\partial p}{\partial x_i}^{n+1/2}\right) \tag{7}$$

After the solution of the pressure $p^{n+1/2}$ from the Poisson equation has been calculated, the final velocity u^{n+1} can be computed with help of equation (6). It should be noted that for a consistent formulation it is essential to use the form given by equation (7) and not to replace the term on the right hand side of equation (7) by the second derivative of $p^{n+1/2}$. The algorithm above is well known and has with an explicit advection and diffusion step in principle third order time accuracy for the velocity and second order time accuracy for the pressure.

The pressure formulation, i.e. the discretization of equation (7) is the only part of the model that is not straightforward. The discretized form of the equation for conservation of mass, using the previously introduced matrices A and B can be written as:

$$\frac{\partial u_i}{\partial x_i} = [A]^{-1}[B]u_i \tag{8}$$

Where $[A_i]$ and $[B_i]$ are the matrices given in the previous section. The discretized version equation (6) reads;

$$u_i^{n+1} = u_i^* - \Delta t[A]^{-1}[B]p^{n+1/2}$$

In our numerical scheme we apply boundary conditions on u_i^{n+1} we apply the same conditions to u_i^*. This choice implies that $[A]^{-1}[B]p^{n+1/2}$ at the boundary should be equal to zero. To achieve this we set the elements of $[A]^{-1}[B]$ which corresponds to the boundary points to zero. Next we apply the discrete divergence operator, equation (8) to the relation above, this gives:

$$[A]^{-1}[B]u_i^{n+1} = [A]^{-1}[B]u_i^* - \Delta t[A]^{-1}[B][[A]^{-1}[B]p^{n+1/2}$$

Enforcing the discrete divergence to zero at time level $n+1$ gives

$$[A]^{-1}[B]u_* = \Delta t[A]^{-1}[B][[A]^{-1}[B]p^{n+1/2} \tag{9}$$

Which is a linear algebraic equation which should be solved to obtain the pressure $p^{n+1/2}$. It should be noted that a direct discretization of the left and right hand side of equation (7), i.e. using formulas for the second derivative, will result in a (small) error in the mass conservation. This error will likely accumulate in time and destroy the favorable properties of the staggered arrangement.

Equation (9) is solved iteratively with the GMRESR algorithm provided by Botchev [6].

3 Results

In this section we will consider turbulent duct flow. This is also a well known flow geometry for which reliable data is available [4], [5]. The discretization in the x and y direction is performed with the 6th order compact scheme outlined above. In the flow (z)-direction, which is assumed to be periodic a Fourier expansion is used. The flow is driven in the z-direction by means of a constant pressure gradient $\partial p/\partial z$, the Reynolds number based on the friction velocity is equal to 360. The computational grid consisted of $128 \times 128 \times 512$ points in the x, y and z direction respectively.

In Figure 3 (left) we show the axial velocity profile, normalized with u_*. The label "DNS" denotes the results of the present study, and the symbols denote the experimental date provided by Nierderschulte [7]. Excellent agreement between our simulation data and the literature data is observed. In Figure 3 (right) we compare the axial and wall normal rms profiles obtained from the DNS with the results provided by Gavrilakis [4] and by Nierderschulte [7]. In these figures we observe a reasonable agreement between our DNS and the results of [4]. The agreement between our data and the experimental data of [7] is much better than the agreement between the results of [4] and [7]. The low numerical accuracy and/or resolution used by Gavrilakis [4] could be responsible for the discrepancy between his results and ours.

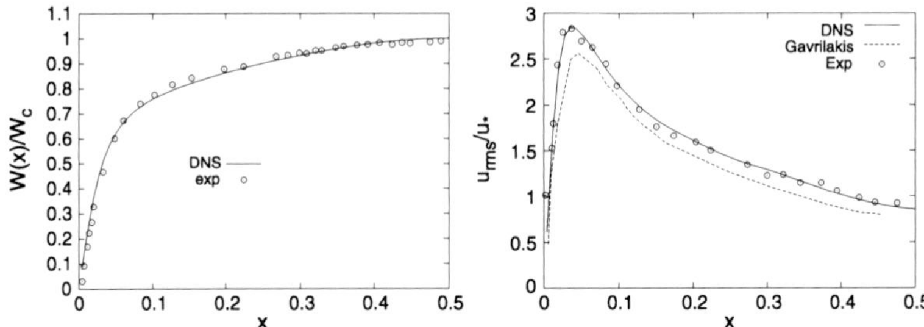

Fig. 3 Left: the mean axial velocity normalized with the friction velocity u_*, right: the axial rms profiles, normalized with u_*. The label "DNS" is for the present work, also included are the results of reference [4] and the experiment by [7].

4 Discussion & conclusion

The developed numerical method is very stable and robust. However, the computational costs are considerable higher than that of a standard finite difference method. On the same mesh size we have observed an increase in CPU time of at least a factor five when we compare the compact method with a standard 2nd order method. The main computational bottleneck is the solution of the Poisson equation. On the other hand the accuracy of the method is substantial higher than that of a standard second order method and for problems where high accuracy is necessary the compact method may be a good alternative to other high order methods.

References

1. Boersma, B.J., 2005, A staggered compact finite difference formulation for the compressible Navier-Stokes equations, *J. of Comp. Phys*, **208**, 675–690.
2. Lele, S.K., 1992, Compact finite difference schemes with spectral like resolution, *J. of Comp. Phys.*, **103**, 16–42.
3. Harlow, F.H., & Welch, J.E., 1965, Numerical calculatios of time-dependent viscous incompressible flow of fluid with a free surface, *Phys. of Fluids*, **8**, 2182–2189.
4. Gavrilakis, S., 1992, Numerical simulation of low-Reynolds number turbulent flow through a straight square duct, *J. of Fluid Mech.*, **244**, 101–129.
5. Huser, A., & Biringen, S., 1993, Direct numerical simulation of turbulent flow in a square duct, *J. of Fluid Mech.*, **257**, 65–95.
6. van der Vorst, H.A., & Vuik, C., 1994, GMRESR: a Family of Nested GMRES Methods, *Num. Lin. Alg. Appl.*, **1**, 369–386.
7. Niederschulte, M.A., 1989, Turbulent flow through a rectangular channel, *PhD thesis*, University of Illinois, Urbana-Champaign.
8. Gear, C.W., 1971, Numerical initial value problems in ordinary differential equations, Prentice Hall, New Jersey, USA.

A multilevel method applied to the numerical simulation of two-dimensional incompressible flows past obstacles at high Reynolds number

François Bouchon, Thierry Dubois and Nicolas James

1 Introduction

Numerical simulation of turbulent flows in complex geometries is one of the most investigated fields in computer science in the last decades. But even though the power of supercomputers has regularly increased for many years, it has been understood that the numerical simulation of realistic flows at high Reynolds number would require too many efforts in term of memory and CPU time if one discretizes directly the Navier-Stokes equation.

In this paper we present an extension of the spectral multilevel method applied for 3D periodic turbulence and the channel flow (see [2], [3]) to the case of numerical simulation of two-dimensional flows past obstacles. In [5], these mutlilevel methods have first been adapted to the context of the finite difference discretization of the Navier-Stokes equations on staggered cartesian grids. Here, this approach is coupled with the immersed boundary (IB) method proposed in [4], which allows to handle complex geometries, and numerical results are presented and compared to "one-level" (or direct) simulations where no multilevel strategy is used. At Reynolds number 9500, comparisons are shown between the mutlilevel method on a 512^2 mesh and a direct simulation on a 3072^2 mesh. At Reynolds number 500 000, the multilevel method is robust, while the corresponding direct simulation (same grid, same time-step) blows up in the first time-iterations of the solution.

François Bouchon, Thierry Dubois and Nicolas James
Clermont Université, Université Blaise-Pascal, Laboratoire de Mathématiques, BP10448, F-63000 Clermont-Ferrand and CNRS, UMR 6620, Laboratoire de Mathématiques, F-63177 Aubière, e-mail: francois.bouchon, thierry.dubois, nicolas.james@math.univ-bpclermont.fr

H. Kuerten et al. (eds.), *Direct and Large-Eddy Simulation VIII*,
ERCOFTAC Series 15, DOI 10.1007/978-94-007-2482-2_12,

2 Description of the problem

2.1 The governing equations and their discretization

We consider an incompressible viscous flow in a domain $\Omega \subset \mathbb{R}^2$ described by the non-dimensional Navier-Stokes equations:

$$\partial_t \boldsymbol{u} - Re^{-1}\,\Delta \boldsymbol{u} + \boldsymbol{\nabla}\cdot(\boldsymbol{u}\otimes\boldsymbol{u}) + \boldsymbol{\nabla} p = \boldsymbol{f}, \tag{1}$$

$$\boldsymbol{\nabla}\cdot\boldsymbol{u} = 0, \tag{2}$$

$$\boldsymbol{u}(\boldsymbol{x}, t=0) = \boldsymbol{u}_0, \tag{3}$$

where Re denotes the Reynolds number, $\boldsymbol{f}$ the external volume force and $\boldsymbol{u}_0$ the initial condition.

This system is completed with boundary conditions. The case of periodic flow will be addressed, as well as Dirichlet boundary conditions for the velocity field.

We now describe the numerical method corresponding to the case of periodic boundary conditions, the numerical treatment near the boundary is detailed in section 2.3. For the points away from $\partial\Omega$, all terms in equations (1) and (2) are discretized in space by using second-order centered finite volume schemes. The discrete unknowns are given on a staggered grid (see [7]): discrete pressure values are located at the center of mesh cells, velocity values are located at the center of edges (see Figure 1).

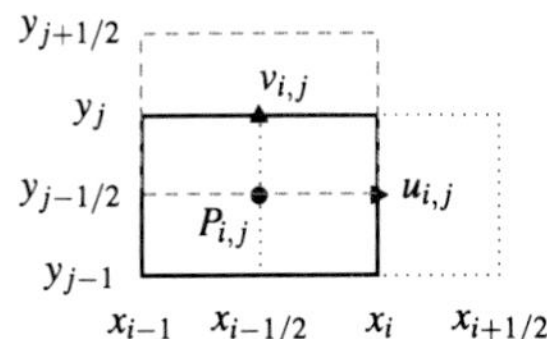

Fig. 1 Cells $K_{i-\frac{1}{2},j-\frac{1}{2}}$ (solid), $K_{i-\frac{1}{2},j}$ (dashed) and $K_{i,j-\frac{1}{2}}$ (dotted) and their corresponding discrete values.

We define the vector $\boldsymbol{u}^k \in \mathbb{R}^{N^2}$ of components $u^k_{i,j}$ where N is the number of unknowns in each direction, and similarly, $\boldsymbol{v}^k \in \mathbb{R}^{N^2}$ and $\boldsymbol{P}^k \in \mathbb{R}^{N^2}$ of components $v^k_{i,j}$ and $P^k_{i,j}$ respectively, corresponding to approximate values of the unknowns at time $t^k = k\delta t$. We introduce $\boldsymbol{U}^k = (\boldsymbol{u}^k, \boldsymbol{v}^k)$ and $\boldsymbol{F}^k = (\boldsymbol{f}^k_u, \boldsymbol{f}^k_v)$ where $\boldsymbol{f}^k_u$ and $\boldsymbol{f}^k_v$ are vectors in $\mathbb{R}^{N^2}$ corresponding to the force $\boldsymbol{f}$. The discrete approximation of (1) writes:

$$\widetilde{U}^{k+1} + \frac{2\delta t}{3Re} A\widetilde{U}^{k+1} = -\frac{2\delta t}{3} GP^k + \frac{1}{3}(4U^k - U^{k-1}) + \frac{2\delta t}{3}(2F^k - F^{k-1}) - \frac{2\delta t}{3}\left(2NL(U^k) - NL(U^{k-1})\right) \quad (4)$$

$$U^{k+1} = \widetilde{U}^{k+1} - \frac{2\delta t}{3} G\phi^{k+1}, \quad (5)$$

where the matrix A is the discrete approximation of the operator $-\Delta$, G is the one of the gradient, NL are the discrete nonlinear operators and $\phi^{k+1} = P^{k+1} - P^k$.

The discretization of the incompressibility constraint is achieved by integrating (2) over the pressure cell $K_{i-\frac{1}{2},j-\frac{1}{2}}$, leading to

$$DU^{k+1} = 0, \quad (6)$$

where $D = G^T$ is an approximation of the divergence operator. Combining (5) and (6), we deduce the linear system satisfied by ϕ, namely:

$$DG\phi^{k+1} = \frac{3}{2\delta t} D\widetilde{U}^{k+1}. \quad (7)$$

Once (7) is solved, the velocity is updated with (5). This implementation ensures that the discrete divergence of the updated velocity is zero up to the computer accuracy.

2.2 The multilevel methodology

The multilevel method for the MAC scheme (marker and cell) has been detailed in [5], it consists in adapting to the physical space the methods of [2] and [3].

We introduce grid operators corresponding to some spatial filtering of the velocity field. Let us consider two embedded grids (see Fig. 2): $G_1 = \{(x_i, y_j), i = 1,\ldots,N, j = 1,\ldots,N\}$ and $G_2 = \{(x_{2i}, y_{2j}), i = 1,\ldots,N/2, j = 1,\ldots,N/2\}$. Let $U = (U,V)$ be a velocity field defined on the fine grid G_1. As in [5], we use the restriction operator $\mathscr{R}$ which defines a filtered velocity field $\mathscr{R}(U)$ on the grid G_2, this velocity field being extended on the grid G_1 to a velocity field $\mathscr{P} \circ \mathscr{R}(U)$ via a prolongation operator $\mathscr{P}$. The velocity field U can be splitted using these operators

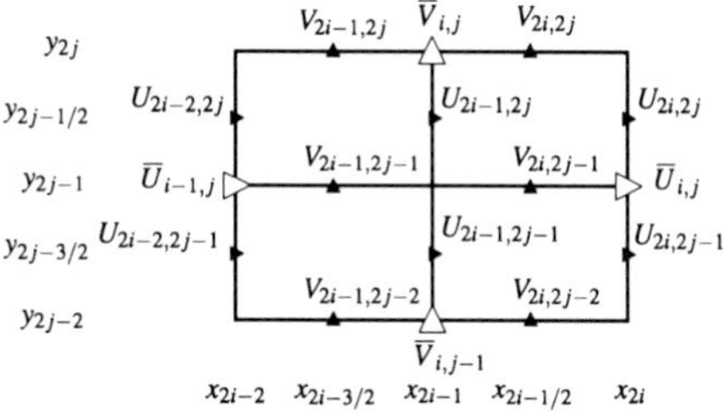

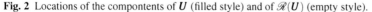

Fig. 2 Locations of the compontents of U (filled style) and of $\mathscr{R}(U)$ (empty style).

$\mathscr{P}$ and $\mathscr{R}$:

$$U = Y + Z = \mathscr{P} \circ \mathscr{R}(U) + Z. \tag{8}$$

It has been shown in [5] that if $DU = 0$, then $D \cdot \mathscr{P} \circ \mathscr{R}(U) = 0$. The component $\mathscr{P} \circ \mathscr{R}(U)$ corresponds to the large scales of the flow, and Z to the smallest ones. Extending this procedure to four embedded grids, we define the following decomposition of the velocity field U:

$$U = U_1 + U_2 + U_3 + U_4. \tag{9}$$

The numerical simulation consists in advancing in time U as described in section 2.1 and dynamically correcting the energy contained in the smallest scales U_4 to maintain a linear discrepency of the energy spectrum in the log-log scale, following the ideas of [2] and [3] (see [5] for further details).

2.3 A second-order immersed boundary method

To take into accound the presence of a solid boundary embedded in Ω, we use an immersed boundary technique where the unknown are placed at the middle of the part of the edges located in the fluid (see Figure 3).

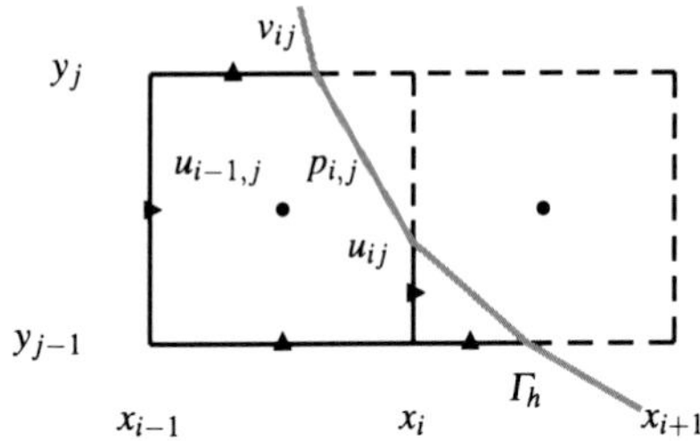

Fig. 3 Arrangement of the unknowns in cut-cells.

The Laplacian operator is then approximate at first order on these points, but due to a super convergence property (see [8]), the numerical method is globally second order.

3 Numerical results

3.1 2D periodic flows

We first present numerical results to show the efficiency of the multilevel methodology for 2D periodic flows, by comparing on Figure 4 the spectra at Reynolds number

500 000 for a direct numerical simulation on 4096^2 points, and a multilevel method on 256^2 points. A finer analysis in this context can be found in [5].

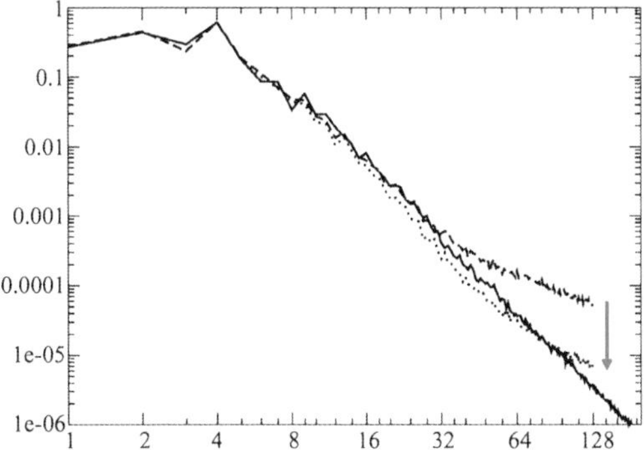

Fig. 4 Energy spectrum. DNS4096 (plain), DNS256 (long-dashed) and ML256 (dashed).

3.2 Flow past a 2D cylinder at Reynolds number 9 500

In the present application, we consider a flow in the domain $\Omega = (-5,5) \times (-2.5,2.5) \setminus \mathscr{D}$, where $\mathscr{D}$ denotes the ball centered in M(0,0) of radius 1. The boundary conditions are then $\boldsymbol{u} = (1,0)$ on $\Sigma = \partial\Omega$ and $\boldsymbol{u} = (0,0)$ on $\Gamma = \partial\mathscr{D}$. DNS have been run and results have been compared to experimental results in [1] and [6]. For $N = 768$, the DNS is stable but the numerical results seem inaccurate. For $N = 512$, a direct numerical simulation is unstable. For the same grid, the multi-level strategy gives numerical results close to those of the direct numerical simulation at $N = 3072$ (see figure 5).

4 Conclusion

A multilevel method has been successively developed for two-dimensional turbulent flows at high Reynolds numbers. Comparisons with DNS on finer meshes show the efficiency of the method. A numerical simulation on a flow past obstacles at Reynolds number 500 000 has also been done on a 3072^2 mesh, while a DNS performed on the same mesh for the same Reynolds number leads to overflows. An analysis of this simulation and many other ones will be the subject of future works.

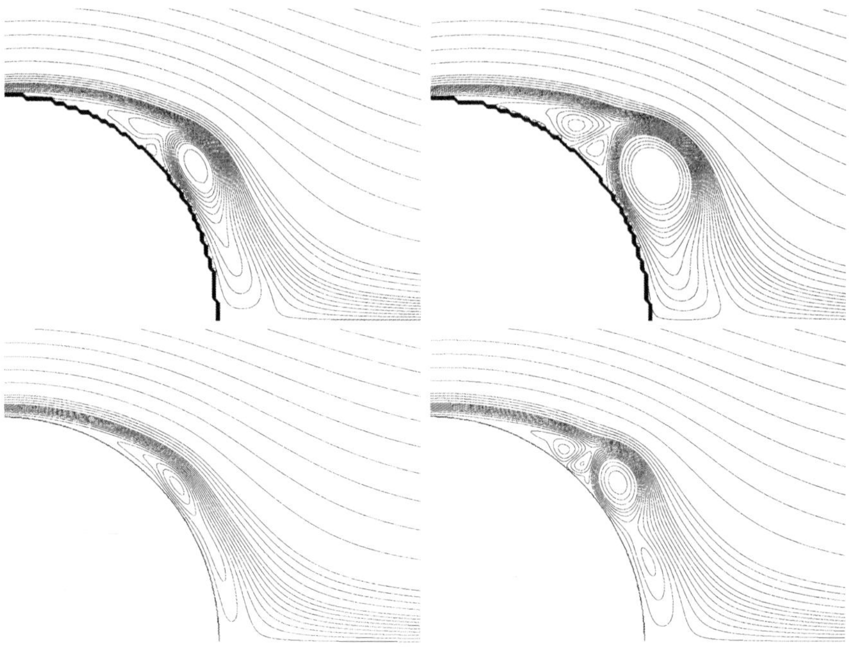

Fig. 5 Streamlines of the velocity at times 0.75 (left) and 1 (right), for the ML512 (top) and the DNS 3072 (bottom).

References

1. R. BOUARD, M. COUTANCEAU, *The early stage of development of the wake behind an impulsively started cylinder for* $40 < Re < 10^4$, J. Fluid Mech., **101**, (1980) 583–607.
2. F. BOUCHON, T. DUBOIS, *A LES model based on the spectral decay of the kinetic energy*, in Advances in Turbulence VIII, CIMNE, Barcelona (2000), 527–530.
3. F. BOUCHON, T. DUBOIS, *A model based on incremental scales applied to LES of turbulent channel flow*, in Direct and Large Eddy Simulation IV, ERCOFTAC SERIES, Kluwer Academic Publishers (2001), 97–104.
4. F. BOUCHON, T. DUBOIS, N. JAMES, *A second-order immersed boundary method for the numerical simulation of two-dimensional incompressible viscous flows past obstacles*, to appear in Computational Fluid Dynamics 2010, Proceedings of ICCFD6, St-Petersbourg, Russia, (2010) Springer.
5. N. JAMES, *Méthodes multi-niveaux sur grilles décalées. Application à la simulation numérique d'écoulements autour d'obstacles.*, Thèse de l'Université Clermont-Ferrand 2, (2009).
6. M. COUTANCEAU, R. BOUARD, *Experimental determination of the main features of the viscous flow in the wake of a circular cylinder in uniform translation. Part 1. Steady flow*, J. Fluid Mech., **79**, (1977) 231–256.
7. F. H. HARLOW, J. E. WELCH, *Numerical calculation of time-dependent viscous incompressible flow of fluid with free surface*, Phys. Fluids, **8**(12), (1965) 2182–2189.
8. N. MATSUNAGA, T. YAMAMOTO, *Superconvergence of the Shortley-Weller approximation for Dirichlet problems*, J. Comp. Appl. Math., **116** (2000) 263–273.

Large eddy simulation with adaptive r-refinement for the flow over periodic hills

Claudia Hertel and Jochen Fröhlich

1 Introduction

Using large eddy simulation (LES) instead of DNS potentially introduces modeling errors and discretization errors both depending on the step size of the grid. Hence it seems promising to optimize the grid in an LES via adaptation to minimize this effect. In contrast to classical grid adaptation, the LES equations depend on the grid via the subgrid-scale term and therefore change whenever the grid is changed, so that criteria for adaptation need to be LES-specific. The present work uses r-adaptive refinement, defined by redistribution of a given number of grid points in the computational domain.

To solve the three-dimensional momentum and mass conservation on moving grids the ALE (Arbitrary Lagrangian Eulerian) formulation was used in its filtered form

$$\frac{\mathrm{d}}{\mathrm{d}t}\int_{V(t)} \bar{u}_i \,\mathrm{d}V + \int_{V(t)} \frac{\partial\left[\bar{u}_i\left(\bar{u}_j - u_{N,j}\right)\right]}{\partial x_j}\mathrm{d}V = \tag{1}$$

$$-\int_{V(t)} \frac{\partial \bar{p}}{\partial x_i}\mathrm{d}V + \int_{V(t)} 2\nu\frac{\partial \bar{S}_{ij}}{\partial x_j}\mathrm{d}V - \int_{V(t)} \frac{\partial \tau_{ij}}{\partial x_j}\mathrm{d}V + \int_{V(t)} \bar{f}_i \,\mathrm{d}V$$

$$\int_{V(t)} \frac{\partial \bar{u}_j}{\partial x_j}\mathrm{d}v = 0 \tag{2}$$

Claudia Hertel
Institut für Strömungsmechanik, TU Dresden, 01062 Dresden, Germany,
e-mail: claudia.hertel@tu-dresden.de

Jochen Fröhlich
Institut für Strömungsmechanik, TU Dresden, 01062 Dresden, Germany,
e-mail: jochen.froehlich@tu-dresden.de

H. Kuerten et al. (eds.), *Direct and Large-Eddy Simulation VIII*,
ERCOFTAC Series 15, DOI 10.1007/978-94-007-2482-2_13,

with the overbar denoting filtered quantities. Nomenclature is as usual with velocity components u_i, pressure p, strain-rate tensor $\bar{S}_{ij} = (\partial_{x_j}\bar{u}_i + \partial_{x_i}\bar{u}_j)/2$ and volume forces f_i. The subgrid-scale stress-tensor τ_{ij} is modeled by the Smagorinsky model using a constant $C_s = 0.1$. When discretizing (1), one has to guarantee that the discretization of the change in volume over time is consistent to that of the grid velocity $u_{N,i}$ by using the so-called Space Conservation Law [2].

The simulations were performed with the in-house code LESOCC2, employing a fractional step method to integrate (2) on curvilinear coordinates in time. A Finite Volume method on cell-centered grids with collocated variable arrangement and central differences is used for spacial discretization.

2 Moving Mesh Approach

The r-adaptive method employed here is based on a Moving Mesh PDE (MMPDE) to position the new grid points $\mathbf{x}$ in physical space [4]. It aims to equidistribute the monitor function G over the domain by solving

$$\tau \frac{\partial \mathbf{x}}{\partial t} = P \left[\sum_{i=1}^{3} \sum_{j=1}^{3} \underbrace{(\mathbf{a}^i \cdot G^{-1} \mathbf{a}^j)}_{A_{ij}} \frac{\partial^2 \mathbf{x}}{\partial \xi^i \partial \xi^j} - \sum_{i=1}^{3} \underbrace{\left(\mathbf{a}^i \cdot \frac{\partial G^{-1}}{\partial \xi^i} \mathbf{a}^i \right)}_{B_i} \frac{\partial \mathbf{x}}{\partial \xi^i} \right] . \tag{3}$$

The monitor function G establishes the connection between the physical solution of the flow field and the grid point distribution and is chosen to be equal in all directions of adaptation, i.e. $G = \omega \mathscr{I}$, with $\mathscr{I}$ the identity matrix. The remaining quantities in (3) identify the contravariant basis vectors $\mathbf{a}^i = \nabla \xi^i$, the time scaling parameter τ and the coordinates of the uniformly discretized computational space ξ. The local scaling parameter P is evaluated by using the coefficients A_{ij} and B_i.

The shape of the computational domain is held fix throughout the adaptation. The tangential movement of boundary points is directly linked to the alteration of the points next to the wall to achieve orthogonal cells.

For the discretization of (3) a Finite Difference Formulation with central differences in space was chosen to determine the cell corner coordinates. This is unavoidable because it is mathematically impossible to determine a valid grid from given cell center values in general cases. An implicit Euler scheme is used for integrating (3) in time.

3 Choice of monitor function

The heart of each moving mesh method is the monitor function ω and hence the so-called Quantity of Interest (QoI). The basic form for the monitor function used here is

$$\omega = \sqrt{1 + \frac{\alpha}{\psi_M} \sum_{k=1}^{N_k} \left(\frac{\psi_k}{\psi_{k,max}} \right)^2}, \tag{4}$$

where $\psi_{k,max}$ is the maximum value of the QoI ψ_k with N_k different QoIs accounted for in the adaptation. The value of this sum for each cell is normalized with its maximum ψ_M over the domain to achieve an equal magnitude of ω during the adaptation. Choosing (4) in the given form guarantees values greater than zero which is needed for a proper solution of (3) whereas the maximum of ω is scaled by the global parameter α. For all quantities used in this section $\langle .. \rangle$ designates averaging in time and homogeneous direction.

As a first QoI, the gradient of the mean streamwise velocity u was chosen, i.e. $\psi_1 = |\nabla \langle u \rangle|$ which should result in a grid that is refined especially near walls.

Using the turbulent kinetic energy (TKE) as a criterion is motivated by the basic idea of LES, and tries to achieve an equal ratio of modeled TKE k_{sgs} and total TKE k_{tot} over the flow field

$$\psi_2 = \frac{\langle k_{sgs} \rangle}{\langle k_{tot} \rangle} \qquad \text{with} \qquad k_{sgs} \approx \left(2^{1/3} - 1\right) 0.5 \left|\bar{\mathbf{u}} - \bar{\bar{\mathbf{u}}}\right|^2 . \tag{5}$$

The evaluation of k_{sgs} in (5) is taken from [1] and $k_{tot} = k_{res} + k_{sgs}$. This model, however, leads to unwanted high values of ψ_2 in areas where k_{res} vanishes. To overcome this drawback the maximum value of the total TKE over the domain $k_{tot,max}$ was chosen as the denominator in (5), instead of the local value.

The third criterion is motivated by the modeled shear-stress τ_{12}^{mod} and the resolved fluctuations $\langle u'v' \rangle$, being dominant in this flow configuration

$$\psi_3 = \frac{\langle \tau_{12}^{mod} \rangle}{\langle \tau_{12}^{mod} \rangle + \langle u'v' \rangle} \qquad \text{with} \qquad \tau_{12}^{mod} = \nu_t \left(\frac{\partial u_1}{\partial x_2} + \frac{\partial u_2}{\partial x_1} \right) . \tag{6}$$

Here ν_t is the turbulent viscosity. Like the second QoI ψ_2, a modification was required to avoid unwanted and unphysical high values of ψ_3 in areas where $\langle u'v' \rangle$ vanishes. To this end, a constant factor was added to the denominator, being one order of magnitude smaller than the maximum of $\tau_{12}^{mod} + \langle u'v' \rangle$.

In addition to the adaptation with one QoI, two results are also presented for a combination of two QoIs ($N_k = 2$) using the gradient of the streamwise velocity ψ_1 combined with (5) in a first adaptation and then with the third criterion (6).

4 Results for the flow over periodic hills

The turbulent flow over periodic hills at $Re = 10595$ (built with the bulk velocity and hill height h) [3, 5] was chosen as a test case. The configuration is shown in Fig. 1 together with the initial grid. The dimension of the computational domain is $L_x = 9h$, $L_y = 3.035h$ and $L_z = 4.5h$ in x, y and $z-$direction, respectively, as in [3].

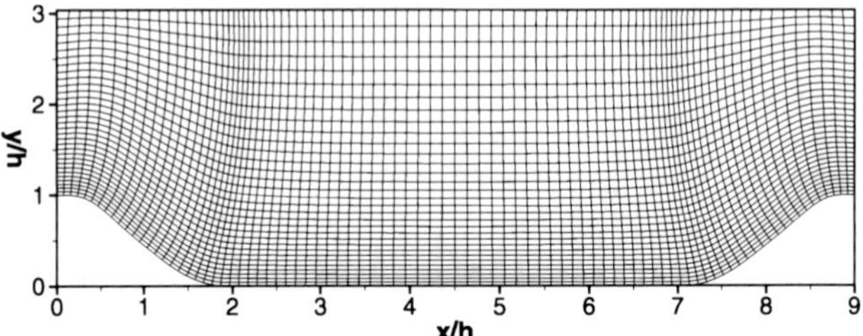

Fig. 1 Centerplane of the geometry and initial grid

The grid resolution of $89 \times 33 \times 49$ points in x, y and $z-$direction is substantially coarser than the grid used in the LES of [3] and even [5] to demonstrate the impact of an r-adaptive method. Each adaptation starts on a well converged solution of the turbulent flow on the initial grid (Fig. 1) using the wall model of Werner-Wengle. The grid was then adapted during one time step using (3). Subsequently, the simulation was continued for 100 time steps with collection of statistics before another step of adaptation was executed. This was repeated 100 times yielding an almost stationary grid. Subsequently, a simulation with 50 flow-through times for averaging on the final grid was performed to evaluate statistics. The values of $\alpha = 50$ and $\tau = 0.1$ were used throughout.

Results for one criterion

The final grids after 100 adaptations resulting from the three different QoI presented in Section 3 are displayed in Fig. 2(a-c). The final grid for ψ_1 shows substantial refinement near the hill crest and the upper wall were the velocity gradient normal to the wall is high. Using (5) yields grid refinement near the separation point and in the shear layer. The third QoI investigated, (6), leads to higher grid point density near the upper wall and the hill crest, generated by the higher value of τ_{12}^{mod} in these areas. In areas where ω is roughly constant an almost uniform grid is obtained.

Table 1 Separation and reattachment points for simulations on final adapted grids compared to the data of [3].

	fine grid [3]	initial grid	ψ_1	ψ_2	ψ_3	ψ_1 & ψ_2	ψ_1 & ψ_3
x_{sep}/h	0.2	0.5	0.3	0.45	0.45	0.3	0.25
x_{rea}/h	4.6	3.1	4.7	3.25	4.15	4.5	4.55

The time-averaged mean flow and shear stress at position $x/h = 2.0$ are shown in Fig. 2(d,e). With all QoI investigated an improvement compared to the solution on the initial grid can be seen. Using ψ_1 and ψ_3 streamwise velocity profiles are obtained that look very similar to that of the reference solution in [3]. However, only criterion ψ_1 predicts the separation and reattachment point of the flow better than

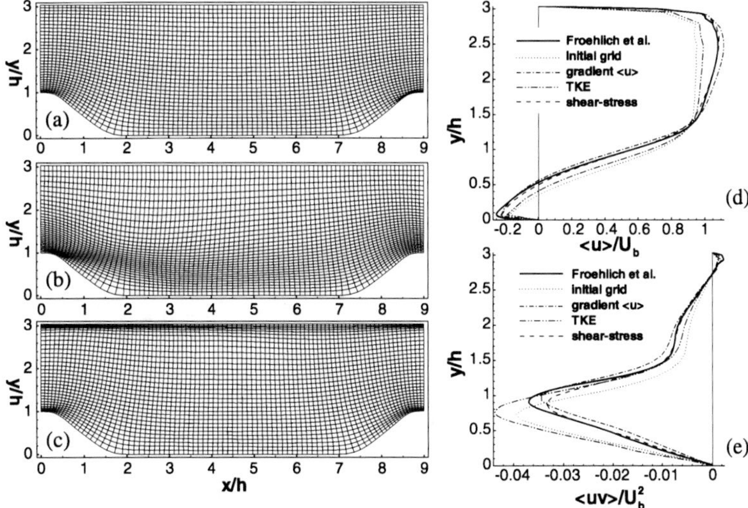

Fig. 2 Final grid after 100 adaptations where only one criterion was used (left). QoI: (a) gradient of streamwise velocity; (b) modeled TKE (5); (c) shear-stress (6). (d) streamwise velocity, (e) Reynolds stress $\langle u'v' \rangle$ at position $x/h = 2.0$.

the simulation on the initial grid (Table 1). With this criterion the grid refinement in wall normal and tangential direction at and upstream of the separation point is stronger than with the other ones (see $x = 8.5 \ldots 9$, $y \approx 1$ in Fig. 2a). This yields a better representation ot the flow at the separation point which is decisive in the present flow (see [5] for a discussion). Hence, the strongest improvement by adaptation is obtained with ψ_1 for the present flow. In the simulations using ψ_2 and ψ_3 the separation point is the same and too far downstream. Due to the different modeling in the separation zone the reattachment point differs between the latter two simulations.

Results for two criteria

The combination of two criteria was motivated by the idea to combine the advantages of different criteria to further improve the simulation. QoI ψ_1 was chosen in both cases to improve the prediction of the separation and reattachment point. The final grids for both simulations are similar to those of ψ_2 and ψ_3, having smaller cells near the hill crest than the final grids for the single QoIs.

As desired the beginning and end of the recirculation area are improved, comparable to those of ψ_1 alone (Table 1). The calculated statistics for the combined criteria can be seen in Fig. 3. The results obtained with the combined ψ_1 & ψ_3 criterion improves upon the one with ψ_1 alone opening a new route to further improvement. In the present case ψ_1 seems to dominate the adaptation, though, as the combined criteria both yield results close to the one with ψ_1 alone. Furthermore the difference in the results with ψ_2 and ψ_3 alone is now substantially decreased when using ψ_1 & ψ_2 and ψ_1 & ψ_3. Further ways to combine criteria, possibly with weighting,

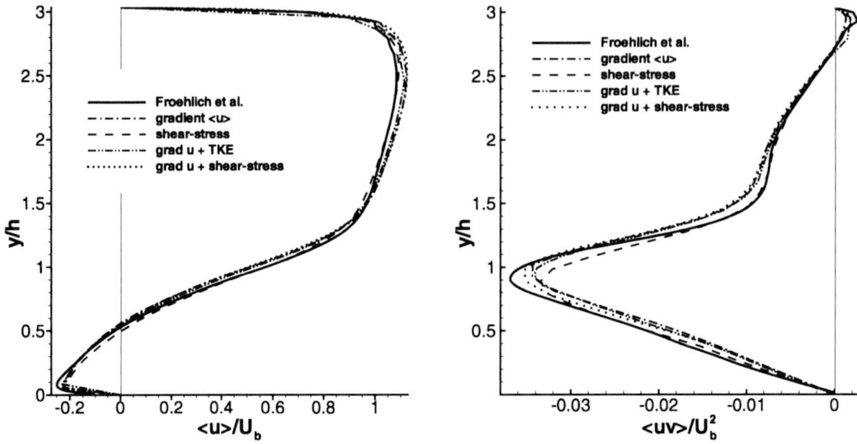

Fig. 3 Averaged flow quantities: streamwise velocity $\langle u \rangle$ (left) with detail plot and Reynolds stress $\langle u'v' \rangle$ (right) at position $x/h = 2.0$ in the middle of the recirculation area.

might be investigated.

On the other hand, the present results are so close to the reference data that hardly any improvement seems possible for the present setup.

Acknowledgements This work is sponsored by the German Research Foundation (DFG) in the framework of Priority Programme SPP 1276, MetStröm. The authors would like to thank their colleagues J. Lang and S. Löbig for instructive discussions and a passionate cooperation.

References

1. Berselli LC, Iliescu W, Layton WJ (2006) Mathematics of large eddy simulation of turbulent flows. Springer, Berlin Heidelberg New York
2. Demirdžić I, Perić M (1988) Space conservation law in finite volume calcualations of fluid flow. Int J Num Meth Fluids 8:1037–1050
3. Fröhlich J, Mellen CP, Rodi W et al. (2005) Highly resolved large-eddy simulation of separated flow in a channel with streamwise periodic constrictions. J Fluid Mech 526:19–66
4. Huang W (2001) Practical aspects of formulation and slution of moving mesh partial differential equations. J Comp Phys 171:753–775
5. Temmerman L, Leschziner M, Mellen CP et al. (2003) Investigation of wall-function approximations and subgrid-scale models in large eddy simulation of separated flow in a channel with streamwise periodic constrictions. Int J Heat Fluid Flow 24:157–180

An eddy-viscosity model based on the invariants of the rate-of-strain tensor

Roel Verstappen

1 Large-eddy simulation

Turbulence can generally not be computed directly from the Navier-Stokes equations, because the flow possess far too many scales of motion. Therefore numerical simulations of turbulence have to resort to models of the small scales of motion for which numerical resolution is not available. Large-eddy simulation (LES) seeks to predict the dynamics of spatially filtered flows. If a spatial filter $u \mapsto \overline{u}$ is applied to the (incompressible) Navier-Stokes equations an expression depending on both full velocity field u and the filtered field $\overline{u}$ results, due to the nonlinearity. The dependence on u can be removed by adopting an eddy-viscosity model, for instance [1]. The governing equation is then given by

$$\partial_t v + (v \cdot \nabla) v + \nabla p = 2\nabla \cdot ((\nu + \nu_e) S(v)) \tag{1}$$

where ν and ν_e denote the fluid viscosity and the eddy viscosity, respectively; $S(v)$ is the symmetric part of the velocity gradient (the rate-of-strain tensor). The solution v is supposed to approximate the filtered Navier-Stokes solution $\overline{u}$.

The very essence of large-eddy simulation is that v contains only eddies of size $\geq \Delta$, where Δ denotes the user-chosen length of the filter. This property enables us to solve Eq. (1) numerically when it is not feasible to compute the full turbulent flow field u. Eq. (1) is formally equivalent to the Navier-Stokes equations with a modified diffusion coefficient; hence, the desired effect thereof is to eliminate all scales of size $< \Delta$. Therefore, we view the eddy viscosity as a function of v that is to be determined such that the dynamically significant scales of motion in the solution v of Eq. (1) are greater than (or equal to) Δ.

Johann Bernoulli Institute for Mathematics and Computer Science
University of Groningen, The Netherlands
e-mail: R.W.C.P.Verstappen@rug.nl

H. Kuerten et al. (eds.), *Direct and Large-Eddy Simulation VIII*, ERCOFTAC Series 15, DOI 10.1007/978-94-007-2482-2_14,

2 When does eddy diffusivity counteract the production of subfilter scales sufficiently?

We consider an arbitrary part Ω_Δ with diameter Δ of the flow domain and take the filtered velocity $\overline{u}$ equal to the average of u over Ω_Δ. Furthermore, we suppose that Ω_Δ is a periodic box, so that boundary terms resulting from integration by parts (in the computations to come) vanish. Poincaré's inequality states that there exists a constant C_Δ, depending only on Ω_Δ, such that for every v,

$$\int_{\Omega_\Delta} ||v-\overline{v}||^2\,dx \le C_\Delta \int_{\Omega_\Delta} ||\nabla v||^2\,dx \tag{2}$$

The optimal constant C_Δ - the Poincaré constant for the domain Ω_Δ - is the inverse of the smallest (non-zero) eigenvalue of the dissipative operator $-\nabla^2$ on Ω_Δ. It is given by $C_\Delta = (\Delta/\pi)^2$ for convex domains Ω_Δ [2].

The residual field $v' = v - \overline{v}$ contains eddies of size smaller than Δ. The eddy viscosity must keep them from becoming dynamically significant. Poincaré's inequality (2) shows that the $L^2(\Omega_\Delta)$ norm of the residual field v' is bounded by a constant (independent of v) times the $L^2(\Omega_\Delta)$ norm of ∇v. Consequently, we can confine the dynamically significant part of the motion to scales $\geq \Delta$ by damping the velocity gradient with the help of an eddy viscosity. To see how the evolution of the $L^2(\Omega_\Delta)$ norm of ∇v is to be damped, we consider the residual field v' first:

$$\frac{\mathrm{d}}{\mathrm{dt}}\int_{\Omega_\Delta} \frac{1}{2}||v'||^2\,dx = \int_{\Omega_\Delta} T\left(\overline{v},v'\right)\,dx - \nu_e \int_{\Omega_\Delta} ||\nabla v'||^2\,dx - \nu \int_{\Omega_\Delta} ||\nabla v'||^2\,dx \tag{3}$$

Here, $\int_{\Omega_\Delta} T\left(\overline{v},v'\right)\,dx$ represents the energy transfer from $\overline{v}$ to v'. Obviously, the energy of v' has to decrease quickly, since Eq. (1) should not produce subfilter scales. Now suppose that the eddy viscosity is taken such that the first two terms in the right-hand side above cancel each other out. Then we have

$$\frac{\mathrm{d}}{\mathrm{dt}}\int_{\Omega_\Delta} \frac{1}{2}||v'||^2\,dx = -\nu \int_{\Omega_\Delta} ||\nabla v'||^2\,dx \tag{4}$$

This equation shows that the evolution of the energy of v' is not depending on $\overline{v}$. Stated otherwise, the energy of subfilter scales dissipates at a natural rate, without any forcing mechanism involving scales larger than Δ. With the help of the Poincaré inequality (2) and the Gronwall lemma, we obtain from Eq. (4) that the energy of the subfilter scales, $\int_{\Omega_\Delta} |||v'||^2(x,t)\,dx$, decays at least as fast as $\exp(-2\nu t/C_\Delta)$, for *any* filter length Δ. Applying Poincaré's inequality and Gronwall's lemma to

$$\frac{\mathrm{d}}{\mathrm{dt}}\int_{\Omega_\Delta} \frac{1}{2}||\nabla v||^2\,dx \le -\nu \int_{\Omega_\Delta} ||\nabla^2 v||^2\,dx \tag{5}$$

results into (at least) the same rate of decay. So, in conclusion, we can keep the subfilter component v' under control by imposing (5). By taking the L^2 innerproduct of Eq. (1) with $\nabla^2 v$ it can be shown that the dissipative condition (5) holds if

$$\nu_e \int_{\Omega_\Delta} q(v)\,dx \geq C_\Delta \int_{\Omega_\Delta} r(v)\,dx \tag{6}$$

where $q(v) = \frac{1}{2}\mathrm{tr}(S^2(v))$ and $r(v) = -\det S(v)$ are the (non-zero) invariants of the rate-of-strain tensor $S(v)$. Condition (6) ensures that subfilter scales are dynamically insignificant, meaning that their energy is bounded by (2) where the upper bound evolves according to (5). In physical terms Eq. (6) expresses that within Ω_Δ the eddy-viscous damping is greater than (or equal to) the production of energy.

3 Modeling consistency

Now, the question is: does the minimal amount of eddy viscosity given by

$$\nu_e = C_\Delta \overline{r(v)}/\overline{q(v)} \tag{7}$$

adequately model the subfilter contributions to the evolution of the filtered velocity? With the help of the filtered Navier-Stokes solution $\overline{u}$ we can analyze the consistency of the eddy-viscosity model $(\overline{u}\overline{u}^T - \overline{uu^T})_{tr} \approx 2\,\nu_e S(\overline{u})$. This is also called *a priori* testing. Here we consider the traceless part (defined by $A_{tr} = A - \frac{1}{3}\mathrm{tr}(A)I$), because the trace of $\overline{u}\overline{u}^T - \overline{uu^T}$ can be incorporated into the pressure. A series expansion gives $(\overline{u}\overline{u}^T - \overline{uu^T})_{tr} = -\frac{\Delta^2}{12}(\nabla\overline{u}\nabla\overline{u}^T)_{tr} + \mathscr{O}(\Delta^4)$. The leading term is known as the gradient or Clark model. Unfortunately, the gradient model cannot be used as a stand-alone LES model, since it produces a finite time blow-up of the kinetic energy [3]. In other words, the gradient model can produce length-scales smaller than Δ. Projecting both Eq. (7) and gradient model onto $S(v)$ yields

$$2C_\Delta \frac{\overline{r}}{\overline{q}} \int_{\Omega_\Delta} S(v) : S(v)\,dx \stackrel{?}{=} -\frac{\Delta^2}{12} \int_{\Omega_\Delta} (\nabla v \nabla v^T)_{tr} : S(v)\,dx \tag{8}$$

The integral in the right-hand side equals $-4\int_{\Omega_\Delta} r(v)\,dx$; see for instance Ref. [4]. This shows that r provides a measure of the alignment of the gradient model and S. By definition we have $S : S = 2q$. Consequently, Eq. (8) shows that the order of the modeling error is optimal if $C_\Delta = \Delta^2/12$. This value is in fair agreement with the Poincaré constant, $C_\Delta = \Delta^2/\pi^2$; yet, it is a slightly lower. The overall situation is sketched in Fig. 1. The horizontal axis in this figure represents all possible eddy-viscosity models; the axis is parameterized by the eddy viscosity. The shaded part of the horizontal axis in Fig. 1 depicts the subset of eddy viscosities that satisfy Eq. (6). The projection of the gradient model onto the horizontal axis falls outside the shaded area; hence it cannot be guaranteed that this projection damps subfilter scales adequately. This reflects that the gradient model can produce subfilter scales. Eq. (7) forms the best approximation of the projection of the gradient model provided the eddy-viscosity model is restricted by Eq. (6).

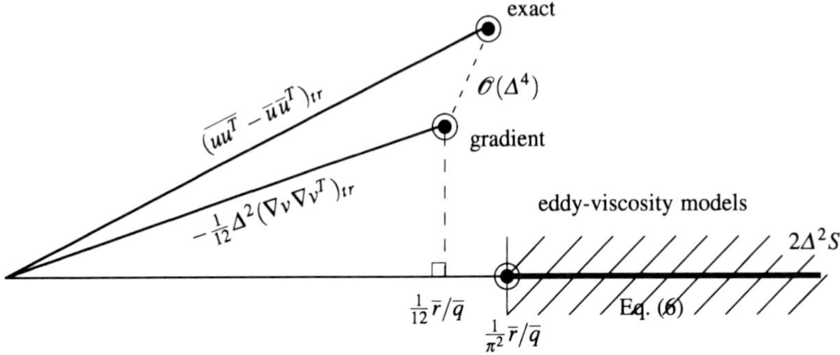

Fig. 1 Some LES-models in the space of trace-less 3x3 tensors.

Furthermore, Eq. (7) yields $\nu_e = 0$ in any (part of the) flow where $r = 0$. That is, the eddy viscosity vanishes if the nonlinear transport to scales $< \Delta$ is absent. At a no-slip wall $r = 0$ too; hence $\nu_e = 0$ at the wall. In homogeneous, isotropic turbulence, we have $r/q \propto Re^{1/2}$. Therefore $\nu_e/\nu \propto Re^{3/2}$ for fixed Δ. Additionally, we obtain that $\nu_e + \nu \to \nu$ if $\nu \propto Re^{-1} \propto \Delta^2 r/q \propto \Delta^2 Re^{1/2}$, that is if $\Delta \propto Re^{-3/4}$. This shows that the eddy viscosity given by Eq. (7) vanishes as Δ is of the order of $Re^{-3/4}$, i.e., if Δ approaches the Kolmogorov scale.

4 Vortex stretching

By taking the curl of Eq. (1) we find the vorticity equation and from that we obtain that the enstrophy is governed by

$$\frac{\mathrm{d}}{\mathrm{dt}} \int_{\Omega_\Delta} \frac{1}{2} ||\omega||^2 \, dx = \int_{\Omega_\Delta} \omega \cdot S\omega \, dx - (\nu + \nu_e) \int_{\Omega_\Delta} ||\nabla \omega||^2 \, dx$$

In the right-hand side we recognize the vortex stretching term that can produce smaller scales of motion. With the help of Eq. (6) we get that this term is bounded by

$$\int_{\Omega_\Delta} \omega \cdot S\omega \, dx = \int_{\Omega_\Delta} 4r(v) \, dx \overset{(6)}{\leq} \frac{\nu_e}{C_\Delta} \int_{\Omega_\Delta} 4q(v) \, dx \leq \nu_e \int_{\Omega_\Delta} ||\nabla \omega||^2 \, dx$$

Notice that the first equality shows that $r(v)$ is a measure for the vortex stretching [4]. Thus, Eq. (6) can also be interpreted as follows: the eddy viscosity is taken such that the corresponding damping of the enstrophy, $\nu_e \int_{\Omega_\Delta} ||\nabla \omega||^2 \, dx$, counteracts the production by means of the vortex stretching mechanism, $\int_{\Omega_\Delta} \omega \cdot S\omega \, dx$. In other words, the eddy viscosity prevents the intensification of vorticity at the scale Δ set by the map $u \mapsto \overline{u}$. Here it may be remarked that Meneveau [5] has also shown that constraints on the eddy viscosity can be based on vortex stretching and enstrophy.

5 Towards a simple qr-model

To compute the eddy viscosity according to Eq. (7), we need know how q and r vary within Ω_Δ. Here, we cannot simply take $\overline{q(v)} = q(\overline{v})$, because the relation between q and v is nonlinear. On the other hand, however, we do not want to compute v' explicitly. In a finite volume approximation, for instance, we resolve only $\overline{v}$ if the grid size is taken equal to Δ, cf. [6]. To express the eddy viscosity in terms of $\overline{v}$, we have to apply an approximate deconvolution method that recovers some of the information lost in the filtering process. To recover an approximation for v' we consider the series expansion of v around $\overline{v}$. Ignoring terms that are of the order Δ^4 (like in the gradient model), we get $v' \approx -\frac{1}{24}\Delta^2\,\nabla^2\overline{v}$. With the help of this approximation it can be shown that $\nu_e = C_S^2\Delta^2\sqrt{4q}$, where the Smagorinsky coefficient is given by

$$C_S^2 = c^2\, r^+ / \sqrt{4q^3} \tag{9}$$

with $c^2 = \frac{1}{\pi^2} + \frac{1}{24}$ and $r^+ = \max\{0, r\}$. Notice that (6) is trivially satisfied by taking $\nu_e = 0$ in case $r < 0$. In homogeneous, isotropic turbulence, the coefficient (9) is (in lowest order) independent of the Reynolds number. The three roots of the characteristic polynomial of $S(v)$ are real-valued. This leads to the constraint $27r^2 \leq 4q^3$. Consequently, the largest value of the Smagorinsky coefficient C_S given by Eq. (9) is (approximately) equal to 0.17. Remarkably this maximum value is identical to Lilly's value, $C_S = 0.17$, which implies that the standard Smagorinsky model with $C_S = 0.17$ satisfies (6). Stated differently, the standard model with $C_S = 0.17$ stops the production of scales $< \Delta$. Interestingly, the value $C_S = 0.17$ has been found too large in many numerical experiments. In turbulent shear flow, for instance, C_S is often reduced to 0.1 to give the standard model a fair change for success.

6 First results

It goes without saying that the performance of the eddy-viscosity model (9) has to be investigated for many cases. As a first step it was tested for turbulent channel flow by means of a comparison with direct numerical simulations. This flow forms a prototype for near-wall turbulence: virtually every LES has been tested for it. The results are compared to the DNS data of Moser *et al.* [7] at $\mathrm{Re}_\tau = 590$. In fact, we should compare the LES-solution v to the filtered DNS-solution $\overline{u}$. Yet, since the filtered DNS-solution is not presented in Ref. [7] we will compare v directly to u. The dimensions of the channel are taken identical to those of the DNS of Moser *et al.* The computational grid used for the large-eddy simulation consists of 64^3 points. The DNS was performed on a 384x257x384 grid, i.e., the DNS uses about 144 times more grid points than the present LES. The LES-results were obtained with an incompressible code that uses a fourth-order, symmetry-preserving, finite-volume discretization. Details about the numerics can be found in Ref. [8]. The Poincaré constant is equal to the inverse of the smallest (non-zero) eigenvalue of the dissipative operator $-\nabla^2$ on Ω_Δ. Since this eigenvalue cannot be represented on the grid, we replace it by the largest, representable eigenvalue, which is given by $4/dx^2 + 4/dy^2 + 4/dz^2$. In this way, we get $\Delta^2 = \frac{3}{1/dx^2+1/dy^2+1/dz^2}$. Further it may be emphasized that Eq. (9) is essentially not more complicated to implement in a LES-code than the standard Smagorinsky model (with C_S constant). Indeed, the eddy-viscosity is expressed in terms of the invariants of the rate-of-strain tensor and does not involve explicit filtering. The invariant $q = \frac{1}{4}|S|^2$ is to be computed in any case; the computation of r is just as difficult. Unlike the standard Smagorinsky model (even with the relatively low value $C_S = 0.1$), the present model showed an appropriate behavior. As can be seen in Figure 2 both the mean velocity and the root-mean-square of the fluctuating velocity are in good agreement with the DNS.

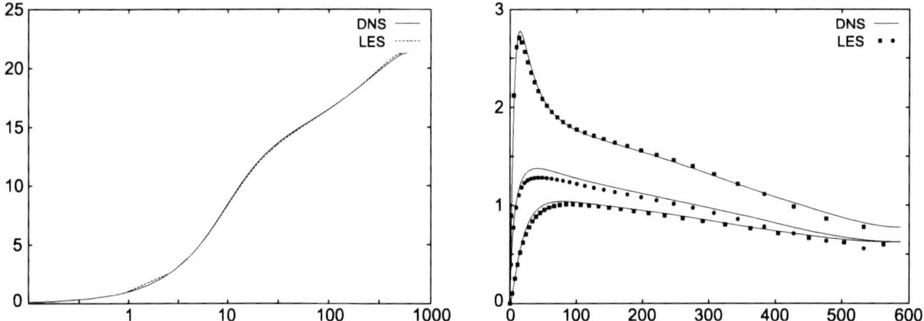

Fig. 2 The left-hand figure shows the mean velocity obtained with the help of the 64^3 LES and the DNS by Moser et al. The right-hand figure displays the root-mean-square of the fluctuating velocities. The boxes and circles represent LES data; every symbol corresponds to data in a grid point.

References

1. P. Sagaut, Large Eddy Simulation for Incompressible Flows, *Springer*, Berlin (2001).
2. L.E.Payne and H.F. Weinberger, An optimal Poincaré inequality for convex domains, *Arch. Rat. Mech. Anal.* **5**, 286–292 (1960).
3. B. Vreman, B. Geurts and H. Kuerten, Large-eddy simulation of the temporal mixing layer using the Clark model, *Theoret. Comput. Fluid Dynamics* **8**, 309–324 (1996).
4. D. Chae, On the spectral dynamics of the deformation tensor and a new a priori estimates for the 3D Euler equations, *Commun. Math. Phys.* **263**, 789–801 (2005).
5. C. Meneveau, Statistics of turbulence subgrid-scale stresses: Necessary conditions and experimental tests, *Phys. Fluids* **6**, 815–833 (1994).
6. U. Schumann, Subgrid scale model for finite difference simulations of turbulent flows in plane channels and annuli, *J. Comput. Phys.* **18**, 376–404 (1975).
7. R.D. Moser, J. Kim and N.N. Mansour, Direct numerical simulation of turbulent channel flow up to $\mathrm{Re}_\tau = 590$, *Phys. Fluids* **11**, 943–945 (1999).
8. R.W.C.P. Verstappen and A.E.P. Veldman, Symmetry-preserving discretization of turbulent flow, *J. Comp. Phys.* **187**, 343–368 (2003).

LES modeling of the turbulent flow over an Ahmed car

O. Lehmkuhl, R. Borrell, C.D. Perez-Segarra, A. Oliva and R. Verstappen

1 Introduction

The Ahmed body car is a semi-rectangular vehicle with a rounded front part and a slant back. Flow over this generic body reproduce the basic fluid-dynamics features of real cars with a typical fastback geometry and its simplified topology allows easy comparisons between experimental and numerical works.

Ahmed *et al.* [1] have carried out experiments of this vehicle with several slant angles (α) and described the characteristics of the flow for various angles. Their conclusions were that the flow is fully unsteady, three-dimensional and shows separations than can be followed by reattachments depending on α. The flow also shows large unsteady phenomena coming from interactions between recirculation bubbles and vortices. Moreover, a critical $\alpha \sim 30^o$ was found to lead an abrupt decrease in drag associated with the merging of separation regions and vortex breakdown. Recently, Lienhart *et al.* [2] have performed laser Doppler anemometry measurements of mean velocity fields and turbulence statistics at $\mathrm{Re} = 7.68 \cdot 10^5$ for a subcritical $\alpha = 25^o$ and a supercritical $\alpha = 35^o$, respectively (being the Reynolds number based on the body height, $\mathrm{Re} = U_{ref} h/\nu$).

Moreover, this test case was selected for the 9th ERCOFTAC Workshop since the complex features of the flow make computations a challenging task, from both, the modeling and the numerical point of view. It was concluded that the unsteady simulation methods like URANS and LES have a stronger potential to accurate

C.D. Perez-Segarra, A. Oliva
Heat and Mass Transfer Technological Center (CTTC), Polytechnical University of Catalonia (UPC), e-mail: cttc@cttc.upc.edu

O. Lehmkuhl, R. Borrell
TermoFluids S.L., Spain, e-mail: termofluids@termofluids.com

R. Verstappen
Institute of Mathematics and Computing Science, University of Groningen, The Netherlands, e-mail: R.W.C.P.Verstappen@math.rug.nl

H. Kuerten et al. (eds.), *Direct and Large-Eddy Simulation VIII*,
ERCOFTAC Series 15, DOI 10.1007/978-94-007-2482-2_15,

capture the details of the flow than the standard RANS techniques, but the case was still not considered to be satisfactory predicted for the subcritical angle ($\alpha = 25^o$).

All these difficulties are mainly related to the partial detachment of the flow at the beginning of the slant. As Krajnovic *et al.* [3] have proved the shear related to the small-scale structures generated by the detachment significantly increases the turbulent stresses. Also, experimental measurements by Lienhart *et al.* [2] shown that at the slant mixing layer, the turbulent kinetic energy is much higher than in the canonical self-similar mixing cases. These features of the flow puts under question the use of standard SGS models (Smagorinsky-like) and its damping/wall-functions (Van-Driest functions, hybrid RANS-LES, DES, etc.), since they have been developed from simple academic flows. However, several LES simulations have been made in the recent years with successful results, as for example, Krajnovic *et al.* [3] or M. Minguez *et al.* [4], among others. In general, these LES results have been carried out using structured spatial methods with high resolution meshes, 15M to 20M control volumes (CV), and Smagorinsky based models in conjunction with wall functions (or similar techniques) to deal with the near-wall region.

The present work aims to assess alternative techniques for the simulation of this type of flow with similar accuracy than the aforementioned works but with less computational cost. To do so, a conservative unstructured mesh discretization [5] will be used in conjunction with the WALE model [6], the qr-model [7] and the $\mathscr{C}_4$ regularization model [8]. As far as the author's knowledge is concerned this is the first time the q-r model is tested on unstructured grids and at high Reynolds numbers.

A reduction of the mesh requirements is expected since the well-known capacity for describing complex geometry of unstructured discretization and the possibility of using local mesh refinement algorithms at both the slant and recirculation regions. Also, our conservative formulation, since it preserves the kinetic energy equation, ensures good stability properties even at high Reynolds numbers and with coarse meshes [9, 5]. In other words, there is no need of fine grids only for stability issues with this discretization.

On the other hand, the three selected models match well with an unstructured formulation since they do not need damping functions to deal with the near-wall region. Furthermore, as far as the authors know, they have not been tested for this flow configuration under unstructured meshes.

2 Numerical Method

The Navier-Stokes and continuity equations can be written as

$$\rho \frac{\partial \boldsymbol{u}}{\partial t} + \mathsf{C}(\boldsymbol{u})\boldsymbol{u} + \mathsf{D}\boldsymbol{u} + \mathsf{G}\boldsymbol{p} = \boldsymbol{0} \quad (1)$$

$$\mathsf{M}\boldsymbol{u} = \boldsymbol{0} \quad (2)$$

where $\boldsymbol{u} \in \mathbb{R}^{3q}$ and $\boldsymbol{p} \in \mathbb{R}^{q}$ are the velocity vector and pressure, respectively. The matrices $\mathsf{C}(\boldsymbol{u})$, $\mathsf{D} \in \mathbb{R}^{3q\times 3q}$ are the convective and diffusive operators, respectively. Note the $\boldsymbol{u}$-dependence of the convective operator (non-linear operator). Finally, $\mathsf{G} \in \mathbb{R}^{3q\times q}$ represents the gradient operator, and the matrix $\mathsf{M} \in \mathbb{R}^{q\times 3q}$ is the divergence operator.

For the temporal discretization of the Navier-Stokes equation (1), a third-order Gear-like scheme [10] is used for the derivative, convective and diffusive term. And for the pressure gradient, a first-order backward Euler scheme is used.

Our spatial discretization schemes are conservative. These conservation properties are held if and only if the discrete convective operator is skew-symmetric ($\mathsf{C}_c(\boldsymbol{u}_c) = -\mathsf{C}_c(\boldsymbol{u}_c)^*$), if the negative conjugate transpose of the discrete gradient operator is exactly equal to the divergence operator ($-(\Omega_c \mathsf{G}_c)^* = \mathsf{M}_c$), and if the diffusive operator D_c is symmetric and positive-definite.

As Felten *et al.* [9] had noted, the total contribution of the pressure gradient term to the evolution of the kinetic energy does not vanish when the fractional step method is used on a collocated arrangement,

$$\varepsilon_{\mathbf{ke}} = (\tilde{\boldsymbol{p}}_c)^* \mathsf{M}_c(\mathsf{G}_c - \mathsf{G}_s)\tilde{\boldsymbol{p}}_c \tag{3}$$

where G_s and G_c are gradient operator at the faces and at the cell centers, respectively. Since this term can not be eliminated it is of interest to evaluate its scaling order. Felten *et al.* [9] had perform an analytical analysis of this term, and from their discussion, it is clear that the spatial term of the pressure error scales like $\mathcal{O}(\triangle_x^2)$ and the temporal term scales like $\mathcal{O}(\triangle_t)$. However, more recently Fishpool *et al.* [10], had developed a third-order Gear-like temporal scheme for collocated meshes, that allows both, a reduction of the dissipation of this type of schemes and the increase of the numerical stability. This last strategy is the one adopted in this work.

Considering the necessity of the use of LES models for studying engineering flows in complex geometries, which in general can not be described with structured grids, it is advisable to test models that do not use the wall-units coordinates. In this sense, the wall-adapting local-eddy viscosity (WALE) [6] and the qr-model proposed by Verstappen [7] are two good choices. In addition, the performance of the $\mathcal{C}_4$ regularization modeling proposed by Verstappen [8] is also considered. The behavior of the three models is compared with the experimental results.

3 Numerical results

The geometry of the car has been defined as in the experiments of Ahmed *et al.* [1] (see figure 1). The values of the body height is $H = 0.288m$. The other geometric values are $L_r = 0.8428m$, $G = 0.2012m$, $W = 0.389m$ and $C = 0.05m$. The slant angle considered is $\alpha = 25^\circ$. The computational domain is a rectangular channel of dimensions $9.1944 \times 1.87 \times 1.4$, with the front of the body located at a distance from the inlet of 2.1024m. The downstream region has a length of 6.048m measured from

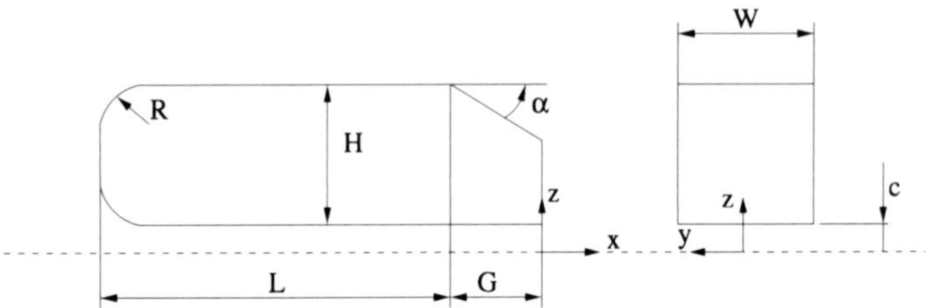

Fig. 1 Ahmed body car geometry

the rear end of the car. The Reynolds number based on the inlet velocity, U_{ref}, and the car height, H, is of $Re = 7.68 \times 10^5$. At the inflow, uniform axial velocity profile is imposed. At the lateral and top walls, slip boundary conditions are prescribed. At the outflow of the domain, a pressure-based boundary condition is applied. In addition, a buffer zone is defined at $x > 5$, in order to suppress non-physical waves which are reflected by the outflow boundary. No-slip conditions at the bottom surface are considered.

The calculations have been performed on an hybrid prism-tetrahedral unstructured grid of about 832000 CVs. A prism-layer to capture the body boundary layer is constructed around the car surface. Results obtained have been compared with the experimental data of Leinhart *et al.* [2] and also with numerical results obtained using Large Eddy Simulations with the wall adapting local eddy viscosity (WALE) [6] model for the sub-grid scale stresses.

In figure 2 a comparison between the three models used and the experimental results for the mean streamwise velocity for the symmetry plane ($y = 0$) at the rear end of the body and at the near wake are shown. As can be seen, there are significant differences concerning the prediction of the mean flow behavior in the slant back for the $\mathscr{C}_4$ regularization model. For this model, the flow separates in the slant corner and forms a large recirculation zone. As a consequence, the flow does not reattaches the slant as in the experiments. Although, *a-priori* it is possible to attribute this behavior to a poor resolution near the wall, when comparing with the performance of the WALE or the qr-model on the same grid, it can be observed that both models show a quite good agreement with the experiments even for this coarse grid. In fact, qr-model performs a little better than the WALE model, reproducing with quite good agreement the experimental measurements also in the near wake of the car.

As expected, larger discrepancies are obtained in the prediction of the Reynolds stresses as can be seen in figure 3. This is more relevant for the regularization model, which gives completely wrong results. In the case of LES models, WALE and qr-model, both follow the profiles of the experimental results, reproducing the position of the peak in the stresses but under-predict their value in some locations about a 50 % of the experimental value. However, it is expected that with an increase in the mesh, both models will approach to the experimental values.

In the case of the regularization model, it seems to be strongly affected by the filtering quality. This is more important in the zone of the slant near the wall, where the mesh is an hybrid between prism and tetrahedral control volumes. In this zone, the mesh rarely satisfy smoothness constraints on the grid spacing. Since the differential filter used has been discretized in an identical manner than the diffusive term in the Navier-Stokes equations, it can be argued that the filter truncation errors are affecting the high-frequency components of the discrete velocity vector. Thus, to avoid such numerical interaction between the filter and the modeled scales it may be interesting to discretize the differential filter with a high order scheme. This aspect will be studied in future works.

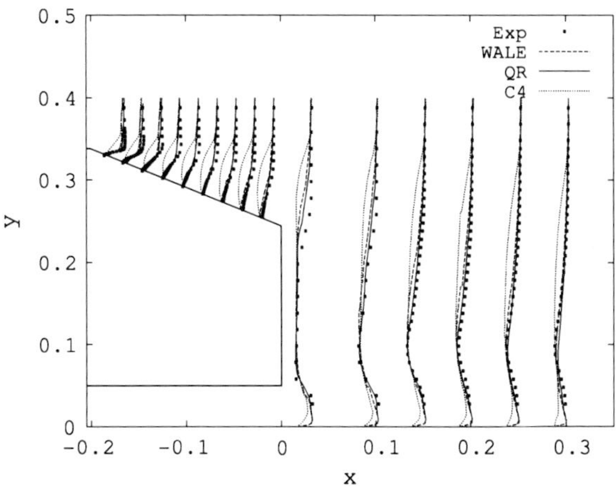

Fig. 2 Mean streamwise velocity profiles in the symmetry plane: solid line qr-model, dashed line WALE model, dotted line $\mathscr{C}_4$, dots experiments [2]

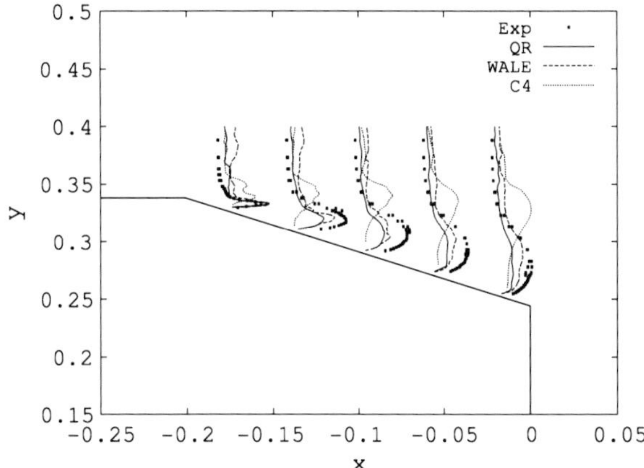

Fig. 3 Root mean square of the streamwise velocity fluctuations u_{rms} profiles in the symmetry plane: solid line qr-model, dashed line WALE model, dotted line $\mathscr{C}_4$, dots experiments [2]

4 Conclusions

The wall-adapting eddy-viscosity model (WALE), the q-r model and the $\mathscr{C}_4$ regularization model have been tested for predicting the flow over an Ahmed car with a slant of 25°. As the governing equations have been discretized by means of a conservative formulation, which has good stability properties even with coarse grids and high Reynolds numbers, it has allowed the use of a grid of about 832000 CVs with quite good results. Both LES models predicts quite well the mean flow while under-predicts the peaks in the Reynolds stresses. However, the results obtained are very promising and it is expected that increasing the grid in the slant and the near-wake zones both models will predict accurately the experimental results. Although the results with the $\mathscr{C}_4$ regularization model were not as expected due to issues with the filtering quality high order schemes for the discrete filter will be tested in the future.

Acknowledgements This work has been financially supported by Ministerio de Educación y Ciencia (Spain). Contract/Grant number ENE2007-67185.

References

1. S. R. Ahmed, G. Ramm, and G. Faltin. Salient features of the time-averaged ground vehicle wake. *SAE Paper*, (840300), 1984.
2. H. Lienhart and S. Becker. Flow and turbulent structure in the wake of a simplified car model. *SAE Paper*, (2003-01-0656), 2003.
3. S. Krajnović and L. Davidson. Flow around a simplified car, Part 1: Large eddy simulation. *Journal of Fluids Engineering*, 127:907–918, 2005.
4. M. Minguez, R. Pasquetti, and E. Serre. High-order large-eddy simulation of flow over the Ahmed body car model. *Physics of Fluids*, 20, 2008.
5. O. Lehmkuhl, R. Borrell, C. D. Pérez-Segarra, J. Chiva, and A. Oliva. Direct numerical simulations and symmetry-preserving regularization simulations of the flow around a circular cylinder at Reynolds number 3900. In *Turbulence, Heat and Mass Transfer 6*, Rome, Italy, September 2009.
6. F. Nicoud and F. Ducros. Subgrid-scale stress modelling based on the square of the velocity gradient tensor. *Flow, Turbulence and Combustion*, 62:183–200, 1999.
7. R. Verstappen. When does eddy viscosity damp subfilter scales sufficiently? In *Quality and Reliability of Large-Eddy Simulations II*, Pisa, Italy, September 2009.
8. R. Verstappen. On restraining the production of small scales of motion in a turbulent channel flow. *Computers and Fluids*, 37(7):887–897, 2008.
9. F. N. Felten and T. S. Lund. Kinetic energy conservation issues associated with the collocated mesh scheme for incompressible flow. *Journal of Computational Physics*, 215:465–484, 2006.
10. G. M. Fishpool and M. A. Leschziner. Stability for fractional-step schemes for the Navier-Stokes equations at high Reynolds number. *Computers & Fluids*, 38:1289–1298, 2009.

Spatially Variable Thresholding for Stochastic Coherent Adaptive LES

AliReza Nejadmalayeri, Oleg V. Vasilyev, Alexei Vezolainen, and Giuliano De Stefano

1 Introduction

The properties of wavelet transform, viz. the ability to identify and efficiently represent temporal/spatial coherent flow structures, self-adaptiveness, and de-noising, have made them attractive candidates for constructing multi-resolution variable fidelity schemes for simulations of turbulence [10]. Stochastic Coherent Adaptive Large Eddy Simulation (SCALES) [6] is the most recent wavelet-based methodology for numerical simulations of turbulent flows that resolves energy containing turbulent motions using wavelet multi-resolution decomposition and self-adaptivity. In this technique, the extraction of the most energetic structures is achieved using wavelet thresholding filter with a priori prescribed threshold level.

SCALES is a methodology, which inherits the advantages of both Coherent Vortex Simulations (CVS) [5] and Large Eddy Simulation (LES) while overcoming the shortcomings of both. Unlike coherent/incoherent and large/small structures decomposition in CVS and LES respectively, in SCALES the separation is between more and less energetic structures. Therefore, unlike CVS, the effect of background flow can not be ignored and needs to be modeled similarly to LES. As a result of using SGS models, the number of degrees-of-freedom is smaller than CVS and consequently a higher grid-compression can be achieved.

Ever since the emergence of the wavelet-based multi-resolution schemes for simulations of turbulence, there has been a major limitation for all wavelet-based techniques: the use of a priori defined global (both in space and time) thresholding-

AliReza Nejadmalayeri, Oleg V. Vasilyev, Alexei Vezolainen
Department of Mechanical Engineering, University of Colorado, Boulder, CO 80309, USA, e-mail: Alireza.Nejadmalayeri@Colorado.edu, Oleg.Vasilyev@Colorado.edu, Alexei.Vezolainen@Colorado.edu

Giuliano De Stefano
Dipartimento di Ingegneria Aerospaziale e Meccanica, Seconda Università di Napoli, I-81031 Aversa, Italy, e-mail: giuliano.destefano@unina2.it

H. Kuerten et al. (eds.), *Direct and Large-Eddy Simulation VIII*,
ERCOFTAC Series 15, DOI 10.1007/978-94-007-2482-2_16,

parameter. In this work the robustness of the SCALES approach is further improved by exploring the spatially and temporally variable thresholding strategy, which allows more efficient representation of intermittent flow structures.

2 Stochastic Coherent Adaptive Large Eddy Simulation

To address the shortcomings of LES and CVS, SCALES uses a wavelet thresholding filter to dynamically resolve and track the deterministic most energetic coherent structures while the effect of less energetic unresolved modes is modeled. The unresolved less energetic structures have been shown to be composed of a minority of deterministic coherent modes that dominate the total SGS dissipation and a majority of stochastic incoherent modes that, due to their decorrelation with the resolved modes, add little to the total SGS dissipation [6, 1]. In the current implementation, similar to the classical LES, only the effect of coherent part of the SGS modes is modeled using deterministic SGS models. The use of stochastic SGS models to capture the effect of the incoherent SGS modes will be the subject of future investigations. The most significant feature of SCALES is the coupling of modeled SGS dissipation and the computational mesh: more grid points (active wavelets) are used for SGS models with lower levels of SGS dissipation. In other words, the SCALES approach compensates for inadequate SGS dissipation by automatically increasing the local resolution and, hence, the level of resolved viscous dissipation. Another noticeable feature of the SCALES method is its ability to match the DNS energy spectra up to the dissipative wavenumber range using very few degrees of freedom.

2.1 Wavelet Thresholding Filter

In the wavelet-based approach to the numerical simulation of turbulence the separation between resolved energetic structures and unresolved residual flow is obtained through nonlinear multi-resolution wavelet threshold filtering (WTF). The filtering procedure is accomplished by applying the wavelet-transform to the unfiltered velocity field, discarding the wavelet coefficients below a given threshold (ε) and transforming back to the physical space. This results in decomposition of the turbulent velocity field into two different parts: a coherent more energetic velocity field and a residual less energetic coherent/incoherent one, i.e., $u_i = \overline{u}_i^{>\varepsilon} + u_i'$, where $\overline{u}_i^{>\varepsilon}$ stands for the wavelet-filtered velocity at level ε

$$\overline{u}_i^{>\varepsilon}(\mathbf{x}) = \sum_{\mathbf{l}\in\mathscr{L}^0} c_{\mathbf{l}}^0 \phi_{\mathbf{l}}^0(\mathbf{x}) + \sum_{j=0}^{+\infty} \sum_{\mu=1}^{2^n-1} \sum_{\substack{\mathbf{k}\in\mathscr{K}^{\mu,j} \\ |d_{\mathbf{k}}^{\mu,j}| > \varepsilon \|u_i\|_{\mathrm{WTF}}}} d_{\mathbf{k}}^{\mu,j} \psi_{\mathbf{k}}^{\mu,j}(\mathbf{x}), \tag{1}$$

where $\psi_{\mathbf{k}}^{\mu,j}$ are wavelets of family μ at j level of resolution, $d_{\mathbf{k}}^{j}$ are the coefficients of the wavelet decomposition, and $\phi_{\mathbf{l}}^{0}$ are scaling functions at zero level of resolution.

The key role in the wavelet-filter definition is clearly played by the non-dimensional relative thresholding level ε that explicitly defines the relative energy level of the eddies that are resolved and, consequently, controls the importance of the influence of the residual field on the dynamics of the resolved motions. In this work we explore the use of spatially and temporary varying thresholding level ε, which follows the evolution of the turbulent velocity field.

2.2 *Wavelet-Filtered Navier-Stokes Equations*

By filtering the Navier-Stokes equations, the following SCALES equations that govern the evolution of coherent energetic structures are obtained:

$$\partial_{x_i}\overline{u}_i^{>\varepsilon} = 0, \tag{2}$$

$$\partial_t\overline{u}_i^{>\varepsilon} + \overline{u}_j^{>\varepsilon}\partial_{x_j}\overline{u}_i^{>\varepsilon} = -\partial_{x_i}\overline{P}^{>\varepsilon} + \nu\partial^2_{x_jx_j}\overline{u}_i^{>\varepsilon} - \partial_{x_j}\tau^*_{ij} + Q\overline{u}_i^{>\varepsilon}, \tag{3}$$

where $\tau_{ij} = \overline{u_iu_j}^{>\varepsilon} - \overline{u}_i^{>\varepsilon}\overline{u}_j^{>\varepsilon}$ are the unresolved "SGS stresses" that need to be modeled, $Q\overline{u}_i^{>\varepsilon}$ is the linear forcing term [8], which is applied in the physical space over the whole range of wavenumbers, and the superscript $(\cdot)^{>\varepsilon}$ denotes wavelet filtered quantities. The SCALES equations are similar to the LES ones with the exception that the nonlinear multiscale band-pass wavelet filter, which depends on instantaneous flow realization, is used. The unresolved SGS stresses represent the effect of "unresolved less energetic deterministic coherent and stochastic incoherent eddies" on the "resolved more energetic coherent structures". In this study the localized kinetic-energy-based model [4] is exploited to close the filtered momentum equations. The SCALES methodology involving both the filtered momentum and the SGS energy equations is implemented by means of the adaptive wavelet collocation method [11].

3 Spatially Variable Thresholding

Previous studies, e.g. [7], demonstrated that in SCALES, the SGS dissipation is proportional to ε^2; therefore, one can enhance SCALES by exploiting spatially-varying ε based on local SGS dissipation $\Pi = -\tau^*_{ij}\overline{S}_{ij}^{>\varepsilon}$. This implies that rate of local-transfer of energy from energetic-resolved-eddies to unresolved-less-energetic structures can be controlled by varying the thresholding-factor. Therefore, the idea is to locally vary ε wherever Π deviates from a priori defined goal-value. A decrease of the thresholding level results in the local grid refinement with subsequent rise of the resolved viscous dissipation, while an increase of ε coarsens the mesh resulting in the growth of the local SGS dissipation. However, in order to vary ε in a physically consistent fashion, it should follow the local flow structures as they evolve in space and time. This necessitates the Lagrangian representation of ε, which is achieved using the Lagrangian Path-Line Diffusive Averaging approach [12]:

$$\partial_t \varepsilon + \overline{u}_j^{>\varepsilon} \partial_{x_j} \varepsilon = -\text{forcing}_{\text{term}} + \nu_\varepsilon \partial^2_{x_j x_j} \varepsilon. \tag{4}$$

For the sake of efficiency, instead of solving Eq. (4) for the evolution of ε, the linear-interpolation along characteristics, similar to the idea of Meneveau et al. [9], is performed

$$\frac{1}{\Delta t}\left[\varepsilon^{\text{new}}\left(\mathbf{x}, t+\Delta t\right) - \varepsilon^{\text{old}}\left(\mathbf{x} - \overline{\mathbf{u}}^{>\varepsilon}\Delta t, t\right)\right] = -\text{forcing}_{\text{term}}. \tag{5}$$

The use of linear interpolation results in sufficient numerical diffusion, thus, eliminating the need for explicit diffusion. The proposed spatially variable thresholding strategy ensures that the wavelet threshold is not *a priori* prescribed but determined on the fly by desired turbulence resolution. In this work two different mechanisms for the forcing term are studied:

FT1 The local fraction SGSD (FSGSD) is defined as $\frac{\Pi}{\varepsilon_{\text{res}}+\Pi}$, where $\varepsilon_{\text{res}} = 2\nu \overline{S}_{ij}^{>\varepsilon} \overline{S}_{ij}^{>\varepsilon}$ is the resolved viscous dissipation. The idea is to maintain FSGD at a "Goal" value which means retain Π at $\varepsilon_{\text{res}} \frac{\text{Goal}}{1-\text{Goal}}$. The first forcing type (FT1) is an attempt to implement this while normalizing the forcing term based on its time average:

$$\text{forcing}_{\text{term}} = \varepsilon^{\text{old}}\left(\mathbf{x} - \overline{\mathbf{u}}^{>\varepsilon}\Delta t, t\right) C_{\text{f}_\varepsilon} \frac{\Pi - \varepsilon_{\text{res}} \frac{\text{Goal}}{1-\text{Goal}}}{\text{TAF}}, \tag{6}$$

where TAF stands for the time average of the forcing, is the time average of $|\Pi - \varepsilon_{\text{res}} \frac{\text{Goal}}{1-\text{Goal}}|$. The forcing constant coefficient, C_{f_ε}, is intentionally set to 400 in order to make the time response of FT1 about three to four times faster than large eddy turnover time which is discussed in the next section.

FT2 In this approach, the variations of local-FSGSD is controlled directly based on the goal-value in conjunction with a relaxation time parameter (time-scale), τ_ε,

$$\text{forcing}_{\text{term}} = \varepsilon^{\text{old}}\left(\mathbf{x} - \overline{\mathbf{u}}^{>\varepsilon}\Delta t, t\right) \frac{1}{\tau_\varepsilon}\left(\frac{\Pi}{\varepsilon_{\text{res}}+\Pi} - \text{Goal}\right). \tag{7}$$

Following the time-varying threshold studies [2], a time-scale associated to the characteristic rate-of-strain is chosen: $\tau_\varepsilon^{-1} = \langle|\overline{S}_{ij}^{>\varepsilon}|\rangle$.

4 Results

The proposed methodology has been tested for linearly forced homogeneous turbulence [3] with linear forcing constant coefficient $Q = 6$ at $Re_\lambda \cong 72$ (Taylor micro-scale Reynolds number) on an adaptive grid with effective resolution 256^3. Figures 1(a,b) demonstrate the preliminary results of this implementation for three

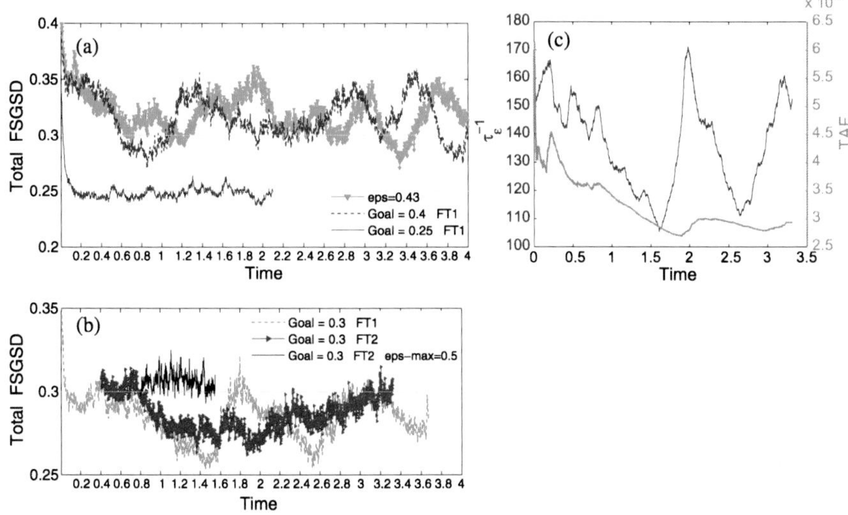

Fig. 1 Time-history of total fraction SGSD (a); Time-history of TAF and τ_{ε}^{-1} (b).

different goal-values (0.25,0.3,0.4) for the local FSGSD with the upper and lower bound for epsilon set as 0.2 and 0.43 ($\varepsilon \in [0.2, 0.43]$). For reference, a constant-thresholding case of $\varepsilon = 0.43$ is included as well. The local and total FSGSD are defined respectively as $\frac{\Pi}{\varepsilon_{\text{res}}+\Pi}$ and $\frac{\langle\Pi\rangle}{\langle\varepsilon_{\text{res}}\rangle+\langle\Pi\rangle}$, where $\langle\Pi\rangle = \langle -\tau^*_{ij}\overline{S}^{>\varepsilon}_{ij}\rangle$ and $\langle\varepsilon_{\text{res}}\rangle = 2\nu\langle\overline{S}^{>\varepsilon}_{ij}\overline{S}^{>\varepsilon}_{ij}\rangle$ are respectively the volume-averaged SGS dissipation and the volume-averaged resolved viscous dissipation.

For the case of Goal=0.4, total-FSGSD never reaches the prescribed goal-value (0.4). The reason is that the total-FSGSD for the case of constant-thresholding with $\varepsilon = 0.43$ is smaller than 0.4 for most of the time. As a result, varying thresholding-factor with a "local-FSGSD goal-value" larger than the average FSGSD of constant-thresholding using the same ε and $\varepsilon_{\max}$ resulted in total-FSGSD, which was bellow the goal-value. Similarly to the previous case, the test case of the goal-value of 0.3 inherits a large-period oscillations due to capping ε at 0.43 level regardless of the forcing method. These oscillations are removed by increasing $\varepsilon_{\max}$ to 0.5. The success of this test with larger $\varepsilon_{\max}$ compared with the abovementioned two tests, where $\varepsilon_{\max}$ was 0.43, revealed that the upper bound of the interval for allowable threshold variations was not large enough to increase the SGS dissipation accordingly, which implies that with $\varepsilon \in [0.2, 0.43]$ flow was over resolved. Therefore, to achieve a FS-GSD greater than the average of FSGSD corresponding to constant-thresholding at a certain $\varepsilon_{\text{constant-thresholding}}$, it is required to set the $\varepsilon_{\max} > \varepsilon_{\text{constant-thresholding}}$. This is further confirmed by considering the case with the goal set to 0.25, which illustrates how precisely the spatially variable thresholding methodology can maintain Π at a priori defined level. In addition, when $\varepsilon_{\max}$ is set up high enough, the SGS dissipation approaches the desired level within few eddy turnover times.

The time history of TAF and τ_ε^{-1} are shown in Fig. 1(c). The relaxation time parameter for FT2, τ_ε, is approximately one-tenth of the large eddy turnover time, $\tau_{\mathrm{eddy}} = \frac{u'^2}{\langle\varepsilon\rangle} = \frac{\frac{2}{3}K}{2KQ} = \frac{1}{3Q} = \frac{1}{18}$. While the relaxation time parameter for FT1, $C_{\mathrm{f}_\varepsilon}\mathrm{TAF}^{-1}$, is between one-third and one-fourth of τ_{eddy}. That is, FT2 has as much as 2 to 3 times faster response compared with FT1. This faster time response was able to partially recover the FSGSD. This improvement reveals the importance of very localized and fast mechanisms for the forcing term. The time-averaged term in FT1 destroys the localized Lagrangian nature of the algorithm; however, to smear out the effect of possible very localized FSGSD values, it is recommended to have some averaging mechanism. Hence, another approach, which is currently under investigation, is to track the forcing term itself within a Lagrangian frame so that the forcing term inherits the history of the flow evolution.

Acknowledgements This work was supported by the United States National Science Foundation (NSF) under grant No. CBET-0756046 and the United States Department of Energy (DOE) under grant No. DE-FG02-07ER64468.

References

1. De Stefano, G., Goldstein, D.E., Vasilyev, O.V.: On the role of sub-grid scale coherent modes in large eddy simulation. Journal of Fluid Mechanics **525**, 263–274 (2005)
2. De Stefano, G., Vasilyev, O.V.: A fully adaptive wavelet-based approach to homogeneous turbulence simulation. Physics of Fluids, submitted (2010)
3. De Stefano, G., Vasilyev, O.V.: Stochastic coherent adaptive large eddy simulation of forced isotropic turbulence. Journal of Fluid Mechanics **646**, 453–470 (2010)
4. De Stefano, G., Vasilyev, O.V., Goldstein, D.E.: Localized dynamic kinetic-energy-based models for stochastic coherent adaptive large eddy simulation. Physics of Fluids **20**, 045102.1–045102.14 (2008)
5. Farge, M., Schneider, K., Kevlahan, N.K.R.: Non-gaussianity and coherent vortex simulation for two-dimensional turbulence using an adaptive orthogonal wavelet basis. Physics of Fluids **11**(8), 2187–2201 (1999)
6. Goldstein, D.E., Vasilyev, O.V.: Stochastic coherent adaptive large eddy simulation method. Physics of Fluids **16**(7), 2497–2513 (2004)
7. Goldstein, D.E., Vasilyev, O.V., Kevlahan, N.K.R.: CVS and SCALES simulation of 3D isotropic turbulence. Journal of Turbulence **6**(37), 1–20 (2005)
8. Lundgren, T.: Linearly forced isotropic turbulence. In: Annual Research Briefs, pp. 461–473. Center for Turbulence Research, NASA Ames/Stanford University (2003)
9. Meneveau, C., Lund, T.S., Cabot, W.H.: A lagrangian dynamic subgrid-scale model of turbulence. Journal of Fluid Mechanics **319**, 353–385 (1996)
10. Schneider, K., Vasilyev, O.V.: Wavelet methods in computational fluid dynamics. Annual Review of Fluid Mechanics **42**, 473–503 (2010)
11. Vasilyev, O.V., Bowman, C.: Second generation wavelet collocation method for the solution of partial differential equations. Journal of Computational Physics **165**, 660–693 (2000)
12. Vasilyev, O.V., De Stefano, G., Goldstein, D.E., Kevlahan, N.K.R.: Lagrangian dynamic SGS model for SCALES of isotropic turbulence. Journal of Turbulence **9**(11), 1–14 (2008)

Stochastic coherent adaptive LES with time-dependent thresholding

Giuliano De Stefano and Oleg V. Vasilyev

1 Introduction

With the recent development of wavelet-based techniques for computational fluid dynamics, adaptive numerical simulations of turbulent flows have become feasible [1]. Adaptive wavelet methods are based on wavelet threshold filtering that makes it possible to separate coherent energetic eddies, which are numerically simulated, from residual background flow structures that are filtered out. By varying the filter thresholding level different approaches with different fidelity are obtained: from the highly accurate wavelet-based direct numerical simulation (WDNS) that does not involve any model to the stochastic coherent adaptive large eddy simulation (SCALES) that needs a closure modeling procedure, *i.e.* [2].

The prescription of a given threshold for SCALES filtering directly links to the desired turbulence resolution. By decreasing the thresholding level more and more eddies with smaller energy are directly simulated so that the effect of unresolved background flow becomes less and less important and, correspondingly, the influence of the modeling procedure as well. To date, the SCALES method has been applied for both decaying and forced turbulence with a specified thresholding level that is based on *a-priori* studies, *i.e.* [3].

In this work, a new original strategy is presented for which the wavelet filtering threshold is not prescribed but determined on the fly for a given level of turbulence resolution. A completely adaptive eddy capturing approach that allows to perform variable fidelity numerical simulations of turbulence is proposed. The new method is based on wavelet filtering with time-dependent thresholding that automatically

Giuliano De Stefano
Dipartimento di Ingegneria Aerospaziale e Meccanica, Seconda Università di Napoli, I-81031 Aversa (Italy), e-mail: giuliano.destefano@unina2.it

Oleg V. Vasilyev
Department of Mechanical Engineering, University of Colorado, Boulder, CO 80309 (USA), e-mail: oleg.vasilyev@colorado.edu

H. Kuerten et al. (eds.), *Direct and Large-Eddy Simulation VIII*,
ERCOFTAC Series 15, DOI 10.1007/978-94-007-2482-2_17,

adapts to the desired level of turbulence resolution. The SCALES governing equations supplemented by a localized dynamic energy-based closure model are solved by means of the adaptive wavelet collocation numerical method.

2 Time-dependent thresholding

The SCALES governing equations for incompressible turbulent flow are represented by the following wavelet-filtered Navier-Stokes equations

$$\frac{\partial \overline{u_i}^{>\varepsilon}}{\partial t} + \overline{u_j}^{>\varepsilon} \frac{\partial \overline{u_i}^{>\varepsilon}}{\partial x_j} = -\frac{1}{\rho}\frac{\partial \overline{p}^{>\varepsilon}}{\partial x_i} + \nu \frac{\partial^2 \overline{u_i}^{>\varepsilon}}{\partial x_j \partial x_j} - \frac{\partial \tau_{ij}}{\partial x_j}, \tag{1}$$

where $\overline{u_i}^{>\varepsilon}$ stands for the wavelet-filtered velocity field while τ_{ij} are the subgrid-scale (SGS) stresses to be modeled [2]. As usual for large eddy simulations, the SGS model is mainly required to provide the right energy dissipation in order to approximate the net effect of unresolved background flow upon the dynamics of resolved eddies. The amount of energy dissipation to be modeled clearly depends upon the wavelet filtering threshold that is used in (1) and, thus, the ratio of SGS to total (resolved plus modeled) dissipation can be practically used as a measure of the turbulence resolution for the SCALES solution. The above ratio is defined as

$$\mathscr{R}(t) = \frac{\mathscr{D}_{\text{sgs}}}{\mathscr{D}_{\text{res}} + \mathscr{D}_{\text{sgs}}}, \tag{2}$$

where $\mathscr{D}_{\text{res}} = 2\nu\langle \overline{S}_{ij}^{>\varepsilon} \overline{S}_{ij}^{>\varepsilon} \rangle$ represents the volume-averaged resolved viscous dissipation and $\mathscr{D}_{\text{sgs}} = \langle -\tau_{ij}^{*} \overline{S}_{ij}^{>\varepsilon} \rangle$ stands for the volume-averaged SGS dissipation, with $0 < \mathscr{R} < 1$. This way, instead of using a prescribed wavelet threshold based upon subjective considerations, a goal value $\mathscr{R}_0$ for the resolved flow parameter (2) can be actually assigned. Correspondingly, the thresholding level ε is evaluated as a time-dependent function according to the simple evolution equation

$$\frac{d\varepsilon}{dt} = -\left(\mathscr{R} - \mathscr{R}_0\right) \frac{\varepsilon}{\tau_\varepsilon}, \tag{3}$$

where τ_ε is a time constant that can be linked to the eddy turnover time of the turbulence. This equation is explicitly discretized and solved step-by-step in time along with the SCALES governing equations. When the grid is too coarse and the SGS dissipation is higher than the goal ($\mathscr{R} > \mathscr{R}_0$), the threshold value is decreased, which leads to the mesh refining. On the contrary, when the turbulence is over-resolved and the SGS dissipation is lower than the goal ($\mathscr{R} < \mathscr{R}_0$), the threshold value is increased, which leads to the mesh coarsening.

With the adoption of the above time-dependent wavelet-filtering threshold, the dynamically adaptive nature of the SCALES method is fully exploited. In fact, the

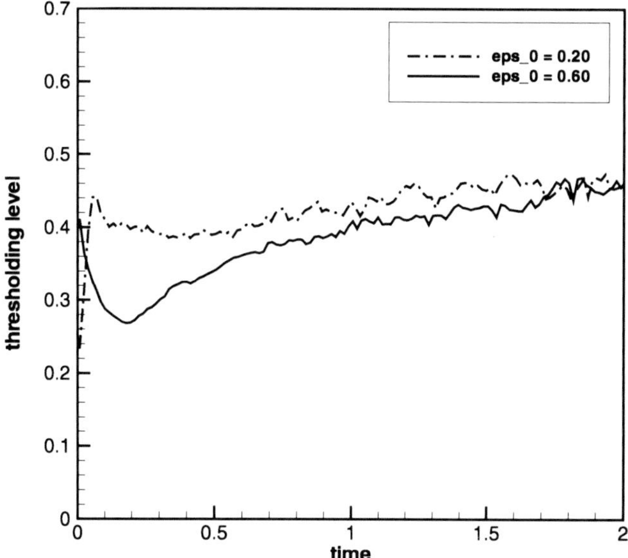

Fig. 1 Thresholding level evolution for two different initial conditions

proposed methodology can be referred to as a "complete" LES approach to the numerical simulation of homogeneous turbulence [4].

3 Results

In this paper, some preliminary results are presented for SCALES of linearly forced homogeneous turbulence at $Re_\lambda = 60$. The filtered governing equations (1) are solved by adding at the right-hand-side a forcing term proportional to the velocity field, viz. $Q\overline{u_i}^{>\varepsilon}$ [5]. The numerical experiments are carried out with $Q = 5.2$ for the sake of comparison with [3].

Given the desired turbulence resolution that corresponds to the prescribed goal value $\mathscr{R}_0 = 0.40$ for the ratio of SGS to total dissipation (2), the solution is initialized with two very different initial thresholding levels that are $\varepsilon_0 = 0.60$ and 0.20, respectively. When starting with a too high threshold like $\varepsilon_0 = 0.60$ the resolution is initially too low and the SGS dissipation provided by the model is too high so that the threshold automatically tends to decrease. The solution with a very low initial threshold like $\varepsilon_0 = 0.20$ shows the opposite behavior because the initial resolution is too high with respect to the goal. In Figure 1 the evolution of the wavelet thresholding level is reported for both solutions for a time corresponding to just two non-dimensional time units to make it possible to examine the short initial transient

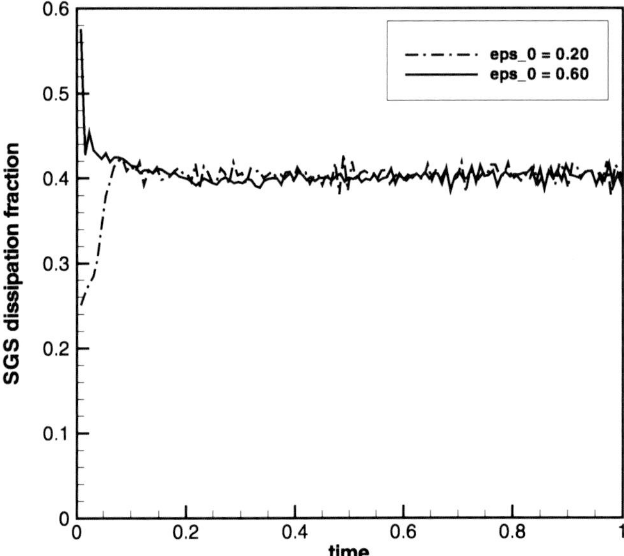

Fig. 2 Fraction of modeled dissipation evolution for two different initial conditions

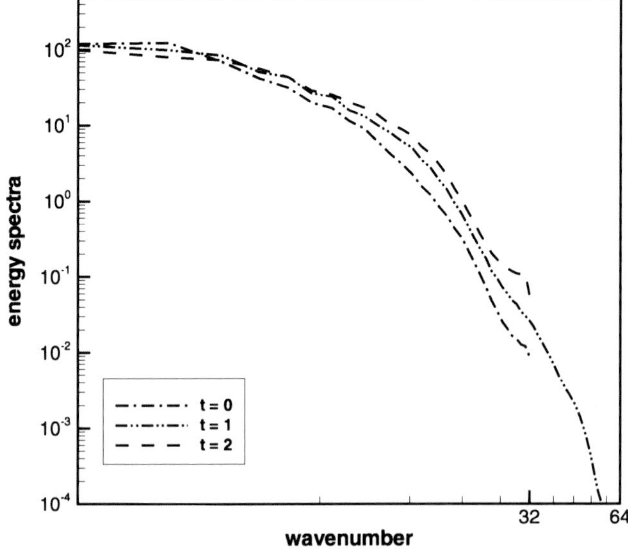

Fig. 3 Energy spectra evolution for $\varepsilon_0 = 0.60$

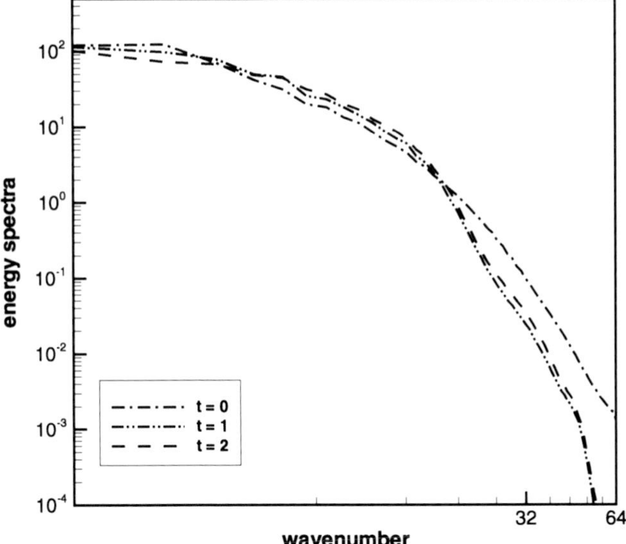

Fig. 4 Energy spectra evolution for $\varepsilon_0 = 0.20$

in some detail. The SCALES solution tends to achieve the prescribed resolution, regardless of the initial thresholding level, as illustrated on Figure 2, where the time history of the SGS to total dissipation ratio is given for one non-dimensional time unit. For both initial conditions the transient time is very short, less than half the eddy turnover time of the turbulence.

The transition from the initial resolution towards the desired one is well represented by the corresponding evolution of the energy spectra, shown in Figures 3 and 4, for the two different initial levels, respectively. For $\varepsilon_0 = 0.60$, since the turbulence resolution is initially insufficient, the energy associated to smaller scales increases as time passes, while the energy spectrum adjusts to the expected shape. On the other hand, for $\varepsilon_0 = 0.20$, where the turbulence is initially over-resolved, the energy associated to smaller scales reduces in time. Note that the energy spectra after two time-units practically correspond to the same solution. In practice, after the initial transient, the two solutions can be thought as two different realizations of the same SCALES solution with the prescribed resolution. By looking at Figure 5, where the different contributions to total dissipation are reported for $\varepsilon_0 = 0.20$, one can see that the constraint $\mathscr{R} \approx 0.40$ is maintained over long time integration, while the total dissipation is in good agreement with the reference spectral solution (SDNS).

As a conclusion, the use of the present time-dependent thresholding strategy makes it possible to achieve the objective separation of resolved energetic and unresolved flow structures in SCALES, for a given desired turbulence resolution. That is particularly promising for the adaptive simulation of complex unsteady turbulent

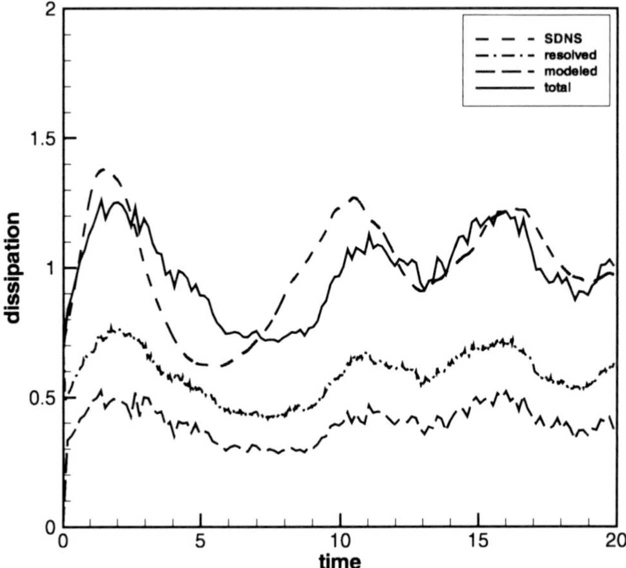

Fig. 5 Resolved, modeled and total dissipation evolution for $\varepsilon_0 = 0.20$, compared to spectral reference solution

flows, where the energetic level of the dominant flow structures can significantly vary in time so that the desired level of turbulence resolution could not be conserved by using a constant wavelet thresholding level.

Acknowledgements This work was supported by the United States National Science Foundation (NSF) under grant No. CBET-0756046 and the United States Department of Energy (DOE) under grant No. DE-FG02-07ER64468.

References

1. Schneider K, Vasilyev O V (2010) Wavelet methods in computational fluid dynamics. Annu Rev Fluid Mechanics 42:473–503
2. De Stefano G, Vasilyev O V, Goldstein D E (2008) Localized dynamic kinetic-energy-based models for stochastic coherent adaptive large eddy simulation. Phys Fluids 20:045102.1–045102.14
3. De Stefano G, Vasilyev O V (2010) Stochastic coherent adaptive large eddy simulation of forced isotropic turbulence. J Fluid Mech 646:453–470
4. Pope S B (2004) Ten questions concerning the large-eddy simulation of turbulent flows. New J of Physics 6:1–24
5. Lundgren T S (2003) Linearly forced isotropic turbulence. Annual Research Briefs CTR 2003:461–473

Mixed subgrid scale models for classical and variational multiscale large-eddy simulations on unstructured grids

Maria Vittoria Salvetti, Hilde Ouvrard, Bruno Koobus and Alain Dervieux

1 Introduction

The scale similarity (SS) idea was first introduced by Bardina et al. [1]. Following this idea the subgrid-scale stress tensor can be modeled through the modified Leonard tensor, which can be computed as a function of the variables resolved in LES by means of the application of an additional explicit filter. In the original idea, this filter should be the same as the grid one. It has been shown in a priori tests in the literature (see [6] for a review) that the SS model correlates very well with the exact SGS stress tensor and represents quite well energy backscatter from the unresolved to the resolved scales. However, when the scale-similarity SGS model is used alone in actual large-eddy simulations, it does not provide enough SGS dissipation and the simulations may undergo numerical instability. Therefore, the modified Leonard tensor (or SS term) is always used in combination with an eddy-viscosity term, leading to the so-called *mixed models* [6].

In this study a family of mixed models is built for finite-element/finite-volume simulations on unstructured grids. Since on unstructured grids explicit filtering implies high computational costs and memory requirements (see e.g. [3]), a generalized SS term is obtained herein through variational projection on the finite-volume macro cells, instead of through the application of an explicit filter. Mixed models are then obtained by combining the SS term with an eddy-viscosity one, given either

M.V. Salvetti
Dip. Ingegneria Aerospaziale, Università di Pisa, Via G. Caruso 8, 56122 Pisa (Italy) e-mail: mv.salvetti@ing.unipi.it

H. Ouvrard, B. Koobus
Département de Mathématiques, Université de Montpellier 2, Case Courrier 051, Place Eugène Bataillon, 34095 Montpellier (France) e-mail: koobus@math.univ-montp2.fr

A. Dervieux
INRIA, 2004 Route des lucioles, BP 93, 06902 Sophia Antipolis (France) e-mail: Alain.Dervieux@sophia.inria.fr

H. Kuerten et al. (eds.), *Direct and Large-Eddy Simulation VIII*, ERCOFTAC Series 15, DOI 10.1007/978-94-007-2482-2_18,

by the Smagorinsky [11] or by the Wall-Adapting Local Eddy-Viscosity (WALE) model [8]. The eddy-viscosity term can be computed as a function of the whole resolved variables and added to the whole range of resolved scales, as classically done in LES. A variational multiscale (VMS) version of the mixed models is also considered, in which the eddy-viscosity part is computed as a function of the smallest resolved scales (SRS) and is applied to these scales only. The separation between the smallest and the largest resolved scales (LRS) in our VMS-LES formulation is obtained through the same variational projection used to define the scale similarity term.

The different SGS models are appraised in classical and VMS large-eddy simulations of the at Reynolds number, based on the freestream velocity and on the cylinder diameter, equal to 3900. This flow has been chosen since it is a classical and well documented benchmark (see e.g. [10] for experimental data and [5, 10] for numerical studies), while containing all the features and all the difficulties encountered in the simulation of bluff-body flows also for more complex configurations and higher Reynolds numbers, at least for laminar boundary-layer separation. Finally, LES and VMS-LES of this flow were previously carried out [9] with the same numerical solver and the same SGS models without the addition of a SS term. These previous results will be herein used for comparison to highlight the effects of the SS-SGS term.

2 Methodology

Classical LES and SGS models - The filtered Navier-Stokes equations for compressible flows and in conservative form are discretized. In our simulations, filtering is implicit, i.e. the numerical discretization of the equations is considered as a filter operator (grid filter).

Since the flows in the applications herein are almost incompressible, the only relevant SGS term to be modeled is the deviatoric part of the SGS stress tensor, T_{ij}. In eddy-viscosity models this is typically expressed through the introduction of a SGS viscosity, μ_{sgs}. Two different eddy-viscosity SGS models are considered in the present work, viz. the classical Smagorinsky model [11] and the WALE model proposed by Nicoud and Ducros [8]. Both models contain a parameter to be a priori set; the value typically used for shear flows of $C_s = 0.1$ is adopted herein for the Smagorinsky model, while, as indicated in [8], the WALE constant C_W is set to 0.5.

As previously discussed, we consider here a scale similarity model in which filtering is replaced by variational projection:

$$T_{ij}^{ss} = \underbrace{P(\overline{\rho}\ \tilde{u}_i\tilde{u}_j) - P(\overline{\rho})P(\tilde{u}_i)P(\tilde{u}_j)}_{M_{ij}^{ss}} - \frac{1}{3}M_{kk}^{ss}\delta_{ij} \qquad (1)$$

δ_{ij} being the Kronecker delta and P the variational projector over the finite-volume macrocells, obtained by a process known as agglomeration [4]. As previously mentioned, the scale similarity model itself is not dissipative enough, thus an eddy viscosity term is added to it to obtain a *mixed model*: $T_{ij}^{m} = T_{ij}^{ss} + T_{ij}^{ev}$ in which, as previously said, the eddy-viscosity term, T_{ij}^{ev} can be expressed either by the Smagorinsky or by the WALE model.

Numerical method - The numerical solver (AERO) used for the present simulations is based on an mixed finite-element/finite-volume discretization of the flow equations on unstructured tetrahedral grids. The diffusive and SGS eddy-viscosity terms are discretized using P1 Galerkin finite elements on the tetrahedra, whereas finite-volumes are used for the convective terms. The numerical approximation of the convective fluxes at the interface of neighboring cells is based on the Roe scheme with low-Mach preconditioning. To obtain second-order accuracy in space, the Monotone Upwind Scheme for Conservation Laws reconstruction method (MUSCL) is used, in which the Roe flux is expressed as a function of reconstructed values of W at each side of the interface between two cells [2]. The numerical dissipation provided by the scheme used in the present work is made of sixth-order space derivatives and thus is concentrated on a narrow-band of the highest frequencies. Moreover, a parameter directly controls the introduced numerical viscosity and is explicitly tuned in order to reduce it to the minimal amount needed to stabilize the simulation. The scale similarity terms are discretized by finite volumes, with centered numerical fluxes at cell interfaces. Time advancing is carried out through a second-order implicit linearized method. More details on the numerical ingredients used in the present work can be found e.g. in [3, 2].

Variational Multiscale LES - A VMS-LES version of the previous mixed models is also investigated. In the VMS approach, the resolved flow variables, W, are decomposed into the largest and the smallest resolved scales, $\overline{W}$ and W' respectively. The effect of the subgrid scales is provided by a closure model as in classical LES, but this model only acts on W', while the Navier-Stokes model is preserved for $\overline{W}$. In the present study, we follow the VMS approach proposed by Koobus and Farhat [4] for the simulation of compressible turbulent flows through a finite volume/finite element discretization on unstructured tetrahedral grids. Starting from the finite-volume and finite-element basis functions associated to the computational grid, ψ_l and ϕ_l, the corresponding basis functions spanning the LRS finite-dimensional space are obtained through the same projector operator, P, previously introduced to obtain the SS term: $\overline{\psi}_l = P(\psi_l)$ and $\overline{\phi}_l = P(\phi_l)$. The basis functions for the SRS space are clearly obtained as follows: $\psi_l' = \psi_l - \overline{\psi}_l$ and $\phi_l' = \phi_l - \overline{\phi}_l$. The SGS eddy viscosity is then computed as a function of the SRS variables(*small-small approach*) and added only to the smallest resolved scales. This is aimed at preventing an excessive dissipation of LRS. As for the SS part, since in classical LES this term is known to well correlate with the *exact* SGS terms and to provide a significant backscatter of energy from unresolved scales, we decided to maintain its classical form, i.e. computed as a function of the whole resolved flow variables, also in VMS-LES. Moreover, it is added to the whole range of resolved scales as in classical LES.

3 Application and discussion

The flow over a circular cylinder at Reynolds number (based on the cylinder diameter, D, and on the freestream velocity) equal to 3900 is simulated. The computational domain, boundary and flow conditions are the same as in [9]. The computational grid considered herein is unstructured, made of tetrahedra and has approximately 1.46×10^6 nodes. The averaged distance of the nearest point to the cylinder boundary is $0.0085D$. Finally, the parameter controlling the amount of numerical viscosity is set in all the simulations equal to the value used in [9], for which it was observed that the numerical viscosity did not introduce undesired effects on the results and did not mask the effect of the SGS term, also in the VMS-LES simulation.

Table 1 Flow bulk coefficients. $\overline{C_d}$ denotes the mean drag coefficient, $\overline{C_{pb}}$ the mean base pressure coefficient and l_r the mean recirculation bubble length.

Turb. model	SGS model	$\overline{C_d}$	-$\overline{C_{pb}}$	l_r
LES	Smagorinsky	0.99	0.85	1.54
LES	Smagorinsky + SS	1.00	0.84	1.65
LES	WALE	0.98	0.94	1.22
LES	WALE + SS	1.02	0.83	1.55
VMS-LES	Smagorinsky	0.93	0.81	1.68
VMS-LES	Smagorinsky + SS	0.94	0.75	1.97
VMS-LES	WALE	0.94	0.83	1.56
VMS-LES	WALE + SS	0.95	0.78	1.73

The simulations considered herein differ for the adopted SGS model, as summarized in Tab. 1, in which a few bulk flow parameters are also shown. As for the mean drag coefficient, $\overline{C_d} = 0.99 \pm 0.05$ was obtained in the experiments by Norberg at $Re = 4020$ (data taken from [5]), which well agrees with those computed in well resolved LES in the literature (e.g. [5] and [10]). The corresponding base mean pressure coefficient is $\overline{C_{pb}} = -0.88 \pm 0.05$. For all the present simulations, the prediction of the mean drag coefficient is within the experimental range; as for the base mean pressure, a good agreement with the experimental data is generally observed, but the scatter of the predictions obtained in the different simulations is larger, with a maximum error of 9% with respect to the lower limit of the experimental range for the VMS-LES simulations using the SS term combined with the Smagorinsky model. Significant differences are present also in the prediction of the negative peak of the mean pressure coefficient (as shown in Figs. 1a and 1b) whose absolute value is overestimated with respect to the experiments in the simulations without SS term, the largest discrepancy being observed in the LES simulation with the WALE model. This is due to the fact that, although mainly acting in the wake, the non-dynamic eddy-viscosity models here used also provide a significant SGS viscosity in the regions that should be laminar. For the considered flow and the adopted values of the model constants (see [9]), the WALE model is more dissipative than the Smagorinsky one, while the small-small VMS approach significantly reduces the introduced SGS viscosity for both the SGS models. We specify that, as in [9], no adaptation of the model constants has been made to the specific test case or to the VMS-LES approach. As a consequence, the largest discrepancy in the negative C_P peak is observed in LES with the WALE model, and for both models the prediction is slightly improved in VMS-LES. The introduction of the SS term brings the predictions obtained both in classical and VMS-LES very close to the experimental curve, indicating that SS term has in this zone the effect of reducing the global introduced SGS dissipation.

As previously found in [9], the amount of introduced SGS viscosity has significant effects on the transition of the shear layers detaching from the cylinder and, consequently, on the mean velocity field. The larger is the introduced SGS viscosity the closer to the cylinder the transition of the shear layers occurs and the vortices forms, this leading to a smaller mean recirculation bubble length. The smallest value of l_r is indeed obtained in the LES simulation with the WALE model, while for a fixed SGS model the VMS-LES approach systematically increases the mean recirculation bubble length. Similarly, the introduction of the SS term has the effect of increasing l_r in both classical and VMS-LES. This indicates again that the SS term has the effect of reducing the global introduced SGS dissipation. For the mean recirculation bubble length, the reference value is the one obtained in [10] ($l_r = 1.51 \pm 10\%$). Thus, for this aspect, the agreement with the experiments is improved by the addition of the SS term only for the LES simulation with the WALE model, in which l_r was underestimated due to the excessive dissipation introduced by this model. This is confirmed also by the profiles of the mean streamwise velocity in the wake (see e.g. Fig. 1c).

Summarizing, it seems that the SS term has the main effect of leading to a reduction of the global introduced SGS dissipation. This is all in all beneficial for the prediction of the pressure distribution on the cylinder surface. Conversely, as far as the prediction of the mean velocity field and of the mean recirculation bubble length is concerned, an improvement of the agreement with experimental data is obtained only for those simulations in which an excessive dissipation is introduced when the eddy-viscosity model is used alone. Thus, this study further confirms that, as also observed in [9], the introduction of the proper amount of SGS dissipation is a crucial issue for this type of flow in classical LES simulations, but also in combination with the VMS-LES formulation. In this perspective, procedures aimed at optimizing the model constants (see e.g. [7]) appear particularly interesting. Finally, thanks to the used projection operator, which is an average on the macro-cells, the additional computational costs brought by the VMS approach are negligible (typically lower than 3% of the total cost of a time step). The use of the same operator is also a key point for maintaining the costs of the generalized SS term computation very low.

Acknowledgements The authors wish to thank Eric Lamballais for kindly sending the experimental data. The authors gratefully acknowledge CINECA (Italy), IDRIS (France) and CINES (France) for providing the computational resources and support.

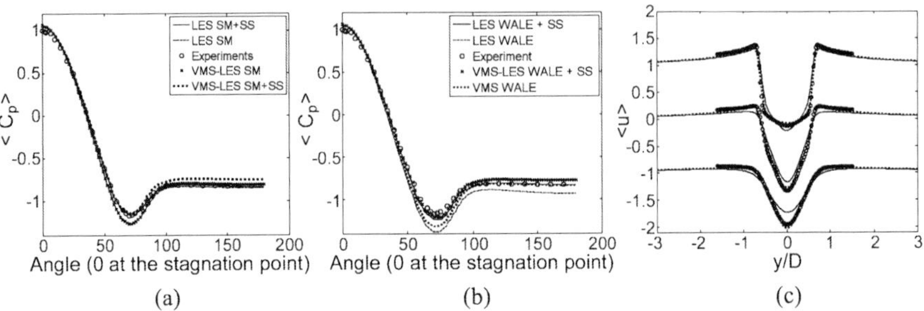

(a) (b) (c)

Fig. 1 Mean pressure coefficient over the cylinder surface in LES and VM-LES simulations: (a) Smagorinsky model with and without SS term; (b) WALE model with and without SS term. (c) Mean streamwise velocity profiles at $x/D = 1.06, 1.54, 2.02$(from top to bottom); open symbols: experiments [10], solid line: LES with WALE, thick dashed line: LES with WALE+SS.

References

1. Bardina, J., Ferziger, J.H. and Reynolds, W.C. (1980). Improved subgrid scale models for large eddy simulation. *AIAA Paper* **80-1357**.
2. Camarri, S., Salvetti, M.V., Koobus, B. and Dervieux, A. (2004). A low diffusion MUSCL scheme for LES on unstructured grids. *Comp. Fluids* **33**: 1101–1129.
3. Camarri, S., Salvetti, M.V., Koobus, B. and Dervieux, A. (2002). Large-eddy simulation of a bluff-body flow on unstructured grids. *Int. J. Num. Meth. Fluids* **40**: 1431–1460.
4. Koobus, B. and Farhat, C. (2004). A variational multiscale method for the large eddy simulation of compressible turbulent flows on unstructured meshes-application to vortex shedding. *Comput. Methods Appl. Mech. Eng.* **193**: 1367–1383.
5. Kravchenko, A.G. and Moin, P. (1999). Numerical studies of flow over a circular cylinder at Re = 3900. *Phys. Fluids* **12**(2): 403–417.
6. Meneveau, C. and Katz, J. (2000). Scale-invariance and turbulence models for large-eddy simulation. *Ann. Rev. Fluid Mech.* **32**: 1–32.
7. Meyers, J., Sagaut, P. and Geurts, B.J. (2006). Optimal model parameters for multi-objective large-eddy simulations, *Phys. Fluids* **18**: 095103.
8. Nicoud, F. and Ducros, F. (1999). Subgrid-scale stress modelling based on the square of the velocity gradient tensor. *Flow Turb. Comb.* **62** (3): 183–200.
9. Ouvrard, H., Koobus B., Dervieux, A. and Salvetti, M.V. (2010). Classical and variational multiscale LES of the flow around a circular cylinder on unstructured grids. *Computers & Fluids* **39**: 1083–1094.
10. Parneaudeau, P., Carlier, J., Heitz, D. and Lamballais, E. (2008). Experimental and numerical studies of the flow over a circular cylinder at Reynolds number 3900. *Phys. Fluids* **20**: 085101.
11. Smagorinsky, J. (1963). General circulation experiments with the primitive equations. *Month. Weath. Rev.* **91**(3): 99–164.

Subgrid-Scale Model and Resolution Influences in Large Eddy Simulations of Channel Flow

Amin Rasam, Geert Brethouwer and Arne V. Johansson

1 Introduction

Subgrid-scale (SGS) modeling and resolution are two important issues that can affect the quality of large eddy simulation (LES) to a large extent. Many SGS models are based on an isotropic description of the SGS motions. However, SGS turbulence is not always isotropic. Near-wall SGS motions in wall-bounded turbulent flows is an example of such. In that case, the quality of LES largely depend on the grid resolution and the SGS modeling used. On the other hand, one could speculate that dependence of LES results on the resolution would be less if the SGS model is able to properly take into account the anisotropy of SGS stresses. In this paper, the influence of resolution on LES is investigated. Three SGS models are used, namely: the standard dynamic Smagorinsky model (DS) based on [3] and modifications of [5] which is an isotropic eddy viscosity model, the high-pass filtered dynamic Smagorinsky (HPF) model of [7] which is based on the variational multi-scale method and the recent explicit algebraic (EA) model of [6] which is capable of properly modeling the anisotropy of the SGS stresses. LES of channel flow is carried out using these SGS models at $Re_\tau = 934$, based on friction velocity and channel half width, and the results are compared to DNS data.

2 Numerical method and simulations

Simulations are performed using a pseudo-spectral Navier Stokes solver for incompressible flows (see [2]). Fourier and Chebychev representations are used in homogeneous and wall-normal directions respectively. The domain size for the compu-

Amin Rasam
Linné FLOW Centre, KTH Mechanics, SE-100 44 Stockholm, Sweden, e-mail: rasam@mech.kth.se

H. Kuerten et al. (eds.), *Direct and Large-Eddy Simulation VIII*,
ERCOFTAC Series 15, DOI 10.1007/978-94-007-2482-2_19,

tations is $L_x \times L_y \times L_z = 8\pi h \times 2h \times 3\pi h$, where h is the channel half width. The box size has the same dimensions as the box used in the DNS of [1] for the same Reynolds number. Periodic boundary conditions are used in the streamwise and spanwise directions. The bulk Reynolds number is taken to be constant and equal to the corresponding DNS value. By imposing the flow rate, a Reynolds number based on the friction velocity and the channel half width, Re_τ, is obtained for each simulation.

The following SGS models are used in this study:
1 - Dynamic Smagorinsky (DS) model [3], where the SGS contribution is:

$$\tau_{ij} - \frac{1}{3}\tau_{kk}\delta_{ij} = -2C_s^2\Delta^2|\tilde{S}|\tilde{S}_{ij}.$$

Tilde denotes a filtered quantity, C_s is the dynamic coefficient, $\tilde{S}$ is the filtered rate of strain tensor ($|\tilde{S}| = \sqrt{2\tilde{S}_{ij}\tilde{S}_{ij}}$) and Δ is the relevant length scale.
2 - Explicit algebraic (EA) SGS model [6]:

$$\tau_{ij} = K_{SGS}\left(\frac{2}{3}\delta_{ij} + \beta_1\tau^*\tilde{S}_{ij} + \beta_4\tau^{*2}\left(\tilde{S}_{ik}\tilde{\Omega}_{kj} - \tilde{\Omega}_{ik}\tilde{S}_{kj}\right)\right),$$

where the first term is the isotropic part, the second term is the eddy viscosity part and the last term is nonlinear and creates proper anisotropy. The filtered rate of rotation tensor is denoted by $\tilde{\Omega}$ and τ^* is the time scale of the SGS motions (see [6]). Model parameters β_1 and β_4 are functions of $\tilde{\Omega}$. SGS kinetic energy is modeled as:

$$K_{SGS} = C\Delta^2|\tilde{S}|^2,$$

where C is determined using Germano's identity and Δ is the filter scale.
3 - High-pass filtered (HPF) dynamic Smagorinsky model [7] where SGS stress τ_{ij} is expressed in terms of high-pass filtered quantities using a suitable filter H:

$$\tau_{ij} - \frac{1}{3}\tau_{kk}\delta_{ij} = -2\nu_t^{HPF}\tilde{S}_{ij}^{HPF}, \quad \nu_t^{HPF} = C^{HPF}\Delta^2\left|\tilde{S}^{HPF}\right|,$$

where the high-pass filtered strain rate tensor is:

$$\tilde{S}_{ij}^{HPF} = \frac{1}{2}\left(\frac{\partial H * \tilde{u}_i}{\partial x_j} + \frac{\partial H * \tilde{u}_j}{\partial x_i}\right),$$

and $*$ denotes a convolution in physical space.

Five different grids have been used for which the details are shown in table 1.

Table 1 Summary of numerical simulations. N_x and N_z are the number of grid points in the streamwise and spanwise directions respectively leading to resolutions Δ_x^+ and Δ_z^+ in wall units in physical space. N_y is the number of Chebychev polynomials in the wall-normal direction leading to a channel center-line resolution of Δy_c^+ in wall units.

Case	SGS Model	$N_x \times N_y \times N_z$	Re_τ	Δx^+	Δy_c^+	Δz^+
DS1		$96 \times 97 \times 72$	789.4	209.0	17.2	104.5
DS2	Dynamic	$128 \times 97 \times 96$	828.7	162.7	18.1	81.4
DS3	Smagorinsky	$192 \times 97 \times 144$	873.5	114.3	19.1	57.2
DS4		$256 \times 129 \times 192$	896.5	88.0	14.7	44.0
DS5		$320 \times 129 \times 320$	912.8	71.7	15.0	26.9
HPF1	High-pass	$96 \times 97 \times 72$	869.1	227.5	19.0	113.8
HPF2	filtered	$128 \times 97 \times 96$	908.0	178.3	19.8	89.1
HPF3	Dynamic	$192 \times 97 \times 144$	948.4	124.1	30.7	62.1
HPF4	Smagorinsky	$256 \times 129 \times 192$	974.3	95.6	16.0	47.8
HPF5		$320 \times 129 \times 320$	998.9	78.4	16.3	29.4
EA1		$96 \times 97 \times 72$	930.7	243.7	29.3	121.8
EA2	Explicit	$128 \times 97 \times 96$	938.7	184.3	30.0	92.2
EA3	Algebraic	$192 \times 97 \times 144$	951.0	124.6	30.6	62.3
EA4		$256 \times 129 \times 192$	966.7	94.9	15.8	47.4
EA5		$320 \times 129 \times 320$	946.3	74.3	15.0	27.9

3 Results and discussions

Fig. 1 shows the convergence of the wall shear stress towards the DNS value. Different behaviors can be observed for various models in this figure. The interesting point is the nonmonotonic behavior of some models. The EA model shows the least variation of the wall shear with resolution while the HPF model has a large overprediction at a certain resolution and does not converge to the DNS value at the resolutions used. However, it is expected that a further increase in the resolution lead to closer convergence to the DNS. This behavior has been observed in LES of channel flow at $Re_\tau = 590$ (results are not shown here).

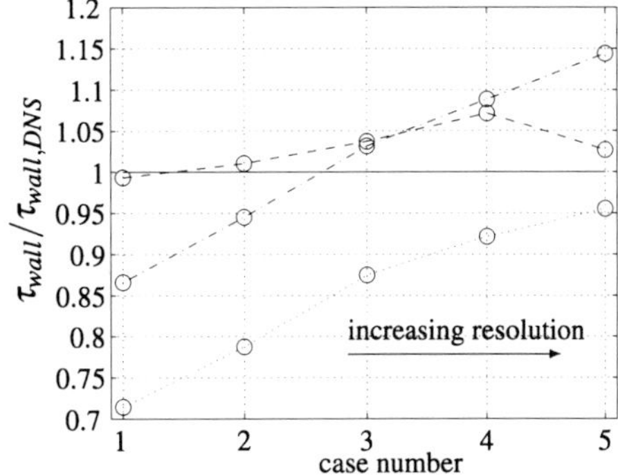

Fig. 1 Variation of average wall shear stress, normalized with the DNS value, with resolution. dash-dotted: HPF model, dashed: EA model and dotted: DS model.

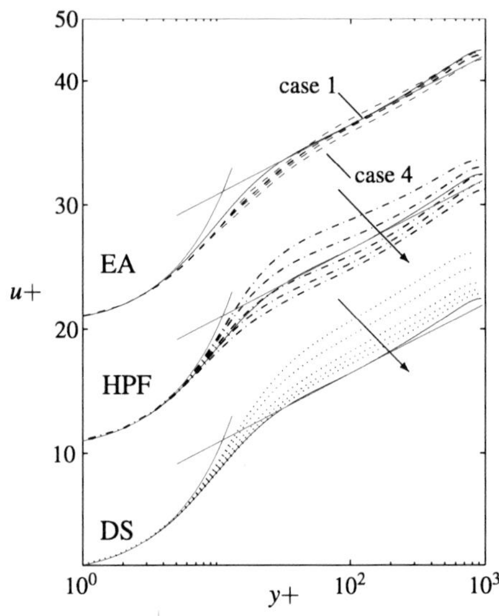

Fig. 2 Mean velocity profiles shifted in the ordinate direction by 10 wall units to separate different model predictions. Solid: DNS, dash-dotted: HPF model, dashed: EA model, dotted: DS and thin solid: $u^+ = y^+$, $y^+ < 11$, $u^+ = \frac{1}{\kappa}\log y^+ + B$, $y^+ > 11$, $\kappa = 0.41, B = 5.2$. Arrows point in the direction of increasing resolution.

Predictions of the mean velocity profiles by different models at various resolutions are shown in Fig. 2. The three SGS model predictions are very different. The DS model predictions converge monotonically to the DNS profile from above. The HPF model predictions does not converge to the DNS profile at these resolutions. It shows a nonmonotonic behavior and it has a pronounced overprediction in the log region at very coarse resolutions and further tend to underpredict the velocity profile at the finest resolution compared to the DNS data. The undesirable behavior of the DS and HPF models at these resolutions is due to their inability to model the near-wall anisotropic SGS motions. The EA model predictions of the mean velocity profile are less resolution dependent and are in good agreement with the DNS data even at the coarsest resolution. This behavior is due to a proper formulation of the anisotropy of the SGS stresses by the EA model. It is worth noting that, although there has been many good predictions using HPF model reported in the literature (e.g. see [7]), the behavior of the model at the finest resolution presented in this study is not as good, which means that a higher resolution is needed for the HPF model to have a good performance at this Reynolds number.

Reynolds stresses are shown in Fig. 3 (a), (b) and (c). The isotropic DS and HPF models overpredict $\langle u'u' \rangle$ at coarse resolutions while the nonisotropic EA model predictions are close to the DNS profile even at the coarsest resolution. The DS model underpredicts $\langle v'v' \rangle$ at low resolutions while the HPF model predictions are in better agreement with the DNS data. The EA model predictions also agree reasonably well with the DNS at all resolutions. The DS model underpredicts $\langle w'w' \rangle$ and the HPF model overpredicts the same quantity. The over-prediction of $\langle w'w' \rangle$ by the HPF model is also reported in the LES computations using the variational multiscale model in [4]. The EA model predictions of $\langle w'w' \rangle$ agree well with DNS at all resolutions. The mis-predictions of different Reynolds stresses by the DS and

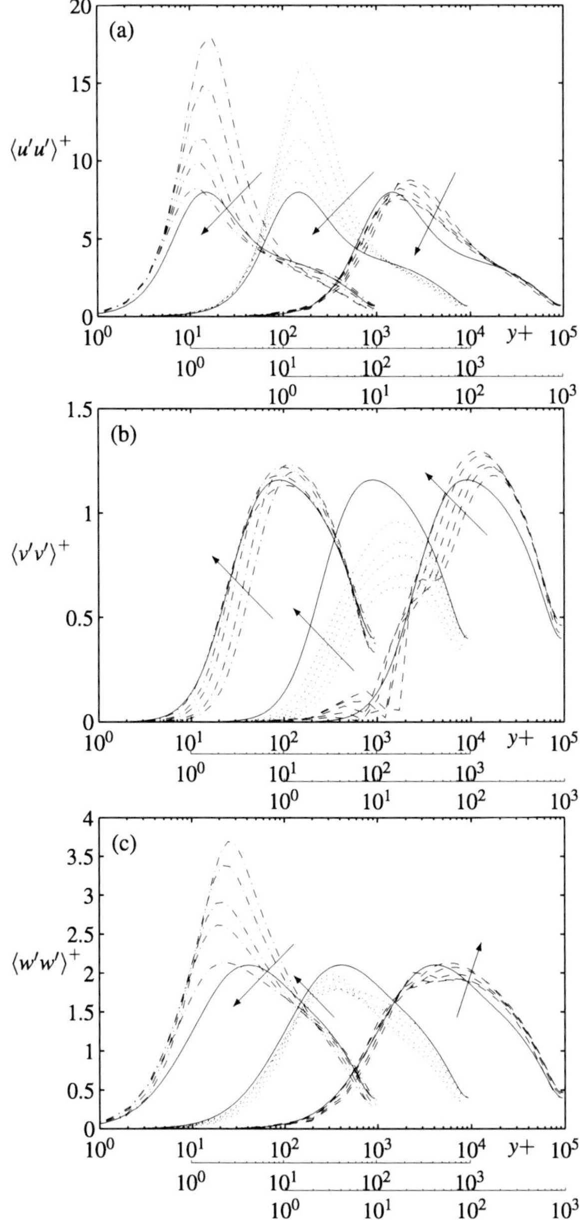

Fig. 3 Reynolds stress profiles, short solid: DNS, dash-dotted: HPF, long solid: EA and dotted DS models. Profiles are shifted in the abscissa direction by 0, 10 and 100 plus units for HPF, DS and EA models respectively. Arrows point in the direction of increasing resolution.

HPF models are due to the isotropic formulation of these models leading to incorrect SGS dissipations provided by these models. The incorrect SGS dissipation lead to under/over predictions of the Reynolds stresses.

4 Conclusions

Resolution effect on LES of fully developed turbulent channel flow at $Re_\tau = 934$ was investigated. A relatively wide range of grid resolutions were used in this study and the results were compared to the DNS data of [1]. LES predictions were found to be very sensitive to the grid resolution and the influence of the SGS model on the predictions can be very large at coarse resolutions. The DS and the HPF models perform poorly at low resolutions for the prediction of the mean velocity and the Reynolds stresses. In contrast, the EA model performs well for the same range of resolutions and its predictions are almost independent of the resolution for prediction of single-point statistics. A conclusion is that in the underresolved resolutions that were considered in this study, where the subgrid-scale stresses are less isotropic, a mixed model that can correctly model the anisotropy of these scales can reproduce accurate single-point turbulence statistics while isotropic models are not able to reproduce correct results.

Acknowledgements The authors acknowledge Dr. Philipp Schlatter and Qiang Li for fruitful discussions on the HPF model. Support from the Swedish research council and computer time provided by the Swedish National Infrastructure for Computing (SNIC) is gratefully acknowledged.

References

1. J. C. del Álamo, J. Jiménez, P. Zandonade and R. T. Moser: Scaling of the energy spectra of turbulent channels. J. Fluid Mech. **500**, 135–144 (2004).
2. M. Chevalier, P. Schlatter and A. Lundbladh and D. S. Henningson: SIMSON a pseudo-spectral solver for incompressible boundary layer flows. KTH Mechanics, Stockholm, Sweden, Trita-MEK. **07** (2007).
3. M. Germano, U. Piomelli, P. Moin and W. E. Cabot: A dynamic subgrid-scale eddy viscosity model. Phys. Fluids A. **3**, 1760–1765 (1991).
4. H. Jeanmart and G. Winckelmans: Investigation of eddy-viscosity models modified using discrete filters: A simplified "regularized variational multiscale model" and "an enhanced field model". Phys. Fluids **19**, 055110 (2007).
5. D. K. Lilly: A proposed modification of the Germano subgrid-scale closure method. Phys. Fluids A. **4**, 633–635 (1992).
6. L. Marstorp, G. Brethouwer, O. Grundestam and A. V. Johansson: Explicit algebraic subgrid stress models with application to rotating channel flow. J. Fluid Mech. **639**, 403–432 (2009).
7. A. Stolz, P. Schlatter and L. Kleiser: High-pass filtered eddy-viscosity models for large-eddy simulations of transitional and turbulent flow. Phys. Fluids **17**, 065103 (2005).

Immersed Boundaries in Large-Eddy Simulation of a transonic cavity flow

C. Merlin, P. Domingo and L. Vervisch

1 Introduction

Complex geometries can be investigated with a non-structured solver. Another approach retained in this paper is to keep a structured mesh with the addition of an immersed boundary concept. Immersed Boundary Methods employ cartesian meshes that do not conform to the shape of the body in the flow and modify the governing equations to incorporate the boundary conditions. First introduced by Peskin [18], IBMs involve either continuous or discrete forcing approaches. Only discrete forcing approach preserves good performance at high Reynolds number. The ghost-cell method (discrete forcing approach) utilizes a sharp boundary with a modification of the computational stencil and an extrapolation scheme to deduce the state variables in the border cells.

To evaluate the accuracy of the boundary layers reconstruction with an immersed boundary, a transonic cavity configuration is simulated for which measurements are available in the literature. Such high velocity subsonic flow exhibits strong aeroacoustic oscillations due to a coupling between pressure disturbances and the vortical structures. This feedback loop is generated by the impact of the shear layer on the downstream wall leading to the subsequent generation of acoustic waves which initiates further instabilities. Numerous experimental studies have been devoted to the study of such an aeroacoustic loop with the pressure spectrum (radiation acoustic analysis). However, only a few of them have investigated the velocity field inside the cavity. A complete data set concerning the flow features is available in the experiment of Forestier *et al.* [6]. Concerning the numerical investigations, many LES computations have been performed to characterize the acoustic radiation of cavities [8, 3]. The major results are obtained at transonic Mach number. An extensive anal-

UMR-CNRS-6614-CORIA & INSA de Rouen, Campus du Madrillet, Avenue de l'Université, BP 8, 76801 Saint Etienne du Rouvray Cedex, France, e-mail: cindy.merlin@coria.fr, pascale.domingo@coria.fr, luc.vervisch@coria.fr

H. Kuerten et al. (eds.), *Direct and Large-Eddy Simulation VIII*, ERCOFTAC Series 15, DOI 10.1007/978-94-007-2482-2_20,

ysis of the cavity flow fields have been provided by Larchevêque *et al.* [14, 15] and Sagaut *et al.* [19].

In the following sections, the numerical method as well as the boundary conditions methodology are briefly described. The cavity flowfield is then investigated in order to assess the relevant use of the Immersed Boundary Method for addressing highly compressible flows.

2 Description of numerical method

The governing equations considered are the unsteady Navier-Stokes equations for a viscous compressible flow. The solver is a parallel one based on an explicit finite volume scheme for cartesian grids. The convective terms are computed resorting to the fourth-order centered skew-symmetric-like scheme proposed by Ducros *et al.* [4]. The diffusive terms are computed using a fourth-order centered scheme. In order to suppress spurious oscillations and damp high-frequency modes, the numerical scheme is augmented by a blend of second- and fourth-order artificial terms [20]. Time integration is performed using a third-order Runge-Kutta scheme [9]. To avoid the appearance of spurious reflections at the open boundaries, the simulations are performed using the three-dimensional Navier-Stokes characteristic boundary conditions (3D-NSCBC approach) to describe non-reflective boundary conditions [16]. This method is particularly effective in the context of compressible flows which prevent the appearance of spurious reflections at the open boundaries. To reproduce a turbulent inflow condition, a correlated random noise [12] is superimposed to the average velocity profile.

The immersed boundaries method has already been combined with high-order scheme in the DNS context by Lamballais and Silvestrini [13] and in compressible LES by De Palma *et al.* [2]. The present approach employs a ghost-cell technique for imposing the boundary conditions on the immersed boundaries (Fadlun *et al.* [5], Iaccarino and Verzicco [11] and Balaras [1]). The interpolation scheme for the cells cut by the immersed boundary is based on the determination of an image-point expressed in terms of a linear, bi-linear or tri-linear interpolation. The boundary conditions are described by a Neumann condition for pressure ($\partial P/\partial n = 0$, where n is the direction normal to the immersed surface) and Dirichlet conditions for velocities (no-slip conditions). The wall temperature is calculated by using an adiabatic hypothesis.

Large-Eddy simulations have been carried out with various subgrid strategies. The first one is based on an implicit subgrid modeling classically denoted MiLES [10]. The next ones rely on traditional SGS modeling approach: the Vreman model [21] and the Lagrangian Dynamic Model denoted LDSM [17]. To introduce the right amount of viscosity in the vicinity of the walls, the filtering operation is performed with a test filter described by a trapezoidal rule [22]. Concerning the filtering procedure near the immersed boundaries, filtered quantities are determined by cutting the filtering volume. More precisely, the scheme is switched to bi-dimensional filtering over the plane parallel to the immersed boundary and ghost cells are not included in the filtering operation. In a similar manner, the lagrangian reconstruction of the fluid path in the LDSM is modified in the vicinity of the immersed and real boundaries. Since the turbulent viscosity on the wall is zero, a new interpolation scheme is proposed. It relies on the Dirichlet condition for the turbulent viscosity to recover the correct ghost viscosity. To determine the SGS viscosity near the boundaries, a tri-linear interpolation scheme is thus used.

3 Large Eddy Simulation of a Cavity Flow

The structure of an unsteady flow over a rectangular open cavity is then investigated using non conforming boundaries Cartesian grid to discretize the governing equations. The cavity studied experimentally by Forestier *et al.* [6] is of parallelepiped shape and has a length-to-depth ratio of $L/D = 2$ and spanwise extent $W/D = 4.8$. The inflow Mach number is M = 0.8 while the Reynolds number is $Re_L = 6.8 \times 10^6$, based on the length of the cavity.

To perform this immersed boundary study, the grid points are clustered near all the walls and in the shear layer above the cavity. The characteristics of the mesh are summed up in the table 1 with typical dimensions of the boundary layer in local wall units. To recover the spanwise inhomogeneity due to lateral walls, the spanwise direction is also refined which yields a mesh of 6.6 millions cells.

Table 1 Mesh properties

Mesh	$290 \times 188 \times 122$
Δx^+	$30 \sim 130$
Δy^+	$20 \sim 70$
Δz^+	$20 \sim 50$
Δt	$3.8 \times 10^{-7} s$

For flows with self-sustained oscillations, open boundaries have to be treated with great care. The previously mentioned 3D-NSCBC approach [16] has been used to avoid any spurious reflections. For the inflow conditions, fluctuations have been added to the main profile. This synthetic turbulence is space and time-filtered correlated using the digital filter approach of Klein *et al.* [12]. Nevertheless, in this particular configuration, the turbulent injection does not play a crucial role in the self-sustained oscillations. The numerical simulations carried out in this study revealed that the turbulence is mainly generated by the Kelvin-Helmholtz instability and the recirculation zones. This result was already mentioned by Sagaut *et al.* [19] for the same configuration. This high subsonic cavity was investigated by Larchevêque *et al.* [15] with MiLES and Selective Mixed Scale modeling strategies. In the present study, the LES investigation is expanded beyond the use of the MiLES approach by using the Localized Dynamic Smagorinsky and the Vreman models.

The key feature of the cavity flow is the presence of specific turbulent structures well described with the simulations. Figure 1 shows the 3D views of the Q-criterion for the cavity considered as immersed on the grid mesh. The interactions of coherent structures with the downstream edge are predominant due to flow impingement and recirculation. The Fig. 2 presents a comparison between an experimental fast Schlieren view of Forestier *et al.* [6] and a numerical strioscopic view at a similar time. The complex waves pattern appearing in the aeroacoustic feedback is well reproduced.

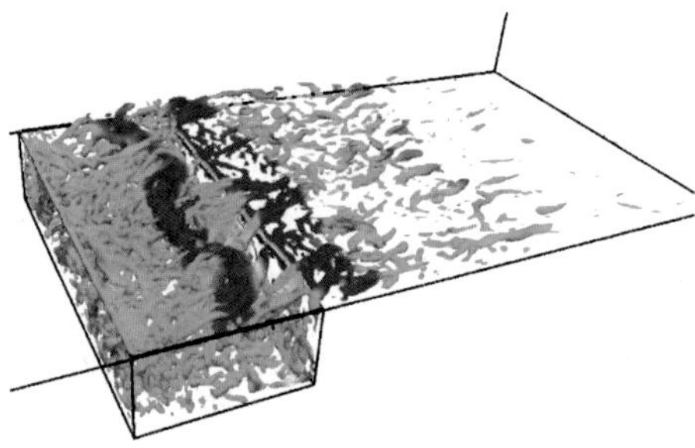

Fig. 1 Three-dimensional vortical structures defined by the Q criterion and obtained with the Lagrangian Dynamic Smagorinsky Model.

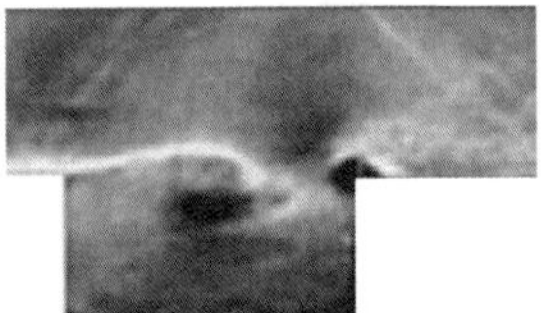

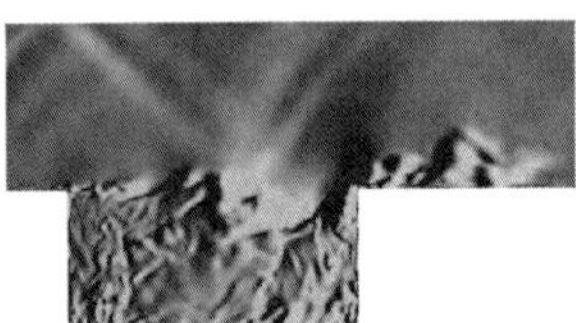

Fig. 2 An experimental fast Schlieren view (left) [6] and a strioscopic view from the MiLES computation (right).

The Fig. 3 shows that the simulations accurately reproduce the first and second order statistics. All the SGS strategies accurately describe the statistical properties of the flow.

In the present study, the feedback process for the self sustained oscillations is the shear layer mode (also called Rossiter mode). In this regime, the oscillations at Rossiter frequencies are generated by the acoustic scattering at the trailing edge and by the receptivity of the shear layer to acoustic disturbances at the leading edge. The pressure power spectrum in sound pressure level (SPL) in units of decibels per Hertz vs Strouhal number highlights the dominant acoustic tones and is presented in Fig. 4. The first Rossiter mode as well as its three first harmonics are well described.

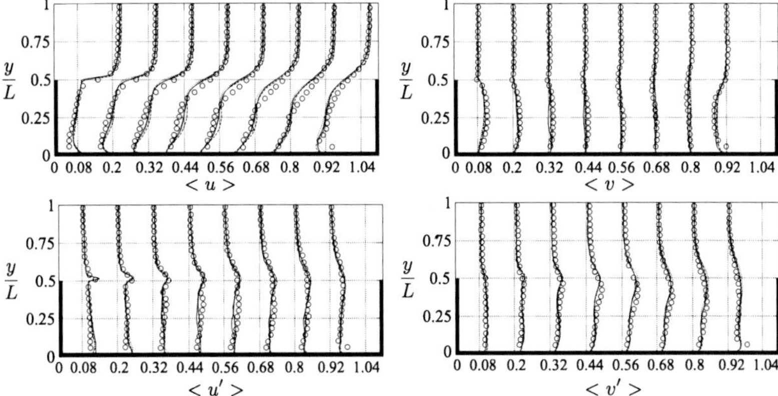

Fig. 3 Velocity statistics in the mid-span plane with immersed boundaries: longitudinal mean velocity, vertical mean velocity, longitudinal fluctuating velocity and vertical fluctuating velocity: solid: LDSM, dashed: MiLES, dotted:Vreman model and ∘: experimental data.

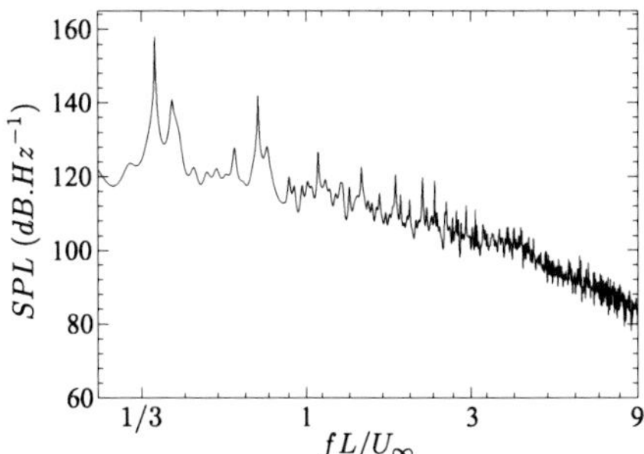

Fig. 4 Pressure power spectrum obtained with the MILES calculation.

4 Conclusions

The essential features of a transonic cavity flow are recovered by IBM. Quantitative and qualitative agreement with experimental data have been obtained with the immersed boundary concept and with various SGS description for the Large Eddy Simulations. This demonstrates the strong potential of LES-IBM to simulate compressible flow and its ability to describe the feedback process which is based on the interaction of small instabilities in the shear layer with the downstream corner and the generation of acoustic waves which propagate upstream. Future work will focus on the description of the turbulent fluid mixing inside complex geometries.

Acknowledgements The investigations presented in this paper have been done as a part of the European research project TECC-AE - FP7 (Technologies Enhancement for Clean Combustion in Aero-Engines), Grant No. 2010-020152. Computing resources were provided by GENCI, IDRIS-CNRS and CRIHAN. Dr. L. Larchevêque is gratefully acknowledged for fruitful discussions.

References

1. Balaras E.: Modeling complex boundaries using an external force field on fixed Cartesian grids in large eddy simulations, Computers and Fluids, **Vol. 33** (3), 375–404 (2004)
2. De Palma P., De Tullio M. D., Pascazio G. and Napolitano M.: An immersed boundary method for compressible viscous flows, Comput. Fluids, **Vol. 35**, 693 (2006)
3. Dubief Y. and Delcayre F.: On the coherent-vortex identification in turbulence, J. Turbulence, **Vol. 1** (1), 11 (2000)
4. Ducros F., Laporte F., Soulères T., Guinot V., Moinat P. and Caruelle B.: High-order fluxes for conservative skew-symmetric-like schemes in stuctures meshes: Application to compressible flows, J. Comput. Phys., **Vol. 161**, 114–139 (2000)
5. Fadlun E. A., Verzicco R., Orlandi P. and Mohd-Yusof J.: Combined immersed-boundary finite-difference methods for three-dimensional complex flow simulations, J. Comput. Phys., **Vol. 161**, 35–60 (2000)
6. Forestier N., Geffroy P. and Jacquin L.: Etude expérimentale des propriétés instationnaires d'une couche de mélange compressible sur une cavité: cas d'une cavité ouverte peu profonde. Rt 22/00153 DAFE, ONERA (in French) (2003)
7. Ghias R., Mittal R. and Dong H.: A sharp interface immersed boundary method for compressible viscous flows, J. Comput. Phys., **Vol. 225**, 528–553 (2007)
8. Gloerfelt X.: Bruit rayonné par un écoulement affleurant une cavité: Simulation aéroacoustique directe et application de méthodes intégrales. PhD thesis, Ecole Centrale de Lyon (2001)
9. Gottlieb S. and Shu C.: Total variation diminishing Runge-Kutta schemes, Mathematics of Computation, **Vol. 67** (221), 73–85 (1998)
10. Grinstein F. F. and Fureby C.: Recent progress on MiLES for high Reynolds number flows, J. Fluids Eng., **Vol. 24**, 848–861 (2002)
11. Iaccarino G. and Verzicco R.: Immersed boundary technique for turbulent flow simulations, Appl. Mech. Rev., **Vol. 56** (3), 331–347 (2003)
12. Klein M., Sadiki A. and Janicka J.: A digital filter based genration of inflow data for spatially developing direct numerical or large eddy simulation, J. Comput. Phys., **Vol. 186**, 652–665 (2003)
13. Lamballais E. and Silvestrini J.: Direct numerical simulation of interactions between a mixing layer and a wake around a cylinder, J. Turbulence, **Vol. 3** (2002)
14. Larchevêque L., Sagaut P., Löe T. H. and Comte P.: Large eddy simulation of a compressible flow in a three dimensional open cavity at high Reynolds number, J. Fluid Mech., **Vol. 516**, 265–301 (2004)
15. Larchevêque L., Sagaut P. and Labbé O.: Large-eddy simulation of a subsonic cavity flow including asymmetric three-dimensional effects, J. Fluid Mech., **Vol. 577**, 105–126 (2007)
16. Lodato G., Domingo P. and Vervisch L.: Three-dimensional boundary conditions for direct and large eddy simulation of compressible viscous flows, J. Comput. Phys., **Vol. 227** (10), 5105–5143 (2008)
17. Meneveau C., Lund T. and Cabot W., A lagrangian dynamic subgrid-scale model of turbulence, J. Fluid Mech., **Vol. 319**, 353–385 (1996)
18. Peskin C. S.: Flow patterns around heart valves: A numerical method, *J. Comput. Phys.*, **Vol. 10**, 252–271 (1972)
19. Sagaut P., Garnier E., Tromeur E., Larchevêque L. and Labourasse E.: Turbulent inflow conditions for large-eddy eimulation of supersonic and subsonic wall flows, AIAA Journal, **Vol. 42**, 469–477 (2004)
20. Tatsumi S., Martinelli L. and Jameson A.: Flux-limited schemes for the compressible Navier-Stokes equations, AIAA Journal, **Vol. 33** (2), 252–261 (1995)
21. Vreman A. W.: An eddy-viscosity subgrid-scale model for turbulent shear flow: Algebraic theory and applications, Phys. Fluids, **Vol. 16** (10) (2004)
22. Zang Y., Street R. L. and Koseff J. R.: A dynamic mixed subgrid-scale model and its application to turbulent recirculating flows, Phys. Fluids A, **Vol. 5** (12), 3186–3196 (1993)

Numerical study of turbulent-laminar patterns in MHD, rotating and stratified shear flows

G. Brethouwer, Y. Duguet and P. Schlatter

1 Introduction

Coexisting laminar and turbulent regions have been observed in several types of wall bounded flows. In Taylor Couette flow, for example, alternating helical shaped laminar and turbulent regions have been observed within a limited Reynolds number range [1] and oblique laminar and turbulent bands have been seen in experiments [1] and simulations [2],[3] of plane Couette flow for Reynolds numbers $Re = U_w h/\nu$ between about 320 and 380. Here $\pm U_w$ is the velocity of the two walls, h is the half width of the wall gap and ν is the viscosity. In this Reynolds number range the turbulent-laminar patterns seem to sustain while at lower Re the flow becomes fully laminar and at higher Re no clear laminar patterns can be distinguished and the flow eventually becomes fully turbulent. Similar oblique laminar-turbulent bands appeared as well in direct numerical simulations (DNS) of plane channel flow for friction Reynolds numbers $Re_\tau = u_\tau h/\nu = 60$ and 80 [4],[5], where u_τ is the friction velocity and h is again the gap half width.

Studies of flows with turbulent-laminar patterns so far were mostly restricted to moderate Reynolds numbers. We show that laminar and turbulent regions can also coexist at higher Reynolds number when there is a mechanism inhibiting turbulence. Three such cases are further examined through DNS: (*i*) magnetohydrodynamic (MHD) channel flow, (*ii*) rotating plane Couette flow and (*iii*) stably stratified open channel flow. It has already been found that turbulent-laminar patterns can emerge in rotating plane Couette flow experiments up to at least $Re = 1000$ [6] while DNS have indicated that they also can appear in stratified channel flow [7].

G. Brethouwer
Linné Flow Centre, KTH Mechanics, Stockholm, Sweden, e-mail: geert@mech.kth.se

Y. Duguet
LIMSI-CNRS, UPR 3251, F-91403 ORSAY CEDEX, France

P. Schlatter
Linné Flow Centre, KTH Mechanics, Stockholm, Sweden

H. Kuerten et al. (eds.), *Direct and Large-Eddy Simulation VIII*,
ERCOFTAC Series 15, DOI 10.1007/978-94-007-2482-2_21,

2 Direct numerical simulations

In order to study turbulent-laminar patterns in wall-bounded we have performed DNSs using a pseudospectral method with periodic Fourier expansions in the streamwise x and spanwise z direction and Chebyshev polynomials in the wall-normal y-direction. The MHD simulations are based on the quasistatic approximations for low magnetic Reynolds numbers [8] with a uniform magnetic field in either the wall-normal or spanwise direction and no-slip and non-conducting boundary conditions at both walls. The plane Couette flow simulations were subject to cyclonic rotation about the spanwise direction. No-slip and free-slip conditions are imposed in stably stratified open channel flow DNSs at the bottom wall and top surface respectively, and constant temperatures are imposed at both sides while the flow field is governed by the Boussinesq equations. The aspect ratio of the computational domains is relatively large since the laminar and turbulent regions are commonly extended [1],[3],[5].

All simulations were initiated by fully turbulent flows. Then either the strength of the magnetic field, rotation rate or stratification were increased until laminar and turbulent patterns emerged in the flow. The simulations were then progressed for a sufficiently long time in order to ensure that the flows with the turbulent-laminar patterns were at least in a quasi-steady state.

Numerical and physical parameters as well as reference names of the DNSs are listed in Tables 1, 2 and 3. The spatial resolution was chosen so that the turbulence was sufficiently resolved in all cases.

Table 1 MHD channel flow simulation parameters. $\mathscr{N}_i$, L_i and N_i are respectively the Stuart number of the magnetic field, the domain size and the number of spectral modes in the respective i-direction. $Re_b = U_b h/\nu$ where U_b is the bulk mean velocity.

run	Re_τ	Re_b	$\mathscr{N}_i$	$L_x \times L_y \times L_z$	$N_x \times N_y \times N_z$
My	140	2400	$\mathscr{N}_y = 0.01$	$80h \times 2h \times 40h$	$864 \times 81 \times 864$
Mz	163	4000	$\mathscr{N}_z = 0.125$	$80h \times 2h \times 40h$	$960 \times 65 \times 960$

Table 2 Rotating Couette flow simulation parameters. $Ro = 2\Omega h/U_w$ is the rotation number and Ω the rotation rate.

run	Re_τ	Re	Ro	$L_x \times L_y \times L_z$	$N_x \times N_y \times N_z$
R750	42	750	-0.025	$250h \times 2h \times 125h$	$800 \times 33 \times 800$
R2000	85	2000	-0.054	$250h \times 2h \times 197.5h$	$1600 \times 49 \times 2048$

Table 3 Stratified open channel flow simulation parameters. $Ri = \Delta\rho g h/\rho_0 U_b$ is the Richardson number with $\Delta\rho$ and h the fluid density difference and distance respectively between wall and surface, g the gravity constant and ρ_0 the reference density.

run	Re_τ	Re_b	Ri	$L_x \times L_y \times L_z$	$N_x \times N_y \times N_z$
S113	113	2020	0.059	$100h \times h \times 50h$	$864 \times 37 \times 864$
S182	182	4080	0.197	$100h \times h \times 50h$	$1600 \times 65 \times 1600$

3 Results

Fig. 1 shows snapshots of the streamwise velocity in a wall parallel plane in the MHD channel flow DNS with either spanwise or wall-normal magnetic field orientation. Regions with intense turbulence and with much calmer, laminar-like flow

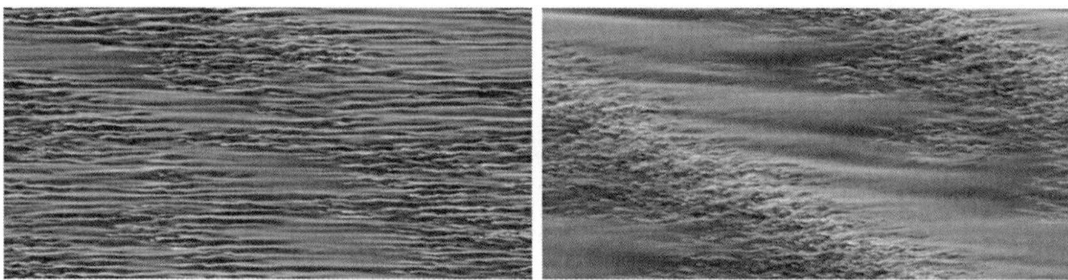

Fig. 1 Snapshots of the instantaneous streamwise velocity at $y = 0.8$ in run My (left figure) and Mz with a wall-normal and spanwise magnetic field respectively. As in Figs. 2 and 3, the mean flow is from left to right.

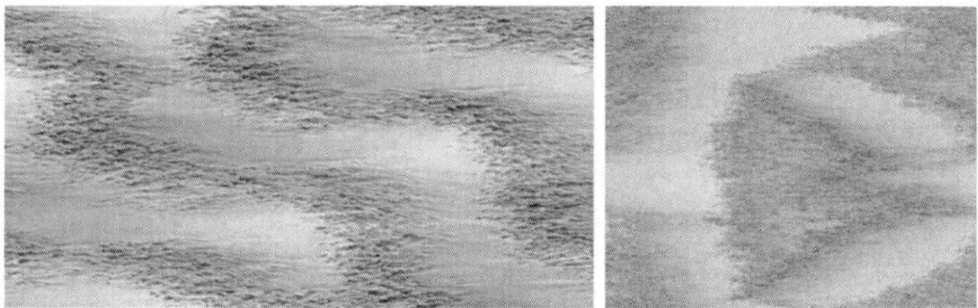

Fig. 2 Snapshots of the streamwise velocity in a plane parallel and close to the wall in run R750 (left figure) and R2000.

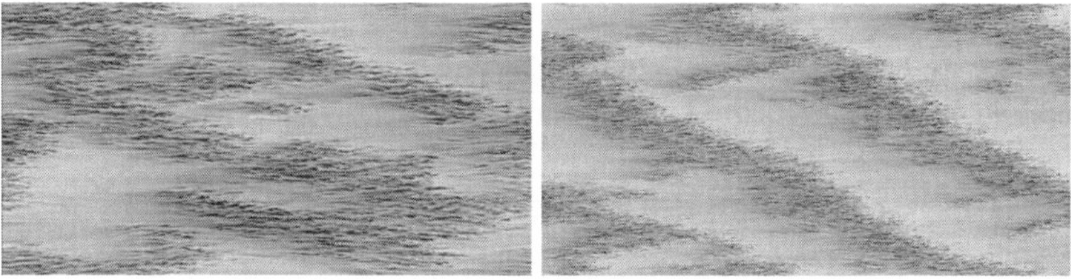

Fig. 3 Snapshots of the streamwise velocity in a plane parallel and close to the wall in run S113 (left figure) and S182.

are seen in both run My and Mz with a strong wall-normal and spanwise magnetic field respectively, but only in the latter case they form clear inclined band-like patterns as in channel flow without a magnetic field [4]. Even in the laminar regions in run My streak-like structures are present whereas they are mostly absent in run Mz. Note that the periodic boundary conditions enforce a certain inclination angle on the turbulent-laminar bands. This restriction is less severe if the computational domain is larger but then the computational costs become prohibitive.

Alternating laminar and turbulent patterns appear as well in run R750 and R2000 of rotating plane Couette flow (Fig. 2). However, in a nonrotating flow at $Re = 350$ the patterns form straight bands (not shown here) while in run R750 they are much more bent and in run R2000 they have various inclination angles. In both cases the turbulent-laminar fractions stay approximately constant while the patterns slowly evolve in time.

Also in run S113 and S182 of stably stratified open channel flow turbulent and large laminar patterns appear (Fig. 3) but especially in run S113 they are less structured than in unstratified channel flow in smaller domains [5],[4]. Similar structures in stratified channel flow DNS in smaller domains were observed by [7]. In both simulations of plane Couette flow and open channel the rotation rate respectively stratification has to be increased at higher Reynolds numbers in order to observe pattern formation.

In plane Couette flows the patterns are not moving due to flow symmetry, but in MHD and stratified channel flow they are convected by the flow. By removing small turbulent scales through low pass filtering and computing the space-time correlation function, we have estimated that the mean convection speed of the turbulent patterns is about equal to the mean bulk velocity in plane channel flow without a magnetic field or stratification while it is about 0.92 respectively 0.8 times the mean bulk velocity in run Mz and run S182. This implies that the patterns travel faster than the mean flow near the wall and slower further away from the wall. Consequently, the leading and trailing edge of the turbulent pattern with respect to the mean flow near the wall and in the core of the flow are reversed.

In order to compute statistics in different parts of the flow a region was defined as turbulent if the median averaged wall normal velocity fluctuations, v_{rms}, are above a threshold. The mean streamwise velocity fluctuations, u^+, and Reynolds stress, uv^+, scaled by mean friction velocities in the laminar and turbulent parts of run S113 are shown in Fig. 4. In the turbulent parts the profile and peak value of u^+ are similar to the ones in fully turbulent open channel flow whereas in the laminar parts they are a bit lower but still significant. On the other hand, the mean wall normal velocity fluctuations (not shown here) and uv^+ are much smaller in the laminar than the turbulent parts.

Mean velocity profiles in the laminar and turbulent parts of the flow differ significantly as well. Fig. 5 shows mean streamwise velocity profiles in the laminar and turbulent regions in rotating Couette flow at $Re = 750$ and stratified open channel flow at $Re_\tau = 113$. In both cases the mean profile in the turbulent part is more similar to mean velocity profiles in fully turbulent flows (steeper gradients near the wall and a flatter profile in the core) than in the laminar parts. This is not surprising, but it implies streamwise gradients in the mean streamwise velocity. Continuity requires then $\partial V/\partial y + \partial W/\partial z \neq 0$, i.e. gradients in the mean wall normal and spanwise velocities. Since the wall normal velocity is restricted by the wall $\partial W/\partial z \neq 0$ and thus

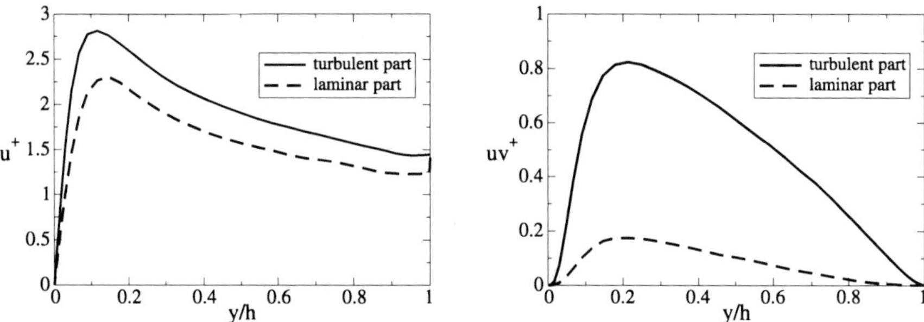

Fig. 4 u_{rms} (left) and $\langle uv \rangle$ (right) in run S113. The statistics in the laminar and turbulent parts are scaled with the mean friction velocity in the laminar and turbulent parts, respectively.

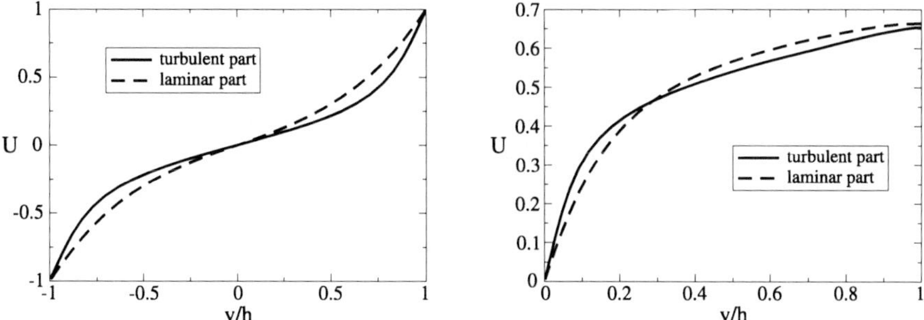

Fig. 5 Mean streamwise velocity in rotating plane Couette flow at $Re = 750$ (left) and stratified open channel flow at $Re_\tau = 113$ (right).

the local mean spanwise velocity has a non-zero component. This is seen in Fig. 6 showing mean velocity vectors obtained by local averaging in a wall parallel plane in run R750. Going through laminar and turbulent flow regions the direction and magnitude of the mean velocity changes. The in-plane mean streamwise velocity has been subtracted to make this clearer. Similar observations have been made by [9] in case of nonrotating Couette flow.

In the case of open and closed channel flow the in-plane mean velocity patterns induced by banded turbulent-laminar structures are even more complex and vary with the distance to the wall since the mean streamwise velocity gradient changes with y (Fig. 5). This is seen in Fig. 6 showing mean velocity vectors in two different wall-parallel planes for run Mz. The local mean spanwise velocity has at some positions an opposite sign at different distance to the wall.

Acknowledgements We acknowledge financial support from the Swedish Research Council. Computational resources at PDC were made available by the Swedish National Infrastructure for Computing (SNIC).

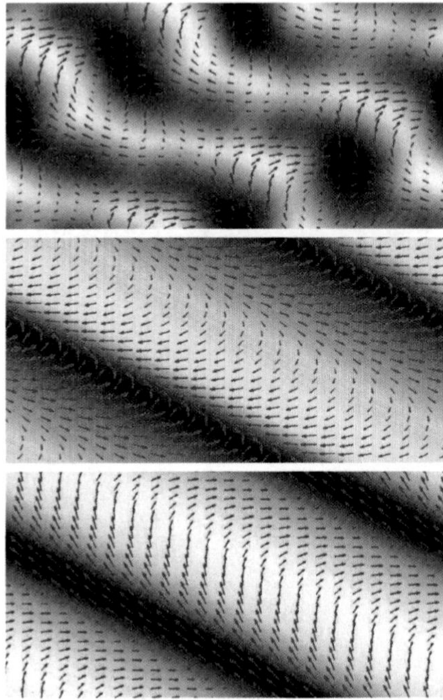

Fig. 6 Mean velocity components in a wall parallel plane for run R750 at $y/h = 0.5$ (top figure), and run Mz at $y/h = 0.1$ (middle figure) and at $y/h = 0.6$ (bottom figure). The mean streamwise velocity is subtracted and the mean flow is from left to right. Dark colours represent turbulent regions.

References

1. Prigent A, Grégoire G, Chaté H, Dauchot O, van Saarloos W (2002) Large-scale finite-wavelength modulation within turbulent shear flows. Phys. Rev. Lett. 89:014501.
2. Barkley D, Tuckerman L S (2005) Computational study of turbulent laminar patterns in Couette flow. Phys. Rev. Lett. 94:014502.
3. Duguet Y, Schlatter P, Henningson D S (2010) Formation of turbulent patterns near the onset of transition in plane Couette flow. J. Fluid Mech. 650:119–129.
4. Fukudome K, Iida O, Nagano Y (2009) The mechanism of energy transfer in turbulent Poiseuille flow at very low Reynolds number. Proc. 6th Intl Symp. on Turbulence and Shear Flow Phenomena:471–476.
5. Tsukahara T, Seki Y, Kawamura H, Tochio D (2005) DNS of turbulent channel flow at very low Reynolds numbers. Proc. 4th Intl Symp. on Turbulence and Shear Flow Phenomena:935–940.
6. Tsukahara T, Tillmark N, Alfredsson P H (2010) Flow regimes in a plane Couette flow with system rotation. J. Fluid Mech. 648:5–33.
7. García-Villalba M, del Álamo J C (2009) Observation of turbulent-laminar patterns in DNS of stably-stratified channel flow. 7th IUTAM Symp. on Laminar-Turbulent Transit.
8. Lee D, Choi H (2001) Magnetohydrodynamic turbulent flow in a channel at low magnetic Reynolds number. J. Fluid Mech. 439:367–394.
9. Barkley D, Tuckerman L S (2007) Mean flow of turbulent-laminar patterns in plane Couette flow. J. Fluid Mech. 576:109–137.

Comparison of SGS Models for Passive Scalar Mixing in Turbulent Channel Flows

Qiang Li, Philipp Schlatter and Dan S. Henningson

1 Introduction and Numerical Methods

In many industrial applications, there is a need to understand scalar mixing, e.g. in gas turbine combustion systems. In such combustion systems, the majority of momentum transport and scalar mixing is driven by the large-scale structures, therefore LES becomes a natural choice. During the last years, promising results are obtained with LES, however, as opposed to LES modeling of the velocities, only a limited body of literature is devoted to scalar modeling, especially in wall-bounded flows. The most commonly used model is the so-called eddy-diffusivity model requiring the predetermination of the turbulent Prandtl number (model coefficient), usually assumed to be a constant. In turbulent channel flow, Moin *et al.* [5] extended the dynamic Smagorinsky model to compressible turbulence and scalar transport. Later, Calmet & Magnaudet [2] performed simulation with three different Schmidt numbers: 1, 100 and 200 using the dynamic mixed model. The mass transfer coefficient is found to agree very well with previous experimental results. Recently, You & Moin [8] generalised their previous SGS model You & Moin [7] to account for scalar transport. The model does not require any spatial nor temporal average of the coefficient and thus can be used to simulate flows in complex geometry. However the results are similar to those obtained with dynamic Smagorinsky model.

This present contribution extends several recent SGS models, i.e. the dynamic Smagorinsky model (DSM) [3,4], dynamic high-pass filtered model (DHPF) [6] and WALE combined with high-pass filtered model (WHPF) which is similar to the RVMs-WALE model proposed by Bricteux *et al.* [1], to include passive scalar and present the statistics pertaining to the scalar field, demonstrating the weakness of the eddy-diffusivity assumption. The predictions of the statistics pertaining to the scalar field from all the models are in general worse than their velocity counterparts. Re-

---

Qiang Li, Philipp Schlatter and Dan S. Henningson
Linné Flow Centre, KTH Mechanics, SE-100 44, Stockholm, Sweden, e-mail: qiang@mech.kth.se

H. Kuerten et al. (eds.), *Direct and Large-Eddy Simulation VIII*, ERCOFTAC Series 15, DOI 10.1007/978-94-007-2482-2_22,

	DNS/EXP data	best reconstruction
mean quantities	$\langle u\rangle$, $\langle\theta\rangle$	$\langle\widetilde{u}\rangle$, $\langle\widetilde{\theta}\rangle$
reduced intensities and scalar flux	$\langle a_{ij}\rangle$, $\langle u'_i\theta'\rangle$ $a_{ij}\equiv u'_iu'_j-\frac{1}{3}\delta_{ij}u'_ku'_k$	$\langle\widetilde{a}_{ij}\rangle+\langle\tau^*_{ij}\rangle$, $\langle\widetilde{u}'_i\widetilde{\theta}'\rangle+\langle\sigma_i\rangle$ $\widetilde{a}_{ij}\equiv\widetilde{u}'_i\widetilde{u}'_j-\frac{1}{3}\delta_{ij}\widetilde{u}'_k\widetilde{u}'_k$, $\tau^*_{ij}\equiv\tau_{ij}-\frac{1}{3}\delta_{ij}\tau_{kk}$
Production ($\mathscr{P}$, $\mathscr{P}_\theta$)	$-\langle u'v'\rangle\frac{\partial\langle u\rangle}{\partial y}$, $-\langle v'\theta'\rangle\frac{\partial\langle\theta\rangle}{\partial y}$	$-\langle\widetilde{u}'\widetilde{v}'+\tau_{12}\rangle\frac{\partial\langle\widetilde{u}\rangle}{\partial y}$, $-\langle\widetilde{v}'\widetilde{\theta}'+\sigma_2\rangle\frac{\partial\langle\widetilde{\theta}\rangle}{\partial y}$
Turb. Diffusion (T, T_θ)	$-\frac{1}{2}\frac{\partial\langle u'_iu'_iv'\rangle}{\partial y}$, $-\frac{1}{2}\frac{\partial\langle\theta'\theta'v'\rangle}{\partial y}$	$-\frac{1}{2}\frac{\partial\langle\widetilde{u}'_i\widetilde{u}'_i\widetilde{v}'\rangle}{\partial y}-\frac{\partial\langle\widetilde{u}'_i\tau^{*\prime}_{i2}\rangle}{\partial y}+\langle\tau_{12}\rangle\frac{\partial\langle\widetilde{u}\rangle}{\partial y}$, $-\frac{1}{2}\frac{\partial\langle\widetilde{\theta}'\widetilde{\theta}'\widetilde{v}'\rangle}{\partial y}-\frac{\partial\langle\widetilde{\theta}'\sigma^{*\prime}_2\rangle}{\partial y}+\langle\sigma_2\rangle\frac{\partial\langle\widetilde{\theta}\rangle}{\partial y}$
Dissipation (ε, ε_θ)	$-\frac{1}{Re}\langle\frac{\partial u'_i}{\partial x_k}\frac{\partial u'_i}{\partial x_k}\rangle$, $-\frac{1}{RePr}\langle\frac{\partial\theta'}{\partial x_k}\frac{\partial\theta'}{\partial x_k}\rangle$	$-\frac{1}{Re}\langle\frac{\partial\widetilde{u}'_i}{\partial x_k}\frac{\partial\widetilde{u}'_i}{\partial x_k}\rangle+\langle\tau^{*\prime}_{ik}\frac{\partial\widetilde{u}'_i}{\partial x_k}\rangle$, $-\frac{1}{RePr}\langle\frac{\partial\widetilde{\theta}'}{\partial x_k}\frac{\partial\widetilde{\theta}'}{\partial x_k}\rangle+\langle\sigma'_k\frac{\partial\widetilde{\theta}'}{\partial x_k}\rangle$

Table 1 Reconstruction of the statistics from LES to compare with DNS data.

sults obtained with dynamic coefficients of the SGS scalar flux are similar to those with fixed coefficients and no improvements is observed. The terms in the budget of the kinetic energy and the scalar variance are reconstructed and compared with the DNS data. In addition, the problems associated with the pressure are discussed.

The filtered Navier-Stokes equations together with the passive scalar transport equation are solved using standard spectral (Fourier/Chebyshev) methods for a periodic channel. The nominal Reynolds number based on the friction velocity u_τ and the channel half width h is 590 with the Prandtl number Pr being 0.71. The computational domain is set to $L_x\times L_y\times L_z=2\pi h\times 2h\times\pi h$ with $384\times 257\times 384$ and $64\times 65\times 64$ spectral collocation points for DNS and LES, respectively. The scalar is introduced on one wall and removed on the other.

2 Data Comparison

In practice, due to the fact that the filtered DNS data is usually not available, one tries to reconstruct the statistics from the available LES data such that they can be compared to the DNS data, see e.g. Winckelmans *et al.* [9]. Following the arguments in Winckelmans *et al.* [9], the comparison between mean quantities, second-order statistics and budget terms is summarized in Table 1.

For the pressure correlation term in the energy budget, due to the fact that the obtained LES solution of pressure is the modified pressure, i.e. the sum of the resolved pressure $\widetilde{P}$ and the sub-grid kinetic energy τ_{kk}, $\widetilde{p}^{LES}=\widetilde{P}+\frac{1}{3}\tau_{kk}$. Therefore, direct comparison of the pressure correlation term is not possible unless a model for τ_{kk} is assumed.

Considering the viscous diffusion term, only the deviatoric part can be reconstructed and then compared to the DNS data. However, for the kinetic energy budget, the deviatoric part of the viscous diffusion term is identically zero since $\widetilde{a}_{ii}=0$.

One can also reconstruct various budget terms for budget of the Reynolds stresses to compare with DNS data. However, the comparison of the pressure correlation term still remains not possible due to the same reason. Note that, for budget of $\langle \widetilde{u}'\widetilde{v}' \rangle$, one can not properly reconstruct since the sub-grid kinetic energy τ_{kk} is needed even for the production term.

The mean contribution from the SGS stresses and SGS scalar fluxes is small for models based on the HPF velocity (scalar) fields. Therefore, one can actually direct compare the resolved quantities with the DNS data without adding the SGS contribution [1]. The near-wall distributions of σ_2 are shown in Figure 1 (a). It is clearly seen that the mean SGS contributions from DHPF and WHPF models are one order of magnitude smaller than those from that from DSM in which the SGS contribution is calculated based on the resolved fields.

Though one cannot compare the pressure statistics, however, by subtracting the mean pressure of LES solution from the DNS solution, an approximation of the SGS kinetic energy can be obtained, i.e. $\tau_{kk} \approx 3(\widetilde{p}^{LES} - p^{DNS})$. The approximations of τ_{kk} obtained from different models are shown in Figure 1 (b) together with data from an *a priori* test. Prediction from DSM is the most closest to the *a priori* test while both DHPF and WHPF have negligible values near the wall.

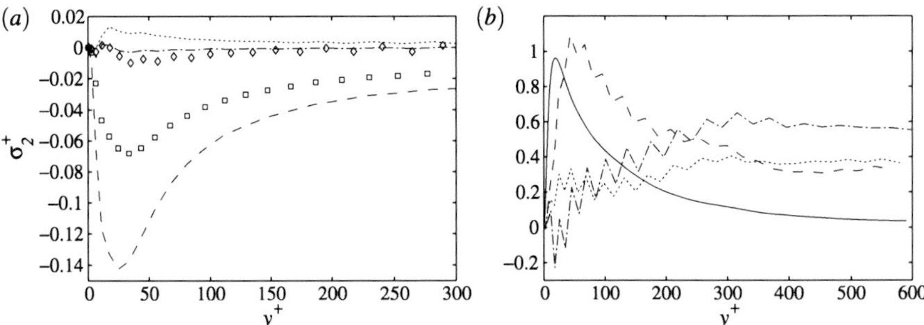

Fig. 1 solid: DNS, dashed: DSM, quare: DDSM, chain-dot: DHPF, diamond: DDHPF, dotted: WHPF. (a) Near wall distributions of $\langle \sigma_2 \rangle^+$. (b) Approximation of $\langle \tau_{kk} \rangle^+$.

3 Results

Only results pertaining to the scalar field are shown. The model coefficient Pr_t is computed both non-dynamically and dynamically. For the non-dynamic case as in DSM and DHPF, Pr_t is set to 0.6 while the dynamic version is termed as DDSM and DDHPF. First the averaged statistics are shown and then the major terms in the budget of the scalar variance are compared with DNS data. The statistics pertaining to the velocity field are consistent with previous studies and therefore not shown.

The mean scalar profiles are shown in Figure 2 (a). DSM and WHPF seem to predict the mean scalar very well up to $y^+ \approx 100$. An underprediction is observed for DHPF close to the wall which is similar to the no-model LES. As the center of the channel is approached, DSM and WHPF overpredict the mean scalar profile

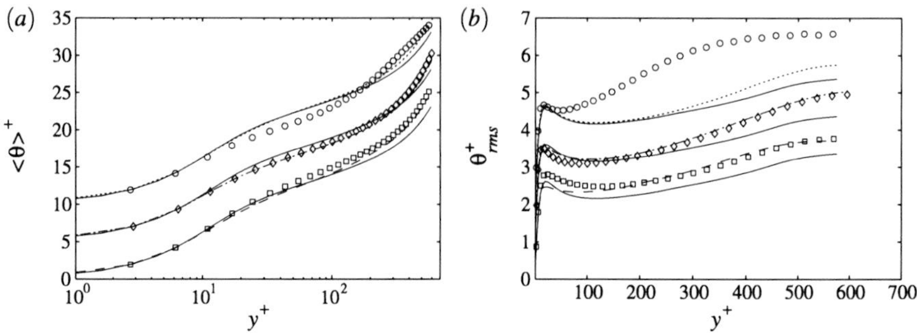

Fig. 2 LES results of channel flow pertaining to the scalar at $Re_\tau \approx 590$; solid: DNS (unfiltered), dashed: DSM, square: DDSM, chain-dot: DHPF, diamond: DDHPF, dotted: WHPF, circle: no-model LES. (*a*) Mean scalar profile, (*b*) RMS of the scalar fluctuation. Note that the plots are shifted by 5 and 2 wall units, respectively.

due to too much dissipation which results in too low wall friction temperature. The DDSM and DDHPF follow DSM and DHPF respectively throughout the channel and no improvement is observed.

The scalar fluctuation is shown in Figure 2 (*b*). A double peak profile is clearly seen and the peak is associated with the local maximum energy production. Moreover, the peak value in the center is Reynolds number dependent [2]. DSM is unable to capture the near-wall peak and due to the dissipative character, an overprediction can be observed throughout the channel. DDSM has even larger overshoot of the near-wall peak. For HPF based models (DHPF and WHPF), both of them can capture the near-wall peak while WHPF slightly overpredicts the peak value due to dissipative character and this is also consistent with the prediction of u^+_{rms}. However, DDHPF slightly underpredicts the peak value. As approaching the center of the channel, all the models deviate from the DNS data and overpredict the peak values. At the center of the channel, the production of the LES prediction is about 50% higher than the DNS value. For the no-model LES, since there is not enough dissipation and this leads to extremely high peak value of the scalar fluctuations. The streamwise and wall-normal heat flux after adding the SGS contribution compare good with DNS data, however, near the center of the channel, all the model predictions deviate from the DNS results for wall-normal component.

The coefficient Pr_t predicted from the dynamic procedure is plotted in Figure 3 (*a*). This coefficient has a maximum value at the wall and decreases away from the wall. For DDSM, Pr_t is around 0.4 in the bulk region of the channel and is close to the value 0.6. For the DHPF model, the non-dynamic value 0.6 is about one order of magnitude larger than the dynamic prediction. However, the resolved scalar statistics of DHPF are quite similar to those with DDHPF which indicates that the model is insensitive to the chosen coefficient. The SGS dissipation of the scalar variance ε_θ^{SGS} is supposed to have the same near-wall behavior as the SGS dissipation of the turbulent kinetic energy ε^{SGS}. However, the DDHPF model gives a different slope than y^3 while DDSM predicts the correct one.

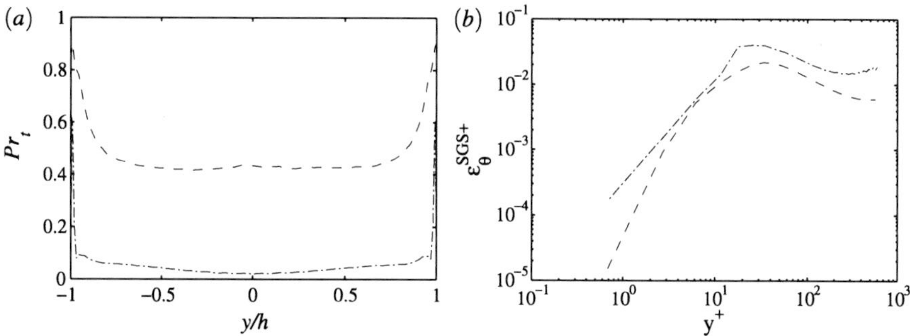

Fig. 3 Distribution of dynamic coefficient Pr_t and the near behavior of ε_θ^{SGS}; dashed: DDSM, chain-dot: DDHPF. (a) Pr_t, (b) ε_θ^{SGS}.

The results comparing the production term in the scalar variance budget are shown in Figure 4 $(a)-(b)$. For DSM, by adding the SGS contribution, a clear improvement is observed, while for DHPF and WHPF the resolved production already agrees well with the DNS data.

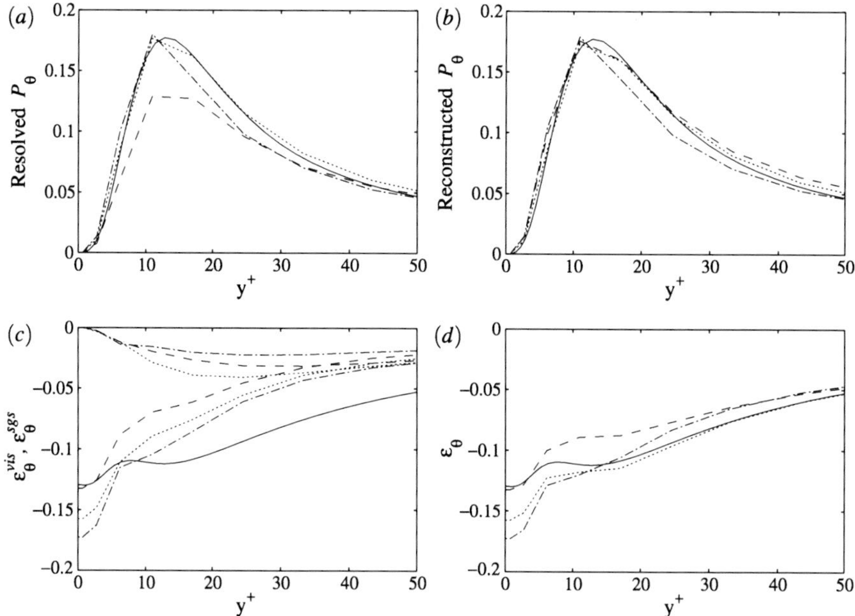

Fig. 4 Production and dissipation of the scalar variance budget; solid: DNS (unfiltered), dashed: DSM, chain-dot: DHPF, dotted: WHPF. (a) Resolved $\mathscr{P}_\theta$, (b) Resolved $\mathscr{P}_\theta$ with SGS contribution. (c) ε_θ^{vis} and ε_θ^{sgs}, (d) ε_θ.

The comparison of dissipation for scalar variance budget are shown in Figure 4 $(c)-(d)$. The dissipation contains two parts: one is the resolved viscous dissipation and the other the SGS dissipation. Very close to the wall i.e. $y^+ \leq 5$, DSM agrees best with DNS data while both DHPF and WHPF predict a larger viscous dissipation. This might be due to the fact that the coefficient decrease too fast close to the wall, as what is also observed in a no-model calculation.

4 Conclusions

Turbulent channel flow with passive scalar transport at $Re_\tau = 590$ has been performed with three SGS models: DSM, DHPF and WHPF using a spectral method. For each model of the velocity, an eddy-diffusivity model of the turbulent passive scalar transport has been proposed. The coefficient of the SGS scalar flux is computed non-dynamically and dynamically. A constant coefficient gives better predictions than a dynamic coefficient while the general behavior is similar. The results from DHPF seem not sensitive to the chosen coefficient of the SGS scalar flux. For DSM, one should add the SGS contribution when comparing the turbulence statistics while for the HPF based models it is good enough to compare the resolved quantity. The same conclusions can be extended when comparing the terms in the budget equations. For DSM, the SGS contribution takes up to 10% of the production while only 1% for HPF based models. For the dissipation, the SGS contributions play an important role and they provide as much as 45% of the total dissipation.

References

1. BRICTEUX, L., DUPONCHEEL, M. & WINCKELMANS, G. (2009). A multiscale subgrid model for both free vortex flows and wall-bounded flows. *Phys. Fluids* **21** (10), 105102.
2. CALMET, I. & MAGNAUDET, J. (1996). Large-eddy simulation of high-schmidt number mass transfer in a turbulent channel flow. *Phys. Fluids*, **9** (2), 438–455, 1997.
3. GERMANO, M., PIOMELLI, U., MOIN, P. & CABOT, W. H. (1991). A dynamic subgrid-scale eddy viscosity model. *Phys. Fluids A* **3** (7), 1760–1765.
4. LILLY, D. K. (1992). A proposed modification of the Germano subgrid-scale closure method. *Phys. Fluids A* **4** (3), 633–635.
5. MOIN, P., SQUIRES, K., CABOT, W.,& LEE, S. (1991). A dynamic subgrid-scale model for compressible turbulence and scalar transport. *Phys. Fluids A*, **3** (11), 2746–2757, 1991.
6. SCHLATTER, P., BRANDT, L., BRUHN, T. & HENNINGSON, D. S. (2006). Dynamic high-pass filtered eddy-viscosity models for LES of turbulent boundary layers. In *Bulletin American Physical Society*, vol. **51**, p. 213.
7. YOU, D. & MOIN, P. (2007). A dynamic global-coefficient subgrid-scale eddy-viscosity model for large-eddy simulation in complex geometries. *Phys. Fluids*, **19** (6), 065110, 2007.
8. YOU, D. & MOIN, P. (2009). A dynamic global-coefficient subgrid-scale model for large-eddy simulation of turbulent scalar transport in complex geometries. *Phys. Fluids*, **21** (4), 045109, 2009.
9. WINCKELMANS, G. S., JEANMART, H. & CARATI, D. (2002). On the comparison of turbulence intensities from large-eddy simulation with those from experiment or direct numerical simulation. *Phys. Fluids*, **14** (5), 1809–1811, 2002.

Improved wall-layer model for forced-convection environmental LES

V. Stocca, V. Armenio and K. R. Sreenivasan

1 Introduction

In environmental flows where the Reynolds numbers are very high it is not possible to solve directly the near-wall region because of the still-limited computer power. Furthermore, the presence of irregular wall-roughness in practical high Reynolds flows would make the direct resolution of the near wall viscous layer somewhat useless. When using Large-eddy simulations (LES), many efforts have been made in the last decades to develop LES wall-layer models (LWM) designed to skip the resolution of the near-wall layer [7]. One of the common approaches used both in channel flows and planetary boundary layers, is to derive the stress to be used as boundary condition by the assumption that the instantaneous near-wall velocity belongs to a logarithmic profile (see among the others [6], [10]). We propose a modification to this kind of approach in which an analytically derived correction to the evaluation of the near wall Smagorinsky eddy viscosity and diffusivity is made. This correction, which is a simplified version of the two part model first proposed by [8] and then adapted by [9] for planetary boundary layers, is computationally simple to implement and does not require homogeneity, thus being a good candidate to simulate complex-terrain configurations typical of environmental flows of practical interest. The quality of the resulting velocity prediction as well as of the temperature profiles is significantly improved. The proposed LWM is validated reproducing with a very coarse mesh a plane channel flow at two Reynolds numbers, $Re_* = 4000$ and $Re_* = 20000$ for a case of forced convection.

V. Stocca
University of Trieste, Trieste, Italy, e-mail: valestocca@gmail.com

V. Armenio
University of Trieste, Trieste, Italy, e-mail: armenio@dica.units.it

K. R. Sreenivasan
New York University, New York, US, e-mail: katepalli.sreenivasan@nyu.edu

H. Kuerten et al. (eds.), *Direct and Large-Eddy Simulation VIII*,
ERCOFTAC Series 15, DOI 10.1007/978-94-007-2482-2_23,

2 Wall model description and numerical set-up

Using the approach first proposed by [6] a stress is imposed as boundary condition at the wall, by assuming that the instantaneous local tangential velocity u_t^ℓ, evaluated the first near-wall computational node belongs to a logarithmic or to a linear profile depending on its non-dimensional distance $z^{+\ell}$ from the wall (see also [7]):

$$u_t^{+\ell} = \begin{cases} \frac{1}{\kappa}\ln(z^{+\ell}) + B & \text{if } z^{+\ell} > 11 \\ z^{+\ell} & \text{if } z^{+\ell} \leq 11 \end{cases} \tag{1}$$

Here x, y, z are, respectively, the streamwise, spanwise and the wall-normal directions and u, v, w the corresponding velocity components. In 1, $u_t^{+\ell} = u_t^\ell / u_*^\ell$ is the local instantaneous non-dimensional tangential velocity, $z^{+\ell} = z u_*^\ell / \nu$, $u_*^\ell = \sqrt{\tau_w^\ell / \rho_0}$ is the local friction velocity with τ_w^ℓ the local wall shear stress, ν is the kinematic viscosity, ρ_0 the reference density, $\kappa = 0.41$ the von Kármán constant and $B = 5.1$. At each time instant of the simulation, for each near-wall node, u_*^ℓ is derived using the log-law in Eq. 1. Then the distance $z^{+\ell}$ is evaluated and if $z^{+\ell} \leq 11$, the linear relation of 1 is used to recalculate the u_*^ℓ accounting for low Reynolds effects as in [7]. However, it should be noticed that, in our simulations, where a coarse grid is used, the limit $z^{+\ell} \leq 11$ is practically never achieved. On the other hand the correction is of importance when studying flow fields with downstream recirculations. Finally τ_w^ℓ is derived from u_*^ℓ and used as boundary condition at the cell boundary. Since the direction of the wall stress is not known *a-priori* we assume it to be equal to the direction of the instantaneous local tangential velocity u_t^ℓ. This set-up is particularly suited for atmospheric boundary layer simulations where the equilibrium layer assumption is valid in most situations. In these cases the grids are usually very coarse in terms of wall units so that the assumption that the near-wall instantaneous local velocity belongs to the log-law is reasonable [6]. Further, the methodology applies to cases of solid walls having small slopes such as gentle hill sides, and its extension to rough walls cases is straightforward.

Beside imposing the stress, we also apply a correction to the evaluation of the eddy viscosity at the near-wall nodes. The use of the WM makes meaningless the no-slip boundary condition for the tangential velocity. It is known that the Smagorinsky model cannot reproduce the correct value of the eddy viscosity near the solid boundaries, since the integral scale of the flow is larger than the grid size used in LES. Moreover, even when one uses a sub-grid scale (SGS) eddy viscosity at the wall in the Smagorinsky model, its evaluation would require the knowledge of the contraction of the resolved strain rate tensor. This quantity is not known due to the fact that the tangential velocity at the wall is not determined. In particular, when the grid is coarse near the solid boundaries, in the computation of $|\overline{S}_{ij}|$, which is necessary to evaluate the eddy viscosity $\nu_T = C_s^2 \Delta^2 |\overline{S}_{ij}|$, the terms $\overline{S}_{xz}$ and $\overline{S}_{yz}$, which are expected to be the leading terms of $\overline{S}_{ij}$, become increasingly incorrect since they are derived assuming no-slip condition (here C_s is a model constant, and Δ is a filter length scale). Apart this conceptual shortcoming, the Smagorinsky model falls short also from the point of view of the implementation at the first grid cell off-the-wall.

We thus modify the way the Smagorinsky eddy viscosity is calculated. In particular the leading terms of the strain-rate tensors are calculated from the derivative of the equations 1. In case the point belongs to the logarithmic or the linear region we end up with equation 2 or 3:

$$\overline{S}_{xz} = \frac{u_*^\ell}{\kappa z}\frac{u^\ell}{u_t^\ell} \quad , \quad \overline{S}_{yz} = \frac{u_*^\ell}{\kappa z}\frac{v^\ell}{u_t^\ell} \tag{2}$$

$$\overline{S}_{xz} = \frac{u_*^{\ell 2}}{\nu}\frac{u^\ell}{u_t^\ell} \quad , \quad \overline{S}_{yz} = \frac{u_*^{\ell 2}}{\nu}\frac{v^\ell}{u_t^\ell} \tag{3}$$

In this way, the resulting eddy viscosity at the near-wall grid-point adjusts consistently with the imposed stress. Assuming a flow in the x-direction, to a first approximation, with the applied correction, $\nu_T^\ell \sim \sqrt{2}C_s^2\Delta^2\overline{S}_{xz} \sim \sqrt{2}C_s^2\Delta^2 u_*^\ell/(\kappa z)$. Further, if the grid has the same spacing in each direction, since the velocity is evaluated at the cell center with $\Delta^2 = (2z)^2$, we have $\nu_T^\ell \sim 4\sqrt{2}C_s^2 z u_*^\ell/\kappa = Dzu_*^\ell$ where $D = 4\sqrt{2}C_s^2/\kappa \sim 13.7C_s^2$ is a constant quantity. Hence, in case of equispaced grids, ν_T reduces to a RANS-like eddy viscosity. This allows us to overcome the main drawback of the use of the Smagorinsky model at the wall, since the correction proposed here allows us to set a proper integral scale. Thus, at the first near-wall point, we use an eddy viscosity which is based on a characteristic length scale, which is the wall-normal distance z and a characteristic velocity which is the friction velocity u_*^ℓ. No averaging is used to obtain ν_T^ℓ since u_*^ℓ is evaluated instantaneously from the solution of eq. 1.

The model also takes into account thermal variations inside the flow field solving a transport equation for temperature. Similar to the procedure applied for the momentum WM, the boundary condition for the temperature is derived by assuming that it instantaneously belongs to a logarithmic profile. Specifically, it employs the logarithmic law proposed by Kader [5] for boundary layers in channel flows:

$$\frac{T^\ell - T_W^\ell}{T_*^\ell} = [2.12\ln(1+z^{+\ell})C + \beta]e^{(-1/\Gamma)} + \Pr z^{+\ell}e^{-\Gamma} \tag{4}$$

where the local friction temperature is defined as $T_*^\ell = -q_W/u_*^\ell$, T_W^ℓ and T^ℓ are the local temperatures at the wall and at the first node off-the-wall respectively, q_W^ℓ is the local heat flux at the wall, $\beta = f(\Pr)$, $C = f(z,\delta)$ and $\Gamma = f(z^{+\ell},\Pr)$. At each time step u_*^ℓ is evaluated, using the T^ℓ of the previous time step, and the value of q_*^ℓ to be imposed at the boundary is derived from (4).

In this way, considering the walls as isothermal and having a constant temperature gap between them, it is possible to obtain the friction temperature $T_*^\ell = q_*^\ell/u_*^\ell$, and hence the wall heat flux q_*^ℓ since u_*^ℓ is already known. Since we treat the temperature as a passive scalar it is possible to account for the SGS temperature fluxes assuming a constant SGS Prandtl number $\Pr_{SGS}$ [2]. Thus the correction on the Smagorinsky model at the wall applies also for the SGS heat flux.

The curvilinear incompressible Navier-Stokes equations are integrated in time using the fractional step formulation proposed by [11], using the explicit second

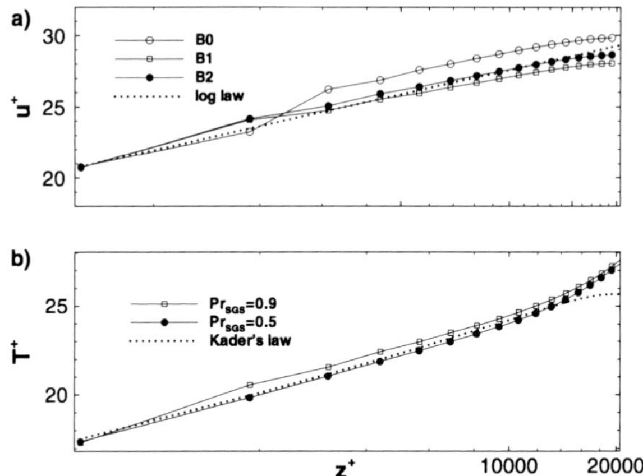

Fig. 1 (a) Mean velocity profile at $Re_* = 20000$ for three values of the SGS constant (dotted line refers to the analytic law 1). (b) Mean temperature profile for three values of Pr_{SGS} number and $C_s = 0.1$ (dotted line refers to the analytic Kader law [5]).

order Adam-Bashforth method for all terms but the diagonal viscous ones which are treated implicitly by means of a second order Crank-Nicolson method. Second order central finite differences are used for the spatial discretization. Simulations have been run with a constant Courant-Friedrichs-Lewy number $CFL = 0.1$. The model is second-order accurate in time and space.

3 Results

To validate the proposed LWM, channel flow simulations at $Re_* = (u_* \delta)/\nu = 4000$ and $Re_* = 20000$ have been carried out. Here we refer to u_* as the average friction velocity, while δ is the channel half height. The domain dimensions are $2\pi/\delta \times 2\pi/3/\delta \times 2$, respectively, in x, y and z directions. We use quite a coarse (32^3) grid having constant grid spacing in each direction, with a resulting resolution for the larger Re_* case of $\Delta x^+ = \delta u_*/\nu \simeq 3927$, $\Delta y^+ \simeq 1309$, $\Delta z^+ = 1250$. Periodic boundary conditions are applied in the horizontal directions while the LWM is used at the solid walls that are kept at constant temperature with a fixed temperature gap between them. The flow is driven by a constant pressure gradient and all statistics have been collected after reaching statistically steady state.

The mean velocity profiles of the $Re_* = 20000$ cases are shown in Fig. 1a. Simulation $B0$ ($C_s = 0.1$), carried out without correction on the near-wall eddy viscosity, gives a wrong velocity profile from the third off-the-wall point up to the channel centerline over-predicting the velocity value. Simulations $B1$ ($C_s = 0.09$) and $B2$

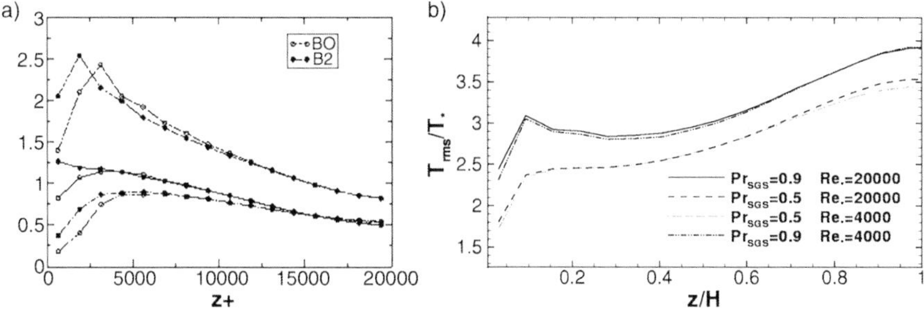

Fig. 2 (a) Root-mean-square profiles of the velocity fluctuations made non-dimensional with u_* for cases $B0$ and $B2$: dashed lines, streamwise component; dash-dotted, wall-normal; solid, spanwise. (b) Root-mean-square components of the temperature fluctuations plotted for $Re_* = 4000$ and $Re_* = 20000$ and for two values of the SGS Prandtl number $\mathrm{Pr}_{SGS} = 0.5$, $\mathrm{Pr}_{SGS} = 0.9$.

($C_s = 0.1$) use the corrected ν_T^ℓ near the wall. The near-wall correction is able to give a reasonably correct velocity profile even in a case of very coarse grid, typical of full-scale environmental application, although it gives a poor prediction at the second off-the wall point where the LES region merges with the near-wall RANS-like model. The value $C_s = 0.1$ gives the most accurate results. It is well known that the value of this constant in the Smagorinsky model depends on numerical issues, as for example the discretization scheme. However, based on the velocity statistics of the $Re_* = 4000$ case not shown in the present work, once the numerical model is set, as in our simulations, it appears that the value of the constant which gives the most accurate results ($C_s = 0.1$) does not depend on the value of the Re_* number.

The temperature WM has been tested in the case $B2$, namely the best-performing configuration for the momentum WM. Although wall modeling for temperature has been extensively investigated within the RANS framework, few attempts have been devoted to this issue in the LES context. A first test has been performed setting $\mathrm{Pr}_{SGS} = 0.9$, a value consistent with DNS studies of passive scalar transport in channel flow simulations and with the passive scalar results of the wall-resolving LES performed with a dynamic mixed model [1]. The resulting averaged temperature profile, made non-dimensional with the average friction temperature T_*, is reported in Fig. 1b. One can again observe the spurious merging between the RANS-like and the LES regions at the second off-wall point. Further, the temperature profile is over-predicted along the whole channel. A second test has been made using $\mathrm{Pr}_{SGS} = 0.5$, which gives a SGS temperature diffusivity $K_{SGS} = 2\nu_T$. The resulting profile, shown in Fig. 1b, collapses satisfactorily over the reference analytic law. Thus, it seems that more SGS diffusion is necessary when coarse grids are used with this temperature WM. The same behavior has been observed in the low-Reynolds-number case also, thus indicating that the WM (for both momentum and temperature) and the values of the constants employed here are robust with respect to variations of the Re_*.

In Fig. 2a the rms velocities are reported for cases $B0$ and $B2$. It is possible to see that, when the SGS correction is applied, the peak of the fluctuations is moved toward the wall and occurs at the second off-wall node for the streamwise component, and at the first off-wall node for the spanwise component—thus being more similar to the real profile. Also, a larger level of fluctuations is detected in the wall normal direction when compared to case $B0$. This is an encouraging result since one of the problems of the coarse LES is the underestimation of the wall-normal fluctuations. In Fig. 2b the rms temperatures are reported for two Reynolds numbers and for two values of Pr_{SGS}. The profiles are very similar for the two values of Re_* while significant differences are found for the two values of Pr_{SGS} considered here. Specifically, $\mathrm{Pr}_{SGS} = 0.5$ gives a lower level of temperature fluctuations producing a profile more consistent with the one reported in [4] for a lower Re_* resolved LES. Although (to the best of the authors' knowledge) there are no T_{rms} profiles available for comparison at this Re_*, from a qualitative comparison of profile shapes with lower Re_* ones, it appears that high fluctuations are present at the channel center. This could be due to the use of a centered interpolation scheme for advective terms of both momentum and temperature variations [4].

4 Conclusions

An improved LWM suited for geophysical flows based on a simple modification of the classical Smagorinsky model is presented. The modification allows a better capturing of the near-wall integral time-scale of the flow. Tests have been carried out considering forced-convection channel-flow simulations at $Re_* = 4000$ and $Re_* = 20000$. The tests used very coarse grids. Nevertheless, encouraging results were found both for the first and second order statistics when $C_s = 0.1$ and $\mathrm{Pr}_{SGS} = 0.5$ for both values of Re_*, thus showing the robustness of this LWM formulation.

References

1. Armenio, V., Sarkar, S., J. F. Mech. **459**, 1–42 (2002)
2. Cabot, W., Moin, P. in *Large Eddy Simulation of Complex Engineering and Geophysical flows*, ed. by Galperin, B., Orszag, S. A., (Cambridge University Press 1993), p. 141.
3. Cabot, W., Moin, P., Flow, Turb. and Comb. **63**, 269–291 (2000)
4. Châtelain, A., Ducros, F., Métais, O., I. J. Num. Meth. in Fluids. **44**, 1017–1044 (2004)
5. Kader, B., Int. J. Heat Mass Transf. **9**, 1541–1544 (1981)
6. Mason, P. J., Callen, N. S., J. F. Mech. **162**, 439–462 (1986)
7. Piomelli, U., Prog. in Aerospace Sci. **44**, 437–446 (2008)
8. Schumann, U., J. Comp. Phys. **18**, 376–404 (1975)
9. Sullivan, P., McWilliams, C., Moeng, C. H., B. L. Met. **71**, 247–276 (1994)
10. Temmerman, L., Leschziner, M. A., Mellen, C. P., Frölich, J., Int. J. Heat Fluid Flow **24**, 157–180 (2003)
11. Zang, Y., Street, R., Koseff, J., J. Comp. Phys. **114**, 18–33 (1994)

The effect of noise on optimal perturbations for turbulent mixing

Sara Delport, Martine Baelmans, Johan Meyers

1 Introduction

The evolution of a mixing layer is very sensitive to upstream flow conditions [3]. We optimize the initial flow conditions of a temporal mixing layer. We use this case as a substitute experiment to study how long flow control can affect the evolution of the mixing layer solution. In earlier work, we showed that optimization of the initial condition can be used to, e.g. increase the rate of kinetic energy dissipation at the end of a selected optimization time window significantly [1]. The current paper focusses on longer time windows and addresses the robustness of the optima in presence of white noise.

We study a temporal mixing layer with initial flow field:

$$u(\boldsymbol{x},0,\boldsymbol{\phi},\alpha) = \frac{\Delta U}{2}\tanh(x_3)\,\mathbf{e}_1 + \boldsymbol{\phi}(\boldsymbol{x}) + \sqrt{\alpha\frac{2L_3}{\delta_\omega}\frac{\|\boldsymbol{\phi}\|}{\|\boldsymbol{\varepsilon}\|}}\,\boldsymbol{\varepsilon}(\boldsymbol{x}) \tag{1}$$

where $\mathbf{e}_1$ is the unit vector in the stream-wise direction x_1, $\boldsymbol{\phi}(\boldsymbol{x})$ are optimized perturbations on the initial mean velocity field, $\boldsymbol{\varepsilon}(\boldsymbol{x})$ is background noise added to the perturbations, and α is the noise level. We optimize the perturbations in the absence of noise, and then investigate how the dissipation is affected when the background noise level is gradually increased in subsequent simulations with initial field (1).

We focus on maximizing the rate of dissipation at the time horizon T. Therefore a cost functional is used that maximizes the enstrophy at the time horizon:

$$\mathscr{J}(u(\boldsymbol{x},t,\boldsymbol{\phi},0) = -\frac{1}{2}\frac{1}{\Omega}\int_\Omega \boldsymbol{\omega}(\boldsymbol{x},T,\boldsymbol{\phi},0)\cdot\boldsymbol{\omega}(\boldsymbol{x},T,\boldsymbol{\phi},0)\,\mathrm{d}\boldsymbol{x} \tag{2}$$

Department of Mechanical Engineering, Katholieke Universiteit Leuven, Celestijnenlaan 300A, B-3001 Leuven, Belgium, e-mail: sara.delport@mech.kuleuven.be; martine.baelmans@mech.kuleuven.be; johan.meyers@mech.kuleuven.be

H. Kuerten et al. (eds.), *Direct and Large-Eddy Simulation VIII*, ERCOFTAC Series 15, DOI 10.1007/978-94-007-2482-2_24,

with $\omega(\boldsymbol{x},T,\boldsymbol{\phi},0)$ the vorticity of the velocity field $u(\boldsymbol{x},t,\boldsymbol{\phi},0)$ at time $t = T$ of the mixing layer with initial perturbations $\boldsymbol{\phi}$ on the mean velocity field.

The optimization of $\boldsymbol{\phi}$ in absence of noise leads to a significant enhancement of the enstrophy. We also observe that the impact of the presence of noise on the enstrophy is small. These results are discussed in sections 3 and 4 respectively. First the method and the case set-up are presented in section 2.

2 Method and case set-up

The optimization procedure combines the Polak-Ribière conjugate-gradient method with the Brent line-search algorithm. The gradient of the cost functional is determined with the continuous adjoint method. The optimization fixes the energy level of the perturbations to 10^{-4} times the mean-field energy $\Delta U^2/8$. For more details about the optimization method we refer to [1].

We consider DNS of a mixing layer with Reynolds number equal to 50 (based on half the velocity difference, and half the initial vorticity thickness, these reference values are also used to make all the properties dimensionless). The box contains 8 of the most unstable wave lengths for this Reynolds number according to the linear stability theory for a tanh-profile, and has dimensions $L_1 = 123.4$, $L_2 = 74.0$, and $L_3 = 240$ (respectively in the stream wise, span wise and the normal direction). The optimized perturbations are a combination of 32×32 Fourier-modes with wave number $(\alpha 2\pi/L_1,\, 0)$, $(0,\, \beta 2\pi/L_2)$, and $(\alpha 2\pi/L_1,\, \beta 2\pi/L_2)$ with $\alpha = \mp 1 \dots 16$ and $\beta = \mp 1 \dots 16$. The energy distribution over the modes, and the shape of the modes in the normal direction are optimized.

Both the Navier-Stokes equations and the adjoint Navier-Stokes equations are discretized using a pseudo-spectral scheme in stream and span-wise directions, while a fourth-order energy conservative scheme is used for the normal direction (for details see [1]). The time integration is a fourth order Runge-Kutta scheme. A grid of $128 \times 128 \times 256$ is used for all simulations.

An important issue in case of gradient-based optimization is the choice of the initial starting point of the optimization. Before going into detail, we introduce the symbol $\widetilde{\mathscr{E}}(t,T)$ for the enstrophy of the mixing layer optimized towards time horizon T, but evaluated at time t:

$$\widetilde{\mathscr{E}}(t,T) = \frac{1}{2}\frac{1}{\Omega}\int_{\Omega} \omega(\boldsymbol{x},t,\boldsymbol{\phi}_T,0)\cdot\omega(\boldsymbol{x},t,\boldsymbol{\phi}_T,0)\,\mathrm{d}\boldsymbol{x}. \tag{3}$$

When optimization is performed, it is expected for two different optimization time horizons $T_2 > T_1$ that $\widetilde{\mathscr{E}}(T_2,T_1) \leq \widetilde{\mathscr{E}}(T_2,T_2) = \mathscr{J}(u(\boldsymbol{x},t,\boldsymbol{\phi}_{T_2},0))$. In previous work [1] we found that this is not always the case while using the conjugate gradient procedure starting all optimizations from the same initial guess for the controls, pointing to the existence of local optima. Also at the outset of the current study, we were faced with these type of local optima, when starting the conjugate-gradient iterations from the same starting point $\boldsymbol{\phi}_A$. Therefore, in order to get a more consistent characterisation of optima as function of the time horizon T, we slightly changed the optimization approach. Instead of starting all optimizations for different time horizons directly from $\boldsymbol{\phi}_A$, we only start the optimization for the lowest time horizon ($T = 60$) from $\boldsymbol{\phi}_A$. Secondly, the optimal parameter set $\boldsymbol{\phi}_{T60}$ is used as a starting point for the optimization to the next time horizon ($T = 80$) (see also [2] Fig. 1), and so forth until the last time horizon $T = 160$. In this way $\widetilde{\mathscr{E}}(T_2,T_1) \leq \widetilde{\mathscr{E}}(T_2,T_2)$ is guaranteed (with $T_2 > T_1$).

3 Enstrophy maximization

The perturbations $\boldsymbol{\phi}$ are optimized to maximize the enstrophy at the time horizon in absence of noise. The cost functional value is monitored during the optimization. When the relative improvement of the cost functional is less than the convergence criterion (set to 10^{-10}), we consider the optimization to be converged. All cases but one converge in less than 70 iterations. One optimization (group A, $T = 140$) is stopped when the iteration number reaches 200 to limit the computational time as approximately 1800 DNS simulations are performed per 200 iterations.

Figure 1 shows the time evolution of the enstrophy for different optimized parameters, and for two sets ($\boldsymbol{\phi}_A$ and $\boldsymbol{\phi}_B$) which are not optimized. Parameters $\boldsymbol{\phi}_A$ are constructed using eigen modes from the linear stability analysis, while parameters $\boldsymbol{\phi}_B$ are based on low-pass filtered random noise. Parameters $\boldsymbol{\phi}_A$ served as starting point for the optimization procedure leading to the optimized parameters used in figure 1(a). Comparison of the enstrophy at the time horizons for $\boldsymbol{\phi}_A$, and the optimal parameters shows that the optimization enhances the enstrophy significantly. For low time horizons ($T = 60$ and 80), the enstrophy evolution strongly peaks near the time horizon for which it is optimized, followed by a sharp decline of enstrophy for $t > T$. For longer time horizons ($T \geq 100$), this is not any more the case. Now peaks around $t = 80$ are observed followed with a gradual decline up to the time horizon ($t = T$). For $t > T$, again a sharper decline in enstrophy is seen (note that, for the current case, the optimal solutions found for $T = 140$ and $T = 160$ are the same).

In figure 1(a), the evolution of the enstrophy starting from the initial field $\boldsymbol{\phi}_B$ (without any optimization) is also displayed. We observe that this (unoptimized) solution yields a higher level of enstrophy than the optimal solutions $\widetilde{\mathscr{E}}(T,T)$ for $T > 120$ optimized starting from $\boldsymbol{\phi}_A$. Consequently, for $T > 120$, these optimized solutions are local optima. To further investigate this, optimization starting from point $\boldsymbol{\phi}_B$ is considered. A similar optimization procedure is used as the procedure described in § 2, but with $\boldsymbol{\phi}_B$ as starting point and $T = 140$ as first time horizon. For the remainder of the work, we will often refer to optima coming from the latter optimization procedure as 'group B' and to the optima of the first procedure as 'group A'.

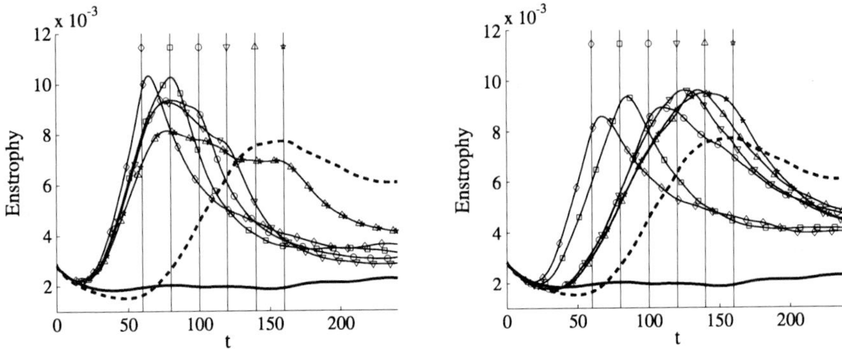

Fig. 1 Enstrophy evolution with parameters $\boldsymbol{\phi}_A$ (—) and $\boldsymbol{\phi}_B$ (−−) and with the optimized parameters (Symbols: $T = 60$ (◇), $T = 80$ (□), $T = 100$ (∘), $T = 120$ (▽), $T = 140$ (△), $T = 160$ (⋆)). Vertical lines are drawn at the different optimization time horizons. a) group A, b) group B

In figure 1(b), the evolution of the enstrophy is shown for group B. They also result in a large increase of enstrophy compared to the unoptimized $\boldsymbol{\phi}_B$ simulation.

The difference between optimizations in group A and group B can be well appreciated. For $T < 100$ optimizations in group A yield the highest enstrophy $\widetilde{\mathscr{E}}(T,T)$, while the inverse holds for $T > 100$. Overall, combining the best optima of group A and group B, we find that optimal controls lead to a maximal enstrophy larger than 8.6×10^{-3}. The optima from group B show a slight drop in the level of maximal enstrophy (about 10%) around $T = 100$. The drop is related to the existence of two groups of optimal solutions within group B.

To illustrate the difference between the groups of optimal solutions, the coherent structures of optimal solutions are depicted in figure 2 using the λ_2-criterion. The 'optimal' mixing layers are evaluated at $t = 40$ (λ_2-visualizations at earlier times do not a yield meaningful visualization of flow structures). The coherent structures are all fairly structured. For group A, we do not observe meaningful differences between visualizations of flows optimized towards different time horizons, and the selected plot in figure 2(a) with $T = 120$ is representative for the type of structures observed for other time horizons. For group B, more pronounced differences are visible when the time horizon is changed. The flow structures for group B can be divided into two groups according to their time horizon: $T \leq 80$ and $T \geq 100$. The typical structures for $T \leq 80$ are presented in figure 2(b). They show a diamond-shape pattern of vorticity in the mixing plane. For $T \geq 100$ typical structures are shown in figure 2(c). The vorticity is now more two-dimensional, mainly distributed in span-wise vorticity. We believe that this feature is strongly related to the better optima found in group B for large time horizons. Since, in this case, the initial energy is mainly stored in two-dimensional structures, vortex stretching, break-up of vortices, and subsequent dissipation, are 'postponed' to later times compared to the flows in figure 2(a,b), such that also the enstrophy peak (e.g. observed in figure 1(b) is significantly shifted to larger time horizons.

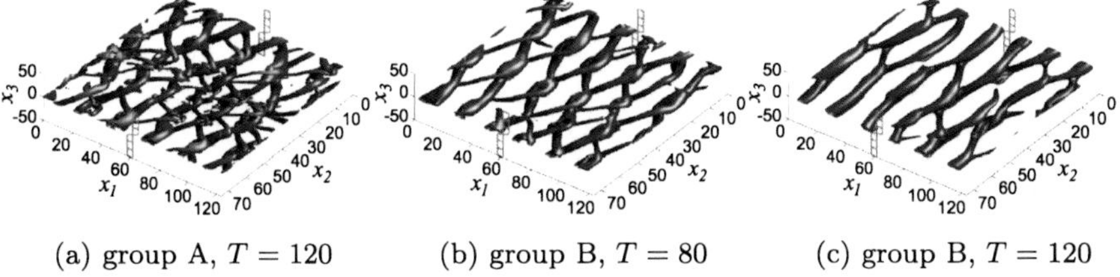

(a) group A, $T = 120$ (b) group B, $T = 80$ (c) group B, $T = 120$

Fig. 2 λ_2 visualization of the velocity-field for optimized parameters; profile $\langle u_1 \rangle$ (—); visualization of $t = 40$

4 Impact of noise

We now investigate the robustness of the optima found in the previous section to noise $\boldsymbol{\varepsilon}$. Three levels of noise are considered with $\alpha = 10^{-1}, 1, 10$. We simulate the mixing layer with initial condition (Eq. 1), using $\boldsymbol{\phi}$ optimized in absence of noise. Twenty different realizations of noise $\boldsymbol{\varepsilon}$ are used, corresponding to white noise uniformly distributed over the box.

The effect of the noise on the enstrophy evolution is shown in figure 3. It is observed that even for noise with 10 times more energy in the box than the perturbations $\boldsymbol{\phi}$, the enstrophy evolution with and without noise shows a similar evolution in the time window $t = [60, 140]$. The mean enstrophy level at the time horizon decreases however with increasing noise level due to non linear interactions between the instabilities triggered by the noise and the optimized perturbations. These interactions suppress only partially the optimized flow evolution.

5 Conclusion

Optimization of the initial condition of a temporal mixing layer showed that flow control can increase the rate of kinetic energy dissipation up to long time horizons. Two different states were found: for the shortest time horizons the optimized parameters lead to diamond shaped coherent structures, for longer time horizons on the other hand structures with a higher 2D component were observed. The mixing layers remained in its state when white noise was added to the optimized parameters, so the optimized states are robust towards the impact of noise.

We would like to acknowledge the financial support of IWT-Vlaanderen The simulations were performed on the HPC-cluster VIC3 of the K.U.Leuven.

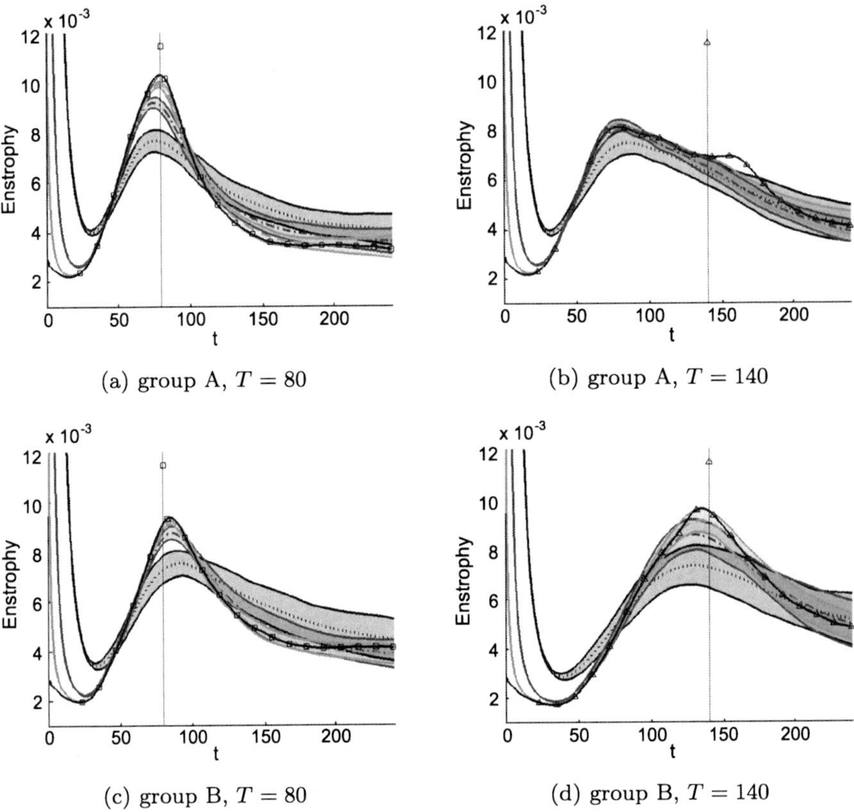

Fig. 3 Enstrophy evolution in time for optimized ϕ without noise (symbols), and plus noise (mean$\mp 2\sigma$ band for $\alpha = 0.1$ (– –), $\alpha = 1$ ($\cdot$ –), $\alpha = 10$ ($\cdots$))

References

1. Delport S, Baelmans M, and Meyers J (2009) Constrained optimization of turbulent mixing-layer evolution. J. Turbul. 10(18).
2. Delport S, Baelmans M, and Meyers J (2010) Optimization of long-term mixing in a turbulent mixing layer. 5th Flow Control Conference, 28 June – 1 July 2010, Chicago, Illinois. AIAA 2010-4422.
3. Moser RD and Rogers MM (1993) The 3-dimensional evolution of a plane mixing layer – pairing and transition to turbulence. J. Fluid Mech. 247:275–320.

Part III
Multiphase Flows

Direct and Large Eddy Simulation of Two-Phase Flows with Evaporation

Josette Bellan

1 Introduction

The modeling of turbulent two-phase flows is a subject of interest both to those who wish to understand and predict natural phenomena (e.g. clouds, tornadoes, volcanic clast dispersion, etc.) and those who wish to design and optimize engineered products (combustion devices based on fuel-spray injection such as gas turbine engines or spark ignition engines, augmenters in military aircraft, spray coating whether for painting or for protection against pests, consumer-product sprays such as those dispensed in cans, medical sprays, etc.). Despite the considerable range of applications and the substantial monetary advantages of successful prediction of turbulent two-phase flows, and despite numerous studies addressing modeling of these flows, there is still a lack of consensus for simulating these flows. The results described below are in the context of volumetrically dilute two-phase flows in which the volume of the condensed phase is negligible with respect to that of the carrier gas (e.g. $O(10^{-3})$) although the ratio of the condensed-phase mass to that of the carrier gas mass can be a substantial fraction (e.g. $O(10^{-1})$) because the density of the condensed phase is larger by a factor of $O(10^3)$ than that of the gas. The mathematical framework for following the gas phase is Eulerian, and because of the volumetrically dilute assumption it is appropriate to follow the drops in a Lagrangian way; in fact, for particles having a diameter, d, much smaller than the Kolmogorov scale, η_K, such as considered in this study, Boivin et al. [1] have shown that the particles can be considered as point sources in the flow and thus that the numerical resolution of the flow is essentially that of an equivalent single-phase flow. The simulation method which is the focus of this work is Large Eddy Simulation (LES), as it has the best potential for high-fidelity portrayal of the temporal and spatial variations of such a flow. LES is a methodology wherein the large flow scales containing the

Jet Propulsion Laboratory, California Institute of Technology, Pasadena, California 91109-8099; Mechanical Engineering Department, California Institute of Technology, Pasadena, CA 91125, USA, e-mail: Josette.Bellan@jpl.nasa.gov

H. Kuerten et al. (eds.), *Direct and Large-Eddy Simulation VIII*, ERCOFTAC Series 15, DOI 10.1007/978-94-007-2482-2_25,

overwhelming amount of energy are computed using the governing equations, and the small scales, being not computed, are instead modeled, with these models being included in the LES equations. Thus, in LES, the potential of the methodology is heavily dependent on the small-scale models. Because these small scales lie at a numerical resolution below that of LES, they are called subgrid-scales (SGS). Although in general LES is thought to only have statistical significance ([2]), we impose here more stringent requirements to achieve both spatial and temporal representation of the flow at the LES scale; clearly, if only statistical significance is sought, the range of models available to the user will be enlarged compared to that recommended by our more stringent study. The description of the methodology and results below adopts that of Bellan and Okong'o [3] and Leboissetier et al. [4] and provides an overall understanding of what has been achieved so far.

In section 2 the basic governing equations ([3]) are recalled, encompassing those for the gas phase (section 2.1) and for the drops (section 2.2). These equations were solved using Direct Numerical Simulation (DNS), a methodology wherein all flow scales overwhelmingly contributing to dissipation are resolved. The LES gas-phase equations are then recalled in section 3.1, and, consistently a methodology to reduce the drop field description is presented in section 3.2. In the Results, section 4, the physically rich information provided by DNS is first used to provide insights into the necessary modeling for LES. Then, the models proposed in section 3 are utilized to examine their performance in LES. Finally, the impact of the results is summarized in section 5.

2 Governing equations

The interest is here in the class of two-phase flows where phase change occurs at the particle surface; this class includes both phenomena of condensation and evaporation. As an example, the focus here is on evaporating drops, although concomitant evaporation and condensation has been treated elsewhere ([5, 6]). The combined assumptions of $d \ll \eta_K$ and volumetrically small loading of the flow, mean that each particle is embedded into a laminar near-field environment, and that the particles communicate only through their effect on the gas phase. Thus, the well known theory of single drop evaporation in laminar flows can be used.

2.1 Gas-phase equations

We define the vector of gas-phase conservative variables $\phi = \{\rho, \rho u_i, \rho e_t, \rho Y_V\}$ and denote the flow field as ϕ, where ρ is the density, u_i is the velocity in the x_i coordinate direction, e_t is the total energy and Y_V is the vapor (subscript V) mass fraction. The carrier gas (subscript C) mass fraction is Y_C, and $Y_C + Y_V = 1$. The gas-phase conservation equations are generically written as

$$\frac{\partial \phi}{\partial t} + \frac{\partial (u_j \phi)}{\partial x_j} = \frac{\partial \theta_j(\phi)}{\partial x_j} + S \tag{1}$$

where the vectors θ_j and S are

$$\theta_j = \{0, (-p\delta_{ij} + \sigma_{ij}), (-pu_j - q_j + \sigma_{ij}u_i), -j_{V,j}\} \tag{2}$$

$$S = \{S_I, S_{II,i}, S_{III}, S_I\} \tag{3}$$

where the first and last components of S are the same, S_I, since the liquid is a single chemical species which through evaporation contributes the vapor that represents the source for both changing the gas density and the vapor-species partial density. p is the pressure, σ_{ij} is the viscous stress

$$\sigma_{ij}(\phi) = 2\mu \left(S_{ij} - \frac{1}{3} S_{kk} \delta_{ij} \right), \tag{4}$$

$$S_{ij}(\phi) = \frac{1}{2} \left(\frac{\partial u_i}{\partial x_j} + \frac{\partial u_j}{\partial x_i} \right) \tag{5}$$

with μ being the viscosity. $j_{V,j}$ and q_j are the species mass and heat fluxes, respectively. Assuming that the Fick contribution dominates the diffusional fluxes, and that the Fourier term dominates the heat flux

$$j_{Vj}(\phi) = -\rho Y_V \left[\frac{D}{Y_V} \frac{\partial Y_V}{\partial x_j} + Y_C \left(Y_V + Y_C \frac{m_V}{m_C} \right) \left[\frac{m_C}{m_V} - 1 \right] \frac{D}{p(\phi)} \frac{\partial p(\phi)}{\partial x_j} \right], \tag{6}$$

$$q_j(\phi) = -\lambda \frac{\partial T(\phi)}{\partial x_j} + (h_V(\phi) - h_C(\phi))\, j_{Vj}(\phi), \tag{7}$$

where D is the species mass diffusivity, m_V and m_C are the molar masses of the vapor and carrier gas, λ is the thermal conductivity and h is the enthalpy $h = e(\phi) + u_i u_i/2$. Particularly, under the assumption of calorically perfect gas $h_C(\phi) = C_{p,C}T, h_V(\phi) = C_{p,V}T + h_V^0$, where $C_{p,C} = C_{p,C}(T^0)$ and $C_{p,V} = C_{p,V}(T^0)$ are the heat capacities at constant pressure, and h_V^0 is the reference vapor enthalpy at (T^0, p^0) obtained from integration or tables, which accounts for the enthalpy difference between the vapor and carrier gas at the reference conditions. Finally, under the perfect gas assumption

$$p(\phi) = \rho R(\phi) T(\phi), \tag{8}$$

where $R(\phi) = Y_V R_V + Y_C R_C$, $R_V = R_u/m_V$, $R_C = R_u/m_C$ where R_u is the universal gas constant.

We also define the Prandtl and Schmidt numbers, $\mathrm{Pr} = \mu C_p/\lambda$ and $Sc = \mu/(\rho D)$.

2.2 Drop equations

To describe in a succinct way the drop conservation equations, we define $Z = \{X_i, v_i, T_d, m_d\}$ as the drop field vector with coordinates representing the position X_i, velocity v_i, temperature T_d, and mass m_d. Thus, the drop evolution equations are

$$\frac{dX_i}{dt}(Z) = v_i, \tag{9}$$

$$\frac{dv_i}{dt}(\psi_f, Z) = \frac{1}{m_d} F_i(\psi_f, Z), \tag{10}$$

$$\frac{dT_d}{dt}(\psi_f, \psi_s, Z) = \frac{1}{m_d C_L}\left[Q(\psi_f, Z) + \dot{m}_d(\psi_f, \psi_s, Z) L_V(Z)\right], \tag{11}$$

$$\frac{dm_d}{dt}(\psi_f, \psi_s, Z) = \dot{m}_d(\psi_f, \psi_s, Z), \tag{12}$$

where F_i is the drag force, Q is the heat flux, $\dot{m}_d$ is the evaporation rate, and C_L is the heat capacity of the drop liquid. L_V is the latent heat of vaporization, which, for calorically perfect gases, is a linear function of temperature, $L_V = h_V^0 - (C_L - C_{p,V})T_d$. Vector $\psi(\phi) = \{u_i, T, Y_V, p\}$ represents the gas-phase primitive variables, evaluated either at the drop surface (subscript s) or at the drop far-field (subscript f). The far-field variables are taken as the gas-phase primitive variables interpolated to the drop locations. The expressions for F_i, Q, and $\dot{m}_d$ include validated correlations for point drops which are based on Stokes drag, with the particle time constant defined by Crowe et al. [7] as $\tau_d = \rho_L d^2/(18\mu)$, where ρ_L is the density of the liquid:

$$F_i(\psi_f, Z) = \frac{m_d}{\tau_d} f_1(u_{i,f} - v_i) \tag{13}$$

$$Q(\psi_f, Z) = \frac{m_d}{\tau_d}\frac{Nu}{3\mathrm{Pr}} C_{p,f} f_2(T_f - T_d) \tag{14}$$

$$\dot{m}_d(\psi_f, Z) = -\frac{m_d}{\tau_d}\frac{Sh}{3Sc}\ln\left[1 + B_M\right], \tag{15}$$

where $m_d = \rho_L \pi d^3/6$. More detail is provided in [3], in particular the expressions for the Nusselt, Nu, and Sherwood, Sh, numbers; Nu is function of Pr and Sh is function of Sc. The mass transfer number is $B_M = (Y_{V,s} - Y_{V,f})/(1 - Y_{V,s})$ (see [8]) where $Y_{V,s}$ is calculated directly from the surface vapor mole fraction, $X_{V,s}$

$$X_{V,s} = \frac{p_{atm}}{p_s}\exp\left[\frac{L_V}{R_V}\left(\frac{1}{T_{B,L}} - \frac{1}{T_d}\right)\right], \qquad Y_{V,s} = \frac{X_{V,s}}{X_{V,s} + (1 - X_{V,s}) m_C/m_V}, \tag{16}$$

where $p_{atm} = 1$ atm, p_s is the saturation pressure and $T_{B,L}$ is the liquid saturation temperature at p_{atm} (i.e. the normal boiling temperature).

2.3 Source terms

As stated above, each drop represents a point source of mass, momentum and energy for the gas phase, with the drop source vector

$$S_d(\psi_f, Z) = \{S_{I,d}, S_{II,i,d}, S_{III,d}\}, \quad (17)$$

$$S_{I,d} = -\dot{m}_d, \quad (18)$$

$$S_{II,i,d} = -[F_i + \dot{m}_d v_i], \quad (19)$$

$$S_{III,d} = -\left[F_i v_i + Q + \dot{m}_d\left(\tfrac{1}{2}v_i v_i + h_{V,s}\right)\right], \quad (20)$$

where $h_{V,s} = C_{p,V}T_d + h_V^0$ is the vapor enthalpy at the drop surface. To construct the Eulerian frame source-terms vector $S(\psi_f, Z) = \{S_I, S_{II,i}, S_{III}, S_I\}$ from the drop sources computed in the Lagrangian frame, $S_d(\psi_f, Z)$, we use

$$S(\psi_f, Z) = \sum_\alpha \frac{w_\alpha}{V}\left[S_d(\psi_f, Z)\right]_\alpha \quad (21)$$

where the summation is over all physical drops α residing within a local numerical discretization volume, V, and the geometrical weighting factor, w_α, is used to distribute the individual drop contributions to the eight nearest neighbor surrounding grid points (i.e. corners of the computational volume V) proportionally to the drop distance from those nodes. Then, to retain numerical stability of the Eulerian gas-phase fields, these source terms are minimally 'smoothed' using a conservative operator. This smoothing is not a filter since it does not remove flow scales, but it is required for successful simulations given the 'spottiness' of the source terms.

3 Large Eddy Simulation equations

Because the intent of LES is to reduce the computational cost, the focus is here on finding a strategy to diminish it by modeling both the gas phase and the drops.

For the gas phase, we filter the equations and remove the small-scales which are no longer computed in LES, but instead are modeled through SGS models. For the liquid phase, we construct from the physical drop field, Z, a reduced, computational, drop field, $\overline{Z}$.

3.1 Filtered gas-phase equations

The filtering operation is defined as:

$$\bar{\psi}(\mathbf{x}) = \int_{V_f} \psi(\mathbf{y})\, G(\mathbf{x}-\mathbf{y})\, d\mathbf{y} \quad (22)$$

where V_f is the filtering volume and G is the filter function. For finite-difference computations, the filter of choice is a top-hat filter which leads to $\bar{\psi}$ being the volume-average. The Favre (density-weighted) filtering is defined as $\tilde{\psi} = \overline{\rho\psi}/\bar{\rho}$. To derive the equations, it is assumed that filtering and differentiation commute, which is correct except near boundaries.

The succinct form of the LES equations is

$$\frac{\partial \overline{\phi}}{\partial t} + \frac{\partial (\widetilde{u}_j \overline{\phi})}{\partial x_j} = \frac{\partial \theta_j(\overline{\phi})}{\partial x_j} + \overline{S} + \frac{\partial \theta_{SGS,j}(\overline{\phi})}{\partial x_j} \tag{23}$$

where

$$\theta_{SGS,j}(\overline{\phi}) = \{0, -\overline{\rho}\tau_{ij}, -\overline{\rho}(\zeta_j + \tau_{ij}\widetilde{u}_i), -\overline{\rho}\eta_j\} \tag{24}$$

$$\overline{S} = \{\overline{S}_I, \overline{S}_{II,i}, \overline{S}_{III}, \overline{S}_I\}, \tag{25}$$

the vector $\overline{S}$ representing the filtered source terms (FSTs). In equation (24), the SGS-flux terms are

$$\tau_{ij} = \widetilde{u_i u_j} - \tilde{u}_i \tilde{u}_j, \quad \zeta_j = \widetilde{h u_j} - \tilde{h}\tilde{u}_j, \quad \eta_j = \widetilde{Y_V u_j} - \tilde{Y}_V \tilde{u}_j, \tag{26}$$

and the correct interpretation of $\overline{S}$ is obtained by considering a drop located at $\mathbf{X}$ within the filtering volume V_f and its contribution within that volume

$$\bar{S}(\mathbf{x}) = \int_{V_f} S_d \delta(\mathbf{y} - \mathbf{X}) G(\mathbf{x} - \mathbf{y})\, d\mathbf{y}, \tag{27}$$

where $S_d \delta(\mathbf{y} - \mathbf{X})$ is the point-source contribution from the drop and δ is the delta function. When G is a top-hat filter

$$\bar{S}(\psi_f, Z) = \frac{1}{V_f} \sum_{\beta} \left[S_d(\psi_f, Z) \right]_{\beta}, \tag{28}$$

which represents a volume-average over the drops β within the filtering volume, where S_d was defined in equation (17).

Equation (23) takes into account the finding that the 'LES assumptions' (see [3]) are negligible in the LES equations. Thus, the only terms to be modeled are the SGS fluxes and the FSTs. For the SGS-flux terms listed in equation (26), we consider three possible models, namely, the Smagorinsky (SM) model ([11]), the Gradient (GR) model ([12]) and the Scale-Similarity (SS) model ([13]), each of which can be used either in a constant-coefficient or a dynamic mode. The description of the FST models is presented next.

3.2 Drop-equations model for LES utilization

The equations for the computational drop field are the same as those for the physical drops, namely eqs. 9 - 13, except that Z is replaced by the modeled drop field $\bar{Z}$.

3.3 Source terms in LES

We define the constant integer N_R as the ratio of the physical drops to the computational drops. Thus, the model (subscript m) of the source terms for the computational drop field is

$$\bar{S}_m\left(\psi_{f,m}\left(\bar{\phi}\right),\bar{Z}\right) = N_R \sum_{\beta=1}^{N_\beta} \frac{1}{V_f}\left[S_d\left(\psi_{f,m}\left(\bar{\phi}\right),\bar{Z}\left(N_R\right)\right)\right]_\beta, \tag{29}$$

where the summation is over the N_β computational drops within V_f, and S_d has the functional form of equation (17) with $\psi_{f,m}$ denoting the model for ψ_f. If the model for the unfiltered flow field at the drop location is that of the filtered flow field, $\psi_{f,m} = \tilde{\psi}_f$, as it is in the present study, then the model for computing the source terms simply becomes

$$\bar{S}\left(\tilde{\psi}_f,\bar{Z}\right) = N_R \sum_{\beta=1}^{N_\beta} \frac{1}{V_f}\left[S_d\left(\tilde{\psi}_f,\bar{Z}\left(N_R\right)\right)\right]_\beta, \tag{30}$$

and the only parameter is N_R. Selecting the N_R value involves a decision regarding the favorable decrease in computational time balanced against an unfavorable decrease in accuracy, when N_R is increased.

4 Results

The goal of LES modeling is to reproduce an accurate picture of the two-phase flow field using reduced computational resources compared to those used for DNS, when DNS is feasible; if accuracy is reached, there is justification to postulate that LES may also be accurate in high Reynolds number regimes where DNS is not feasible, although this must always be verified by comparing with experiments. Since the SGS-flux and FST models and their interaction with the resolved scales is what distinguishes LES from DNS, one must first understand how such models help achieve accuracy. That is, one must examine the DNS database and gain insights on how can the SGS-flux and FST models reproduce the governing effects in DNS.

The DNS has been performed for the configuration of a temporal mixing layer shown in figure 1. Details on the initial and boundary conditions are in Okong'o

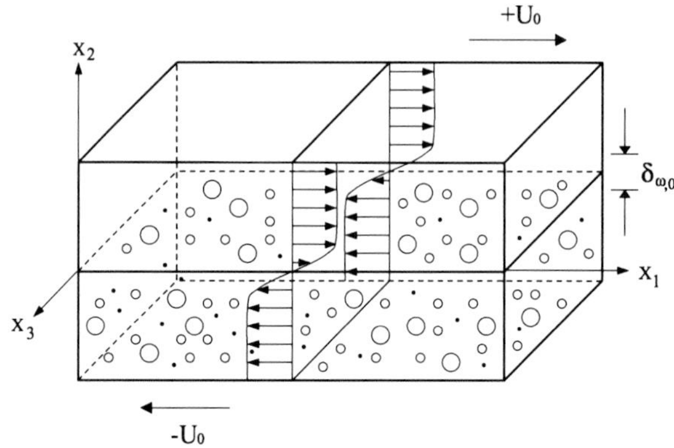

Fig. 1 Sketch of the configuration from [3].

Run	ML_0	Number of drops	t^*_{tr}	$\mathrm{Re}_{m,tr}$
TP600a2	0.2	2.99×10^6	105	1576
TP600a5	0.5	7.48×10^6	105	1568

$\mathrm{Re}_0 = 600$, $\Delta U_0 = 271.7$m/s, $\delta_{\omega,0} = 6.859\times10^{-3}$m,
Domain size ($L_1 \times L_2 \times L_3$) 0.2m×0.22m×0.12m; 16.22×10^6 grid nodes,
Carrier gas (air): $M_{c,0} = 0.35$; $T_0 = 375$K; $\rho_0 = 0.9415$kg/m^3,
μ calculated from Re_0; $\mathrm{Pr} = Sc = 0.697$,
Drops (n-decane): $T_{d,0} = 345$K; $\rho_L = 642$kg/m^3,
St_0 has a Gaussian (3, 0.5) distribution.

Table 1 List of initial conditions for the database analyzed. Re_m is the momentum thickness based Reynolds number defined in [3].

and Bellan [3]; some values are provided in table 1, where $\mathrm{Re}_0 \equiv \rho_0 \Delta U_0 \delta_{\omega,0}/\mu$ is the initial vorticity-thickness ($\delta_{\omega,0}$) based Reynolds number, $M_{c,0}$ is the initial Mach number, and $St_0 = \tau_d \Delta U_0/\delta_{\omega,0}$ is the initial Stokes number. The computation is initiated with four spanwise vortices hosted in the layer; a vorticity perturbation is imposed on the initial field to accelerate double-vortex pairing, leading to the formation of an ultimate vortex in which small scales proliferate. The boundary conditions for the gas are periodic in the streamwise (x_1) and spanwise (x_3) directions, and reflecting in the cross-stream (x_2) direction. In LES, we have chosen grid spacings that credibly reproduced the initial conditions and have used the same numerical method and boundary conditions as in DNS so as to be able to assign success, or lack thereof, uniquely to the SGS-flux and FST models.

4.1 Insights from the DNS results

We use here only a subset of the DNS computational runs, all of which are listed in Okong'o and Bellan [3]. The complete set includes runs performed for two initial

Reynolds number values, $\mathrm{Re}_0 = 500$ and 600, and three initial mass loading values, $ML_0 = 0, 0.2$ and 0.5, where the mass loading is the initial ratio of liquid mass to mass of carrier gas in the drop-laden part of domain. The initial drop temperature is smaller than that of the carrier gas (air) and than the boiling point of the liquid, so that the drops heat up and evaporate. All runs were conducted up to a transitional state identified as occurring when the energy spectra based on gas-velocity fluctuations are smooth (except for the peak at the forcing frequency, used to perturb the layer in order to accelerate vortex pairing). Time is tracked in non-dimensional units, $t^* = t\Delta U_0/\delta_{\omega,0}$, where ΔU_0 is the initial velocity difference between the two streams of the layer (see figure 1); the transitional time is indexed by subscript tr. Our goal in examining the DNS database is to elucidate the hierarchy of phenomena, the result of which will indicate where should the emphasis be for reproducing the dissipation (i.e. the unresolved small-scale effects) in LES.

To prioritize the allocation of resources, it is important to understand which are the modeling aspects that are most responsible for the features of the flow field. This understanding can be achieved in many ways (see [3]), but we choose to report here on one of the less conventional ways: an analysis of irreversible entropy production.

The entropy equation is

$$\frac{\partial(\rho s)}{\partial t} + \frac{\partial(\rho s u_j)}{\partial x_j} = -\frac{\partial \Sigma_j}{\partial x_j} + g \tag{31}$$

where s is the entropy, Σ_j represents the flux of reversible entropy and g is the rate of irreversible entropy production

$$\Sigma_j = \frac{1}{T}\left(q_j - n_C V_{Cj}\mu_C - n_V V_{Vj}\mu_V\right) \tag{32}$$

$$g = g_{III} + g_{II} + g_{I,kine} + g_{I,chpot} + g_{visc} + g_{temp} + g_{mass}, \tag{33}$$

where n_C and n_V are molar densities, V_{Cj} and V_{Vj} are the carrier-gas and vapor diffusion velocities, and μ_C and μ_V are the chemical potentials associated with carrier gas and vapor. Also,

$$g_{III} = \frac{S_{III}}{T}, \quad g_{II} = -\frac{u_i S_{II,i}}{T}, \quad g_{I,kine} = \frac{\frac{1}{2}u_i u_i S_I}{T}, \quad g_{I,chpot} = -\frac{\mu_V S_I}{T}, \tag{34}$$

$$g_{visc} = \frac{\mu}{T}\left(2S_{ij}S_{ij} - \frac{2}{3}S_{kk}S_{ll}\right) = \frac{2\mu}{T}\left(S_{ij} - \frac{1}{3}S_{kk}\delta_{ij}\right)\left(S_{ij} - \frac{1}{3}S_{ll}\delta_{ij}\right), \tag{35}$$

$$g_{temp} = \frac{\lambda}{T^2}\frac{\partial T}{\partial x_j}\frac{\partial T}{\partial x_j}, \quad g_{mass} = \frac{R_C R_V}{Y_C Y_V\left(R_V Y_V + R_C Y_C\right)}\frac{j_{Vj} j_{Vj}}{\rho D}. \tag{36}$$

The quantity g is by definition the dissipation ([10]), and it is this quantity that must be reproduced by subgrid-scale modeling. For single-phase flow, g is only represented by the last three terms of equation (33) and it is positive or null. For two-phase flows, g may be positive, null or negative, depending on the contribution of the first four terms in equation (33) representing the source terms. We wish to evaluate the ranking of the seven terms in equation (33), both on the average and rms basis. The average is performed over the entire domain at t_{tr}^* and encompasses both $ML_0 = 0.2$ and 0.5 databases. The results show that for the average

$$\underbrace{g_{III}}_{>0} \underset{\|\cdot\|\times 2}{>} \underbrace{g_{I,chpot}}_{<0} \underset{\|\cdot\|\times 4}{>} g_{visc} \underset{\times 10}{\gg} (g_{mass}, \underbrace{g_{I,kine}}_{>0}, g_{temp}) \underset{\|\cdot\|\times 10}{\gg} \underbrace{g_{II}}_{\substack{<0 \text{ at } ML_0=0.2 \\ >0 \text{ at } ML_0=0.5}} \tag{37}$$

and for the rms

$$g_{III} \underset{\times 2}{>} g_{I,chpot} \underset{\times(4\text{ to }7)}{\gg} g_{visc} \begin{array}{c} > \text{at } ML_0 = 0.2 \\ < \text{at } ML_0 = 0.5 \end{array} g_{II} \underset{\times 6}{\gg} g_{mass} \underset{\times 3}{>} (g_{I,kine}, g_{temp}). \tag{38}$$

Clearly, g_{III} and $g_{I,chpot}$ are the largest contributions, and therefore it is evident that drop evaporation and the energy source/sink resulting from this process are the major phenomena which must receive attention when performing subgrid-scale modeling. Thus, the modeling of source terms in LES, described in section 3.3, is crucial for success in LES. Moreover, a similar analysis performed at the LES and SGS scales (see [3]), for two different filter widths, shows that independent of filter size, the g rms increases with increasing ML_0. Also, while at the LES scale the average of g has same variation with ML_0 as in the DNS, at the SGS scale, the average has opposite variation with ML_0 compared to DNS or LES. When restricting the analysis to g_{III} and $g_{I,chpot}$ only, at the LES scale the average has same sign as DNS, however, at the SGS scale the average has opposite sign to DNS or LES. Thus, the different SGS-scale behavior from that at the DNS and LES scales must be captured.

It also must be emphasized that in order to capture the drop distribution, which is highly non-uniform as experimentally observed by Eaton and Fessler [9], the correct prediction of the vortical flow aspects is essential since drops accumulate in regions of high strain and low vorticity.

4.2 LES results

The models' ultimate performance can only be assessed in LES, as it substantially depends on the interaction between resolved and unresolved scales; the template for examining the models' performance is here the filtered-and-coarsened DNS (FC-DNS), where the coarsening reproduces the LES grid. Since the focus is here on two-phase flows, the interest is on duplicating in LES the drop number density and vapor mass fraction, although other aspects of the flow will be discussed as well. We wish to examine both the effect of the SGS-flux model and that of the drop field model through the modeling of the FSTs.

Effect of the SGS-flux model

The effect of the SGS-flux model used in LES is illustrated in figure 2 for the time-wise evolution of selected quantities depicting important characteristics of the flow.

To understand the effect of flow filtering, in figure 2 we show additional results from DNS. To complement these results, figure 3 depicts a visualization of the drop number density field at t_{tr}^* for the FC-DNS template and LES simulations using the dynamic Smagorinsky (SMD), dynamic Gradient (GRD) and constant-coefficient Scale-Similarity (SSC) models.

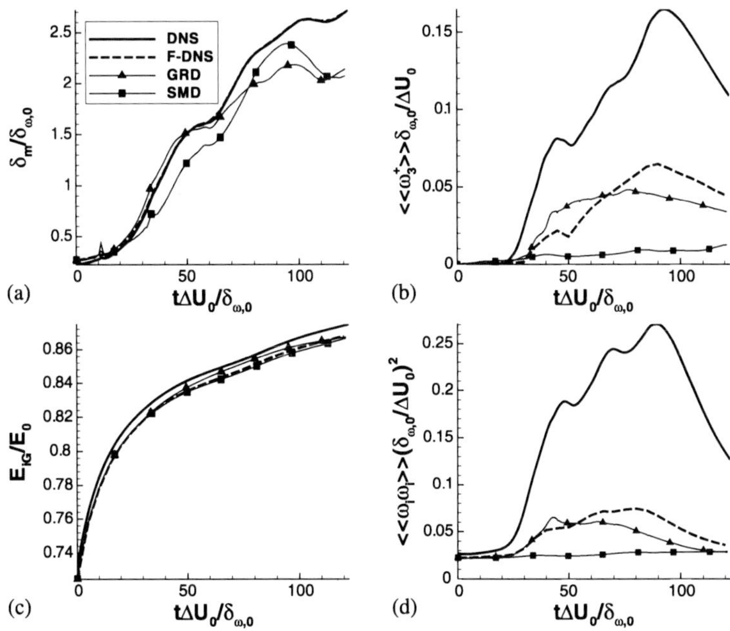

Fig. 2 Time evolution of the mixing layer TP600a2 in FC-DNS and LES conducted with several dynamic SGS-flux models and $N_R = 8$: (a) momentum thickness, (b) positive spanwise vorticity, (c) internal energy and (d) enstrophy.

Figure 2a shows that filtering has no impact on the layer growth represented by the momentum thickness, $\delta_m/\delta_{\omega,0}$, evolution; indeed, the curves are virtually indistinguishable. Up to the first pairing, manifested by a plateau in $\delta_m/\delta_{\omega,0}$, the GRD simulation only slightly overestimates the template, while the LES conducted with SMD substantially underestimates it. However, approximately half-way through the second pairing, the LES with the SMD model shifts closer to the template, whereas the GRD model considerably underpredicts the layer growth. The non-dimensionalized internal energy of the flow, E_{iG}/E_0, experiences a definite reduction when the flow is filtered, as shown in figure 2c, and the GRD model overestimates its evolution while the SMD model slightly underestimates it. The most dramatic effect of flow filtering is apparent in figures 2b and 2d portraying the non-dimensional positive spanwise vorticity, $\langle\langle \omega_3^+ \rangle\rangle \delta_{\omega,0}/\Delta U_0$, and the non-dimensional enstrophy, $\langle\langle \omega_i \omega_i \rangle\rangle (\delta_{\omega,0}/\Delta U_0)^2$, where ω is the vorticity vector and $\langle\langle\rangle\rangle$ denotes domain averaging. Initially, ω_3 is negative, and thus formation of positive ω_3 is an indication of small-scale formation. Filtering the flow clearly removes

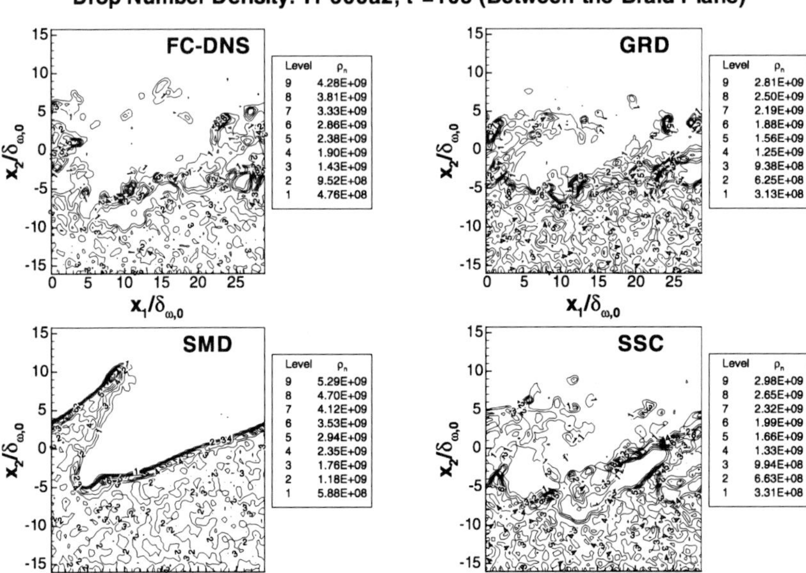

Fig. 3 Comparison in the prediction of the drop number density (m^{-3}) for TP600a2 at $t^*_{tr} = 105$. The between-the-braid plane is at $x_3/L_3 = 0.5$. In all LES, $N_R = 8$.

the majority of positive spanwise vorticity, and it is this reduced field which represents the template for LES. The LES simulation using GRD overestimates this template up to the beginning of the second vortex pairing, and thereafter underestimates it. The LES simulation utilizing the SMD model substantially underestimates $\langle\langle \omega_3^+ \rangle\rangle\, \delta_{\omega,0}/\Delta U_0$ throughout the layer evolution. A similar outcome of LES with these two SGS-flux models is evident even more dramatically when assessing the $\langle\langle \omega_i \omega_i \rangle\rangle\, (\delta_{\omega,0}/\Delta U_0)^2$ evolution; the SMD model is incapable of generating enstrophy which is a manifestation of the stretching and tilting, an important mechanism for turbulence production. Further inquiry into the shortcomings of the SMD model were conducted by Leboissetier et al. [4] who traced it through energy spectra to the overwhelmingly dissipative aspect of the SMD model. This aspect is also apparent in the visualizations presented in figure 3 where the LES conducted with GRD and SSC were able to reproduce small-scale structures of the drop number density field, with drop accumulation in various regions of the flow, but the SMD LES incorrectly predicts that drops accumulate only at the edge of a smooth vortex.

Effect of filtered source terms model

To assess the effect of the FST model, we focus on the reduction from the physical drop field to the computational drop field; no examination is here made of the impact of the reconstruction model for the gas field at drops' locations (for such a study, see [3]). Although not illustrated for the sake of brevity, both the drop number density

Simulation	Grid	N_d	N_R	CPU-hours
DNS	288×320×176	2,993,630	-	2252
LES (GRD)	72×80×44	2,993,630	1	113
LES (GRD)	72×80×44	374,203	8	20
LES (GRD)	72×80×44	181,101	16	12
LES (GRD)	72×80×44	93,550	32	9
LES (GRD)	72×80×44	46,775	64	8

Table 2 CPU hours for various simulations when ML_0=0.2. N_d is the number of drops. The simulations were performed on an Origin 2000 supercomputer.

and the vapor mass fraction as obtained from the FC-DNS, and LES conducted with the GRD and $N_R = 16, 32$ and 64 have been examined. Reducing the physical drop field by a factor of 16 results in a reduction of the structure of the lower-stream drop number density, however, the structure in the mixing layer is reminiscent of that of the FC-DNS. Most important, the Y_V field with $N_R = 16$ gives a reasonably good representation of that obtained in the FC-DNS. This representation deteriorates as N_R increases to 32 and further to 64, as does the reproduction of the drop number density. Ultimately, the user of a LES model must choose between computational cost and accuracy of the computation.

A comparison between computational time for various simulations is listed in table 2. As can be seen, most of the reduction in computational time, a factor of 20, is due to the coarser LES grid. When reducing the physical drop field by a factor of 8 to a computation drop field, only a factor of 5.65 reduction in computational time is achieved. A further reduction by a factor of 2 ($N_R = 16$) in the drop field only leads to a modest additional reduction by a factor of 1.67 in computational time. Further reduction in the drop field by a factor of 2 seems unwarranted as the fidelity of the results is considerably diminished while the computational time is only very slightly reduced, by a factor of 1.33. We can thus provisionally conclude (i.e. additional studies are needed) that the computational benefits of the strategy of performing LES with computational drops reaches an asymptote beyond $N_R = 16$.

5 Summary and conclusions

A Direct Numerical Simulation (DNS) database study was conducted for the purpose of determining which aspects should be emphasized in modeling Large Eddy Simulation (LES) subgrid effects, so as to achieve a high-fidelity representation of drop-laden flow with evaporation. The study identified the modeling of the source terms in the gas governing equations, representing the interaction of gas and drops, as being crucial to model accuracy, as well as the vortical features of the flow. Further, several LES computations were analyzed to examine both the influence of the subgrid-scale-flux models, the role of which is to reproduce the vortical features of the flow, and the impact of the model for the source terms from the standpoint of the physical drop field reduction to a computational drop field. Comparisons of these simulations were performed with the filtered-and-coarsened DNS (FC-DNS) which represents the template for LES. The results show that the very popular Smagorinsky model has poor capability in reproducing the vortical features of the flow, but that the Gradient and Scale-Similarity models each leads to a reasonable approximation of the FC-DNS flow field. By reducing the physical drop field by factors of 8, 16, 32 and 64, it was determined that the computational efficiency benefits reached an asymptote beyond the reduction by a factor of 16, indicating that the further loss in accuracy may not be balanced by computational savings.

Acknowledgements This work was conducted at the Jet Propulsion Laboratory (JPL) of the California Institute of Technology, and was sponsored by the U. S. Department of Energy and the U. S. Air Force Office of Scientific Research under an agreement with the National Aeronautics and Space Administration. Computations were performed on the SGI Origin2000 at the JPL Supercomputing Center.

References

1. Boivin, M., Simonin, O. and Squires, K. 1998 Direct numerical simulation of turbulence modulation by particles in isotropic turbulence. *J. Fluid Mech.* **375**, 235–263.
2. Pope, S. B. 2004 Ten questions concerning the large-eddy simulation of turbulent flows, *New Journal of Physics* **6**, 35–59.
3. Okong'o, N. and Bellan, J. 2004 Consistent large eddy simulation of a temporal mixing layer laden with evaporating drops. Part 1: Direct numerical simulation, formulation and *a priori* analysis. *J. Fluid Mech.* **499**, 1–47.
4. Leboissetier, A., Okong'o, N. A. and Bellan, J. 2005 Consistent Large Eddy Simulation of a temporal mixing layer laden with evaporating drops. Part 2: *A posteriori* modeling. *J. Fluid Mech.*, **523**, 37–78.
5. Selle, L. C. and Bellan, J. 2007 Characteristics of transitional multicomponent gaseous and drop-laden mixing layers from Direct Numerical Simulation: Composition effects. *Phys. Fluids* **19**(6), doi: 10.1063/1.2734997, 063301-1-33.
6. Bellan, J. and Selle, L. C. 2009 Large Eddy Simulation composition equations for single-phase and two-phase fully multicomponent flows. *Proc. Combust. Inst.* **32**, 2239–2246.
7. Crowe, C., Chung, J. and Troutt, T. 1998 Particle mixing in free shear flows. *Progress in Energy and Combustion Science* **14**, 171–194.
8. Williams, F. 1965 *Combustion Theory* Addison-Wesley.
9. Eaton J. K. and Fessler J. R. 1994 Preferential concentration of particles by turbulence. *Int J Multiphase Flow* **20**, 169–209.
10. Hirshfelder, J., Curtis, C. and Bird, R. 1954 *Molecular Theory of Gases and Liquids*. John Wiley and Sons.
11. Smagorinksy, J. 1993 Some historical remarks on the use of nonlinear viscosities. In *Large Eddy Simulation of Complex Engineering and Geophysical Flows* (eds. B. Galperin & S. Orszag), chap. 1, pp. 3–36. Cambridge University Press.
12. Clark, R., Ferziger, J. and Reynolds, W. 1979 Evaluation of subgrid-scale models using an accurately simulated turbulent flow. *Journal of Fluid Mechanics* **91**(1), 1–16.
13. Bardina, J., Ferziger, J. and Reynolds, W. 1980 Improved subgrid scale models for large eddy simulation. *Tech. Rep.* 80–1357. AIAA.

Influence of Particle-Wall Interaction Modeling on Particle Dynamics in Near-Wall Regions of Turbulent Channel Down-Flow

A. Kubik and L. Kleiser

1 Background and Methodology

Particle-laden channel flows are found in a variety of natural settings and are also of great practical importance due to their frequent occurrence in engineering applications. Numerous experimental and computational studies related to particle transport in turbulent flows were reported in the literature. There have been a few applications of Direct Numerical Simulations to particle-laden channel flows, such as in [7] and [9]. Those studies necessarily focus on rather low Reynolds numbers. Kulick *et al.* [6] reported results of experiments on particle-laden flows in a vertical channel down-flow at $Re_\tau \approx 644$ (based on friction velocity and channel half-width). Large-Eddy Simulation results for a channel flow at approximately the same Reynolds number are also available, e.g. [4].

In the present study, a vertical turbulent channel flow (with gravity pointing in the mean flow direction) at Reynolds number $Re_\tau \approx 210$ is investigated by means of DNS. The aim is to study the influence of wall-particle interactions on the particle statistics.

The Eulerian-Lagrangian approach was adopted for the calculations in which the fluid phase is described by 3D time-dependent Navier-Stokes equations whereas the particles are tracked individually. The two phases are coupled, as the fluid phase exerts forces on the particles and experiences a feedback force from the dispersed phase. The feedback force is added as an effective body force to the Navier-Stokes equations for the fluid phase. The equations are solved together with the incompressibility constraint using a spectral–spectral-element Fourier–Legendre code [12] with periodic boundary conditions in the wall-parallel directions and no-slip condition at the channel walls. The coordinates in the channel are labeled as x for the streamwise, y for the wall-normal and z for the spanwise direction.

A. Kubik and L. Kleiser
Institute of Fluid Dynamics, ETH Zurich, Switzerland, e-mail: kubik@ifd.mavt.ethz.ch

H. Kuerten et al. (eds.), *Direct and Large-Eddy Simulation VIII*,
ERCOFTAC Series 15, DOI 10.1007/978-94-007-2482-2_26,

The trajectories of the particles are calculated simultaneously in time with the fluid phase by integrating the equation of motion for each particle. Maxey and Riley provided a modified Basset-Boussinesq-Oseen (BBO) equation for describing the particle motion. The lift force [8] is used as an *ad hoc* added component of the total hydrodynamic force. The BBO equation for particles in channel flow can be simplified to include only drag and gravity; lift force, however, becomes significant close to the walls [5]. Empirical and analytical corrections for the drag and lift were necessary to accommodate for moderate Reynolds numbers [2], [11] and the proximity of walls [3], [1], [11]. Particle-wall collisions are modeled taking into account the elasticity of the impact and particle deposition for low-velocity particles in regions of low shear [5]. Particle-particle collisions are omitted in this study and the parameter range for the calculations is restricted such as to keep this assumption valid.

2 Results and Discussion

The experiment by Kulick *et al.* [6] was often used as a benchmark for particle-laden channel flows due to detailed data it provides. It will be consulted here to study the influence of modeling on accumulation of particles near the wall and the importance of wall-particle interactions. Although the discrepancy in the Reynolds numbers between the experiment and the present DNS is significant, comparisons in near-wall regions are still possible if the scaling of the particle parameters is appropriate.

For the simulations, copper particles with a diameter of 70 μm, corresponding to a Stokes relaxation time τ_p^+ of 2000 wall units, were chosen. The lowest mass loading (i.e. particle-to-fluid mass ratio) in the experiments is $\phi = 0.02$. Particle-laden flow with this parameters is assumed to be dilute enough to neglect inter-particle collisions even in the low-speed streaks where particles tend to accumulate. (The mass loading in near-wall regions is approximately six times higher than the average.) Different cases were examined employing different models and parameters for particle-wall impact: restitution coefficient e, dynamic friction μ_d, Hamaker constant A for copper particles and a glass wall. Case C1 describes a model for a slightly non-elastic, sliding collisions with no deposition ($e = 0.97$, $\mu_d = 0.65$ and $A = 0\ J$); C2 ($e = 0$, μ_d not applicable, $A = 18 \cdot 10^{-20}\ J$) characterizes the case where all particles deposit upon impact and C3 is a synthesis of the preceding cases ($e = 0.97$, $\mu_d = 0.65$ and $A = 18 \cdot 10^{-20}\ J$).

Figures 1*(a)–(c)* show the mean and r.m.s. values of particle velocities. For all cases the mean particle velocity u_x^{p+} is in good agreement with the experiments in the channel center. However, the abrupt increase of u_x^{p+} in the wall-near region for $y^+ > 10$ in the experimental data was not reproduced in any of the present simulations although the tendency to an increase of the particle velocity close to the walls can be seen. The level of streamwise particle velocity fluctuations $u_{x,rms}^p / u_\tau$ shows similar values as those of the experimental data in the $30 < y^+ < 100$ region in all

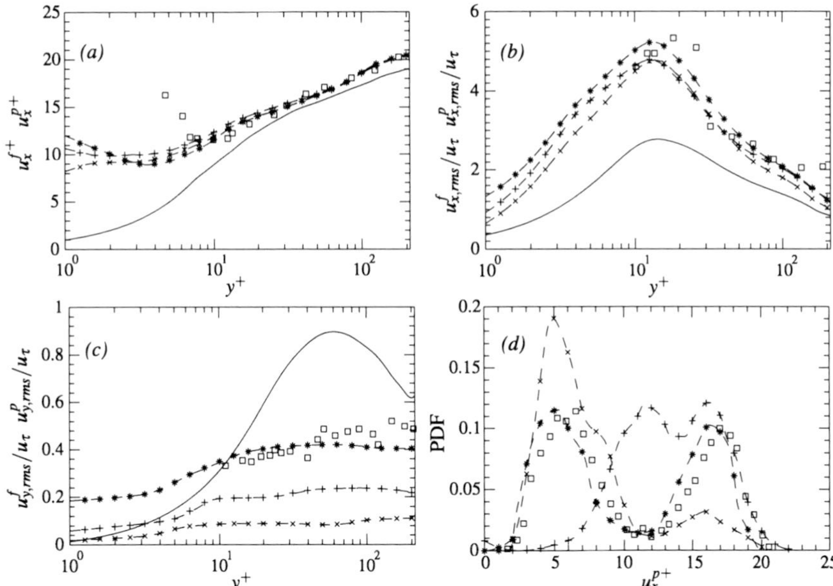

Fig. 1 Particle velocities: *(a)* mean velocity; *(b)* streamwise velocity fluctuations *(c)* wall-normal velocity fluctuations *(d)* PDF of streamwise particle velocity at $y^+ = 12$ plane. solid: fluid flow; $-+-$: C1; $-\times-$: C2; $-*-$: C3; squares: experiment [6].

cases. The profiles in cases C1 and C2 are similar and have a local maximum value slightly lower as in the case of the experimental data. The maximal value in case C3 is almost identical to the measurements [6]. The spatial location of the maximum is lower in all simulations, i.e. closer to the wall, than in [6]. However, all cases predict significantly lower values in the center region. The level of the wall-normal particle velocity fluctuations $u^p_{y,rms}/u_\tau$ in case C1 is almost half of that in the experiment and falls significantly in case C2 while good agreement is exhibited in case C3.

The probability distribution function (PDF) of the streamwise particle velocity around the $y^+ = 12$ plane is shown in fig. 1*(d)*. A clear bimodal distribution similar to the experimental data is only observed in Case C3. The profile for case C2 also shows a weak bi-modality. Assuming that these two modes can be separated at $u^{p+} = 12$, the mean value of the lower and the higher mode, u^{p+}_L and u^{p+}_H

$$u^{p+}_L = \frac{1}{P_L}\int_{-\infty}^{12} u^{p+}P(u^{p+})\mathrm{d}u^{p+} \quad \text{and} \quad u^{p+}_H = \frac{1}{P_H}\int_{12}^{+\infty} u^{p+}P(u^{p+})\mathrm{d}u^{p+} \qquad (1)$$

and the probability of each mode, P_L and P_H

$$P_L = \int_{-\infty}^{12} P(u^{p+})\mathrm{d}u^{p+} \quad \text{and} \quad P_H = \int_{12}^{+\infty} P(u^{p+})\mathrm{d}u^{p+} \qquad (2)$$

can be calculated using the data in fig. 1*(d)*, as shown in table 1. It was found that the mean values and the probabilities of modes calculated from the simulation data are in good agreement with the values of the experimental data when the wall potential was taken into consideration (case C2). A further improvement was achieved when the models for particle reflection and particle deposition were combined (case C3).

Table 1 Mean velocities and probabilities for lower and higher modes.

	u_L^{p+}	u_H^{p+}	P_L	P_H	P_H/P_L
C2	6.75	15.33	0.79	0.21	0.27
C3	5.14	15.80	0.58	0.42	0.72
Experimental data	6.26	16.52	0.57	0.43	0.75

The cause of the occurrence of this bi-modality in cases C2 and C3 can also be investigated by studying a snapshot of correlations of (u_x^{p+}, u_y^{p+}) taken in the region $10 < y^+ < 20$, which is shown in fig. 2. Different groups of particles can be identified here. The first group consists of particles with high values of u_x^{p+} and negative values u_y^{p+}. These particles are those coming from the center of the channel and moving toward the wall. The second group consists of the particles with intermediate velocity and almost zero wall-normal velocity. In case C1, another family of high velocity particles can be identified, this time with a positive value u_y^{p+}, which corresponds to particles leaving the near-wall region after an impact at the wall. In case C2, this last group of particles moving away from the wall have a low streamwise velocity, typically less than six, i.e. high-velocity particles moving from the wall are missing. In case C3 both scenarios are united: low-velocity particles are retained at the wall and high velocity particles are reflected from the wall after the impact.

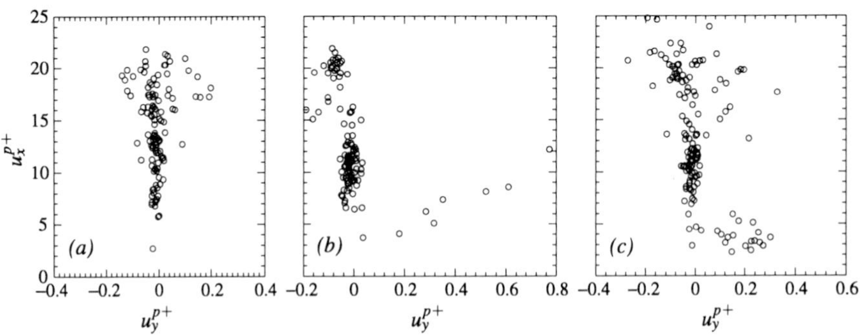

Fig. 2 Phase space distribution of particle velocity at $y^+ = 12$. *(a)* C1; *(b)* C2 *(c)* C3.

Given the information above, the presence of a bimodal probability distribution can be explained as follows. Assuming purely elastic collisions and long relaxation time, particles entering the near-wall region are reflected at the wall and leave this region after an impact. A particle loses part of its energy only due to viscous effects. How large this loss will be depends on the relaxation time of the particle and the direction of its motion relative to the wall. In such case, particles of different streamwise velocities would be present in the near-wall region with a mono-modal probability distribution function. On the other hand, if there is any mechanism that can prevent all or some of the particles to move freely away from the wall, a cloud of low-velocity particles will be formed in the vicinity of the wall. Some of these particles will be injected into the flow. This, in fact, happens when a correction for drag and lift forces is applied: particles are decelerated strongly due to aerodynamic interaction with the wall (e.g. the displacement of the fluid between the particle and wall). This inter-relation depends on the particle's residence time in the viscous sublayer and becomes more pronounced for slow particles. This creates a group of particles with a probability distribution function strongly dependent on the conditions at the wall. For the latter flow case, with certain combination of parameters, the two families of particles, i.e., the one with low streamwise velocity created near the wall and the one with high streamwise velocity coming from the outer region, can co-exist at a certain distance to the wall. If their mean directional velocity is sufficiently different, a bimodal probability distribution function will be observed. It should be mentioned that the above explanation of the cause of bi-modality of the probability distribution function is a detailed version of that given by Kulick *et al.* [6] with a slight modification. Here, the particle-wall interaction mechanism is subdivided into individual effects which are manipulated independently, which is of course not possible in a physical experiment. Finally, the creation of the low-velocity cloud of particles near the wall gives a clear contribution to non-directional velocities which can be observed in the whole channel. As a result, sufficiently higher r.m.s. levels in the wall-normal component of the particle velocity were predicted when an attractive force was present near the wall, in good agreement with experimental data.

In order to test the propositions made above, case C1 was re-computed without the near-wall drag correction. Figure 3*(a)* and *(b)* show the wall-normal particle velocity fluctuations and PDF of streamwise particle velocity with and without near-wall drag correction. The r.m.s. levels of wall-normal particle velocity seem not to be overly sensitive to the choice of formula of near-wall drag correction. The PDF of streamwise particle velocity, however, shows a clear dependence on the model. Without the wall-correction factor, the probability distribution is mono-modal, as expected.

Fukagata *et al.* [4] showed that in addition to the hydrodynamic effects inter-particle collisions play a decisive role in the mechanism to trap particles near the wall. Even for very low mass loadings the particle-particle collisions partially convert the streamwise to wall-normal momentum. There are other parameters, such as wall roughness which have an influence on turbulence as shown by Vreman [10]. The conclusions from this study cannot be definite without an inquiry into these effects.

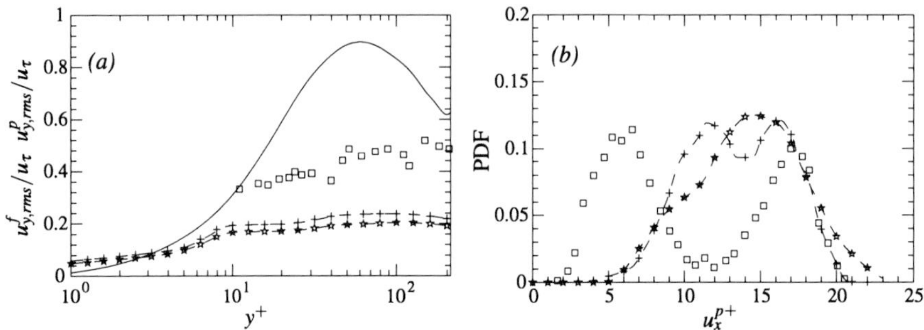

Fig. 3 Wall-normal particle velocity fluctuations *(a)* and PDF of streamwise particle velocity at $y^+ = 12$ plane *(b)*. solid: fluid flow; $-+-$: C1 with correction factor for near-wall forces; $-\times-$: C1 without correction factor; squares: experiment [6].

3 Conclusions

In this study a vertical turbulent channel down-flow at Reynolds number $Re_\tau \approx 210$ was investigated. An excellent agreement between simulation results and experimental findings was achieved when models for particle reflection and particle deposition were combined.

References

1. Brenner, H.: The slow motion of a sphere through a viscous fluid towards a plane surface. Chemical Engineering Science **16**, 242–251 (1961)
2. Clift, R., Grace, J. R., Weber, M. E.: Bubbles, Drops, and Particles. Academic Press (1978)
3. Faxén, H.: Die Bewegung einer starren Kugel längs der Achse eines mit zäher Flüssigkeit gefüllten Rohres. Arkiv Mat. Astron. Fys. **17**(27), 1–28 (1923)
4. Fukagata, K., Zahrai, S., Kondo, S., Bark, F.H.: Anomalous velocity fluctuations in particulate turbulent channel flow. Int. J. Multiphase Flow **27**, 701–719 (2001)
5. Kubik, A.: Numerical simulation of particle-laden, wall-bounded attached and separated flows. ETH Dissertation No. **17205**, Zurich (2007)
6. Kulick, J.D., Fessler, J.R., Eaton, J.K.: Particle response and modification in fully turbulent channel flow. J. Fluid Mech. **277**, 109–134 (1994)
7. Li, Y., McLaughlin, J.B., Kontomaris, K., Portela, L.: Numerical simulation of particle-laden turbulent channel flow. Phys. Fluids **13**(10), 2957–2967 (2001)
8. Saffmann, P.G.: The lift on a small sphere in a slow shear flow. J. Fluid Mech. **22**, 385–400, (1965) and J. Fluid Mech. **31**, 624, (1968)
9. Soltani, M., Ahmadi, M.: Direct numerical simulation of particle entrainment in turbulent channel flow. Phys. Fluids **7**(3), 647–657 (1995)
10. Vreman, A.W.: Turbulence characteristics of particle-laden pipe flow. J. Fluid Mech. **584**, 235–279 (2007)
11. Wang, Q., Squires, K.D., Chen, M., McLaughlin, J.B.: On the role of the lift force in turbulence simulations of particle deposition. Int. J. Multiphase Flow **23**, 749–763 (1997)
12. Wilhelm, D., Härtel, C., L. Kleiser: Computational analysis of the two-dimensional–three-dimensional transition in forward-facing step flow. J. Fluid Mech. **489**, 1–27 (2003)

On the Error Estimate in Sub-Grid Models for Particles in Turbulent Flows

E. Calzavarini, A. Donini, V. Lavezzo, C. Marchioli, E. Pitton, A. Soldati and F. Toschi

1 Introduction

The use of Large Eddy Simulation (LES) has emerged in recent years as a powerful simulation technique with the specific goal of achieving a good statistical accuracy while retaining a computational cost lower than Direct Numerical Simulations (DNS) [1]. In LES, only large-scale motions are directly computed (resolved on the computational grid) while small scale motions are not computed explicitly but modeled via Sub-Grid Scale (SGS) models. Due to the complex statistical properties of turbulence, many models and methodologies have been proposed in the past. Although none of the proposed models can be considered a perfect substitute to DNS, their performance can be sometimes considered fairly accurate for what concerns the most common Eulerian turbulent flow statistics. The problem of particle transport in turbulence demands much more to LES than just reproducing low order Eulerian statistics (e.g. spectra, average profiles etc) [2, 3]. Here we propose a way to quantify the effect of (the error due to) sub-grid modeling on particle properties.

2 Description of the numerical methods

Objective of this paper is to provide an accurate quantification of the effect of sub-grid modeling on particle dynamics. To this purpose DNS of both Homogeneous

E. Calzavarini
École Normale Supérieure de Lyon, CNRS UMR 5672, 46 Allée d'Italie, 69007 Lyon, France

A. Donini · V. Lavezzo · F. Toschi
Eindhoven University of Technology, P.O. Box 513, 5600 MB Eindhoven, The Netherlands, e-mail: f.toschi@tue.nl

C. Marchioli · E. Pitton · A. Soldati
Università degli Studi di Udine, Via delle scienze 208, 33100 Udine, Italy

H. Kuerten et al. (eds.), *Direct and Large-Eddy Simulation VIII*, ERCOFTAC Series 15, DOI 10.1007/978-94-007-2482-2_27,

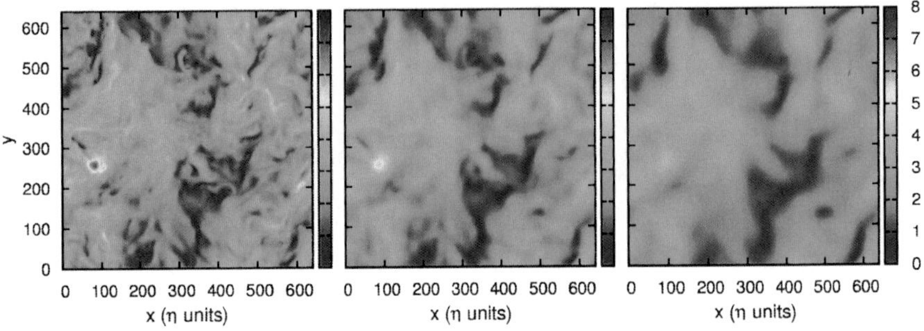

Fig. 1 2D view of the absolute value of the velocity field coarse-grained with larger filter width corresponding to, from left to right, 1η, 5η and 10η respectively.

Isotropic Turbulence (HIT) and Turbulent Channel Flow (TCF) are carried out to generate sets of velocity fields that are subsequently filtered at every time step using a Gaussian filter. This operation allows to smooth out all flow scales smaller than the filter width, Δ. Several filter widths were considered to produce a full temporal evolution of different filtered Eulerian fields, each mimicking a "perfect" LES, i.e. a reduced simulation where the largest scales evolve closely matching DNS ones. Particles dynamics are then determined using a Lagrangian approach: first, the motion of particles initially released as close-by pairs [2] is computed in the (unfiltered) DNS fields; then, a Gaussian filter is applied to generate coarse-grained fields in which the same particles are tracked again. In this way we can generate perfectly comparable LES simulations. For both HIT and TCF, simulations were performed using a pseudo-spectral method. The Eulerian field is governed by the incompressible Navier-Stokes equations discretized on a regular grid of 512^3 grid points for HIT and on an irregular grid of 128^3 grid points stretched in the wall-normal direction for TCF. Governing equations are integrated using a Fourier-Chebyshev representation of field variables and advanced in time with a second-order Adams-Bashforth integrator. For HIT, the solution domain is a cube, subjected to periodic boundary conditions. Turbulence scales are fixed in time by forcing the flow at the small wavenumbers, with constant power, in order to achieve a statistically stationary turbulent flow ($Re_\lambda = 363$). Lagrangian statistics are then extracted tracking large ensembles of particles and of particle pairs with prescribed initial separation, roughly equal to the Kolmogorov length scale, η [2, 3]. For TCF, simulations were performed on a reference domain with periodic boundary conditions in the streamwise and spanwise directions and with no-slip boundary condition enforced at the walls. In this case turbulence is sustained by an imposed pressure gradient along the streamwise direction ($Re_\tau = 150$). Particle pairs are released in wall-parallel planes with three different initial orientations, aligned with each flow direction to investigate also the influence of shear on pair dispersion. In this study we tracked both tracers and particles with inertia. Differently from HIT, the Gaussian filter is employed in TCF only along the two homogeneous directions.

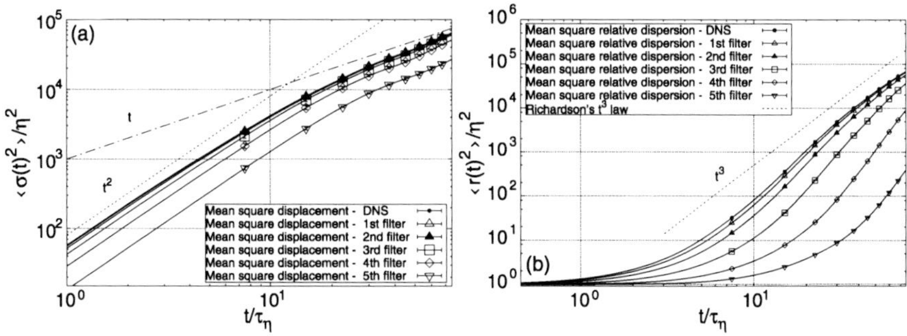

Fig. 2 Mean square displacement (a), and mean square relative dispersion (b) as a function of time t, normalized by Kolmogorov time-scale. Profiles correspond to different degrees of filtering: no filter is applied in DNS, then filter widths of $1.55\eta, 3.1\eta, 6.2\eta, 12.42\eta$ and 24.84η respectively.

3 Results

In Fig. 1 a typical 2D view of homogeneous isotropic turbulence is shown. For increasing filter widths, the field becomes more and more coarse grained. Yet, large scale structures remain unchanged for all filtered evolutions. In isotropic turbulence, mean velocity is zero so the first non-vanishing statistical moment is the variance of particle displacement, σ^2, which measures the displacement of a fluid particle relative to its initial position. In Fig. 2 the variance of particle displacement obtained applying filters of increasing width to the DNS field is shown. For comparison the two asymptotic trends $\langle r(t)^2 \rangle \sim t$ and $\langle r(t)^2 \rangle \sim t^2$ predicted by the theory of G.I. Taylor [4] are also shown. It is observed that particle displacement variance increases with time and exhibits the expected asymptotic scaling behaviors. The results show a good agreement for all cases, yet it can be seen that an increase in filter width leads to lower dispersion. This result allows to conclude that LES under-predicts single particle dispersion. Particle pairs in a turbulent flow will on average move away from each other, leading to mean separations that increase in time. The relative dispersion depends on the properties of the turbulent velocity field and on the initial separation distance. In DNS, the velocity field is fully resolved, and the contributions to relative dispersion are well captured at all scales of motion. However, in our "perfect" LES, the large-scale velocity fields are fully resolved whereas small-scale fluid motions are completely missing. We thus expect a significant effect on the relative dispersion of particle pairs, especially at early dispersion stages. This effect can be measured statistically by looking at relative separations and velocities. An important quantity to measure is particle pair dispersion. For homogeneous isotropic turbulence, two regimes can be identified for relative dispersion: the instantaneous separation of the positions of the two particles $\mathbf{r}(t) = \mathbf{r}^{(1)}(t) - \mathbf{r}^{(2)}(t)$ and the separation magnitude $\mathbf{r} = |\mathbf{r}|$. For short times, the mean square separation between two fluid elements grows either exponentially or quadratically. Furthermore, when the particle separation distance falls in the inertial range, it can be described using a diffusion equation for the probability density function (pdf) of the pair separation

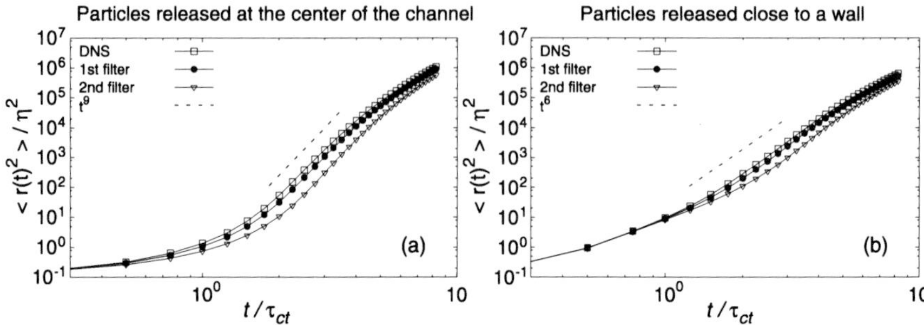

Fig. 3 Particle pair mean square relative dispersion in turbulent channel flow. Time t is normalized by the crossing time τ_{ct}, namely the time taken by the flow to cross the channel in the streamwise direction. Slopes corresponds to different filter widths (□: DNS, •: 1st filter, ▽: 2nd filter).

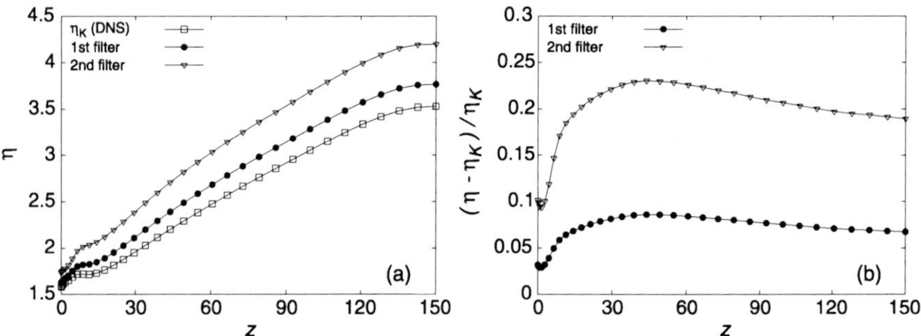

Fig. 4 Left panel: smallest length-scale as a function of wall normal coordinate in turbulent channel flow (wall located at $z = 0$, center of the channel located at $z = 150$). Right panel: the effect of increase of the smallest length-scale by filtering (comparison with η_K obtained with DNS).

$p(r,t)$, following the ideas of Richardson [5]. This gives rise to the celebrated scaling for the second order moment: $\langle r^2 \rangle = g\varepsilon t^3$. Here g is the Richardson constant, and ε is the energy dissipation rate per unit mass. For long times, a diffusion limit exists similar to single-particle dispersion. Regimes depend on the initial particle separation, and only the inertial regime is universal since at long times particles belonging to some pairs become uncorrelated. The t^3 regime is derived theoretically for high Reynolds number flows with a clear inertial range. From Fig. 2 we can see that filters of increasing width shift profiles toward lower values. Based on these results, one can conclude that, similarly to what was observed for single particles, LES under-predicts particle pair dispersion. In this case under-prediction is even higher.

Particle pair dispersion in turbulent channel flow is shown in Fig. 3 for two different locations of release. Although the values of Reynolds number and grid resolution are too small to provide a clear inertial range, we define it as $10\,\overline{\eta_K} < r(t) < h$ (where $\overline{\eta_K}$ is the Kolmogorov length-scale averaged over the all directions and h is the channel half-height) for a comparison with HIT. In inertial range the slope of $\langle r^2 \rangle$ is steeper than in homogeneous isotropic turbulence. This is due to the strong influence of the shear near the channel walls. In this regions the streamwise velocity gradient dominate pair separation. As in HIT LES under-predicts particle pair dispersion. In particular, the degree of under-prediction changes along the wall-normal direction, being stronger for particles released in the center of the channel. Note that initial separation between two particles is always smaller than the Kolmogorov length-scale $\eta_K(z)$. A possible explanation to this behavior is provided in Fig. 4, which shows the wall-normal behavior of the smallest length-scale $\eta(z)$ computed when filters of increasing width are applied to the DNS fields. The smallest length-scale, equal to the Kolmogorov length-scale $\eta_K(z)$ in DNS, increases upon filtering, and the increase is higher in the center of the channel than near the walls, both in absolute and relative terms. This means that filtering removes more "information" in the center of the channel: particles become exposed to a narrower spectrum of structures and under-prediction of pair separation is thus stronger over short times (e.g. $\frac{t}{\tau_{ct}} < 1$).

Particles released in fields with different filtering will have a different evolution because of two effects (see Fig. 5): (i) lack of high frequency oscillations and (ii) error accumulation. Due to the filtered nature of the fields which lack high frequency oscillations, particles will also miss these frequencies from their dynamics. The second effect is the result of particles evolving in the filtered field along a trajectory which differs from the DNS one. This leads particles to experience different values of the velocities (see Fig. 5) and to accumulate an error during their motion. It is of paramount importance to discriminate between these two effects: while the first can be addressed quite easily (e.g. by adding an appropriate stochastic high frequency noise to the particle velocity), the latter poses great modeling challenges. It is hence fundamental to understand in which conditions and for which observables the second effect plays a role. In order to analyze and quantify these two effects, we introduce a new investigation tool. Referring to the velocity field we can calculate its material derivative with respect to the filter width, obtaining

$$\frac{d}{d\Delta}\mathbf{v}_\Delta(\mathbf{x}_\Delta(t),t) = \frac{\partial \mathbf{u}_\Delta}{\partial \Delta} + \frac{\partial \mathbf{x}_\Delta(t)}{\partial \Delta} \cdot \nabla \mathbf{u}_\Delta(\mathbf{x}_\Delta(t),t), \tag{1}$$

where Δ is the filter width. This simple and exact relation is helpful to our purpose, given that we can calculate its two different hand sides and then compare them in order to obtain an error. In our databases, comprehensive of several Eulerian and Lagrangian data, we integrated also the filtered partial derivative and gradients of the evolution of the particles. The left panel of Fig. 5 shows the trend of the error between the two hand sides, defined as their difference divided by their semi-sum.

4 Conclusions

Small turbulent scale effects on the dynamics of individual particles and of particle pairs are analyzed systematically, to validate the use of LES in two-phase flows. To this aim DNS of homogeneous isotropic turbulent flow and of turbulent channel

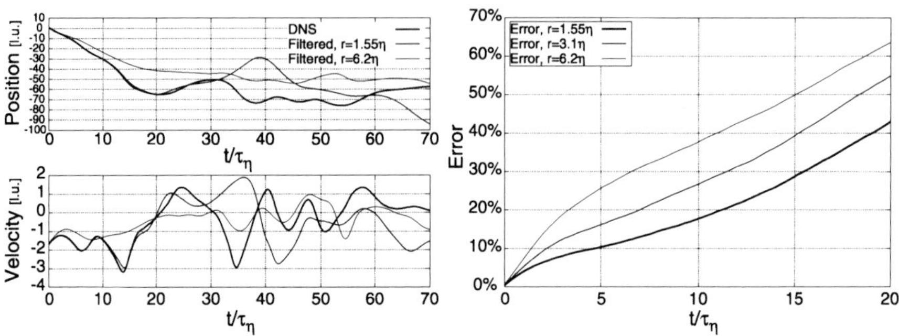

Fig. 5 Left panels: example of the evolution of particles in velocity fields filtered with different filters. Both position and velocity are shown as a function of time. Right panel: error for three different filter widths, averaged over all the particles.

flow are carried out, and the motion of dispersed particles is followed in time. The DNS velocity fields are then filtered with a Gaussian filter of increasing width, to generate a-priori LES fields in which particles are tracked again. In this way it is possible to isolate the effect of the small scales with respect to the real flow field (represented by the DNS one). It is also possible to highlight the behavior of flow field statistics with filter of increasing width. A general under-prediction of particle dispersion is found in both flow configurations. However, while for single particles dispersion under-prediction seems to be small, for particle pairs it becomes more significant. This is due to the stronger influence of the small scales of motion on pair dispersion in the dissipation range. Finally a new analysis tool has been introduced to investigate deviation of particles evolving in the filtered field with respect to "exact" DNS trajectory.

References

1. Sagaut, P.: Large eddy simulation for incompressible flows: an introduction. Springer, (2006)
2. Salazar, J.P.L.C., Collins, L.R.: Two-particle dispersion in isotropic turbulent flows. Ann. Rev. Fluid Mech. **41**, 405–432 (2009)
3. Toschi, F., Bodenschatz, E.: Lagrangian properties of particles in Turbulence. Ann. Rev. Fluid Mech. **41**, 375–404 (2009)
4. Taylor, G.I.: Diffusion by continuous movements. Proc. R. Soc. Lond. 196–211 (1921)
5. Richardson, L.F.: Weather prediction by numerical process. Cambridge (1922)

Benchmark test on particle-laden channel flow with point-particle LES

C. Marchioli, A. Soldati, M.V. Salvetti, J.G.M. Kuerten, A. Konan, P. Fede, O. Simonin, K.D. Squires, C. Gobert, M. Manhart, M. Jaszczur, L.M. Portela

1 Introduction

Dispersion of particles in a turbulent wall-bounded flow is crucial in many practical applications. For numerical simulation of particle-laden turbulent flow various approaches are available. Among these, LES is perhaps the most promising because its computational cost is lower than that of DNS and its predictive capability is much higher than Reynolds-Averaged Navier-Stokes methods especially in case of particles-turbulence interaction in boundary layers. Various subgrid models are available which have proved their validity for several types of flow. However, the treatment of particles in LES is still a relatively new topic with open questions regarding e.g. Sub-Grid Scales (SGS) effects on particle behavior and the modeling of particle-particle, particle-fluid or particle-wall interactions. To address these issues, an international collaborative benchmark test has been proposed as part of the activity of the COST Action P20 LESAID. The objective is to gather a large database of results obtained with different numerical methods, SGS models and physical models in order to resolve questions about the validity of these models. In this paper the first statistics of the benchmark for a base Eulerian-Lagrangian simulation of particle-laden channel flow, are presented. The specific simulation parameters have been chosen also to allow estimate of the quality of the LES results upon comparison with available DNS results [1] for the same test case. The groups participating in the benchmark are: UUD-UPI (Marchioli, Soldati, Salvetti); TUE (Kuerten); IMFT-ASU (Konan, Fede, Simonin, Squires); TUM (Gobert, Manhart); TUK-TUD

CM & AS: Dept. Energy Technologies, University of Udine, Italy
MVS: Dept. Aerospace Engineering, University of Pisa, Italy
JGMK: Dept. Mechanical Engineering, Eindhoven University of Technology, The Netherlands
AK, PF & OS: Université de Toulouse; INPT, UPS; IMFT; CNRS; Toulouse, France
KDS: Dept. Mechanical and Aerospace Engineering, Arizona State University, Tempe (AZ), USA
CG & MM: Dept. Civil Engineering, Munich University of Technology, Germany
MJ: AGH University of Science and Technology, Krakow, Poland
LP: Dept. Multiscale Physics, Delft University of Technology, The Netherlands

H. Kuerten et al. (eds.), *Direct and Large-Eddy Simulation VIII*, ERCOFTAC Series 15, DOI 10.1007/978-94-007-2482-2_28,

(Jaszczur, Portela). Results provided by each group refer to a statistically stationary situation in which the particle concentration has reached a steady state. The time taken to reach steady concentration is very long (up to $2 \cdot 10^4$ in wall units [1]) thus making the required computational effort quite high even for a LES-based calculation.

2 Base Simulation and Common Parameters

Physical problem - The base simulation is LES of turbulent incompressible flow in a channel of size $L_x \times L_y \times L_z = 4\pi H \times 2\pi H \times 2H$ laden with particles of sizes $d_p/H = 1.02 \cdot 10^{-3}, 2.28 \cdot 10^{-3}$, and $5.10 \cdot 10^{-3}$ where $H = 0.02\ m$ is half the channel height. The friction Reynolds number is $Re_\tau = u_\tau H/\nu = 150$, based on friction velocity $u_\tau = 0.11775\ m/s$, and on fluid viscosity $\nu = 1.57 \cdot 10^{-5}\ m^2/s$. The particle-to-fluid density ratio is $\rho_p/\rho = 769.23$. One-way coupling is considered, and collisions between particles are neglected. The particle equation of motion is: $\frac{d\mathbf{v}(t)}{dt} = \frac{\mathbf{u}(\mathbf{x},t)-\mathbf{v}(t)}{\tau_p}\left(1+0.15Re_p^{0.687}\right)$, where $\mathbf{v}(t)$ is the particle velocity and $\mathbf{u}(\mathbf{x},t)$ is the fluid velocity at the location of the particle, $\mathbf{x}$. The particle relaxation time is $\tau_p = \rho_p d_p^2/(18\rho\nu)$. The standard drag correction for particles with Reynolds number $Re_p = |\mathbf{u}(\mathbf{x},t) - \mathbf{v}(t)|d_p/\nu < 10^3$ is applied. The particle Stokes number $St = \tau_p/\tau_f$ ($\tau_f = \nu/u_\tau^2$ being the characteristic time scale of the flow) is equal to 1, 5 and 25.

Boundary and initial conditions - In the streamwise and spanwise directions the fluid flow is periodic. At the walls the no-slip condition is applied. The initial field is a fully-developed statistically stationary solution of the LES equations. Initially particles are randomly distributed with initial velocity equal to the fluid velocity at particle location. Particles moving outside of the domain in the streamwise and/or spanwise directions are reintroduced via periodicity. Particles collide elastically with the walls. The time step is chosen sufficiently small compared to the particle relaxation time. In order to obtain accurate statistics, at least 10^5 particles of each kind were tracked.

Subgrid models - Two simulations are considered: one without SGS model (no-model simulation) and one with an eddy-viscosity model (either Smagorinsky with wall damping or a variant of the dynamic eddy-viscosity model), both on a 64^3-point grid. No SGS model is applied in the particle equation of motion. This basic test case was designed to see the effects of the different numerical methods used in the LES and in the Lagrangian tracking on fluid and particles. The no-model simulation serves as a zero measurement: any useful SGS model should yield better results.

Specific methodology details - They are summarized in Table 1. The possibility of using different numerical schemes and/or different values for some simulation parameters (like the time step size, for instance) allows clearcut evaluation of how the accuracy of the LES results depends on the choice made.

3 Results and Discussion

In this Section, fluid and particle statistics are discussed to benchmark the performance of the different numerical approaches. Particle statistics were computed summing the desired variable over all particles in a sampling volume, constituted by wall-parallel fluid slab, and averaging by the number of particles in the sampling volume.

Base LES simulation statistics (with SGS model) - The mean streamwise fluid velocity, U_x^+, and the Root Mean Square of streamwise and wall-normal fluid velocities, $RMS(U_x^+)$ and $RMS(U_z^+)$, are shown in Figs. 1a) and in Fig. 1b). For comparison purposes, the reference DNS profile obtained by group UUD-UPI (thick solid line) and the von Karman log law (dotted line) are also shown. Overall, all codes produce fluid velocity statistics that are in reasonable agreement. Maximum deviations for mean velocity are limited to few percents in the center of the channel (see close-up view of Fig. 1a). We remark here that these quantitative differences are of the same order as those obtained in the DNS test case [1]. This was somehow expected given the fine grid resolution adopted in base simulation (grid coarsening factor of 2 with respect to DNS). Results for the RMS are also in rather good agreement. Only small spread in the quantitative numbers arise, particularly in the near-wall region where profiles reach their peak value. The above mentioned differences are, of course, due solely to differences in modeling and discretizing the flow field, which will add to differences in modeling the particle motion and will show up also in the statistical moments for the particle velocity, shown in Figs. 1c-d). For brevity, statistics are shown here only for the $St = 1$ particles, which are the most responsive to LES filtering. The agreement among mean quantities is again satisfactory in the log law region. Particle RMS near the centerline is also well predicted by all groups (see Fig. 1d). Near the wall, however, the uncertainty associated with the calculation of the peak value is higher. This is observed for all Stokes numbers. Fig. 2 shows (statistically-steady) concentrations as a function of the wall-normal coor-

Group	Flow solver	SGS model for the fluid phase	Time integration of fluid (non-linear + viscous terms)	Time integration of particles	Fluid velocity interpolation	Wall-normal distribution of collocation points
UUD-UPI	PS	DS	AB2 + CN	RK4	L6	Chebyschev
TUE	PS	DS	RK3 + CN	H2	LH4	Chebyschev
IMFT-ASU	FD2	SWD	AB2 + CN	AB2	L4	HT (SF=2)
TUD-TUK	FV2	SWD/DS	AB2 + CN	AB2	TL	HT (SF=1.7)
TUM	FV2	–	RK3	RK2	L4	HT (SF=1.7)

Table 1 Summary of numerical methodologies. Nomenclature: *Flow solver* - PS: pseudospectral, FD2: 2nd order finite differences, FV2: 2nd order finite volumes; *SGS model* - SWD: Smagorinsky with wall damping, DS: dynamic Smagorinsky; *Time integration* - AB2: 2nd order Adams-Bashforth, CN: Crank-Nicolson, RK2/RK3/RK4: $2^{nd}/3^{rd}/4^{th}$ order Runge-Kutta, H2: 2nd order Heun method; *Fluid velocity interpolation* - L2/L4/L6: $2^{nd}/4^{th}/6^{th}$ order Lagrange polynomials, LH4: 4th order Lagrange-Hermite polynomials; *Collocation points* - HT: hyperbolic tangent (SF: stretching factor).

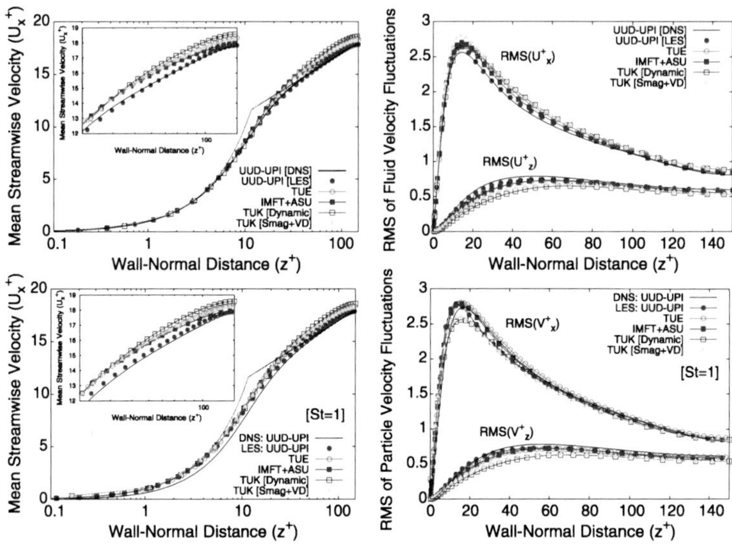

Fig. 1 Mean and RMS velocities in the base LES. Panels: a-b) fluid; c-d) $St = 1$ particles.

dinate, z^+, at time $t^+ = 20{,}000$ for $St = 1$ and $St = 25$. Particle concentration was obtained exactly as in [1]. Discrepancies are observed particularly in the near-wall region where particle accumulation builds-up in time, and may increase or decrease depending on the Stokes number. Deviations arise not only because of different choices in modeling the flow field but also, if not mainly, because of the numerical errors associated (i) with the interpolation technique used to obtain the fluid velocity at particle location and (ii) with the time-step size chosen to integrate the equation of particle motion. These errors sum up over time and give the accumulated profile deviations of Fig. 2. A significant spread in the prediction of concentrations was observed also for DNS [1]. In LES, however, concentration is systematically underestimated with respect to DNS, as can be seen by comparing LES and DNS results of groups UUD-UPI and TUE in Fig. 2.

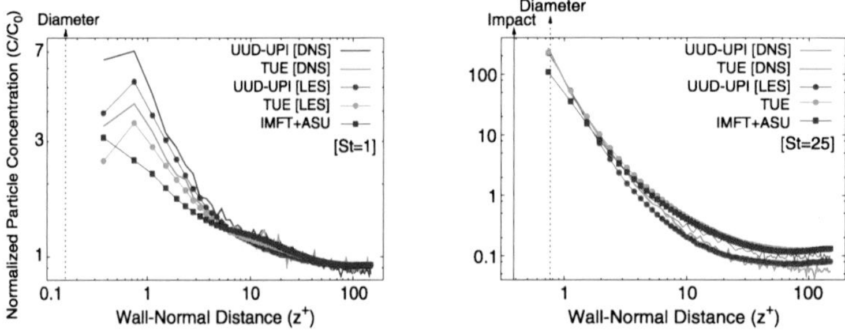

Fig. 2 Particle concentrations in the base LES: (a) $St = 1$; (b) $St = 25$.

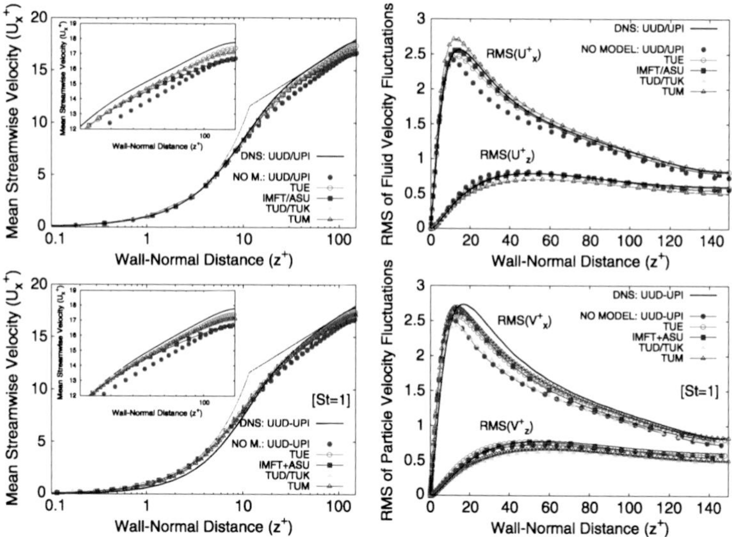

Fig. 3 Mean and RMS velocities in the no-model simulation. Panels: a-b) fluid; c-d) $St = 1$ particles.

No-model simulation statistics - In this section, we focus on the modifications of velocity and concentration statistics obtained in the no-model simulation, where simulation parameters are exactly the same as in the LES base simulation except that now the SGS model for the fluid has been switched off. Fig. 3 shows the mean streamwise fluid velocity (Fig. 3a), and the RMS of streamwise and wall-normal fluid velocities (Fig. 3b). As expected results become less accurate. Both the mean velocity and the streamwise RMS are now underestimated, as can be readily seen comparing DNS and no-model results of group UUD-UPI. The velocity deficit is more evident outside the buffer layer. Also, the uncertainty associated with the calculation of both the peak value and the peak location of $RMS(U_x^+)$ appears increased. It is interesting to observe that, compared with the other groups, the profiles of groups TUD/TUK and TUM, which are both using a finite-volume flow solver, tend to produce a relative overshoot of $RMS(U_x^+)$ and a slight relative undershoot of $RMS(U_z^+)$. Similar trends may be observed in the particle velocity statistics, shown in Fig. 3c-d) for the $St = 1$ particles. These statistics, however, also depend on additional low-pass filtering effects associated to particle inertia. The coarse-grained mean velocity, for instance, is again lower than the DNS one outside the buffer layer. Yet inertial effects produce an increase in the velocity with which particles are advected in the near-wall region, making the fluid lag behind. The streamwise RMS is also significantly affected: peak values are smaller than those of the fluid and their location appears to be shifted towards the wall. To conclude our analysis, in Fig. 4 we show the concentration profiles for $St = 1$ and 25 in the no-model simulation. Compared with DNS but also with the base LES, near-wall concentration is overestimated for all Stokes numbers.

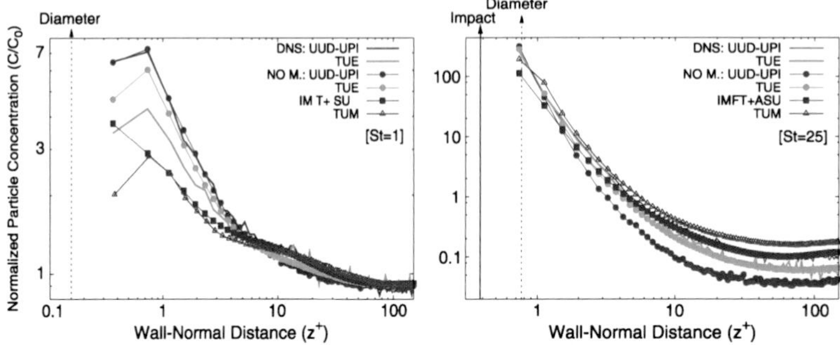

Fig. 4 Particle concentrations in the no-model simulation. Panels: (a) $St = 1$; (b) $St = 25$.

4 Conclusions and Possible Test Case Extensions

Analysis of mean and RMS particle velocities indicates that single-point statistics are not much affected by filtering and by differences in modeling or numerics, provided that a suitable SGS model is used for the fluid. Conversely, particle concentrations are much more sensitive to differences in the numerical set-up, and significant discrepancies are observed between LES and DNS values of concentration. This confirms that a SGS model should be introduced in the particle motion equations to obtain a reliable prediction of particle concentration (see [2] and references therein). Starting from the base simulation, test case extensions will concern issues related to physical modeling and computational parametrization. In particular, other LES grid resolutions and SGS models for both fluid and particles will be studied. Inclusion of gravity or lift in the particle motion equation motion, and of two-way coupling (effect of particles on fluid) and four-way coupling (particle-particle interactions) will also be explored.

References

1. Marchioli C., Soldati A., Kuerten J.G.M., Arcen B., Taniere A., Goldensoph G., Squires K.D., Cargnelutti M.F. and Portela L.M. (2008) Statistics of particle dispersion in direct numerical simulations of wall-bounded turbulence: Results of an international collaborative benchmark test. *Int. J. Multiphase Flow*, **34**, 879–893.
2. Marchioli C., Salvetti M.V., and Soldati A. (2008) Some issues concerning Large-Eddy Simulation of inertial particle dispersion in turbulent bounded flows. *Phys. Fluids*, **20**, 040603.

On large eddy simulation of particle laden flow: taking advantage of spectral properties of interpolation schemes for modeling SGS effects

Christian Gobert and Michael Manhart

1 Introduction

This contribution deals with Large-Eddy simulation (LES) of particle-laden flow. State of the art methods are capable to predict the dynamics of small particles in dilute suspensions as long as direct numerical simulation (DNS) is possible. However, as soon as the Reynolds number is too high, DNS is not an alternative and often LES is the method of choice. For LES of particle-laden flow, the effect of the unresolved subgrid scales (SGS) needs to be modeled, a least for small or moderate Stokes numbers of the particles. This means that two turbulence models are necessary. One model for SGS effects on the resolved scales of the carrier fluid flow (such as the Smagorinsky model) and another model for SGS effects on the particle dynamics. Hereinafter, the former type of models is refered to as *fluid-LES models* and the latter type as *particle-LES models*.

Several particle-LES models have been proposed, most intended for a Eulerian-Lagrangian representation on which we focus. One promising particle-LES model is the approximate deconvolution method (ADM). ADM was developed for single phase flow by Stolz & Adams [10]. Later, ADM was adopted for particle-laden flow by Kuerten [5] and Shotorban *et al.* [9]. The idea of ADM is to improve barely resolved scales in LES but not to 'create' scales which are smaller than the LES grid (like in stochastic models).

However, all these models do not take into consideration the effect of interpolation of the fluid velocity on the particle position. It is well known that this effect is much more significant in LES than in DNS. For ADM, Kuerten [5] even observed that 'Compared to fourth-order interpolation of the fluid velocity to the particle position, second-order interpolation approximately cancels the effect of the subgrid model in the particle equation'.

Technische Universität München, e-mail: ch.gobert@bv.tum.de, m.manhart@bv.tum.de

H. Kuerten et al. (eds.), *Direct and Large-Eddy Simulation VIII*,
ERCOFTAC Series 15, DOI 10.1007/978-94-007-2482-2_29,

In the present work we focus on this effect and we take advantage of the interpolation error in order to construct a novel particle-LES model. The idea is similar to the implicit LES approach of Adams *et al.* [1]. Implicit LES stands for numerical approximation of the Navier–Stokes equations such that the numerical error acts as fluid-LES model. Likewise, the idea of our new particle-LES model is to interpolate the fluid velocity seen by the particles such that the interpolation error acts as particle-LES model.

2 Numerical Simulation of the carrier flow

In the present work we analyse particle dynamics in forced isotropic turbulence by DNS and LES. For the simulation of the carrier fluid, we use a second order Finite-Volume method together with a third order Runge-Kutta scheme [13] for advancement in time. The conservation of mass is satisfied by solving the Poisson equation for the pressure using an iterative solver [11]. More details on the flow solver can be found in [6].

The flow is driven using a slightly modified version of the deterministic forcing scheme proposed by Sullivan *et al.* [12]. The flow was computed at three Reynolds numbers, namely $Re_\lambda = 52$, 99 and 265 based on the transverse Taylor microscale λ and the rms value of one (arbitrary) component of the fluctuations u_{rms}. In all computations the flow was solved in a cube on a staggered Cartesian equidistant grid. The size of the computational box and the cell width was chosen in dependence of the Reynolds number, see table 1.

The new particle-LES model was assessed by a priori analysis at $Re_\lambda = 52$ and a posteriori analysis at $Re_\lambda = 52$, 99 and 265. For the a priori analysis, we filtered the DNS field $\mathbf{u}$ by a box filter with filter width $\Delta = 7\Delta x$, Δx being the DNS cell width. The kinetic energy of the filtered field $\hat{k}_f = \left\langle \mathscr{G} u_i^2 \right\rangle /2$ is 87% of the energy of the unfiltered field. For the a posteriori analysis, $\mathscr{G}\mathbf{u}$ was computed by LES. As fluid-LES model we used the Lagrangian dynamic Smagorinsky model proposed by Meneveau *et al.* [7].

Table 1 Parameters for DNS and LES of forced isotropic turbulence.

	DNS			LES		
Re_λ	52	99	265	52	99	265
N	256^3	512^3	1030^3	42^3	64^3	42^3
$\Delta x/\lambda$	0.093	0.078	0.047	0.567	0.623	1.15
$\Delta x/\eta_K$	1.34	1.54	1.54	8.17	12.3	37.7

3 Discrete particle simulation

In this study we consider dilute suspensions of small particles and neglect effects of the particles on the fluid and particle-particle interactions (one way coupling).

We set the density of the particles to $\rho_p = 1800\rho$ where ρ is the density of the fluid. Several fractions of different diameter d were considered, all of which are smaller than the Kolmogorov length scale. Consequently, the particles can be treated as point particles. The particle relaxation times are adjusted that the Stokes numbers based on fluid Kolmogorov time scale range from $St = 0.1$ to $St = 100$.

We assumed that in the given configurations the acceleration of a particle $\frac{d\mathbf{v}}{dt}$ is given by Stokes drag only,

$$\frac{d\mathbf{v}}{dt} = -\frac{c_D Re_p}{24\tau_p}(\mathbf{v} - \mathbf{u}_{f@p}). \tag{1}$$

Here, $\mathbf{v}(t)$ denotes the particle velocity and $\mathbf{u}_{f@p}$ the fluid velocity at the particle position. The particle Reynolds number Re_p is based on particle diameter and particle slip velocity $\|\mathbf{u}_{f@p} - \mathbf{v}\|$ which leads to a nonlinear term for the Stokes drag. The drag coefficient c_D was computed in dependence of Re_p according to the scheme proposed by Clift *et al.* [2].

The fluid velocity $\mathbf{u}_{f@p}$ is interpolated to the particle position $\mathbf{x}_p(t)$, i.e. $\mathbf{u}_{f@p} = \mathbf{u}(\mathbf{x}_p(t),t)$ by a standard fourth order interpolation scheme except for the simulations with the new model which defines its own interpolation scheme.

Equation (1) is solved by a Rosenbrock-Wanner method [4]. This method is a fourth order method with adaptive time stepping. The stiff term in equation (1) is linearized in each time step and treated by an implicit Runge-Kutta scheme. Note that due to this method, particle time step sizes can be smaller than fluid time step sizes and are interpolated in time as well.

4 Effect of interpolation methods on the spectrum seen by particles

The present section addresses the effect of interpolation on the fluid velocity seen by the particles. This effect is quantified in spectral space. Focus is on the high wave number effects of interpolation. The particle-LES model, which is presented in section 5, takes advantage of these observations.

Consider interpolation schemes that can be formulated as a linear combination of the sample values u_j and some weighting functions $w(x)$, the interpolation kernel

$$u_{f@p}(x) = \sum_j w(x - x_j)\, u_j. \tag{2}$$

With formula (2), the fluid velocity seen by particles $u_{f@p}$ is a continuous function and its Fourier transform can be computed for an arbitrary wave number κ, by multiplying the spectrum of the interpolation kernel with the continuous spectrum of the sample data,

$$\left|\mathscr{F}\mathscr{T}(u_{f@p})(\kappa)\right|^2 = |\mathscr{F}\mathscr{T}(w)(\kappa)|^2 \left|\mathscr{F}\mathscr{T}(u_f)(\kappa)\right|^2. \quad (3)$$

It should be noted that $\mathscr{F}\mathscr{T}(u_f)(\kappa)$ is not zero beyond the cutoff wave number κ_c. It can be computed by piecewise reflection of the spectrum for $\kappa < \kappa_c$, cf. figure 1. This result is well known under the term aliasing.

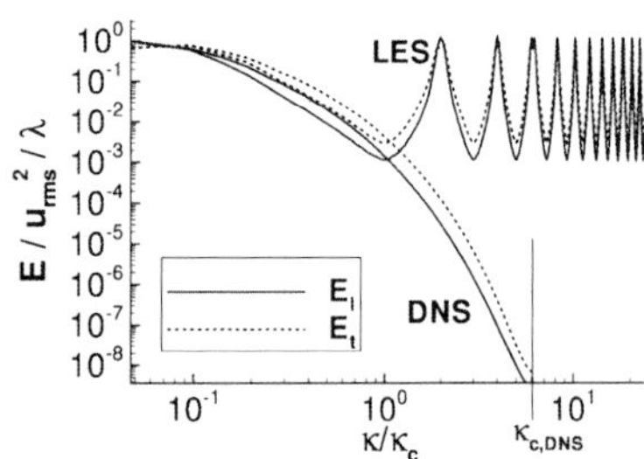

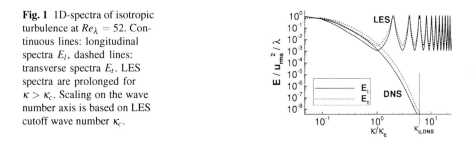

Fig. 1 1D-spectra of isotropic turbulence at $Re_\lambda = 52$. Continuous lines: longitudinal spectra E_l, dashed lines: transverse spectra E_t. LES spectra are prolonged for $\kappa > \kappa_c$. Scaling on the wave number axis is based on LES cutoff wave number κ_c.

In three dimensions, the interpolation formula must be extended to tensorial form considering interpolation in longitudinal and transverse direction, respectively. These are computed for trilinear and cubic interpolation for LES of isotropic turbulence in $Re_\lambda = 52$ in figure 2 for both schemes. In addition, the LES and DNS spectra computed from the flow field are shown.

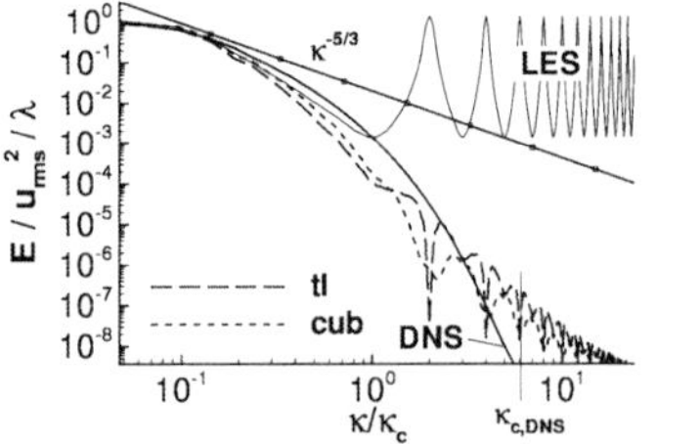

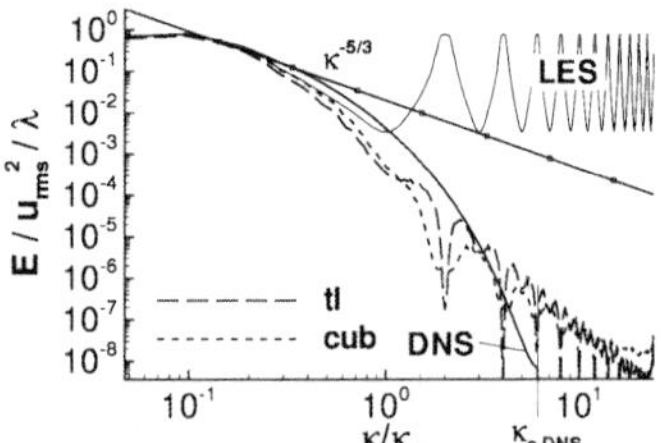

Fig. 2 Longitudinal (left) and transverse (right) spectra seen by particles in isotropic turbulence at $Re_\lambda = 52$, computed by LES. Long-dashed line (tl): trilinear interpolation, short-dashed line (cub): fourth-order interpolation. For reference, the spectra computed from the grid points in DNS and LES (plus reflections) are also shown (continuous lines).

5 Construction of the SOI model

In the present section, the findings of section 4 are combined to construct a new particle-LES model. The idea of the model is to reconstruct the scales beyond the cutoff of the LES by making use of the fact that the particles see a continuous spectrum which can be controlled by the interpolation kernel. In 1D, the modeling strategy involves the following:

1. Compute the LES spectrum $|\mathscr{F}\mathscr{T}(\langle u_f \rangle)(\kappa)|^2$ for $\kappa < \kappa_c$. Extend this spectrum by reflection for higher values of κ (cf. figure 1).
2. Define a target spectrum $E^{target}(\kappa)$ for the fluid velocity seen by the particles.
3. Search for an interpolation kernel w such that $|\mathscr{F}\mathscr{T}(\langle u_f \rangle)|^2 |\mathscr{F}\mathscr{T}(w)|^2 \approx E^{target}$.
4. Apply w in order to interpolate the fluid velocity seen by the particles.

We have tested two different target spectra: (i) spectra from a DNS of the carrier flow and (ii) a model spectrum for homogeneous isotropic turbulence proposed by Pope [8]. The target spectrum can only be approximated because of additional admissibility conditions for the interpolation, such as compact support and order of the interpolation. However, we seek for the optimal stencil in a sense that the target spectrum is approximated as exact as possible. The model can be regarded as an interpolation scheme that is optimized with respect to the spectrum seen by the particles and is therefore referred to as 'Spectrally Optimized Interpolation' (SOI).

The optimized stencil is represented by a series of cubic splines. Therefore, SOI requires only about 7% more CPU time than LES without particle-LES model and fourth-order interpolation. More details on the optimization algorithm can be found in [3].

SOI is assessed by numerical simulation of forced isotropic turbulence at $Re_\lambda = 52, 99$ and 265. At all Reynolds numbers, DNS, LES with ADM, LES with SOI and LES without particle-LES model were conducted. This allows a direct comparison of SOI against ADM with respect to the accuracy of the models.

In Figure 3 the spectra seen by the particles are shown for $Re = 265$. SOI is able to produce spectra very close to the model spectra even at that high Reynolds number. This increases the kinetic energy of the fluid velocity to the desired level which in

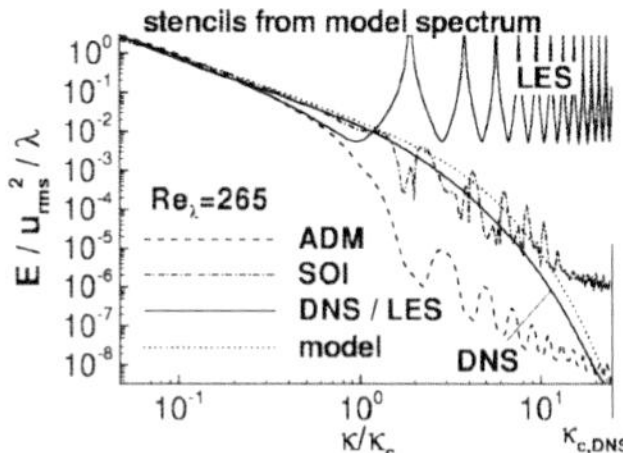

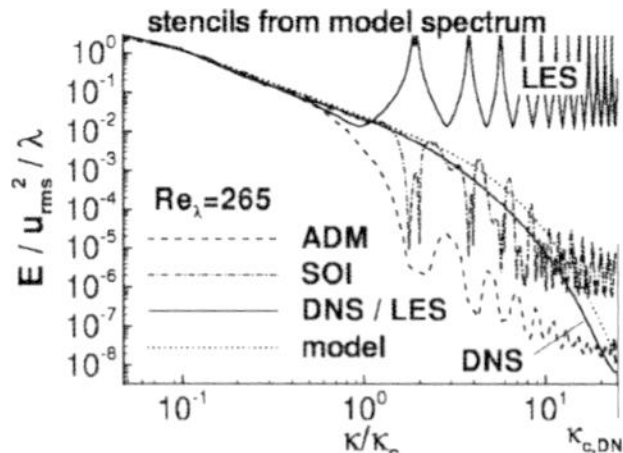

Fig. 3 Longitudinal (left) and transverse (right) spectra seen by particles in isotropic turbulence at $Re_\lambda = 265$, computed by LES with ADM and SOI.

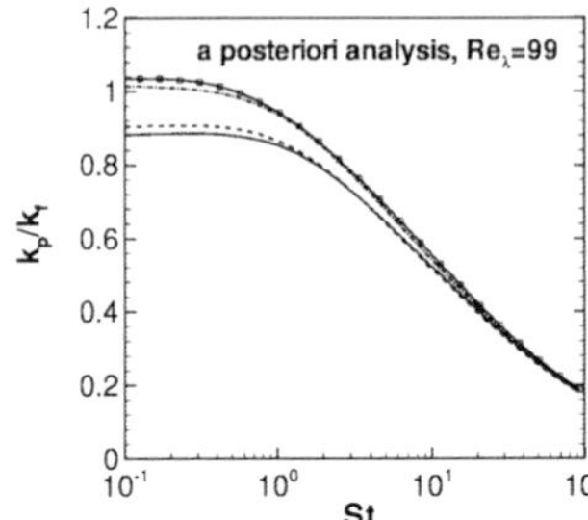

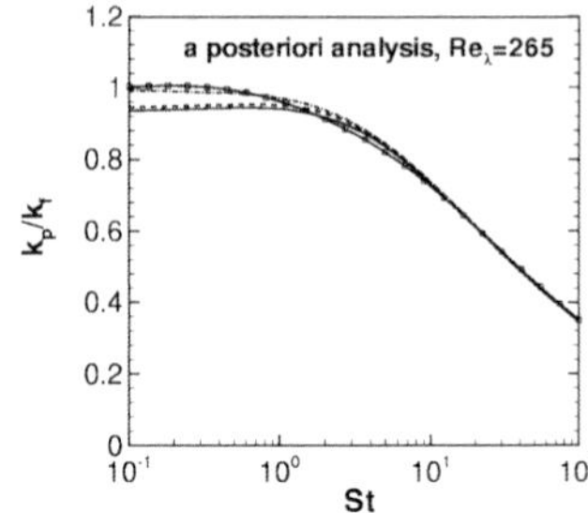

Fig. 4 Kinetic energy of the particles. Continuous lines with symbols: DNS, continuous lines without symbols: LES. Dashed lines: ADM, dash-dotted lines: SOI.

turn increases the kinetic energy of the particles themselves to the right level, even at low Stokes numbers (see Figure 4). Improvement of the particle statistics is also obtained for the integral time scales that are reduced by a reasonable factor (not shown here).

References

1. N. A. Adams, S. Hickel, and S. Franz. Implicit subgrid-scale modeling by adaptive deconvolution. *J. Comput. Phys.*, 200(2):412–431, November 2004.
2. R. Clift, J. R. Grace, and M. E. Weber. *Bubbles, Drops and Particles*. Academic Press, New York, 1978.
3. C. Gobert and M. Manhart. Subgrid modelling for particle-les by spectrally optimised interpolation (SOI). *submitted to J. Comput. Phys.*, 2010.
4. E. Hairer and G. Wanner. *Solving Ordinary Differential Equations II. Stiff and Differential-Algebraic Problems. Springer Series in Computational Mathematics*. Springer, New York, 1990.
5. J. G. M. Kuerten. Subgrid modeling in particle-laden channel flow. *Phys. Fluids*, 18:025108, 2006.
6. M. Manhart. A zonal grid algorithm for DNS of turbulent boundary layers. *Comput. Fluids*, 33(3):435–461, 2004.
7. C. Meneveau, T. S. Lund, and W. H. Cabot. A Lagrangian dynamic subgrid-scale model of turbulence. *J. Fluid Mech.*, 319:353–385, 1996.
8. S. B. Pope. *Turbulent Flows*. Cambridge University Press, Cambridge, UK, 2000.
9. B. Shotorban, K. Zhang, and F. Mashayek. Improvement of particle concentration prediction in large-eddy simulation by defiltering. *Int. J. Heat Mass Transfer*, 50(19-20):3728–3739, September 2007.
10. S. Stolz and N. A. Adams. An approximate deconvolution procedure for large-eddy simulation. *Phys. Fluids*, 11(7):1699–1701, 1999.
11. H. L. Stone. Iterative solution of implicit approximations of multidimensional partial differential equations. *SIAM J. Num. Anal.*, 5(3):530–558, 1968.
12. N. P. Sullivan, S. Mahalingam, and R. M. Kerr. Deterministic forcing of homogeneous, isotropic turbulence. *Phys. Fluids*, 6(4):1612–1614, 1994.
13. J. H. Williamson. Low-storage Runge-Kutta schemes. *J. Comput. Phys.*, 35:48–56, 1980.

DNS of a free turbulent jet laden with small inertial particles

F. Picano, G. Sardina, P. Gualtieri and C.M. Casciola

1 Introduction & Methodology

Turbulent jets with a dispersed phase are widely found in technological applications or in natural flows. In Plinian volcano eruptions a multiphase jet-column is produced. In this process the mixing of the entrained fresh air into the hot stream of gas is crucial in establishing the conditions for pyroclastic flows [4].

On the other hand, the droplet-laden jet is also the prototypal flow for fuel injectors, which are frequently found in internal combustion engines, turbine engine or rockets. Also in these conditions the entrainment of the surrounding fluid in the jet core is crucial to allow the evaporation of the droplets.

Real particle trajectories differ from those of pure Lagrangian tracers because of inertia. Actually, the particle inertia introduces a time lag–particle relaxation time– in the particle response to fast fluid velocity fluctuations, leading to many anomalous phenomena. If the particle relaxation time is of the order of the dissipative time scale of a turbulent flow–Kolmogorov time–small scale clustering occurs. Hence, the turbulent mixing never fully homogenizes a suspension of inertial particles, which will be constituted mainly by multi-scale voids and clusters of particles. The issue has

Francesco Picano
Dipartimento di Meccanica e Aeronautica, "Sapienza" University, via Eudossiana 18, 00184, Rome, Italy, e-mail: francesco.picano@uniroma1.it

Gaetano Sardina
Dipartimento di Meccanica e Aeronautica, "Sapienza" University, via Eudossiana 18, 00184, Rome, Italy, e-mail: gaetano.sardina@uniroma1.it

Paolo Gualtieri
Dipartimento di Meccanica e Aeronautica, "Sapienza" University, via Eudossiana 18, 00184, Rome, Italy, e-mail: paolo.gualtieri@uniroma1.it

Carlo M. Casciola
Dipartimento di Meccanica e Aeronautica, "Sapienza" University, via Eudossiana 18, 00184, Rome, Italy, e-mail: carlomassimo.casciola@uniroma1.it

H. Kuerten et al. (eds.), *Direct and Large-Eddy Simulation VIII*,
ERCOFTAC Series 15, DOI 10.1007/978-94-007-2482-2_30,

been thoroughly examined in isotropic (see e.g. [15], for a review) and in homogeneous shear turbulence ([14, 2]).

Inhomogeneity adds new effects, particularly evident in wall flows where a mean drift of particles towards the wall–the so-called turbophoresis–is apparent [6, 1].

Similar effects are expected also in jets, where the milder inhomogeneity in the streamwise direction as well as the intermittent entrainment region separating the turbulent jet core and the irrotational environment, may induce important new phenomenologies.

Several papers investigate the behavior of particles (e.g. [12, 5]) and of evaporating droplets (e.g. [13]) in the transitional, non-universal, near field region of jets. Inertial particles are found to concentrate in the shear-layer, outside the large coherent vortical structures of the near field, consistently with the general trend observed in other flows. Few works [5, 3] concern the far-field behavior, where the back-reaction of particles on the fluid stream is shown to influence the jet spreading rate. The outward particle mean radial velocity is found to be larger than that of the fluid [11], suggesting localization effects associated to preferential sampling of outward fluid motions, see [9] for similar effects in the wall layer of wall turbulence. The effect of inertia on particle dynamics decreases moving downstream of the jet [11], consistently with the increased time scale of the flow.

In this framework, the present paper aims at analyzing the particle behavior from a direct numerical simulation (DNS) of a free jet laden with different particle populations to understand the behavior of the system in the far field.

The carrier fluid is assumed to obey to the incompressible ($\nabla \cdot \mathbf{u} = 0$) Navier-Stokes equations,

$$\frac{\partial \mathbf{u}}{\partial t} + \mathbf{u} \cdot \nabla \mathbf{u} = -\frac{\nabla p}{\rho} + \nu \nabla^2 \mathbf{u} \tag{1}$$

$$\frac{\partial Y}{\partial t} + \mathbf{u} \cdot \nabla Y = D \nabla^2 Y \,,$$

where ρ and ν are density and kinematic viscosity of the fluid, respectively, $\mathbf{u}$ is the fluid velocity, Y is the concentration of a passive scalar injected with the flow and D its diffusion coefficient ($Sc = \nu/D = 0.7$). Since a very dilute suspension of tiny particles is assumed no force-feedback on the fluid is considered in equation (1) [1].

The inflow boundary condition for the turbulent round jet is assigned at each time-step by a companion DNS of a turbulent pipe flow. Traction-free conditions on the side mantle and a convective Orlanski-type outlet condition is used to allow the correct entrainment rate and to preserve the constant momentum flux of the jet [8].

The algorithm discretizes equations (1) in cylindrical coordinates by a conservative second order finite difference scheme on a staggered grid, with time integration performed by an explicit third order low-storage Runge-Kutta scheme, see [8, 9] for additional details on numerics and code validation. A bounded central difference scheme is adopted for the discretization of the convective term of the passive scalar in order to avoid spurious oscillations.

The bulk Reynolds number of the DNS is $Re_R = U_b R/\nu = 2000$, with U_b the bulk velocity and R the nozzle radius. The domain dimensions are $2\pi \times 22R \times 83R$ in the azimuthal, θ, radial, r, and axial, z, directions, respectively, with a corresponding mesh of $128 \times 145 \times 784$ nodes that is stretched in the radial and axial directions to keep a resolution comparable to the local Kolmogorov length η_k [8].

Tiny rigid and spherical particles (diameter $d_p \ll \eta_k$) are treated as material points with finite inertia evolving according to Newton's law. The mass density of the solid phase is taken much larger than that of the fluid, $\rho_p/\rho \approx 1000$. In these conditions, the only significant force acting on the particles is the viscous Stokes drag, and each particle evolves according to the equations [7]

$$\frac{d\mathbf{v}}{dt} = \frac{\mathbf{u}(\mathbf{x}) - \mathbf{v}}{\tau_p} \qquad \frac{d\mathbf{x}}{dt} = \mathbf{v}, \tag{2}$$

where $\mathbf{v}$ denotes the particle velocity and $\tau_p = \rho_p d_p^2/(\rho \nu 18)$ is the particle response time (Stokes time). The Stokes number, defined as the ratio of τ_p to the characteristic time scale of the carrier fluid, controls the particle dynamics for a given flow field.

Seven particle populations, differing in nominal Stokes number $St = \tau_p U_b/R = 2 \div 128$, are injected at the fixed rate of 600 particles per eddy turn-over time R/U_b for each population. A mixed linear-quadratic formula based on Lagrange polynomials is used to interpolate the fluid velocity $\mathbf{u}$ at the particle position $\mathbf{x}$, see eq. (2). The same three-stages third order low storage Runge-Kutta method used for the fluid phase evolves the particle populations. Further numerical details can be found in [9, 10] where the same algorithm is used for the DNS simulations of particle-laden turbulent pipe and jets.

After the particle populations achieved the statistical steady state, more than 160 time independent fields with temporal separation of $2.5R/U_b$ were collected.

2 Results & Discussions

The general behavior of the particle-laden jet can be observed in figure 1 where snapshots of three particle populations, $St = 2$; 16; 128, are provided together with a cut of the passive scalar concentration field. Lightest particles (left panel of fig. 1) behave almost as the passive scalar, persisting in the region where the concentration of the passive scalar (jet core) is appreciable. Nonetheless the signature of the inertia is apparent in terms of small-scale clustering [15]. On the contrary, particles with $St = 16$ (middle panel) show an almost even distribution up to $z \approx 50R$, location where they seem to concentrate, exhibiting an apparent small scale clustering further downstream. The heaviest particles ($St = 128$, right panel) seem more evenly distributed in the whole domain and, in contrast to the smaller ones, partially intrude the irrotational entrainment region well outside of the jet core.

To address the statistical features of the particle-laden jet, figure 2 provides the axial behavior of the centerline mean particle concentration (left panel) and of the

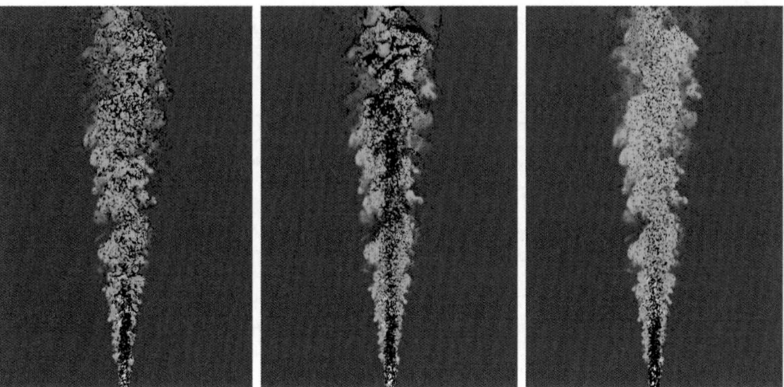

Fig. 1 Snapshots of particle positions by symbols and the corresponding passive scalar field by contours. Left: particles with $St = 2$; middle: particles with $St = 16$; right: particle with $St = 128$

mean centerline velocity (right panel). Particle populations show unexpected centerline concentration peaks that move downstream increasing the particle relaxation time. The location of these humps is controlled by the local large scale Stokes number $St_L = \tau_p/T_L$ that is based on the time scale given by $T_L = r_{1/2}/U_c$, where $r_{1/2} = S(z - z_0)$ is the jet half-width [10] and $U_c = U_b B 2R/(z - z_0)$ is the average centerline fluid velocity (in this case $S = 0.089$, $B = 6.77$ $z_0 = -0.8R$, see [8, 10]). Hence the local Stokes number $St_L = \tau_p/T_L \propto (z - z_0)^{-2}$ decreases quadratically in the downstream evolution. Peaks occur in locations where $St_L = \mathscr{O}(1)$, while particles behave almost as tracers further downstream ($St_L \ll 1$), showing a centerline concentration similar to that of a passive scalar [10]. The mean centerline particle velocity, right panel of fig. 2, is larger than that of the fluid, collapsing on it in locations where the large scale Stokes number becomes smaller than unity. The large-scale particle dynamics is actually controlled by the local Stokes number St_L.

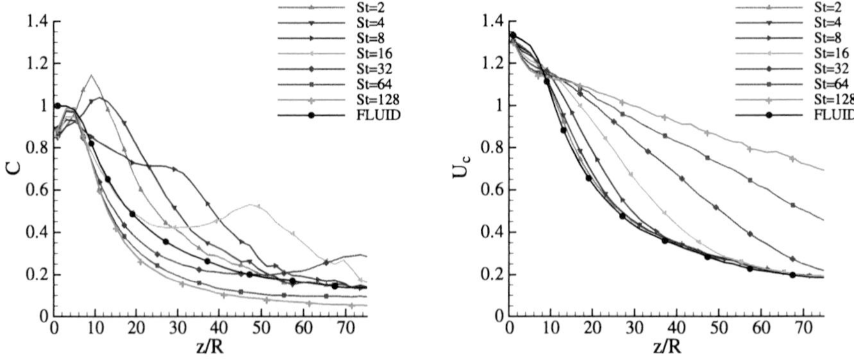

Fig. 2 Left panel: axial development of centerline mean concentration C of particles and passive scalar. Right panel: axial development of centerline mean velocity of particles and of the fluid

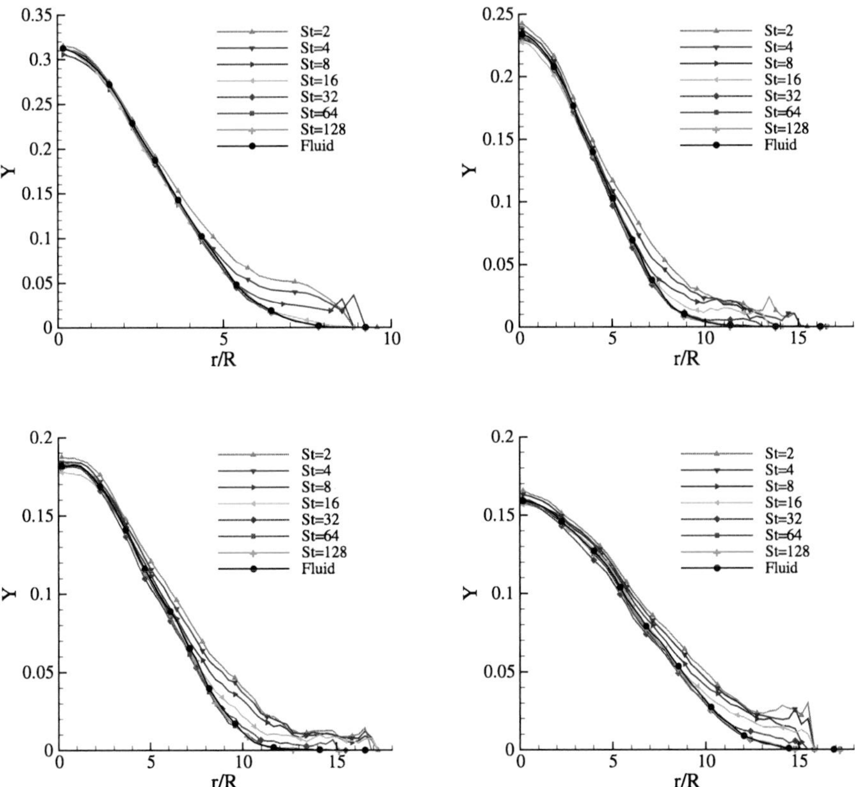

Fig. 3 Profiles of passive scalar mean concentration *sampled* by the particles vs r/R at different axial distances: top-left $z/R = 30$, top-right $z/R = 40$, bottom-left $z/R = 50$, bottom- right $z/R = 60$

Inertial effects are particularly relevant in the region where $St_L(z) \geq 1$, inducing unexpected mean centerline concentration peaks where $St_L(z) \approx 1$.

Despite particles tend to behave as passive tracers where $St_L(z) < 1$, they still exhibit some peculiarities, such as the preferential "sampling". Jets are characterized by two distinct zones: the turbulent jet core and the irrotational entrainment region, separated by an intermittent interface (the viscous super-layer). A passive scalar injected within the fluid becomes more and more diluted in the axial evolution due to the entrainment of the ambient fluid through this interface. A tracer issued by the inflow never reaches the irrotational ambient fluid, passing through the intermittent interface. This mechanism induces a sampling bias in the Eulerian statistics sampled by particles with negligible inertia which will sample only events pertaining to the turbulent jet core. Figure 3 displays the radial profiles of the mean passive scalar concentration as seen by the particles at different z/R. As apparent, the mean flow profile (unconditioned) is reproduced only by particles with $St \geq 8$ at $z/R = 30$, which becomes $St \geq 32$ at $z/R = 60$. Smaller particles, instead, experience a larger mean scalar concentration especially in the outer jet region spanned by the viscous super-layer. This behavior can be interpreted observing the instantaneous snapshots of figure 1. Since trajectories of small particles ($St = 2$) almost match those of the tracers (passive scalar), they are located in the jet core in coincidence of the passive scalar, while no particles are present in the irrotational entrainment region where the passive scalar concentration is null. On the contrary, the largest particles ($St = 128$) whose trajectories differ from those of tracers, are able to overcome the interface and are more uniformly found in both regions with no relevant sampling bias. The critical parameter, separating the two dynamics, should be given in terms of the local $St_L(z)$ and the critical value is $St_L \approx 1$, e.g. both particles with $St = 8$ at $z/R = 30$ and those with $St = 32$ at $z/R = 60$ have $St_L \simeq 1.35$. It is expected that the above sampling features of the scalar concentration field may play a role in the evaporation of small droplets in sprays.

References

1. Balachandar, S., Eaton, J.: Turbulent dispersed multiphase flow. Ann. Rev. Fluid Mech. **42**, 111 (2010)
2. Gualtieri, P., Picano, F., Casciola, C.: Anisotropic clustering of inertial particles in homogeneous shear flow. J. Fluid Mech. **629**, 25 (2009)
3. Hardalupas, Y., Taylor, A., Whitelaw, J.: Velocity and particle-flux characteristics of turbulent particle-laden jets. Proc. R. Soc. London A **426**(1870), 31 (1989)
4. Kaminski, E., Tait, S., Carazzo, G.: Turbulent entrainment in jets with arbitrary buoyancy. J. Fluid Mech. **526**, 361 (2005)
5. Longmire, E., Eaton, J.: Structure of a particle-laden round jet. J. Fluid Mech. **236**(1), 217 (1992)
6. Marchioli, C., Soldati, A., Kuerten, J., Arcen, B., Taniere, A., Goldensoph, G., Squires, K., Cargnelutti, M., Portela, L.: Statistics of particle dispersion in Direct Numerical Simulations of wall-bounded turbulence: Results of an international collaborative benchmark test. Int. Journal of Multiphase Flows **34**(9), 879 (2008)
7. Maxey, M., Riley, J.: Equation of motion for a small rigid sphere in a nonuniform flow. Phys. Fluids **26**, 883 (1983)
8. Picano, F., Casciola, C.: Small-scale isotropy and universality of axisymmetric jets. Phys. Fluids **19**, 118106 (2007)
9. Picano, F., Sardina, G., Casciola, C.: Spatial development of particle-laden turbulent pipe flow. Phys. Fluids **21**, 093305 (2009)
10. Picano, F., Sardina, G., Gualtieri, P., Casciola, C.: Anomalous memory effects on transport of inertial particles in turbulent jets. Phys. Fluids **22**, 051705 (2010)
11. Prevost, F., Boree, J., Nuglisch, H., Charnay, G.: Measurements of fluid/particle correlated motion in the far field of an axisymmetric jet. Int. J. Multiphase Flow **22**(4), 685 (1996)
12. Sbrizzai, F., Verzicco, R., Pidria, M., Soldati, A.: Mechanisms for selective radial dispersion of microparticles in the transitional region of a confined turbulent round jet. Int. J. Multiphase Flow **30**(11), 1389–1417 (2004)
13. Selle, L., Bellan, J.: Characteristics of transitional multicomponent gaseous and drop-laden mixing layers from direct numerical simulation: Composition effects. Phys. Fluids **19**, 063301 (2007)
14. Shotorban, B., Balachandar, S.: Particle concentration in homogeneous shear turbulence simulated via Lagrangian and equilibrium Eulerian approaches. Phys. Fluids **18**, 065105 (2006)
15. Toschi, F., Bodenschatz, E.: Lagrangian properties of particles in turbulence. Ann. Rev. Fluid Mech. **41**, 375 (2009)

A numerical simulation of the passive heat transfer in a particle-laden turbulent flow

Marek Jaszczur

1 Introduction

Non-isothermal turbulent flows laden with large number of particles or droplets occur in numerous situations e.g., combustion, catalytic cracking, droplet growth in clouds, etc. Due to interactions between continuous and dispersed phase the distribution of the particles and its statistical properties can be highly non-uniform [2]. This non-uniformity can have important consequences on the chemical processes affecting the efficiency of the combustion or chemical reactions. Experiments and DNS demonstrates that shear flow, mean gradient in the fluid and velocity of the particles have complex effect on the particle fluctuations.

For multiphase flows, Eulerian-Lagrangian point-particle DNS has been applied in the studies of the particle-turbulence dynamics and transport phenomena in isothermal flows. Wang and Maxey [10] performed DNS in isotropic turbulence with particles, and showed that particles with time constant of the order of the smallest turbulence scale are preferentially concentrated in the region of low vorticity and high strain-rate. For wall-bounded shear flow Pedinotti [7], Rouson and Eaton [9] have shown that particle are also preferentially concentrated in low speed streaks. McLaughlin [5] has shown that particles accumulate in the near wall region, Portela [8] shows that particles concentration in the near wall region in the low speed streaks is high. However, there do not exist similar studies dealing with heat/mass transfer. Hetsroni [3] makes an analysis of the heat transfer and thermal pattern only around a single sphere. Chagras [1] in his work focuses on heat transfer enhancement as a effect of collision but no information on the particle behaviour is provides.

Direct Numerical Simulation of the flow, combined with Lagrangian particle tracking technique has been performed to study the problem. In presented configuration small solid and dense particles carrying by fluid forces are influence by turbulent non-isothermal flow. The simulations have been performed at $Re_\tau = 180$ and

AGH - University of Science and Technology, Department of Fundamental Research in Energy Engineering, Al. Mickiewicza 30, 30059 Kraków, Poland, e-mail: jaszczur@agh.edu.pl

H. Kuerten et al. (eds.), *Direct and Large-Eddy Simulation VIII*, ERCOFTAC Series 15, DOI 10.1007/978-94-007-2482-2_31,

395, and with Pr=1.0. The main focus is on the interactions between particles and turbulence and their effect on the temperature of the particles. The influence of the particles on the fluid and inter-particle interactions have been neglected. Presented data obtained with Direct Numerical Simulation show new effects related to particle temperature and particle turbulent heat fluxes.

2 Formulation of the problem

For current wall-bounded configuration small solid and dense particles carried by fluid forces are influenced by turbulent non-isothermal flow. The interactions between the fluid and the particle are done through exchange of momentum and thermal energy. Dealing with very small volume fraction of the particles the effect of the particles on the continuity equations can be neglected. For that the Eulerian-Lagrangian point-particle approach [8] has been used. The transfer of momentum between the particle and the fluid is considered through a force located at the particle center, which is determined from the velocities of the particle and of the surrounding fluid. The Nusselt Number required to heat transfer computation has been calculated from the Ranz-Marshall correlation and is base on velocities of the particle and the surrounding fluid. The approach is valid if the particles are significantly smaller than the smallest flow scales and the Biot number is less than 0.1. The continuous-phase is solved using standard direct numerical simulation techniques for incompressible flow. The effect of particle diameter and density on the statistical quantities has been examined. The problem under consideration is shown in Fig. 1.

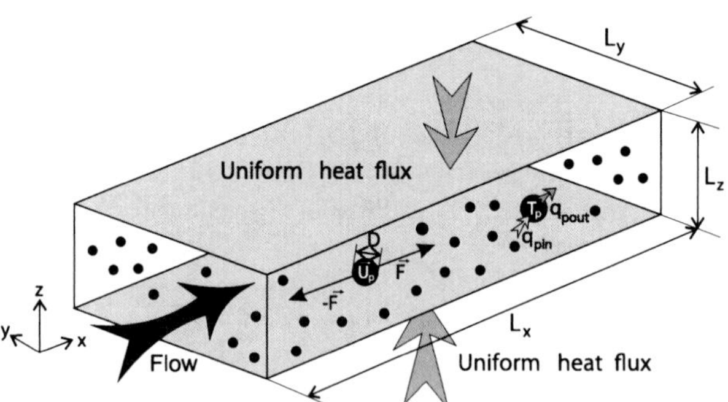

Fig. 1 Sketch of the computational domain.

The streamwise wall normal and spanwise coordinate are denoted x,y,z. The wall heat flux q_w is assumed to be uniformly distributed on the walls. The working fluid is assumed to be Newtonian fluid with constant properties. Temperature is treated as passive scalar. For small heavy particles the significant force is only the drag force [2] and the equation of motion for a particle can be written as:

$$\frac{d\mathbf{v}}{dt} = C_d \frac{Re_p}{24} \frac{1}{\tau_v} (\mathbf{u} - \mathbf{v}) \tag{1}$$

where u is the velocity of the fluid interpolated at the center of the particle. The particle Reynolds number Re_p, and the hydrodynamic particle-relaxation-time τ_v, are defined as:

$$Re_p = \mid (\mathbf{u} - \mathbf{v}) \mid \frac{D_p}{\nu} \qquad \tau_v = \frac{\rho_p}{\rho} \frac{D_p^2}{18\nu} \qquad C_d = \frac{24}{Re_p} \tag{2}$$

where ρ_p and ρ are the particle and fluid densities, D_p is the diameter of the particles and C_d is the drag coefficient for Stokes flow. The equation for the particle temperature can be written as:

$$\frac{dT_p}{dt} = \frac{Nu}{2} \frac{1}{\tau_T} (T - T_p) \qquad \tau_T = \frac{\rho_p c_p D_p^2}{12k} \tag{3}$$

where T is the temperature of the fluid interpolated at the center of the particle, and τ_T is the thermal particle-relaxation-time. The Nusselt number can be calculated from the Ranz-Marshall correlation. The position, flow, and particle quantities are normalized by the channel half-width δ, the friction velocity u_τ, and the friction temperature T_τ. The convective and diffusive terms in all the equations are discretized using a second-order central scheme. The particle motion and particle temperature algorithms are obtained with using a second-order Adams-Bashforth scheme for the time-advancement, and a tri-linear interpolation for the velocity and temperature. The channel walls are considered to be perfectly smooth and the particles are assumed to contact wall when the center is one radius from the wall, then they are bounced-back using elastic specular-reflection (elastic collisions), and when they leave the domain they are re-introduced at the opposite side of the channel with the same velocity and temperature. To force the fluid motion, a constant pressure-gradient has been imposed along the streamwise direction. The flow is heated by a uniform heat-flux from both walls; hence there is no restriction on wall-temperature fluctuations. Periodic boundary conditions are imposed in streamwise and spanwise directions.

3 Numerical results

The simulations have been performed for $Re_\tau = 180$ and 395 with a 256x128x256 computational grid, uniform in the streamwise and spanwise directions and with an hyperbolic-tangent stretching in the normalwise direction. The computation of the flow start from initial random flow field and has been time advanced to get a statistically-steady state for velocity and temperature. Then particles are assigned uniformly to the channel with initial velocity assumed to be the same as the fluid, eventually the particles get into statistically-steady state independent from the initial velocity and position. In present computations time before particles start to be averaged takes at least t*=200. The statistics for the fluid and particles were averaged for $100\delta/u_\tau$ at Re_τ=180 and for $40\delta/u_\tau$ at Re_τ=395.

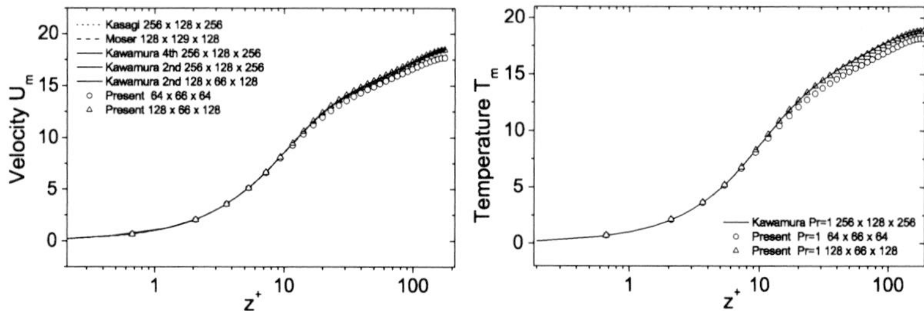

Fig. 2 The mean velocity profile for streamwise component and mean temperature for continuous phase (comparison with data from literature) at $Re_\tau = 180$.

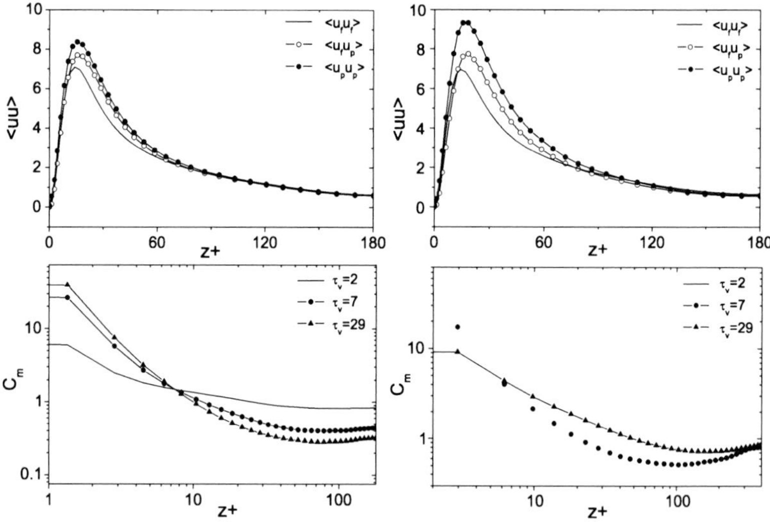

Fig. 3 Fluid streamwise velocity component fluctuation for St=2 (left) and St=7 (right) and particle concentration for St=2, 7, 29 at $Re_\tau = 180$ (left) and 395 (right).

The mean velocity profile and temperature profiles for continuous phase obtained from the DNS computations are shown in Fig. 2 for different computational grid. Results are compared with data from literature [4, 6]. Fluid streamwise velocity fluctuation for St=2 and 7 together with particle concentration for all considered here Stokes number at $Re_\tau = 180$ and 395 has been show on Fig. 3.The mean temperature profile for particle and fluid has been shown in Fig. 4. For all thermal particle response times (and momentum time $\tau_v^+ = 2$) and for most of channel height temperature of particle is smaller then temperature of the fluid. Which is different from

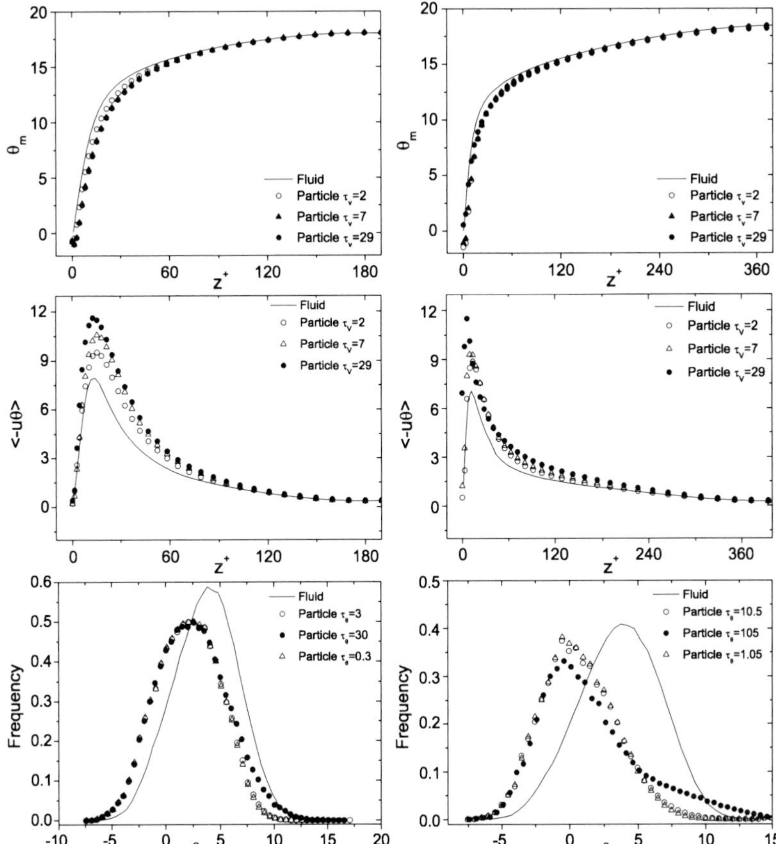

Fig. 4 Particle temperature, streamwise velocity-temperature correlation (for the plane z^+=3.6) at $Re_\tau = 180$ (left) and 395 (right). At the bottom frequency at $Re_\tau = 180$ and St=2 (left) and St=7 (right).

typical profile of streamwise velocity where in the core of channel the velocity of the particle overlaps velocity of the fluid. But the main difference can be seen close to the wall where temperature of the particle is much smaller than temperature of the fluid. This effect occurs also for relatively light particles and is more pronounced for high Reynolds number. Fig. 4 also presents velocity-temperature correlation for fluid statistics and for particles with $\tau_v^+ = 2, 7, 29$ and τ_T^+ from 0.3 up to 105 respectively. Correlations of streamwise velocity fluctuations and temperature for particle are much larger than for the fluid and the difference is increasing when particle response time increases. At the bottom of Fig. 4 frequency distribution for temperature (for particle and fluid volume elements) for the plane z^+=3.6 and $\tau_v^+ = 2$ has been presented. It can be clearly seen that this distribution is shifted to the left compared to the fluid one. For higher Reynolds number this shift is even higher.

4 Conclusion

The Eulerian-Lagrangian point-particle DNS approach has been extended in order to include heat transfer. A numerical simulation has been carried out to study particle transport and heat transport in wall-bounded turbulent flow at Re_{τ}=180, and 395. Several statistical quantities for particle velocity, temperature and concentrations and flow-particle correlations for the correlations between continuous-phase and dispersed phase have been obtained from the calculations. Mean temperature for the fluid in the boundary layer region is larger than the mean temperature for the particles. Similar effect can be found for mean streamwise velocity but the difference for temperature is larger. The mean streamwise velocity fluctuation in the boundary layer is different for the fluid and for the particle, and decorrelates with increasing Stokes number. This affects the velocity-temperature correlations which are proportional to the turbulent heat fluxes. High concentrations of the particles close to the wall in low speed streaks where temperature differ from mean temperature affect strongly mean temperature of the particles.

Acknowledgements This research was supported by the European Commission under project Dev-BIOSOFC, MTKD-CT-2006-042436.

References

1. V. Chagras, B. Oesterlé, P. Boulet, Int. J. Heat Mass Transfer **48**, 1649–1661 (2005)
2. J.K. Eaton, J. Fessler, Int. J. Multiphase Flow **20**, 169–209 (1994)
3. G. Hetsroni, C.-F. Li, A. Mosyak, I. Tiselj, Int. J. Multiphase Flow **27**, 1873–1894 (2002)
4. N. Kasagi, Int. J. of Heat and Fluid Flow **19**, 125–134 (1998)
5. B. McLaughlin, Phys. Fluid A **1**, 186–203 (1975)
6. R.D. Moser, J. Kim, N.N. Mansour, Phys. Fluids **11**, 943–945 (1999)
7. S. Pedinotti, G. Mariotti, S. Banerjee, Int. J. Multiphase Flow **18**, 927–941 (1992)
8. L.M. Portela, R.V.A. Oliemans, Int. J. Num. Fluids **43**, 1045–1065 (2003)
9. D. Rouson, J. Eaton, in Proc. of 7th Workshop on Two-Phase flow Predictions, Erlangen 1994
10. L. Wang, M. Maxey, J. Fluid Mech. **256**, 27–68 (1993)

Effect of evaporation and condensation on droplet size distribution in turbulence

Briti S. Deb, Lilya Ghazaryan, Bernard J. Geurts, Herman J.H. Clercx, J.G.M. Kuerten and Cees W.M. van der Geld

1 Introduction and Motivation

The interaction of droplets that undergo phase transition with a turbulent flow is encountered in many areas of engineering and atmospheric science as described in [4]. In the context of cloud physics the evaporation and condensation of water vapor from and to the droplets is the governing process for the growth of the droplets from sub micron size up to a size of around 20 μm, after which they grow mostly by coalescence until they become large enough to fall as rain drops under gravity. Much pioneering work has been done in [3, 4, 6] on the theoretical and numerical investigation of the influence of turbulence on evaporation and condensation associated with aerosol droplets. In this paper we consider the situation of water droplets undergoing phase change and moving in air. Air also advects the vapor concentration field. We compute the natural size distribution of the droplets that arises as a result of the interaction between the droplets and the transporting turbulent flow. We assume the turbulent flow to be homogeneous and isotropic. We will perform DNS of the velocity field and the passively advected vapor and temperature field. The droplet trajectories are computed time-accurately in a domain with periodic boundary conditions.

From our simulation results we inferred the dynamical behavior of inertial particles moving in a homogeneous, isotropic turbulent flow. We have computed the probability density function (PDF) of the droplets radii at various points in time.

Briti S. Deb, Lilya Ghazaryan, Bernard J. Geurts
Multiscale Modeling and Simulation, Faculty EEMCS, J.M. Burgers Center, University of Twente, P.O. Box 217, 7500 AE Enschede, The Netherlands,
e-mail: {b.s.deb,l.ghazaryan,b.j.geurts}@utwente.nl

Herman J.H. Clercx, Hans G.M. Kuerten, Cees W.M. van der Geld
Fluid Dynamics Laboratory, Faculty of Applied Physics.
Process Engineering, Faculty of Mechanical Engineering.
Eindhoven University of Technology, P.O. Box 513, 5600 MB Eindhoven, The Netherlands,
e-mail: {H.J.H.Clercx,J.G.M.Kuerten,C.W.M.v.d.Geld}@tue.nl

H. Kuerten et al. (eds.), *Direct and Large-Eddy Simulation VIII*,
ERCOFTAC Series 15, DOI 10.1007/978-94-007-2482-2_32,

Starting from an ensemble of equal-sized small droplets, it was observed that the distribution becomes more wider as time increases, while the location of the peak gradually increases to large droplets. It is seen that these PDFs resemble locally a more-or-less Gaussian distribution. The full convergence toward a statistically stationary state is seen to be a very slow process.

The paper is organized in the following way. In Section 2 we will introduce the mathematical models that we have used for the air flow, droplets trajectory and the passive scalars. In Section 3 we focus on the details of the numerical tools that we have used and the statistics of the results that we have got from our simulation. The concluding remarks are presented in Section 4.

2 Mathematical Models in Dimensionless Form

In this Section we first present the Lagrangian description of particle trajectories and the treatment of the evaporation and condensation processes in Subsection 2.1. The fluid mechanical model for the flow field and the passive scalars is given in Subsection 2.2.

2.1 Dispersed Phase

Various forces affect the trajectory of droplets in the flow field. The dominant contribution to the force in the flow we consider is the drag force or Stokes drag. We consider the volume fraction of the droplets in the domain to be small so collisions and coalescence among the droplets is neglected. Under these assumptions, from [5] the equation governing the particle position $\mathbf{x}$ and velocity $\mathbf{v}$ becomes:

$$\begin{aligned} \frac{d\mathbf{v}}{dt} &= \frac{\mathbf{u}-\mathbf{v}}{\mathrm{St}} \\ \frac{d\mathbf{x}}{dt} &= \mathbf{v} \end{aligned} \tag{1}$$

where we evaluate the fluid velocity $\mathbf{u}$ at the instantaneous location of the particle. Here St is a dimensionless number called the Stokes number of the particle, which is defined in the following way:

$$\mathrm{St} = \frac{\rho_d}{18\rho_g}\left(\frac{\mathrm{R}}{\eta}\right)^2$$

where η is the Kolmogorov length scale, ρ_d (ρ_g) the droplet (gas) mass density and R is the radius of the droplet. In this paper we will consider the particles as water droplets and the gas to be air with the initial Stokes number around 0.2.

The focus in this paper is on the transport of dispersed droplets by a turbulent flow, in which phase transition through evaporation and condensation takes place. We are considering the situation where the radii of the droplets are larger than the mean free path of the carrier gas. The droplet size is quite large in the cases considered, so the change in size of the droplets by evaporation and condensation, which takes place as a result of exchange of vapor between the droplet surface and the surrounding, is governed by diffusion laws. Then from [1] the rate of change of the radius R of the droplet obeys the following equation,

$$\frac{1}{\mathrm{R}}\frac{\mathrm{dR}}{\mathrm{dt}} = \frac{\mathrm{Sh}}{72\ \mathrm{Sc}\ \mathrm{St}}(\mathrm{C}_{v,g} - \mathrm{C}_{v,d}) \tag{2}$$

In this expression Sh and Sc are respectively the Sherwood number and Schmidt number; $\mathrm{C}_{v,d}$, $\mathrm{C}_{v,g}$ are the mass fractions of the vapor on the droplet surface and of the gas at the location of the droplet. The mass fraction on the droplet surface $\mathrm{C}_{v,d}$ can be expressed in terms of the ratio of the partial vapor pressure to the total pressure on the droplet surface. Now considering the total pressure to be constant, the partial vapor pressure on the droplet surface can be evaluated in terms of the temperature of the droplet by Antoine's equation; which is actually derived from the Clausius-Clapeyron saturation law used in [3].

The temperature evolution of the droplet is influenced both by the local temperature difference with the gas and change in mass of the droplet due to phase change. Using a characteristic temperature T_0 of the flow field as the reference temperature, we can express the droplet temperature equation in dimensionless form as;

$$\frac{\mathrm{dT}_d}{\mathrm{dt}} = \frac{\mathrm{Nu}}{3\mathrm{Pr}}\frac{c_g}{c_d}\frac{\mathrm{T}_g - \mathrm{T}_d}{\mathrm{St}} + \frac{h_L}{\mathrm{T}_0 c_d}\frac{1}{\mathrm{m}}\frac{\mathrm{dm}}{\mathrm{dt}} \tag{3}$$

where h_L is the latent heat and c_d and c_g are the specific heat of the droplet and the gas respectively. Nu and Pr are respectively the Nusselt and Prandtl numbers. In our simulation Nu and Sh are taken to be equal to be 2 and Pr is equal to 1. The value of the ratio $\frac{\mathrm{Nu}}{3\mathrm{Pr}}\frac{c_g}{c_d}$ is around 0.159 and that of $\frac{h_L}{\mathrm{T}_0 c_d}$ is 540.66.

2.2 Continuum Phase

Our principal goal here is to model the turbulent velocity field $\mathbf{u} = (u, v, w)$ which is advecting both the droplets and the vapor. The flow field we consider is assumed to be homogeneous and isotropic. We model the three dimensional velocity field $\mathbf{u}$ in dimensionless form by the incompressible Navier-Stokes equation

$$\frac{\partial \mathbf{u}}{\partial \mathrm{t}} + (\mathbf{u}\cdot\nabla)\mathbf{u} = -\nabla p_g + \frac{1}{\mathrm{Re}}\nabla^2\mathbf{u} + \mathbf{f} \tag{4}$$

$$\nabla\cdot\mathbf{u} = 0$$

where p_g is the pressure of the gas, **f** is the external forcing term and Re is the Reynolds number based on a reference length, a reference velocity and the kinematic viscosity of the fluid.

Both the temperature of the gas T_g and vapor mass fraction of the gas $C_{v,g}$ are modeled as passive scalars in the flow field and their transport is governed by the advection-diffusion equation. This implies that these quantities do not affect the flow. In addition we decompose the temperature T_g of the gas into its mean value, denoted by $\langle T \rangle$, and its fluctuating part $\tilde{T}$. The mean temperature varies linearly and for our purpose it suffices to assume that $\langle T \rangle = \alpha x$. Under these assumptions the resulting equations becomes

$$\frac{\partial \tilde{T}}{\partial t} + (\mathbf{u} \cdot \nabla)\tilde{T} - \kappa_t \nabla^2 \tilde{T} = -\alpha u \tag{5}$$

$$\frac{\partial C_{v,g}}{\partial t} + (\mathbf{u} \cdot \nabla) C_{v,g} - \kappa_v \nabla^2 C_{v,g} = -\Sigma_i \, \delta(x - x_i) \frac{dm_i}{dt} \tag{6}$$

where κ_t and κ_v are respectively the diffusivity of the temperature and vapor field which are equal to 10^{-3} in our case. Here δ is the Dirac delta function and x_i is the instantaneous location of the ith particle. The term on the right hand side of the equation accounts for the fact that evaporation and condensation locally affect the vapor concentration field.

3 Numerical Schemes and Results

We use a de-aliased pseudo-spectral method for both the velocity and the passive scalars in order to perform direct numerical simulation for the turbulent flow. In this method the velocity and the scalar fields in the Navier-Stokes and the advection diffusion equations are represented by their Fourier series. The numerical integration of the equations is done as in [2] by the four-stage, second order, compact storage, Runge-Kutta method. We consider the computational domain to be a cubic box of length one with periodic boundary conditions in each direction. We choose large scale forcing of the velocity field in such a way that the kinetic energy in the collection of forced modes remains constant. In our simulations the Reynolds number is taken as 1061 and the corresponding Taylor Reynolds number as 50.

The numerical treatment of the discrete droplet phase proceeds in a number of steps. We distribute the droplets initially randomly in space and provide as initial velocity the fluid velocity as found at the particle location. We assume that all the particles have initially the same radius. After a sufficiently long evolution the ensemble of droplets evolves into a statistically stationary distribution which is the result of evaporation and condensation in the forced turbulent flow. The initial temperature of the particle is set to be equal to that of the initial temperature of the air at the particle position. In order to get information of the flow field such as velocity, temperature and vapor mass fraction at the location of a particle we perform tri-linear interpolation. The time integration of the particle trajectories was performed using the forward Euler method.

DNS of the flow of a dispersed ensemble of droplets undergoing phase transition in a homogeneous, isotropic turbulent flow requires the specification of a number of parameters. In Figure 1(a) we have presented the PDF of the droplets radii for a sample of 10^4 droplets at time $130\tau_L$, where $\tau_L = 0.4$ is the large eddy turn over time assuming that initially all droplets have the same radius which is $5 \cdot 10^{-4}$. We have shifted the origin of the radii to zero using the relation $(\mathrm{R} - \mathrm{R}_{peak})$ where $\mathrm{R}_{peak} = 7.07 \cdot 10^{-4}$ is the radius at which the PDF has a peak at time $130\tau_L$. We focus on the evolving radius of all the droplets and compute the corresponding distribution function. In Figure 1(b) we display the location of the peak of the distribution function (R_{peak}) as a function of time for a sample of 10^2 droplets. Here by location we mean the radius at which the size distribution curve corresponding to a particular time has the highest value. The initial condition corresponds to a sharply peaked PDF in which all droplets have the same radius. The location of the peak of this distribution is increasing with time due to condensation. This growth gradually decreases as the statistically stationary state is approached. In order to achieve a more precise approximation of this ultimate PDF a longer simulation time is required. The evolution of the PDF of the droplets size-distribution as a function of

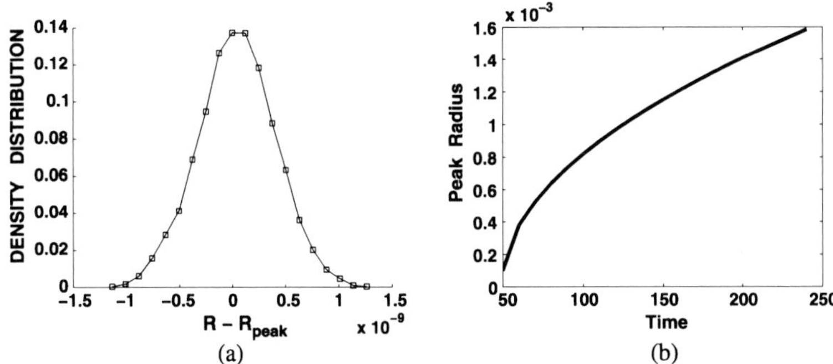

Fig. 1 A typical PDF for a sample of 10^4 droplets is plotted against $(\mathrm{R} - \mathrm{R}_{peak})$ shown in Figure 1(a) where R_{peak} is the radius at which the PDF has a peak. While in Figure 1(b) we see the evolution of R_{peak} corresponding to different PDFs as a function of time

time is displayed in Figure 2. One can see from the plot that the distribution curve spreads more with the increase in time. This broadening of the size distribution is due to the fact that as time increases the droplet move each through different regions in the flow field experiencing different vapor concentration fields and temperatures, which ultimately results in different radii of the droplets. The shape of the PDF and its dependence on physical parameters of our model will be investigated numerically in the future and published elsewhere.

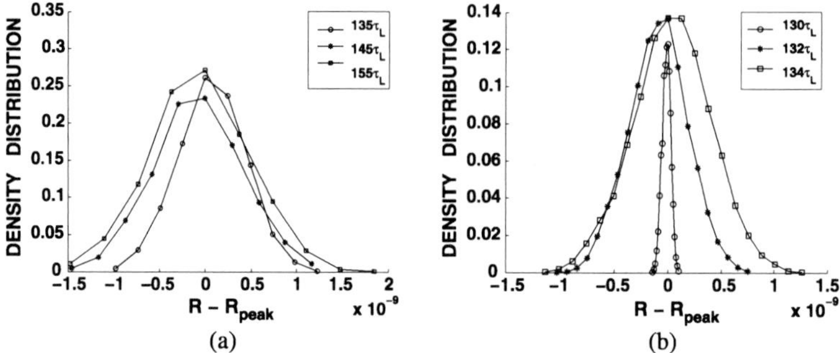

Fig. 2 The figures shows the broadening of the size distribution function for $3 \cdot 10^3$(Figure 2(a)) and 10^4(Figure 2(b)) droplets with increase in time. Here we have shifted the origin to zero using the relation $\mathrm{R} - \mathrm{R}_{peak}$, where R_{peak} is the radius at which the PDFs has a peak.

4 Concluding remarks

In this paper we have studied the turbulent transport of water droplets in air, that undergo phase transition represented by evaporation and condensation. The simulation was based on DNS using pseudo-spectral discretization. We have developed a two-way coupled mathematical model of the vapor concentration field with the change in mass of the droplets and observed the effect of turbulence on the size distribution of the droplets. From our simulation we observed that the PDF of the size distribution becomes wider as time increases. In the future we will incorporate the effect of the droplet temperature on the gas temperature equation and we will also compare our simulation results with experimental data.

Acknowledgements The authors gratefully acknowledge financial support from the Dutch Foundation for Technical Sciences, STW. This project is part of the Multiscale Simulation Techniques program. The numerical simulations have been made possible through a grant from NCF - SH061.

References

1. Crowe, C., Sommerfeld, M., Tsuji,Y.: Multiphase flows with droplets and particles, *CRC Press* (1998)
2. Geurts, B. J.: Elements of Direct and Large-Eddy Simulation. R.T. Edwards (2004)
3. Luo, K., Desjardins, O., Pitsch, H.: DNS of droplet evaporation and combustion in a swirling combustor. Center for Turbulence Research, Annual Research Briefs (2008)
4. Lanotte, A. S., Seminara, A., Toschi, F.: Condensation of cloud microdroplets in homogeneous isotropic turbulence. arXiv, **2**, 0710.3282 (2008)
5. Maxey, M. R., Riley, J. R.: Equation of motion for a small rigid sphere in a non uniform flow. Phys. Fluids, **26**(4) (1983)
6. Sidin, R. S. R., IJzermans, R. H. A., Reeks, M. W.: A Lagrangian approach to droplet condensation in atmospheric clouds. Phys. Fluids, **21**, 103303 (2009)

Reduced turbophoresis in two-way coupled particle-laden flows

D.G.E. Grigoriadis and B.J. Geurts

1 Introduction

Direct numerical simulation of turbulent channel flow is employed to show that two-way coupling effects in particle-laden flows leads to reduced preferential clustering and turbophoresis [9] even for low values of volume and mass fraction. The effect of including two-way coupling on the phenomenon of turbophoresis and preferential clustering is studied for particles of different response times at a variety of loading ratios.

We adopt the Euler-Lagrange formalism in which a large number of discrete point particles is embedded in a continuous flow description. We consider particle-laden turbulent flows for different values of the volume fraction $\Phi_v = N_p V_p / V_{tot}$ and mass fraction $\Phi_m = (\rho_p/\rho_f)\Phi_v$ defined in terms of the number of particles N_p, each having volume V_p and flowing in the flow domain of volume V_{tot}. The mass density of the particles is denoted by ρ_p, while ρ_f represents the fluid density.

For values of the volume fraction smaller than $\Phi_v \leq O(10^{-6})$, the interaction between the two phases is classified as a *one-way coupling* [4] since momentum exchange only occurs from the fluid to the particulate phase. For volume fractions values higher than $\Phi_v \geq O(10^{-6})$, the interaction is classified as a *two-way coupling interaction* where the exchange of momentum between the two phases can no longer be neglected.

Dimokratis G.E. Grigoriadis
Department of Mechanical and Manufacturing Engineering, University of Cyprus, 75 Kallipoleos Avenue, P.O. Box 20537 Nicosia, 1678, Cyprus, e-mail: grigoria@ucy.ac.cy

Bernard J. Geurts
Multiscale Modeling and Simulation, Faculty EEMCS, University of Twente, P.O. Box 217, 7500 AE Enschede, The Netherlands
Anisotropic Turbulence, Laboratory for Fluid Dynamics, Faculty Applied Physics, Eindhoven University of Technology, P.O. Box 513, 5600 MB Eindhoven, The Netherlands, e-mail: b.geurts@utwente.nl

H. Kuerten et al. (eds.), *Direct and Large-Eddy Simulation VIII*, ERCOFTAC Series 15, DOI 10.1007/978-94-007-2482-2_33,

Turbulence modification due to the presence of small dispersed particles is expected to be strengthened in regions where preferential concentration effects are important. As shown by [3] from DNS simulations at $\Phi_v = 6.8 \times 10^{-5}$ and $\Phi_m = 0.5$, the inclusion of two-way coupling modifies the local coherency of the flow and the particulate phase, thereby contributing to a modification of preferential clustering effects.

Several studies have been dedicated to understanding the influence of dispersed particle motion on turbulence. As pointed out by [2] in a recent review of turbulence modulation this mechanism is still purely understood in particle laden flows, and several contradictory results have been presented.

For a turbulent channel flow, the damping of fluid turbulence with increasing volume fraction for loading ratios up to $\Phi_m = 0.8$ was reported experimentally by [10] and [13] but such an attenuation was not captured by LES simulations performed by [14]. In contrast, [11], found significant feedback effects for a vertical channel at $Re_\tau = 125$. Using a coarse DNS, at loading ratios in the range of $\Phi_m = 0.2 - 2$, they reported that the phase coupling reduces the near-wall concentration and increases the anisotropy of turbulence in the carrier phase. Apparently, the capturing of small-scale turbulence structures in a flow is essential to represent the dynamics of small particles as emphasized in [8].

To the Authors knowledge, the effect of two-way coupling and modified particle clustering in the lower range of two-way coupling interactions has not been studied before; it forms the objective of the present paper. We report DNS simulations of two-way coupled suspended particles at volume and mass ratios of $10^{-6} < \Phi_v < 10^{-4}$ and $0.0007 < \Phi_m < 0.077$ respectively. This range of loading for a two-way coupled suspension is much lower than previously reported in two-way coupling studies [11, 3] because so far it was considered to have a negligible effect on preferential clustering and turbophoresis.

2 Mathematical modeling of particle-laden flow

2.1 Fluid phase

The fluid motion is governed by the continuity equation $\nabla . u_i = 0$ and the Navier-Stokes equations for an incompressible fluid formulated in a dimensionless form as,

$$\frac{\partial u_i}{\partial t} + \frac{\partial u_i u_j}{\partial x_j} = -\frac{\partial p}{\partial x_i} + \frac{1}{Re_\tau} \frac{\partial^2 u_i}{\partial x_j^2} - f_{i,R}, \tag{1}$$

where $Re_\tau = (u_\tau h)/\nu_f$ is the characteristic Reynolds number based on friction velocity u_τ, the kinematic viscosity ν_f and the half width of the channel h. The last term $f_{i,R}$ in the *rhs* of Equation (1) accounts for the two-way coupling between the two phases. It represents the feedback effect of the particles on the fluid carrier, i.e. the reactive force density exerted by the motion of particles on the fluid.

A cartesian flow solver was used for the present study to numerically solve the Navier-Stokes equations using the fractional time-step approach and a fully explicit projection scheme with pressure correction. The spatial discretisation was based on a second-order finite-difference scheme on orthogonal grids with a staggered variable arrangement [6].

2.2 Particulate phase

Following a Lagrangian approach for the particulate phase, an equation of motion is solved for each of the embedded particles. These are treated as point-particles assuming that they are solid, rigid and spherical, colliding fully elastically with the wall boundaries. The importance of particle inertia is expressed by the particle relaxation time τ_p is defined as,

$$\tau_p = \frac{{d_p}^2}{18\nu_f}\frac{\rho_p}{\rho_f}, \tag{2}$$

where the subscripts f and p denote the fluid and particulate phase respectively and d_p denotes the particle diameter. Using $\tau_f = h/u_\tau$ as the characteristic time scale of the flow, the Stokes number St_τ which is defined as the ratio between the particulate time scale τ_p and the fluid time scale τ_f becomes,

$$St_\tau = \frac{\rho_p}{\rho_f}\left(\frac{d_p}{h}\right)^2\frac{Re_\tau}{18} = \frac{St^+}{Re_\tau}, \tag{3}$$

where St^+ is the Stokes number defined with respect to the fluid time scale $\tau_f^+ = \nu/u_\tau^2$.

Due to the density ratio ($\rho_p/\rho_f = 769.23$) and the size of the released particles ($d_p/h < 3.3\times10^{-3}$) used here, only the drag force was considered to contribute to the equation of motion [1, 12]. Particle lift, buoyancy, Basset force, added mass and rotation effects were thus neglected for the present analysis. Under these assumptions, the motion of a dispersed particle located at $\mathbf{x}_p$ and travelling at velocity $\mathbf{u}_p$ is governed by the dimensionless equation,

$$\frac{d\mathbf{u}_p}{dt} = \frac{\mathbf{u}_f(\mathbf{x}_p) - \mathbf{u}_p}{St_\tau}(1 + 0.15{Re_p}^{0.687}) \equiv \mathbf{H}_p, \tag{4}$$

where we introduced the short-hand notation $\mathbf{H}_p$ for the right hand side and $Re_p = \frac{d_p|\mathbf{u}_p - \mathbf{u}_f|}{\nu_f}$ is the particle Reynolds number. The local fluid velocity $\mathbf{u}_f(\mathbf{x}_p)$ at the location of the particle $\mathbf{x}_p$ is computed based on the interpolation scheme presented in [7]. The location of each particle $\mathbf{x}_p$ is updated by integrating $\frac{d\mathbf{x}_p}{dt} = \mathbf{u}_p$, using a simple Euler scheme.

Case	St^+ $\frac{\tau_p}{(\nu/u_\tau{}^2)}$	Φ_v $\times 10^6$	d_p/h $\times 10^3$	Φ_m (%)	N_p
A1	2.5	1	1.612	0.077	71938
A2	2.5	10	1.612	0.769	719383
A3	2.5	100	1.612	7.692	7193834
B1	5	1	2.280	0.077	25434
B2	5	10	2.280	0.769	254340
B3	5	100	2.280	7.692	2543405
C1	10	1	3.225	0.077	8992
C2	10	10	3.225	0.769	89922
C3	10	100	3.225	7.692	899229

Table 1 Test cases considered for the case of two-way coupled particle-laden channel flow at $Re_\tau = 150$.

2.3 Force coupling

The 'point force approximation' is used to account for the coupling between the two phases. The point force exerted by each particle on the surrounding fluid is entirely allocated to the finite fluid element that contains the particle [5]. Thus, the reactive force per unit mass in the $N-S$ Equation (1) contributing to the total flux at location **x**, is given by,

$$f_{i,R} = \frac{1}{m_f} \sum_{p=1}^{M} F_{i,p}, \tag{5}$$

where m_f is the fluid mass contained in a grid cell and $F_{i,p}$ is the actual force exerted by particle p along direction i. The summation is over the M particles contained in the control volume around **x**. With the adopted definition of the particle's acceleration (equation 4), and assuming monodisperse particles of equal mass m_p, the total force exerted by the particles on each control volume becomes,

$$f_{i,R} = \frac{m_p}{m_f} \sum_{p=1}^{M} H_{i,p} = \frac{\phi_m}{M} \sum_{p=1}^{M} H_{i,p}, \tag{6}$$

where $\phi_m = \frac{M m_p}{m_f}$ is the local mass fraction.

3 Results

DNS of particle-laden fully developed turbulent channel flows were conducted at a constant flow rate for a Reynolds number of $Re_\tau = 150$ [8] as shown in Table 1. The computational domain used here has dimensions $4\pi h \times 2\pi h \times 2h$ using a numerical resolution of $128 \times 96 \times 90$ cells.

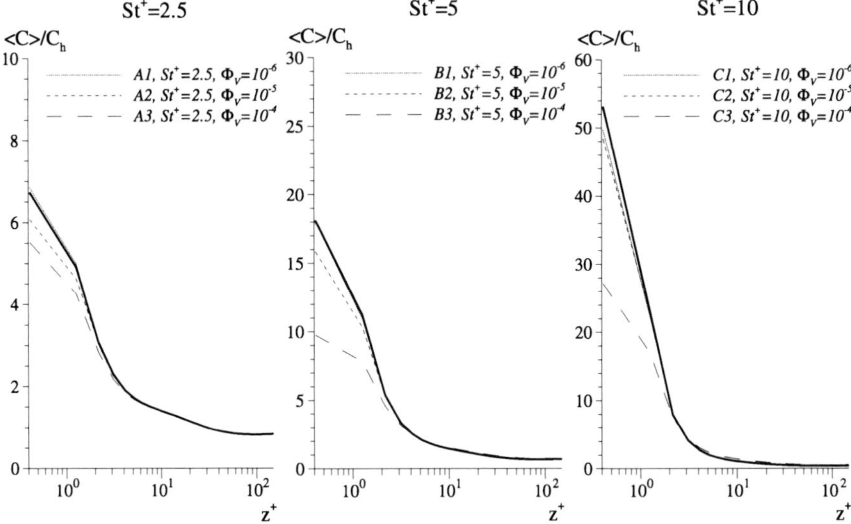

Fig. 1 Time-averaged particle concentration normalized with the homogeneous concentration C_h for various St numbers. (Solid lines): one-way coupling, (dashed lines): two-way coupling.

After a fully developed turbulent state was reached for a particle-free flow, the particles were introduced at randomly selected locations within the domain. The simulation was performed for a total time of 72 flow-through times or 60 time units h/u_τ. Statistics were collected and presented for the second half of this time period, i.e., for $30 < t < 60$.

All the present results indicate that the degree of particle clustering in the near-wall region and the strength of turbophoretic effects, are significantly modified by adding two-way momentum exchange between the dispersed particulate phase and the carrying fluid flow. The overall importance of including a two-way coupling approach was a strong function of the St^+ number. For light particles, even for the highest loading ratio of $\Phi_v = 10^{-4}$ where a two-way coupling regime would be justified [4], two-way coupling had a much smaller effect on turbophoresis and the centerline concentration when compared to those obtained for larger particles at the same volume and mass fractions (Cases B3 and C3).

For the highest particle loading at a mass fraction $\Phi_m = 7.7\%$ considered here, the inclusion of two-way coupling effects was found to reduce turbophoresis by 18%, 46% and 50% for particles with dynamic response characterized by Stokes numbers $St^+ = 2.5, 5$ and 10 respectively (Figure 1).

4 Conclusions

Direct numerical simulations of particle laden turbulent channel flow is employed to show that two-way coupling in particle-laden flows leads to reduced preferential clustering. It was demonstrated that with the inclusion of long-range two-way coupling effects, the average volume and mass loading fractions Φ_v and Φ_m only offer a rough indication on the degree of turbulence modulation and the associated preferential clustering and turbophoresis effects.

By demonstrating the relevance of two-way coupling in relation to preferential clustering, we can conclude that not only this coupling can significantly attenuate clustering and turbophoretic effects for heavier particles, but also that because of these preferential clustering mechanisms, two-way coupling expresses itself at rather lower average mass fractions than previously reported.

References

1. Armenio, V. and Fiorotto, V. (2001). The importance of the forces acting on particles in turbulent flows. *Phys. Fluids*, **13**(8), 2437.
2. Balachandar, S. and Eaton, J.K., (2010). Turbulent dispersed multiphase flow. *Annual Review of Fluid Mechanics*, **42**, 111–133.
3. Dritselis, C.D. and Vlachos, N.S. (2008). Numerical study of educed coherent structures in the near-wall region of a particle-laden channel flow. *Phys. Fluids*, **20**, 1790–1801.
4. Elghobashi, S. (2006). An Updated Classification Map of Particle-Laden Turbulent Flows. S. Balachandar and A. Prosperetti (eds), Proceedings of the *IUTAM Symposium on Computational Multiphase Flow*, Springer.
5. Elghobashi, S. and Truesdell, G.C. (1993). On the two-way interaction between homogeneous turbulence and dispersed solid particles I: Turbulence modification. *Phys. Fluids A*, **5**, 1790–1801.
6. Grigoriadis, D.G.E., Bartzis, J.G. and Goulas, A. (2004). Efficient treatment of complex geometries for large eddy simulations of turbulent flows. *Computers and Fluids*, **33**(2), 201–222.
7. Grigoriadis, D.G.E. and Kassinos, S.C. (2009), Lagrangian particle dispersion in turbulent flow over a wall mounted obstacle. *Int. J. of Heat and Fluid Flow*, **30**(3), 462–470.
8. Kuerten, J.G.M. (2006). Subgrid modeling in particle-laden channel flow. *Phys. Fluids*, **18**, 025108.
9. Kuerten, J.G.M. and Vreman, A.W. (2005). Can turbophoresis be predicted by large-eddy simulation? *Phys. Fluids*, **17(1)**, 011701.
10. Kulick, J.D., Fessler, J.R. and Eaton, J.K., (1994). Particle response and turbulence modification in fully developed channel flow. *J. Fluid Mech.*, **277**, 109–134.
11. Li, Y., McLaughlin, J.B., Kontomaris K. and Portela, L. (2001). Numerical simulation of particle-laden turbulent channel flow. *Phys. Fluids*, **13**, 2957–2967.
12. Marchioli, C., Picciotto, M. and Soldati, A. (2007). Influence of gravity and lift on particle velocity statistics and transfer rates in turbulent vertical channel flow. *Int. J. Multiphase Flow*, **33**(3), 227–251.
13. Paris, A.D., Eaton, J.K. (2001). Turbulence attenuation in a particle-laden channel flow. Report TSD-137, Stanford University, CA.
14. Segura, J.C., Eaton, J.K., Oefelein, J.C. (2004). Predictive capabilities of particle-laden large eddy simulation. Report TSD-156, Stanford University, CA.

Development of a particle laden pipe flow: implications for evaporation

G. Sardina, F. Picano, P. Gualtieri, C.M. Casciola

1 Introduction

Turbulent motions of a multiphase fluid are present in many natural processes and technological applications. In clouds the interactions between turbulence and dispersed droplets lead to the growth of water micro-particles by means of collisions and coalescence that may accelerate the rain. In engineering, transport and evaporation phenomena of micro-droplets in pipes are crucial in the mixing between the fuel and oxidizer in combustion chambers. A good mixing efficiency between the two phases enables smaller and lighter burners.

Particles/droplets dynamics in turbulent fields is a complex process that involves a great number of degrees of freedom and properties such as inertia, vapor diffusivity and collisions. In the limit of small, diluted particles heavier than the carrier fluid, the leading force acting on a particle is the viscous Stokes drag [2]. The particle velocity V evolves according to: $\dot{V} = (U - V)/\tau_p$, where $\tau_p = d_p^2 \rho_p/(18\rho_f \nu)$ is the Stokes response time (d_p, ρ_p are the particle diameter and density respectively; ρ_f and ν are the fluid density and viscosity). Normalizing the response time with a characteristic flow time scale, e.g. the viscous time (ν/U^{*2}, with U^* friction velocity) we obtain the so called friction Stokes number St^+. Fluid particles are recov-

G. Sardina
Dipartimento di Meccanica e Aeronautica, via Eudossiana 18, 00184 Rome, Italy, e-mail: gaetano.sardina@uniroma1.it

F. Picano
Dipartimento di Meccanica e Aeronautica, via Eudossiana 18, 00184 Rome, Italy, e-mail: francesco.picano@uniroma1.it

P. Gualtieri
Dipartimento di Meccanica e Aeronautica, via Eudossiana 18, 00184 Rome, Italy, e-mail: paolo.gualtieri@uniroma1.it

C.M. Casciola
Dipartimento di Meccanica e Aeronautica, via Eudossiana 18, 00184 Rome, Italy, e-mail: carlomassimo.casciola@uniroma1.it

H. Kuerten et al. (eds.), *Direct and Large-Eddy Simulation VIII*, ERCOFTAC Series 15, DOI 10.1007/978-94-007-2482-2_34,

ered in the limit of zero Stokes time, while the opposite limit of ballistic particles is achieved for large St^+. The difference between particle velocity V and fluid velocity U gives origin to various anomalous phenomena such as small-scale clustering [1] and preferential accumulation at the wall (turbophoresis) even for incompressible flows [8]. In fact, in the limit of Lagrangian tracers, the particle distribution is uniform, while in the case of finite inertia, particles tend to accumulate in the periphery of vortical structures at scales smaller than Kolmogorov length. This behavior is crucial in droplet dynamics because the collisions and consequent coalescence process is strongly enhanced by turbulent fluctuations. While small scale clustering is present also in homogeneous flows, turbophoresis is peculiar of wall flows. In these conditions particles tend to migrate towards the walls where they remain entrapped for very long times. Also this process can play a fundamental role for the dynamics of evaporation, since, droplets tend to reach the wall with the turbophoretic drift velocity, entering in a region with high level of vapor saturation that can slow down the evaporation process.

2 Numerical methodology

Our objective is to perform several simulations of droplet-laden turbulent pipe flows in order to study some aspects of the evaporation mechanism. The particles/droplets are very diluted in order to neglect momentum feedback on the carrier phase and inter-particle collisions, one-way coupling regime. The solver adopts staggered second order finite differences for the spatial discretization and a third order Runge-Kutta scheme for the temporal advancement in order to solve the Navier-Stokes equations for the carrier fluid. The dispersed phase is evolved in a Lagrangian formulation and a mixed linear/second order accurate interpolation for the fluid velocity at the particle position is implemented. While the fluid is evolved in a base domain element with axial length larger than the correlation length, the particles/droplets occupy a larger domain given by several replications of the base computational element. The same code coupled with a scalar field representing the vapor phase is used to perform simulations of the evaporation of droplets. In particular, the vapor is evolved in the same domain of the particles and a TVD scheme is adopted for the nonlinear terms of the equation. The vapor and droplets are coupled by a source term accounting for droplet mass depletion during evaporation. Further details about the numerical algorithm can be found in [6, 7]. Two kinds of DNSs have been performed, the former deals with the spatial development of inertial particles in a long pipe flow without evaporation, the latter adopts the same geometrical configuration but in presence of evaporation of the droplets. DNS simulations concerning evaporation problems are very rare, one of the more recent can be [4]. The present simulations are characterized by a nominal Reynolds number $Re = U_b R/\nu = 3000$, where U_b is the bulk velocity, R is the pipe radius and ν the kinematic viscosity, while the friction Reynolds number is $Re_\tau = U^* R/\nu = 200$. Concerning the first simulation, seven different particle populations are injected at a fixed location at the axis of the pipe and their evolution is analyzed for a streamwise extension of $200R$ (with R the pipe radius).

The axial length of the second kind of simulation is reduced to $20\pi R$ due to computational time limitations. Droplets are injected at the axis with the same inlet mechanism of inertial particles with vapor scalar inlet condition set to zero. Three different runs are performed considering three initial droplet sizes (Stokes number). Droplets counteract with their vapor as a source term. At the beginning of the simulation, there is no vapor inside the field (the initial field is dry) and the scalar production is only due to droplets evaporation. The equation for the droplet radius is obtained with the following model [3]:

$$\frac{dr_p}{dt} = -\frac{Sh}{9Sc}\frac{r_p}{\tau_p}(Y_s - Y^*) \tag{1}$$

where r_p, Sh, Sc, Y_s, Y^* are the droplet radius, the Sherwood number, the Schmidt number, the saturation level and the vapor field at particle position respectively. The model assumes the balance of the vapor mass fraction at the droplet surface under the hypothesis that the droplet is insoluble to the gas phase species. The range of applicability of the present model is confined to very small evaporation rates, see [3]. We remark that the droplet relaxation time τ_p depends on the square of droplet radius so that the evaporation rate in equation (1) is inversely proportional to the droplet radius. The Schmidt number used in the simulation is $Sc = 0.7$ while the Sherwood number, the following expression given by [5] is assumed, $Sh = 2 + 0.552 Re_p^{1/2} Sc^{1/3}$, where $Re_p = (U - V)d_p/\nu$ is the droplet Reynolds number. Saturation level Y_s is fixed to 0.5 while the vapor value at particle position Y^* is the interpolation of the vapor Eulerian field evolved in the simulation as a scalar.

3 Results

Concerning the particle laden flow, we can observe that the turbophoresis is apparent by the mean particle concentration in the viscous sublayer, see figure 1. All particles are introduced in the domain at the inlet section near the pipe axis, the input rate is fixed to 900 particles per eddy-turnover-time for each population. They are dispersed by turbulent fluctuations and tend to eventually reach the wall starting the accumulation process. Statistically, axial homogeneity is reached at different distances from the origin depending on particle inertia. Lightest particles assume uniform concentration in the whole cross section, order $1200R^3$ particles per unit volume. Particles with $St^+ = 10, 50$ reach instead a near wall concentration 100 times larger.

In the case of evaporating droplets, we have performed three DNSs for three different initial Stokes number, $St^+ = 1, 10, 100$, respectively. The liquid mass flux in the three cases has been kept constant. In figure 2 droplet instantaneous configurations and vapor field (contours) are shown for the three initial Stokes number. The

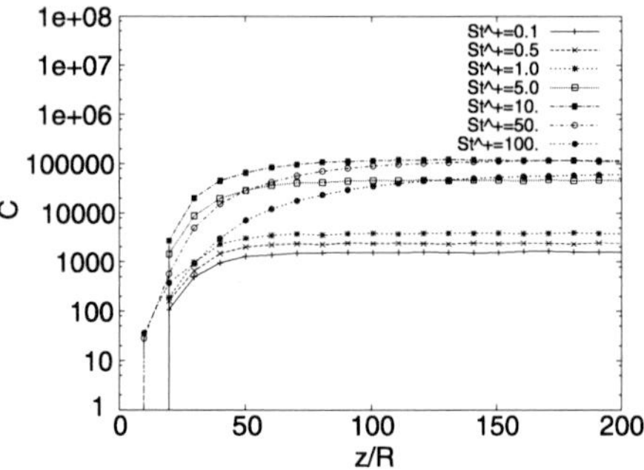

Fig. 1 Mean particle concentration in the viscous sublayer along the pipe axis.

evaporation length L_e, defined as the length required to evaporate 95% of the initial liquid phase (indicated by the red arrow in figure), increases with initial droplet inertia. Concerning the vapor phase, the mixing is still incomplete at the end of the computational domain.

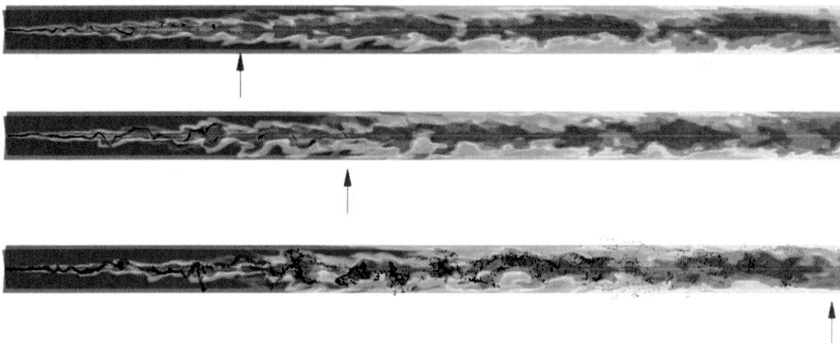

Fig. 2 Instantaneous droplets configurations inside the pipe, contours represent the vapor concentration. After the red arrows 95% of the injected liquid mass is evaporated. From top to bottom the initial viscous Stokes number of $St^+ = 1, 10, 100$.

To quantitatively extract the value of evaporation length, the plots of liquid and vapor mass fluxes are shown in figure 3. Both fluxes are normalized with the inlet value which, as stated before, is the same for the three cases. The estimated evaporation lengths for the three simulations are $10R$, $12R$ and $30R$, ordered by increasing initial droplet inertia.

The effects of turbulent fluctuations can be spotted by looking at the left panel of figure 4. The ratio of laminar to turbulent evaporation length is plotted in the graph–$(L_e)_{lam}/(L_e)_{tur} = 1.4, 1.6, 2.3$–for droplets with initial Stokes number of $St^+ = 0.1, 1, 10$ respectively. Turbulence promotes a faster evaporation especially for larger droplets. The reason is the turbophoretic drift that causes the removal of

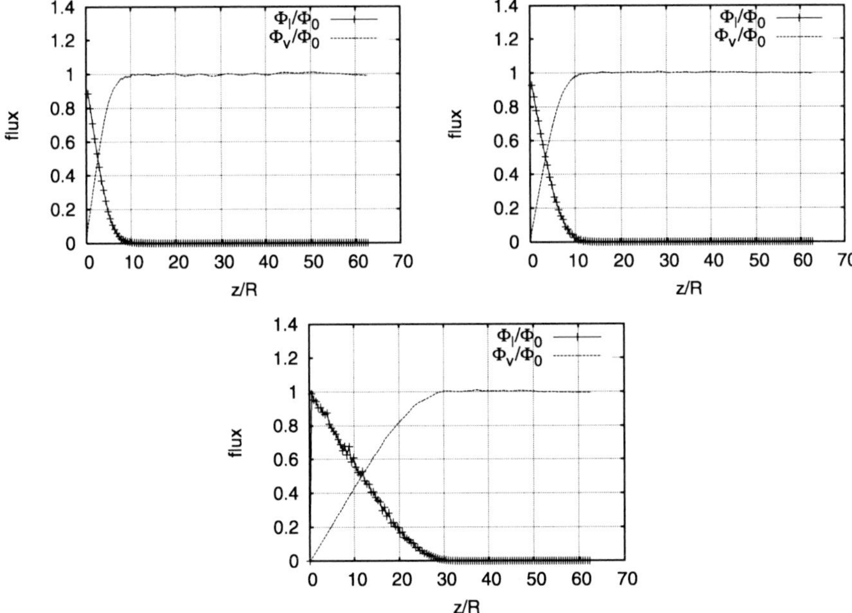

Fig. 3 Liquid and vapor fluxes versus axial distance for the three cases of evaporation configuration.

the droplets from the pipe center. Actually, this region is characterized by a high vapor level. It is a saturate zone, as opposed to the near wall region where the vapor concentration is considerable smaller as qualitatively shown in figure 2. The right panel of figure 4 represents the variance of the droplets radius pdf. This particle size distribution has been calculated as a function of axial distance by accounting for all particles in a transversal section of the pipe. After an initial transient characterized by the increase of the variance, there is a plateau where the variance is constant. In the meantime the average droplet radius decreases (not shown) suggesting that to a first approximation the shape of the pdf does not change, but simply translates towards smaller droplet sizes.

4 Conclusions

The strong combination of turbulent fluctuations and turbophoresis promotes the evaporation efficiency in pipe flow configurations. This is particularly accentuated for large inertia droplets in comparison to the laminar regime. A complete vapor mixing requires a very long pipe distances in the streamwise directions. In view of closure models for droplet evaporation it is interesting to note that the variance of droplet radii remains constant for a substantial pipe extension. Improvements are

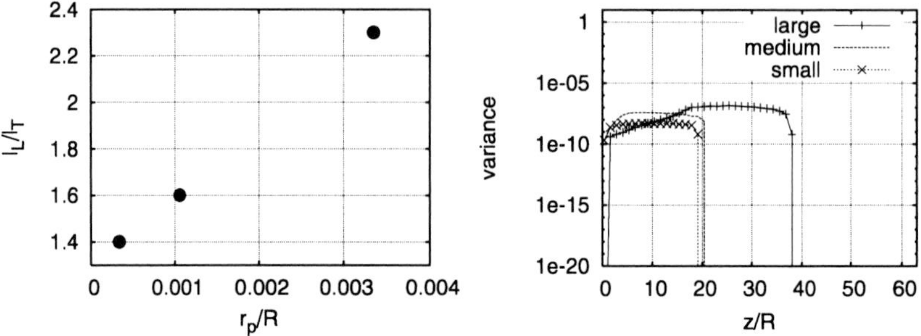

Fig. 4 Left panel: ratio between the evaporation length in the laminar and turbulent regimes. Right panel: variance of the pdf of droplet radius versus axial distance.

however needed for a more realistic description of the process by implementing the temperature transport equation and droplet collisions/coalescence.

References

1. Bec, J., Biferale, L., Cencini, M., Lanotte, A., Musacchio, S. and Toschi, F.: Heavy particle concentration in turbulence at dissipative and inertial scales. Physical Review Letters. **98/9**, 8452. (2007)
2. Maxey, M.R. and Riley, J.J.: Equation of motion for a small rigid sphere in a nonuniform flow. Physics of Fluids. **26**, 883. (1983)
3. Miller, R.S., Harstad, K. and Bellan, J.: Evaluation of equilibrium and non-equilibrium evaporation models for many-droplet gas-liquid flow simulations. International Journal of Multiphase Flow. **24/6**, 1025–1055. (1998)
4. Okongo, N., Leboissetier, A. and Bellan, J.: Detailed characteristics of drop-laden mixing layers: Large eddy simulation predictions compared to direct numerical simulation. Physics of Fluids. **20**, 103305. (2008)
5. Ranz, W.E. and Marshall, W.R.: Evaporation from drops. Chem. Eng. Prog. **48/3**, 141–146. (1952)
6. Picano, F., Sardina, G. and Casciola, C.M.: Spatial development of particle-laden turbulent pipe flow. Physics of Fluids. **21**, 093305. (2009)
7. Picano, F., Sardina, G., Gualtieri, P. and Casciola, C.M.: Anomalous memory effects on transport of inertial particles in turbulent jets. Physics of Fluids. **22**, 051705. (2010)
8. Reeks, M.W.: The transport of discrete particles in inhomogeneous turbulence. Journal of aerosol science. **14/6**, 729–739. (1983)

Direct numerical simulation of binary-species mixing layers

M. Pezeshki, K.H. Luo and S. Gu

1 Introduction

Mixing layers are a fundamental phenomenon that occurs in many more complex flows such as jets, counter-flows and recirculating flows. The importance of mixing layers as a building-block in fluid mechanics is evident in the large number of computational, experimental and theoretical studies devoted to the topic. With its simple configuration and easy control of flow parameters, mixing layers were one of the first flows that became amenable to direct numerical simulation (DNS). Examples of DNS of incompressible [1] and compressible [2] can be found. However, these studies and most others have considered only single-species (usually air) mixing layers, in which the two streams have different flow properties but the same fluid properties. Relatively few studies by DNS have considered binary-species [3] and multi-species [4–6] mixing layers in which the fluid properties of different streams differ. Such binary- and multi-species mixing layers are of special importance in the process and energy (i.e. combustion) industry.

A few studies using realistic diffusion coefficients for species are conducted in the context of combustion. Among those, studies performed by Pitsch [7] and Hilbert and Thevenin [8] reveal some aspects of the importance of differential diffusion and its strong effects on the flow field properties. Some researchers, though, studied on both reactive and non-reactive flows considering different diffusivity. Jaberi et al. [9] considered a problem of passive scalar mixing as well as binary chemical reaction in constant density, isothermal, homogeneous turbulent flow. Yeung and Pope [10] and Yeung et al. [11] were the groups who studied the effects of differential diffusion of passive scalars with different molecular diffusivity on the flow and scalar statistics.

In the current study, DNS results of non-reacting binary-species compressible mixing layers are presented. Pure hydrogen (upper stream) and pure oxygen (lower

---

M. Pezeshki, K.H. Luo, S. Gu
Energy Technology Research Group, School of Engineering Sciences, University of Southampton, Southampton SO17 1BJ, United Kingdom, e-mail: m.pezeshki@soton.ac.uk

H. Kuerten et al. (eds.), *Direct and Large-Eddy Simulation VIII*, 219
ERCOFTAC Series 15, DOI 10.1007/978-94-007-2482-2_35,

stream) flow in opposite directions to form a temporally evolving mixing layer. Considering the unique properties of hydrogen such as high diffusivity (Lewis number 0.3) and lightness (molecular weight 2), a lot of challenges are encountered from the numerical simulation point of view. The situation is made worse with oxygen in the other stream, which has a diffusivity 4 times less than hydrogen (Lewis number 1.11) and molecular weight 16 times more. These may partly explain why few numerical studies have simulated pure hydrogen-oxygen mixing layers. In the DNS presented here, real fluid properties for H_2 and O_2 are specified, which is contrasted with the results under the unity Lewis number assumption.

2 Methodology and computational setup

Three-dimensional compressible time-dependent Navier-Stokes equations have been solved in conjunction with transport equations for non-premixed hydrogen and oxygen streams. Details about numerical details can be found in Luo [13]. In addition, sixth order filtering is invoked to remove the high wave number fluctuations, which is necessary when real fluid properties are used in the DNS.

Simulations have been performed using a box-type geometry with nearly 26 million grid points. The flow field is initialized using an asymmetric error function profiles for species mass fraction, temperature and velocity with the interface at $z = 10$ in cross-streamwise direction which causes hydrogen to occupy two third of the computational domain. The initial mean pressure is equal to the ambient pressure. Equal temperature has been applied for two cases studied here; 300K and 2000K for hydrogen and oxygen, respectively. For the second case, unity Lewis number assumption has been made to enable us simulate equal diffusivity mixing layer flow. Convective Mach number of 0.4 is chosen to have small effects of compressibility and Reynolds number used is 750 and is based on the mean initial velocity of free streams and initial vorticity thickness. Mean initial velocity for upper stream is equal to that of the lower stream. The initial disturbances are introduced using the method used in [3], which will cause two stages of pairing to occur.

3 Diffusivity effects on flow development and characteristics

Diffusivity of hydrogen is extremely large compared with most common gases. Few previous studies have considered pure hydrogen and the effects of hydrogen's high diffusivity on other transport phenomena have not been fully clarified. In this study, the DNS case with real Lewis numbers for H_2 and O_2 is compared with a different case in which the Lewis numbers of two streams are artificially set to unity. For the sake of isolating the Lewis number effects, all the other input parameters are chosen to be equal. Figure 1 shows a snapshot of the instantaneous distribution of the hydrogen mass fraction in the mid-plane in the spanwise direction comparing results of Lewis number 1.0 and 0.3. It can be seen that hydrogen (with real Lewis number) diffuses more effectively than an artificial fluid with unity Lewis number. In figure 1-A the area in which mass fraction of hydrogen is zero, is much smaller than its counterpart in figure 1-B, showing more diffusion of upper stream (H_2) toward lower stream (O_2). In figure 1-B, one can observe that some hydrogen is entrained by the oxygen stream without much mixing (region with H_2 mass fraction close to 1.0 at $z = 5$), whereas similar region in figure 1-A has values less than 1.0 that shows much better mixing. This is again due to the higher diffusivity of real hydrogen.

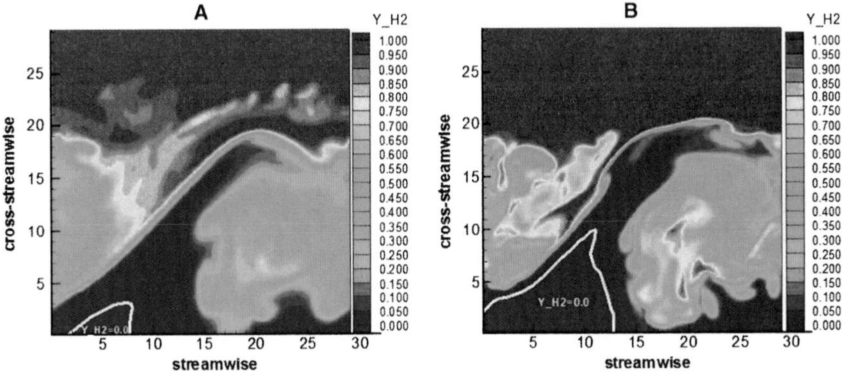

Fig. 1 H_2 mass fraction distributions at time 110 in the spanwise mid-plane for: A. Real molecular diffusivity; B. Equal molecular diffusivity (Le=1).

To study the effect of diffusivity on scalar statistics, the transport equation for scalar variance $\overline{\theta''\theta''}$ is analyzed in which θ is any scalar (fuel or oxidizer) mass fraction with its Favre fluctuations denoted by θ''. Figure 2 depicts the variance for H_2 and O_2 mass fractions across the mixing layer at time 75 and time 110. As shown in this figure, variance of hydrogen with real diffusivity is less than its counterpart, oxygen. In addition, variance of scalar with unity Lewis number is larger than both scalars with real values for diffusion coefficient. To explain this behavior one should refer to scalar variance transport equation. Among the terms in this equation, the scalar dissipation rate is picked up to be evaluated as this term is found to be analyzed more in the literatures focused on diffusivity effects, although other terms will also have their own effects on the scalar variance. Looking at the variation of this quantity (not shown here), it can be seen that dissipation rate of scalar (H_2 or O_2 mass fractions) with Le=1 is less than the dissipation rate of the scalars with different Lewis number. So, as a smaller sink term, it will cause increase in variance compare to the other case. For H_2 and O_2 with real diffusivity, dissipation rate of hydrogen with smaller Lewis number is more than oxygen which will cause variance of hydrogen mass fraction gets lesser value compare to oxygen. Similar results have also been found and published by Jaberi et al. [9] and Yeung and Pope [10]. They studied the differential diffusion effects under homogeneous turbulence and showed that the variance of a scalar with higher diffusivity (smaller Schmidt or Lewis number) is less than the variance of a scalar with higher Lewis number. Another feature of these two figures is 'over-prediction' of the scalar variance for the case in which unity Lewis number assumption has been made. This 'over-prediction' is intensified at later time stage (t=110). Also, both scalars show the same trend and have equal variances which results in neglecting their different diffusivity effects.

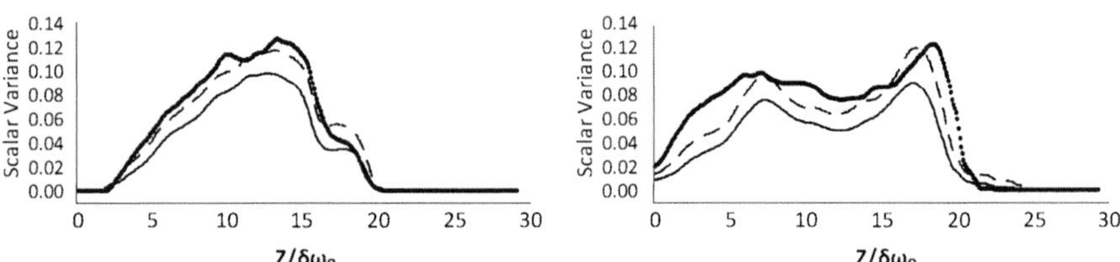

Fig. 2 Variance of the scalars mass fraction at time 75 (left) and 110 (right). Solid line: $\overline{Y''_{H_2}Y''_{H_2}}$, Dashed line: $\overline{Y''_{O_2}Y''_{O_2}}$, both with real diffusion coefficient; Dotted line: either $\overline{Y''_{H_2}Y''_{H_2}}$ or $\overline{Y''_{O_2}Y''_{O_2}}$ with unity Lewis number.

4 Differential diffusion in mixing layers

A mixture fraction in the conventional sense cannot be defined for a binary-fluid mixing layer with different Lewis numbers. This is one manifestation of differential diffusion. According to definition presented in [12], differential diffusion can be defined as the difference between diffusion term of the mixture fraction transport equation with real properties for the fluid and diffusion term of the mixture fraction transport equation with the assumption of equal diffusivity for the fluid:

$$DD \equiv \frac{-1}{\beta_1 - \beta_0} \sum_{\ell=1}^{N_e} \gamma_\ell \Big(\sum_{i=1}^{N_s} \frac{a_{\ell i} W_\ell}{W_i} \nabla(\rho D_i \nabla Y_i) \Big) - \nabla(\rho D_{mix} \nabla f) = \gamma_{exact} - \gamma_{approx.} \quad (1)$$

in which γ_l is weighting factor, $a_{\ell i}$ is number of atoms of element ℓ in species i, W_ℓ is the molecular weight of element ℓ, W_i is species molecular weight, D_i is the diffusivity of species i, Y_i is mass fraction of species i, ρ is the mixture density, γ_{exact} is the diffusion term with real diffusivity of species, γ_{approx} is the diffusion term with equal diffusivity of species, D_{mix} is the mixture fraction diffusivity and is a constant and taken to be the average of species diffusivity [14] and finally, β and f are coupling function and mixture fraction based on coupling function, respectively, which are defined by:

$$\beta = \sum_{\ell=1}^{N_e} \gamma_\ell \left(\sum_{i=1}^{N_s} \frac{a_{\ell i} W_\ell Y_i}{W_i} \right) \quad , \quad f = \frac{\beta - \beta_o^0}{\beta_f^0 - \beta_o^0} \tag{2}$$

Differential diffusion (DD) shows the difference between the exact diffusion term γ_{exact} (first term on the r.h.s.) and that assuming equal diffusivities $\gamma_{approx.}$ (second term on the r.h.s.). This is essentially the error caused by the equal-diffusivity assumption in the form of unity Lewis number. To quantify this parameter in mixing layer flows, cases 1 and 2 have been selected to be analyzed. To facilitate comparison at equivalent stages of development of the mixing layers, results are obtained at the same momentum thickness magnitude of 3 (δ_m evolution is not shown in here). Simulation time corresponding to δ_m=3 is t=120 for the case with non-equal Lewis number and t=105 for the other one. This difference refers to different evolution of mixing layers caused solely by the effect of molecular diffusivity as all other initial parameters are the same. It must be emphasized that the value of DD is constant (and is equal to zero) for the mixing layers with Le=1 and it is just shown for the sake of comparison with the other case to highlight the magnitude of the error caused by equal diffusivity assumption. Figure 3 shows the DD normalized by γ_{exact} across the mixing layer (left) as well as the value of γ_{exact} (right). As it was mentioned above, DD shows no variations for mixing layers with equal diffusivity as both terms of equation 1 cancel each other. In contrast, for the case with non-unity Lewis numbers, DD represents rigorous fluctuations across the mixing layers. At some locations, the error has large peaks with the values up to 27. To be able to show the smaller fluctuations, this plot has only been shown within the range of [-2,2]. One of the peaks has been appeared at the interface of two layers at $z = 10$ while the majority have been located in the hydrogen side from $z = 20$ toward upper boundary proving the fact that elimination of diffusivity of hydrogen and approximation with unity value for Lewis number would cause significant error.

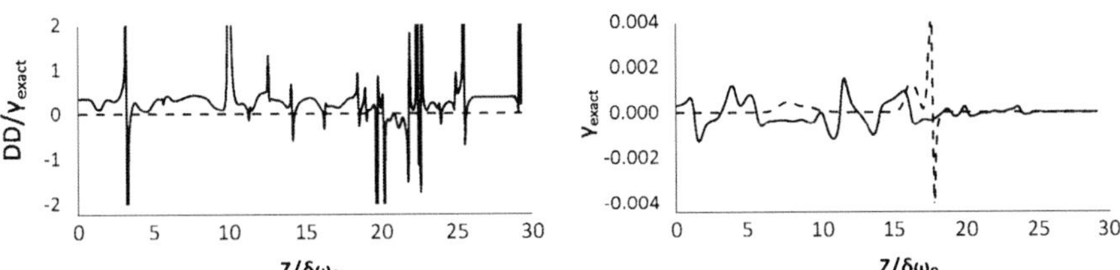

Fig. 3 Differential diffusion (left) and γ_{exact} (right) across the mixing layers for both cases in the mid-plane in spanwise direction. Solid line: case with real Lewis number, Dashed line: case with unity Lewis number assumption; T_{H2}=300K and T_{O2}=2000K for both cases.

5 Concluding remarks

A series of DNS simulations have been conducted to study the effects of real fluid properties, mainly diffusivity of hydrogen and oxygen on the development of the mixing layers, and scalar statistics. It is shown that non-unity Lewis numbers of hydrogen and oxygen have a large impact on scalar transport as measured by scalar mass fraction and scalar variance. Differential diffusion (*DD*) is analyzed by comparing the transport equations for a generalized "mixture fraction" and the mixture fraction defined under the unity Lewis number assumption. The level of *DD* shows the error while the effect of different species diffusivity is considered negligible. The magnitude of this error can be significant especially when the mixing layers consists of species with large difference between diffusion coefficients, such as hydrogen and oxygen studied here.

References

1. Moser, R.D., Rogers, M.M.: Mixing transition and cascade to small scales in a plane mixing layer. Phys. Fluids A **3(5)**, 1128–1134 (1991)
2. Vreman, W., Sandham, N.D., Luo, K.H.: Compressible mixing layer growth rate and turbulence characteristics. J. Fluid Mech. **320**, 235–258 (1996)
3. Okong'o, N., Harstad, K., Bellan, J.: Direct numerical simulations of O_2/H_2 temporal mixing layers under supercritical conditions. AIAA J. **40**, 914–926 (2002)
4. Knaus, R., Pantano, C.: On the effect of heat release in turbulence spectra of non-premixed reacting shear layers. J. Fluid Mech. **626**, 67–109 (2009)
5. Echekki, T., Chen, J.H.: Direct numerical simulation of autoignition in non-homogeneous hydrogen-air mixtures. Combust. Flame **134**, 169–191 (2003)
6. Zheng, X.L., Yuan, J., Law, C.K.: Nonpremixed ignition of H_2/air in a mixing layer with a vortex. Proc. Combust. Inst. **30**, 415–421 (2004)
7. Pitsch, H.: Unsteady flamelet modeling of differential diffusion in turbulent jet diffusion flames. Comb. Flame. **123**, 358–374 (2000)
8. Hilbert, R., Thevenin D.: Influence of differential diffusion on maximum flame temperature in turbulent nonpremixed hydrogen/air flames. Comb. Flame. **138**, 175–187 (2004)
9. Jaberi, F.A., Miller, R.S., Mashayek, F., Givi, P.: Differential diffusion in binary scalar mixing and reaction. Comb. Flame. **109**, 561–577 (1997)
10. Yeung, P.K., Pope, S.B.: Differential diffusion of passive scalars in isotropic turbulence. Phys. Fluids A. **5(10)**, 2467–2478 (1993)
11. Yeung, P.K., Sykes, M.C., Vedula, P.: Direct numerical simulation of differential diffusion with Schmidt numbers up to 4.0. Phys. Fluids A. **12(6)**, 1601–1604 (2000)
12. Sutherland, J.C., Smith, P.J., Chen, J.H.: Quantification of differential diffusion in nonpremixed systems. Comb. Theory Modelling. **9(2)**, 365–383 (2005)
13. Luo, K.H.: Combustion effects on turbulence in a partially premixed supersonic diffusion flame. Combust. Flame. **119**, 417–435 (1999)
14. Kronenburg, A., Bilger, R.W.: Modelling of differential diffusion effects in nonpremixed nonreacting turbulent flow. Phys. Flow. **9(5)**, 1435–1447 (1997)

Direct Numerical Simulation of a Buoyant Droplet Array

Marcel Kwakkel, Wim-Paul Breugem and Bendiks Jan Boersma

1 Introduction

In many industrial and natural processes turbulent dispersion of immiscible phases occur. An industrial example is the process of steel making, where bubble flotation is used to remove inclusions which downgrade the quality of steel. An example from nature is the formation of droplets in clouds, where turbulent air influences the collision-coalescence rate. To gain a better understanding of droplet dynamics and turbulence modification in the clustering regime at Stokes number $St \sim 1$, proper modeling of coalescence and break-up is crucial. To be able to investigate these effects the number of deformable droplets should be relatively high and the problem has to be solved in an accurate and efficient way. The goal of our research is therefore to perform Direct Numerical Simulation (DNS) of a large number ($\sim 10^3$) of inertial droplets in a turbulent carrier fluid, where coalescence and break-up is treated in a physical way. In general such simulations are expensive in terms of CPU, but in this work we show that our code scales well with an increasing number of deformable droplets and grid sizes.

2 Governing Equations

The motion of two immiscible fluid phases, where one fluid phase is dispersed in a surrounding continuous fluid phase, can be described by combining the equations for the fluid motion with some appropriate interface conditions. The motion of the interface Γ depends on the local velocity field of the fluid at the location of the interface, which makes this a moving boundary problem. By demanding continuity of

Marcel Kwakkel
Delft University of Technology, Laboratory for Aero Hydrodynamics Leeghwaterstraat 21, NL-2628 CA Delft, The Netherlands, e-mail: m.kwakkel@tudelft.nl

H. Kuerten et al. (eds.), *Direct and Large-Eddy Simulation VIII*,
ERCOFTAC Series 15, DOI 10.1007/978-94-007-2482-2_36,

velocity and total stress, both fluid phases are coupled. At the interface, the density ρ, viscosity μ and surface tension forces are discontinuous.

2.1 Fluid motion

If both fluid phases are considered to be isothermal incompressible Newtonian fluids, the fluid motion can be described by

$$\nabla \cdot \mathbf{u} = 0, \tag{1}$$

$$\rho \left(\frac{\partial \mathbf{u}}{\partial t} + \mathbf{u} \cdot \nabla \mathbf{u} \right) = -\nabla p + \nabla \cdot \mu \left(\nabla \mathbf{u} + \nabla \mathbf{u}^T \right) + \mathbf{f}, \tag{2}$$

where $\mathbf{u} = (u, v, w)^T$, p and $\mathbf{f}$ respectively denote the velocity vector, pressure and a body force term (gravity, interfacial tension). The density ρ and viscosity μ are assumed constant within each phase and are defined by ρ_0, μ_0 in fluid phase '0' and by ρ_1, μ_1 in fluid phase '1'. If a color function χ is introduced as

$$\chi = \begin{cases} 0 & \text{in fluid phase '0'} \\ 1 & \text{in fluid phase '1',} \end{cases} \tag{3}$$

then ρ and μ can be expressed as

$$\rho = \rho_0 (1 - \chi) + \rho_1 \chi, \tag{4}$$

$$\mu = \mu_0 (1 - \chi) + \mu_1 \chi. \tag{5}$$

2.2 Interface conditions

Since the fluids are considered immiscible, their mutual interface Γ is a material property of the flow and its motion can be described by

$$\Gamma_t + \mathbf{u} \cdot \nabla \Gamma = 0, \tag{6}$$

where $\mathbf{u}$ is the velocity of the fluid at the location of the interface. At the interface continuity of velocity and continuity of stresses hold, which are given by

$$[\mathbf{u}]_\Gamma = 0, \tag{7}$$

$$\left[p\mathbf{n} + \mu \left(\nabla \mathbf{u} + \nabla \mathbf{u}^T \right) \cdot \mathbf{n} \right]_\Gamma = \sigma \kappa \mathbf{n}, \tag{8}$$

where $[.]_\Gamma$ denotes a jump across the interface Γ, $\mathbf{n}$ the interface normal vector, σ the surface tension coefficient (assumed constant), and κ the value of the interface curvature. Interface condition (8) shows that the jump conditions for the pressure and velocity are coupled. [5] derived that when the viscosity is continuous at the interface the derivatives of the velocity are also continuous and (8) reduces to

$$[p\mathbf{n}]_{\Gamma} = \sigma \kappa \mathbf{n}. \tag{9}$$

By using a continuous viscosity the pressure and velocity jump conditions are decoupled and derivatives of the velocity can be equally approximated in the whole computational domain. The only resulting interface condition is a jump in the pressure at the location of the interface due to the surface tension force.

3 Computational Method

The deformable droplets are described by the multiple marker Coupled Level-Set Volume-of-Fluid (CLSVOF) method as used in [1, 2]. The full dimensionless Navier-Stokes equations [1] are solved on a fixed, uniform grid using a finite-volume/front-capturing method with the inclusion of surface tension. These equations are integrated in time using a pressure-correction method and solved by a preconditioned Conjugate Gradient (PCG) solver. Parallelization of the numerical method is performed using a domain decomposition approach. The Message Passing Interface (MPI) library is used for communication with neighbors. To achieve a very efficient and scalable method, we introduce the localized multiple marker method and the deflated PCG method.

3.1 CLSVOF Method

The boundary between the two phases is regarded as a deformable interface. This interface is captured by two so-called marker functions, the Level-Set (LS) and Volume-of-Fluid (VOF) functions. The LS function is used as the color function χ, as described in Section 2.1. The LS function is required to be (and remain) a distance function to the interface. The discrete LS function implicitly defines the position of the interface on the computational grid, but does not ensure conservation of the volume enclosed by the interface. The VOF function represents the volume fraction in a computational cell and does conserve the enclosed volume. Since the VOF function does not give any information about the distribution of the fluid inside a cell, the LS function is used to determine the amount of fluid that flows across the boundaries of a cell. By coupling both functions, the second order accurate CLSVOF method is obtained [7]. After the advection of both functions, the VOF function is used to reinitialize the advected LS function to represent a distance function to the advected VOF function.

3.2 Local Multiple Marker Method

By using the multiple marker method the problem of (artificial) numerical coalescence is circumvented, but in [1, 2] each marker is defined globally. Disadvantages of this implementation are the large memory footprint and the sequential advection of interfaces. This sequential advection is a result of the parallelization of the markers, which is also performed by the domain decomposition approach. Since the marker functions are only needed around the interface, this implementation does not lead to an optimal load balancing. In our implementation the marker functions are localized in a box around the interface, leading to a smaller memory footprint. The parallelization is replaced by a master/slave technique, where one processor (master) advects the marker functions. Other processors that physically contain part of the local box (slaves) only send their part of the velocity field to the master. By decoupling computation and communication a better load balancing can be achieved.

3.3 Deflated PCG Method

In [1] it was reported that the variable Poisson equation resulting from Equation (2) required more than 80% of the total CPU time. Due to large density ratios in multiphase flows, the Poisson equation becomes ill-conditioned leading to a slow convergence. In general it is expected that more droplets lead to a worse conditioning, so longer solve times. The conditioning of the Poisson equation can be improved by the application of the deflation method. The deflation method was introduced independently by [6] and [3] and is a two-level PCG (2L-PCG) method whose two-level preconditioner is based on a deflation technique. [8] applied the method to bubbly flows, where the density contrast is relatively large. He showed that 2L-PCG methods are less sensitive to the density contrast, the number of bubbles and the size of the grid compared with standard PCG methods. The cost per iteration is slightly higher if the deflation method is applied, but it also reduces the number of iterations of the CG solver. The efficiency of the deflation method depends on the number of deflation vectors. More deflation vectors lead to less iterations, but more work per iteration. See [8] for more details about the deflation method and its parameters.

4 Results

To investigate the effect of the local marker and deflation methods, simulations of a buoyant droplet array in a periodic box have been performed. Our configuration and parameter set complies with those of [4, 2]: Reynolds number $Re = 29.907$, Eötvös number $Eo = 2$, density ratio $\zeta = 0.1$, viscosity ratio $\gamma = 0.1$ and volume fraction $\alpha = 0.1256$. Only the number of droplets and the grid size have been varied. The domain decomposition is set to $8 \times 1 \times 1$, so the box is devided into slices.

To compare different settings, the wall clock time over the first 100 time steps is averaged.

4.1 Effect of Local Marker Method

To investigate the effect of the local marker method with the master/slave technique, the average wall clock time of one iteration is measured. The time spend in the pressure solver is subtracted, since this part differed between the used codes. In Fig. 1 it can be seen that for a fixed grid size the global marker method does not scale well with an increasing number of droplets. For the local marker method however, the wall clock time is almost constant with an increasing number of droplets. For fixed number of droplets the local marker also scales better than the global marker method with respect to the grid size.

4.2 Effect of Deflation Method

To investigate the effect of the deflation method, the number of iterations of the pressure solver and the wall clock time spend in the pressure solver are compared. Fig. 2 shows that deflation successfully decreases the number of iterations. The number of iterations with the deflation method is also almost independent of grid size, which is clearly not the case for simulations without deflation. The reduction in the number of iterations is a factor of about 4 to 7. The amount of work per iteration will be higher if deflation is used, so the real point of interest is the amount of time spend in the pressure solver. Fig. 3 shows that the reduction of the time spend in the pressure solver is not as large as the reduction of iterations, but we still achieve a speed-up of about 3 to 5. Since more droplets lead to a worse conditioning, it was expected that 27 droplets would need more iterations than 8 droplets to solve the pressure equation. This is however not the case here, probably the distribution of the droplets over the periodic box and the choice of deflation vectors has some influence here.

5 Conclusion

To be able to perform DNS of a large number of droplets, a very accurate and efficient method is necessary. Accuracy was already achieved by using the multiple marker CLSVOF method [1, 2], but this implementation still lacked efficiency (especially if the number of droplets was increased). Our implementations of the local marker method with master/slave technique combined with the deflated PCG

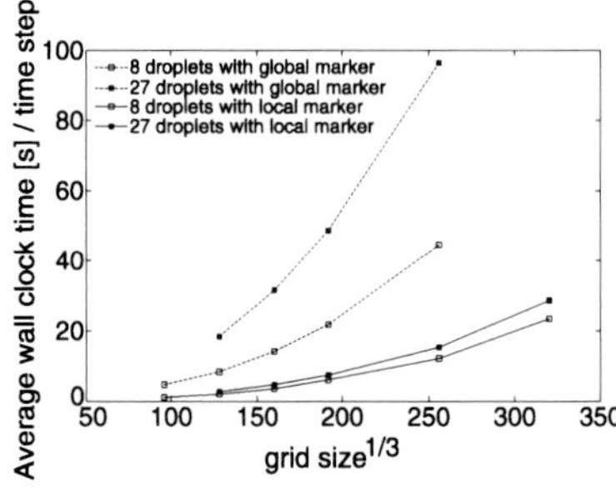

Fig. 1 Average wall clock time per time step without pressure solver.

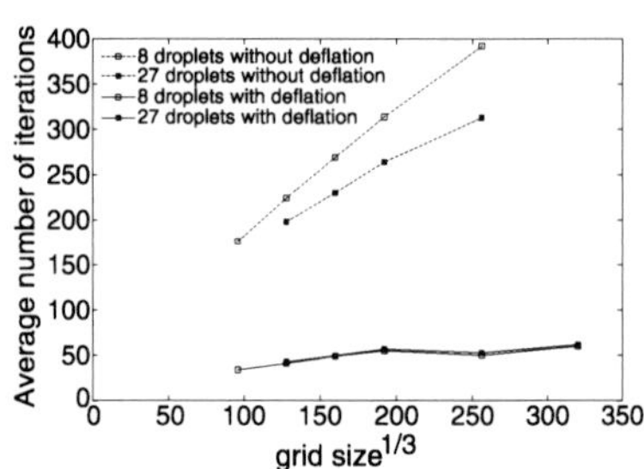

Fig. 2 Number of iterations for the pressure solver.

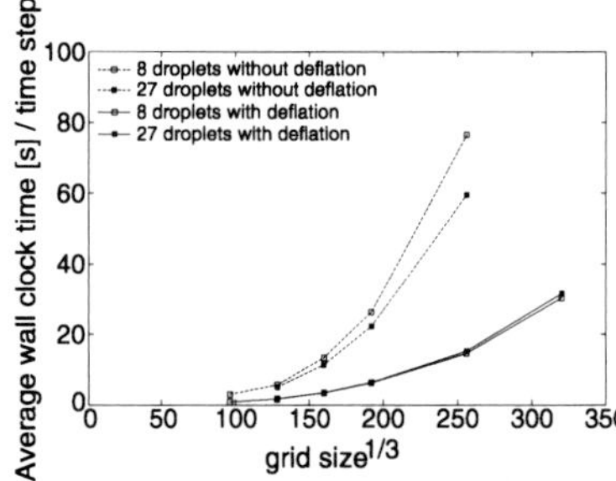

Fig. 3 Average wall clock time spend in the pressure solver.

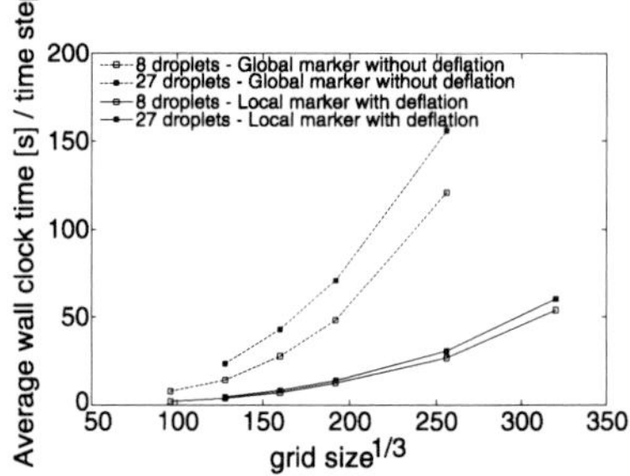

Fig. 4 Average wall clock time per time step (overall effect).

method removes this bottleneck, see Fig. 4. It enables us to investigate the interesting physical phenomena that occur in turbulent multiphase flows.

References

1. E.R.A. Coyajee. *A front-capturing method for the numerical simulation of dispersed two-phase flow*. PhD thesis, TU Delft, 2007.
2. E.R.A. Coyajee and B.J. Boersma. Numerical simulation of drop impact on a liquid-liquid interface with a multiple marker front-capturing method. *J. Comp. Phys.*, 228:4444–4467, 2009.
3. Z. Dostal. Conjugate gradient method with preconditioning by projector. *Int. J. Comput. Math.*, 23:315–323, 1988.
4. A. Esmaeeli and G. Tryggvason. Direct numerical simulations of bubbly flows part 2. moderate reynolds number arrays. *J. Fluid Mech.*, 385:325–358, 2000.
5. M. Kang, R. Fedkiw, and X.-D. Liu. A boundary condition capturing method for multiphase incompressible flow. *J. Sci. Comput.*, 15:323–360, 2000.
6. R.A. Nicolaides. Deflation of conjugate gradients with applications to boundary value problems. *SIAM J. Numer. Anal.*, 24:355–365, 1987.
7. M. Sussman and E. Puckett. A coupled level set and Volume-of-Fluid method for computing 3D and axisymmetric incompressible two-phase flows. *J. Comp. Phys.*, 162:301–337, 2000.
8. J.M. Tang. *Two-level preconditioned conjugate gradient methods with applications to bubbly flow problems*. PhD thesis, TU Delft, 2008.

Part IV
Environmental Flows

LES modeling and experimental measurement of boundary layer flow over multi-scale, fractal canopies

Jason Graham, Kunlun Bai, Charles Meneveau and Joseph Katz

1 Introduction

In many regions the atmospheric surface layer is affected substantially by vegetation canopies. Most previous work has focused on effects of vegetated terrain characterized by a single length scale, e.g. a single obstruction of a particular size, or canopies consisting of plants, often modeled using a prescribed leaf-area density distribution with a characteristic dominant scale. It is well known, however, that typical flow obstructions such as canopies are characterized by a wide range of length scales, branches, sub-branches, etc. Yet, it is not known how to parameterize the effects of such multi-scale objects on the lower atmospheric dynamics. This work aims to study boundary layer flow over fractal, tree-like shapes. Fractals provide convenient idealizations of the inherently multi-scale character of vegetation geometries, within certain ranges of scales. Preliminary results from a large-eddy simulation (LES) and experimental study of a fractal tree canopy in a turbulent boundary layer are reported. The LES use Renormalized Numerical Simulation (Chester et al., 2007, J. Comp. Phys.) to provide subgrid parameterizations of drag forces from unresolved small-scale branches. Experiments aiming at understanding drag forces acting on fractal trees are performed in a water tunnel facility. Drag force measurements are obtained on a set of "pre-fractal" trees containing 1-5 branch generations.

The canopy consists of a periodic array of equal trees. The trees in the canopy have at each generation a splitting into 3 branches with a scale contraction factor of 1/2. The similarity fractal dimension of the resulting object (strictly speaking a "pre-fractal", since it does not contain an infinite number of branchings) is: $D_s = logN/log(1/r) = log3/log2 = 1.585$. This tree geometry is a compromise between realism and simplicity. It shares with real vegetation elements the co-existence of multiple scales and elements of scale-invariance. Its fractal similarity dimension is

Department of Mechanical Engineering, and Center for Environmental and Applied Fluid Mechanics, Johns Hopkins University, 3400 North Charles Street, Baltimore MD 21218, USA, CM e-mail: meneveau@jhu.edu, JG e-mail: jgraha34@jhu.edu, KB e-mail: kbai2@jhu.edu, and JK e-mail: katz@jhu.edu

H. Kuerten et al. (eds.), *Direct and Large-Eddy Simulation VIII*,
ERCOFTAC Series 15, DOI 10.1007/978-94-007-2482-2_37,

consistent with known vegetation fractal dimensions that range mostly between $D = 1.4$ and 2. Fig. 1a shows the layout of the canopy simulated during LES (only two of the trees are simulated and the rest represented using periodic boundary conditions). The experimental canopy setup is shown in Fig. 1b.

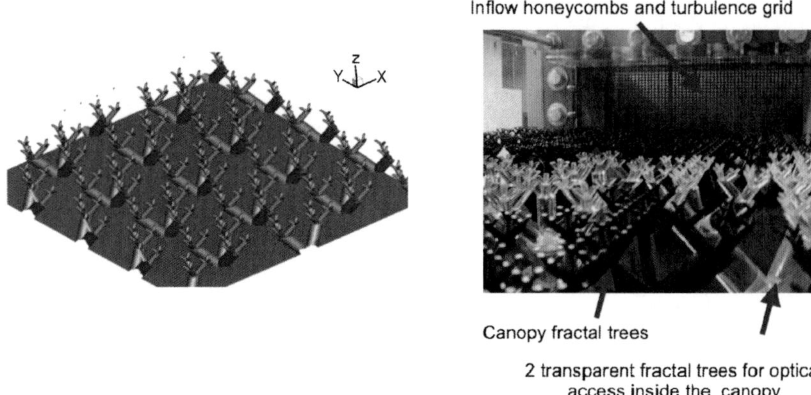

Fig. 1 (a) Layout of simulated canopy with periodic arrangement of fractal trees. (b) Photograph of fractal tree canopy inside water tunnel, showing transparent target trees inside canopy and flow conditioning grids upstream of the test section.

2 Description of numerical methods

Simulations are performed using a variant of the JHU-LES code [3, 1]. The LES code solves the filtered, incompressible Navier-Stokes equations for a neutrally buoyant and high-*Re* flow such that,

$$\frac{\partial \widetilde{u}_i}{\partial t} + \widetilde{u}_j \left(\frac{\partial \widetilde{u}_i}{\partial x_j} - \frac{\partial \widetilde{u}_j}{\partial x_i} \right) = -\frac{\partial \widetilde{p}}{\partial x_i} - \frac{\partial \tau_{i,j}}{\partial x_j} + \delta_{i,1} \Pi, \qquad \frac{\partial \widetilde{u}_i}{\partial x_i} = 0 \tag{1}$$

where $i = 1,2,3$ and corresponds to the x,y,z directions respectively, $\widetilde{u}_i$ the filtered velocity, $\widetilde{p}$ the filtered pressure, $\tau_{i,j}$ the deviatoric component of the sub-grid scale stress tensor, and Π the forcing term in the x-direction. The classical Smagorinksy model is employed for sub-grid scale stresses where $\tau_{i,j} = -2c_s^2 \Delta^2 |S_{i,j}| S_{i,j}$ and c_s, the Smagorinsky coefficient, specified as 0.16 (except near solid objects where standard wall damping [6] is implemented), Δ is the filter width (specified here as the grid spacing $\Delta = dx = dy = dz$), and $S_{i,j} = (\partial \widetilde{u}_i / \partial x_j + \partial \widetilde{u}_j / \partial x_i)/2$.

The governing equations are discretized using a pseudo-spectral method where spectral discretization is used in the x and y directions and 2nd order finite differencing in the z (or vertical) direction. Periodic boundary conditions are used along the sides of the domain, so flow simulated is over an array of trees. A stress-free boundary condition is imposed at the top of the domain, while a rough-wall (log-law) boundary condition is imposed at the bottom surface. The resolved tree branches are represented using the immersed boundary method (IBM, for details, see [2]) whereas the effects of the unresolved tree branches are imposed using the Renormalized Numerical Simulation (RNS) framework from [2]. See Fig. 2a for an illustration of the resolved and unresolved (sub-grid) branches.

As briefly noted previously, the tree canopy is broken into resolved and unresolved branches. The drag forces due to the unresolved branches are imposed using RNS. The RNS procedure is a down scaling technique that uses drag forces from the resolved flow field to parameterize the drag forces due to the unresolved scales. The formulation of the RNS framework begins by equating the drag force due to branch b at generation g (i.e. the last resolved generation) and its descendants to the sum of the resolved and unresolved branches such that,

$$\boldsymbol{F}_b^n = \boldsymbol{R}_b^n + \sum_{\beta \in sub(b)} \boldsymbol{F}_\beta^n, \quad \text{where} \quad \boldsymbol{R}_b^n = \int \boldsymbol{r}^n(\boldsymbol{x}) d^3\boldsymbol{x}; \quad \boldsymbol{F}_\beta^n = \int \boldsymbol{f}_\beta^n(\boldsymbol{x}) d^3\boldsymbol{x}, \tag{2}$$

and $\boldsymbol{r}$ is the IBM force. The total drag force of the resolved branches in (2), $\boldsymbol{R}_b^n$, is directly evaluated from the immersed boundary method whereas $\boldsymbol{F}_b^n$ and the unresolved descendant branch force, $\boldsymbol{F}_\beta^n$, are parameterized using a form drag model. The current work employs the explicit time formulation of [2] for the form drag model such that $\boldsymbol{F}_b^n = \boldsymbol{F}_b^n(c_d^{n-1})$ and scale invariance is assumed (i.e. c_d is independent of the generation number). Then

$$\boldsymbol{F}_b^n = -c_d^n \frac{\rho}{2} |V_b^n| \boldsymbol{V}_b^n A_b, \quad \boldsymbol{F}_b^n(c_d^{n-1}) = \boldsymbol{R}_b^n - c_d^{n-1} \frac{\rho}{2} \sum_{\beta \in sub(b)} |V_\beta^n| \boldsymbol{V}_\beta^n A_\beta \tag{3}$$

and ρ is the density of the fluid, $\boldsymbol{V}_b^n$ and A_b are the reference velocity and area, respectively, of branch b and its descendants, and $\boldsymbol{V}_\beta^n$ and A_β are the reference velocity and area, respectively, of the unresolved descendant branches of branch b. Using a least squares error minimization, the error between equations in (3) is minimized when

$$c_d^n = -\frac{2 \sum_b \boldsymbol{F}_b^n(c_d^{n-1}) \cdot |V_b^n| \boldsymbol{V}_b^n A_b}{\rho \sum_b |V_b^n|^4 A_b^2} . \tag{4}$$

It is still necessary, however, to correctly distribute the unresolved forces on the computational mesh in order to preserve the local flow structure. The unresolved force distribution is accomplished by expressing the drag force as a body force for any point in space as using a local force field, $f_{\beta_i}^n(\boldsymbol{x}) = -\kappa_\beta^n |\widetilde{\boldsymbol{u}}(\boldsymbol{x})| \widetilde{u}_i(\boldsymbol{x}) \widetilde{\chi}(\boldsymbol{x})$ where κ_β^n is a constant determined such that the desired total drag force is obtained (see Ref. [2]), $\widetilde{u}_i(\boldsymbol{x})$ the filtered velocity, and $\widetilde{\chi}$ the filtered indicator function (see Fig. 2b). The filtered indicator function is defined as $\widetilde{\chi} = \mathscr{G} * \chi$ where χ is the true indicator function (1 inside a branch; 0 otherwise) and $\mathscr{G}$ a Gaussian filter kernel of width 2Δ.

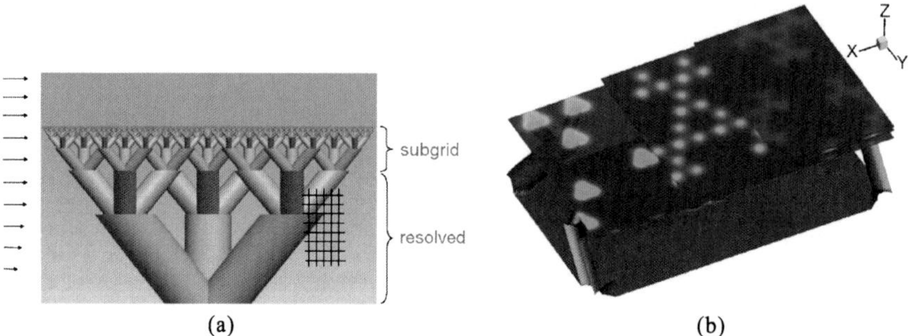

Fig. 2 (a) Description of resolved and unresolved (sub-grid) branch groups. (b) Contours of $\widetilde{\chi}$ at the mid-height of generations 3, 4, and 5 for the computation domain used in the canopy simulations. The resolved branches are shown as the grey iso-surface.

3 Experimental set up and measurements

Experiments are carried out in a water tunnel facility in the Laboratory of Experimental Fluid Dynamics (LEFD) at the Johns Hopkins University [5, 4]. The new test section is 1.2 m long, 0.56 m wide and 0.23 m deep. Flow conditioning is provided by a set honeycombs and passive grids which are located 0.25 m upstream of the test section. The incoming velocity profile is designed to mimic the velocity profile in a forest canopy. The reference velocity used for the drag coefficient calculations is obtained using a Pitot-tube in the center of the test section, 0.1 m behind and 0.05 m above the tree canopy. For this work the water flow rate was controlled to provide a range of five reference velocities between 0.8 m/s and 1.3 m/s.

The canopy model has a modular design that allows for the target tree to have between one and five generations. Located in the center of the canopy the target tree bears a load cell at its base for drag measurements. The load cell is regularly calibrated in order to account for any creep or hysteresis that may be caused from prior tests. Voltage signals from the load cell are conditioned by an analog transmitter before entering a data acquisition board connected to a desktop PC. For each reference velocity, trees with five different generation levels are considered for measurements.

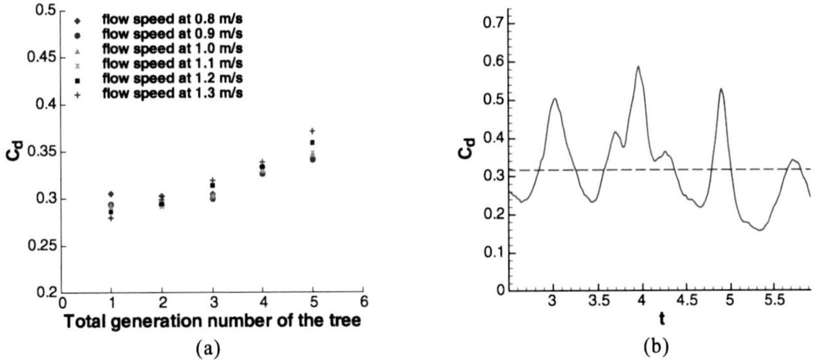

Fig. 3 (a) Experimentally determined drag coefficient for the target tree. (b) Drag coefficient computed using RNS (dashed line indicates averaged $c_d = 0.316$).

4 Numerical and experimental results comparison

The drag coefficient, c_d, for the entire tree is defined as $c_d = \frac{F_{tot}}{\frac{1}{2}\rho V^2 A}$ where F_{tot} is the total drag force, ρ the density of the fluid, V the reference velocity and A the projected frontal area [7]. The area is computed as the true "line-of-sight" projected area of the target tree whose normal direction is along the dominant flow path. In Fig. 3a the experimentally determined c_d are shown with varying generation number of the fractal tree. These results indicate increases in c_d with increasing generation number. Dependence on velocity (Reynolds number) is also visible.

The computed c_d from the LES using RNS is shown in Fig. 3b and shows significant time fluctuations. The time-average value (not converged yet) is shown as horizontal line. It is slightly above 0.3, smaller than the 5th-generation drag coefficient measured in the experiments by about 10%.

5 Conclusions

The RNS framework was utilized to provide the sub-grid parameterizations for the unresolved branches of the fractal tree canopy. Experimental force measurements were obtained for a set of varying water flow rates and number of generations for the target tree. The experimental results indicate scale dependence of the drag coefficient for the current geometry. Although the current RNS simulations employ a scale-invariant model for the drag coefficient and are not well converged yet, the numerical results match the experimental drag coefficient for higher generation numbers to about 10%.

Acknowledgements This research is supported by the National Science Foundation (IGERT Project # 0801471 and ATM grant # 0621396).

References

1. Bou-Zeid, E., C. Meneveau, and M.B. Parlange, A scale-dependent Lagrangian dynamic model for large eddy simulation of complex turbulent flows. *Phys. Fluids* **17**, 025105 (2005).
2. Chester S., C. Meneveau, and M.B. Parlange, Modeling turbulent flow over fractal trees with renormalized numerical simulation, *J. of Comp. Phys.* **222**, 427–448 (2007).
3. Porté-Agel, F., C. Meneveau, and M.B. Parlange, A scale-dependent dynamic model for large-eddy simulation: application to a neutral atmospheric boundary layer. *J. Fluid Mech.* **415**, 261–284 (2000).
4. Soranna, F., Y.C. Chow, O. Uzol, and J. Katz, The effect of inlet guide vanes wake impingement on the flow structure and turbulence around a rotor blade, *J. Turbomachinery* **128**, 92–95 (2006).
5. Uzol, O., C.Y. Chow, J. Katz and C. Meneveau, Unobstructed PIV measurements within an axial turbo-pump using liquid and blades with matched refractive indices, *Experiments in Fluids* **33**, 909–919 (2002).
6. Mason, P. and D. Thompson, Stochastic backscatter in large-eddy simulations of boundary layers, *J. Fluid Mech.* **242**, 51–78 (1992).
7. Finnigan, J.J., Turbulence in Plant Canopies, *Annual Review of Fluid Mechanics* **32**, 519–571 (2000).

Large Eddy Simulation study of a fully developed thermal wind-turbine array boundary layer

Marc Calaf, Charles Meneveau and Marc Parlange

1 Introduction

When wind turbines are arranged in large wind farms, their efficiency decreases significantly due to wake effects and to complex turbulence interactions with the atmospheric boundary layer (ABL) [1]. For large wind farms whose length exceeds the ABL height by over an order of magnitude, a "fully developed" flow regime may be established [1, 2, 3]. In this asymptotic regime, the changes in the stream-wise direction are small compared to the more relevant vertical exchange mechanisms. Such a fully developed wind-turbine array boundary layer (WTABL) has recently been studied [2] using Large Eddy Simulations (LES) under neutral stability conditions. The simulations showed the existence of two log-laws, one above (characterized by: u_*^{hi}, z_o^{hi}) and one below (u_*^{lo}, z_o^{lo}) the wind turbine region. This enabled the development of more accurate parameterizations of the effective roughness scale for a wind farm. Now, a suite of Large Eddy Simulations, in which wind turbines are modeled as in [2] using the classical drag disk concept are performed, again in neutral conditions but also considering temperature. Figure 1 shows a schematic of the geometry of the simulation.

The aim is to study the effects of different thermal ABL stratifications, and thus to study the efficiency and characteristics of large wind farms and the associated land-atmosphere interactions for realistic atmospheric flow regimes. Such studies

Marc Calaf
Department of Environmental Engineering, Ecole Polythecnique Fédérale de Lausanne (EPFL), GR A0 445, Station 2, 1015 Lausanne, Switzerland, e-mail: marc.calaf@epfl.ch

Charles Meneveau
Department of Mechanical Engineering, and Center for Environmental and Applied Fluid Mechanics, Johns Hopkins University, 3400 North Charles Street, Baltimore MD 21218, USA, e-mail: meneveau@jhu.edu

Marc Parlange
Department of Environmental Engineering, Ecole Polythecnique Fédérale de Lausanne (EPFL), GR A0 412, Station 2, 1015 Lausanne, Switzerland, e-mail: marc.parlange@epfl.ch

H. Kuerten et al. (eds.), *Direct and Large-Eddy Simulation VIII*, ERCOFTAC Series 15, DOI 10.1007/978-94-007-2482-2_38,

help to unravel the physics involved in extensive aggregations of wind turbines, allowing us to design better wind farm arrangements. As a first step, temperature is treated in a passive mode, allowing us to focus the study on the influence of a large WFABL into the scalar fluxes. By considering various turbine loading factors, surface roughness values and different atmospheric stratifications, it is possible to analyze the influence of these parameters on the induced surface roughness, and the sensible heat roughness length. These last two parameters can be used to model wind turbine arrays in simulations of atmospheric dynamics at larger (regional and global) scales [4], where the coarse meshes used do not allow to account for the specifics of each wind turbine. Results from different sets of simulations are presented, for which also the corresponding effective roughness length-scales can be determined. The results also help our understanding of how wind turbines affect scalar transport processes in the turbine wakes.

By using a simple drag disk approach for modeling the wind turbines, it is found that the surface heat flux inside the thermal wind-turbine array boundary layer is increased. This is the result of two competing effects: (1) a major increase on $u_{*,hi}$; (2) a smaller decrease due to lower $u_{*,lo}$ near the ground.

2 Description of numerical method

In this work we consider flow that is neutrally stratified, and driven by an imposed pressure gradient with temperature as a passive scalar.

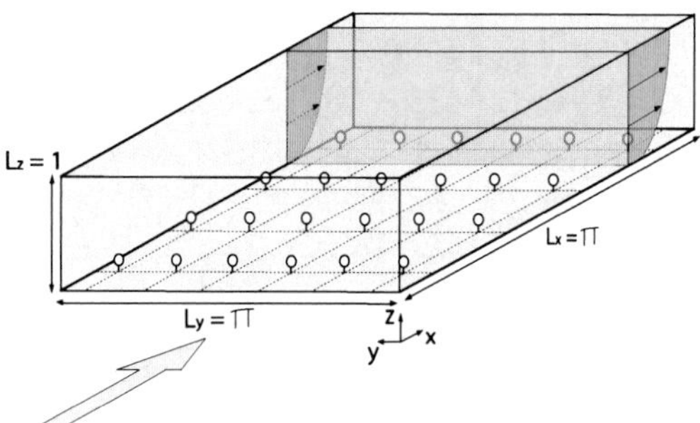

Fig. 1 Schematic sketch of the modeled wind farm, with dimensions: $(L_x,L_y,L_z) = (\pi,\pi,1) \times 1000$ m and 128^3 grid cells. With the periodic boundary conditions the simulations reproduce the conditions of a fully developed Wind Turbine Array Boundary Layer (WFABL).

Therefore, the LES is based on the filtered incompressible Navier-Stokes equations, the continuity equation and an equation for potential temperature, i.e.

$$\partial_i \widetilde{u}_i = 0 \tag{1}$$

$$\partial_t \widetilde{u}_i + \partial_j (\widetilde{u}_i \widetilde{u}_j) = -\partial_i \widetilde{p}^* - \partial_j \tau_{ij} + f_i - \delta_{i1} \partial_1 p_\infty / \rho \tag{2}$$

$$\partial_t \widetilde{\theta} + \widetilde{u}_j \partial_j \widetilde{\theta} = -\partial_j \pi_j \tag{3}$$

where $\widetilde{u}_i$ is the filtered velocity field, $\widetilde{\theta}$ is the filtered temperature field, and $\widetilde{p}^*$ is the filtered modified pressure equal to $\widetilde{p}/\rho + \tau_{kk}/3 - p_\infty/\rho$. Further, τ_{ij} is the subgrid-scale stress term. Its deviatoric part $(\tau_{ij} - \delta_{ij}\tau_{kk}/3)$ is modeled using an eddy viscosity subgrid-scale model, as discussed further below; the trace of this term $(\tau_{kk}/3)$ is combined into the modified pressure, as is common practice in incompressible LES. Equivalently, π_j is the scalar SGS flux term. (Note that if temperature was included as an active scalar, an extra term would be added on the right hand side of the momentum equation. Using Boussinesq's approximation this would be given by: $g\frac{\widetilde{\theta}-\langle\widetilde{\theta}\rangle}{\langle\widetilde{\theta}\rangle}$). Figure 1 shows a sketch of the computational domain with representative dimensions: $(L_x,L_y,L_z) = (\pi,\pi,1)$ x1000 m and 128^3 grid cells.

The force f_i is added for modeling the effects of the wind turbines in the momentum equation using the "drag disk" approach in LES [5], with a new local variant [2, 6]. Since simulations are done at very large Reynolds numbers and the bottom surface as well as the wind-turbine effects are parameterized, viscous stresses are neglected. In the real case of wind turbines in the atmospheric boundary layer, the flow is forced by geostrophic wind and in the outer layer is affected by Coriolis accelerations. The flow changes direction near the ground, and for a given geostrophic wind direction, the turning depends upon the shear stresses (momentum exchanges) at the bottom surface. Since these are not known ahead of time, and we wish to have a mean wind that is perpendicular to the wind-turbine disks in the array to be simulated, in the simulations we prefer to use forcing with an imposed pressure gradient $\partial_1 p_\infty$ in the x_1 direction. The results of the simulations, especially in the surface layer region, can still be interpreted in the context of geostrophic wind forcing.

The skew-symmetric form of the NS equation is implemented. The numerical discretization follows the approach used by Moeng [7] and Albertson & Parlange [8], which combines a pseudo-spectral discretization in the horizontal directions and a centered second-order finite differencing in the vertical direction. With the periodic boundary conditions the simulations reproduce the conditions of a fully developed Wind Turbine Array Boundary Layer (WFABL). A second order accurate Adams–Bashforth scheme is used for time integration. The subgrid model used is the dynamic Smagorinsky model [9] using the Lagrangian scale-dependent version as described in [10]. The nonlinear convective terms and the SGS stress are dealiased using the 3/2 rule.

In the streamwise direction, we use fully periodic boundary conditions (in accordance with the spectral discretization). The top boundary uses zero vertical velocity and zero shear stress boundary condition (same for temperature). At the bottom surface, we use a classic imposed wall stress boundary condition relating the wall stress to the velocity at the first grid-point and the surface heat flux to the temperature at the first grid-point using the standard log (Monin-Obukhov) similarity law [4].

$$\tau_{w1} = -\left(\frac{\kappa}{\ln z/z_{0,lo}}\right)^2 \left(\widehat{\tilde{u}}^2 + \widehat{\tilde{v}}^2\right)^{0.5} \widehat{\tilde{u}} \tag{4}$$

$$\tau_{\theta w} = \frac{(\theta_s - \tilde{\theta})u_* \kappa}{ln(z/z_{0,lo})} \tag{5}$$

where the hat on $\widehat{\tilde{u}}$ and $\widehat{\tilde{v}}$ represents a local average obtained by filtering the LES velocity field with filter width 2Δ (see [9] for more details about such filtering).

3 LES results

Figure 2 shows vertical profiles of the horizontally averaged heat fluxes. Lines with symbols represent the case with no wind turbines. Lines without symbols are the reference case without wind turbines. The dot-dashed lines represent the subgrid-scale component of the heat flux, which for both cases, with and without wind turbines, are very similar. The dashed lines account for the turbulent Reynolds heat flux components, these being $\sim 15\%$ larger near the ground (between 0 to $\sim 0.1/H$ height) for the case with wind turbines. Such differences can also be observed in the dispersive (canopy) heat flux term, but at a height above the actual wind turbines (between $\sim 0.2/H$ to $\sim 0.7/H$ height). Overall, an increase of about $\sim 15\%$ on the total heat flux (solid lines) is observed in the scenario where the wind turbines are present. This increase in the heat flux is the result of two competing effects: one, a major increase of $u_{*,hi}$ ($\sim 50\%$); two, a decrease due to lower $u_{*,lo}$ near the ground. The increase of $u_{*,hi}$ is due to the increase in turbulence and mixing induced by the wind turbines blades; while the reduction of $u_{*,lo}$ is explained due to the slow down of the mean flow near the ground because of the wind turbines presence. For further details see [2].

Figure 3 compares vertical profiles between cases with different thrust coefficients $C_t' = \{0, 0.6, 1.33, 2\}$ of the horizontally averaged heat flux. The solid line represents the case where there are no wind turbines. $C_t' = 2$, represents the corresponding Betz limit case, while $C_t' = 1.33$ is closer to a real scenario. $C_t' = 0.6$ represents a weakly loaded case. Results show that the increase of the heat flux is proportional to the thrust coefficient, although this proportionality seems to reach an asymptote for the more loaded cases.

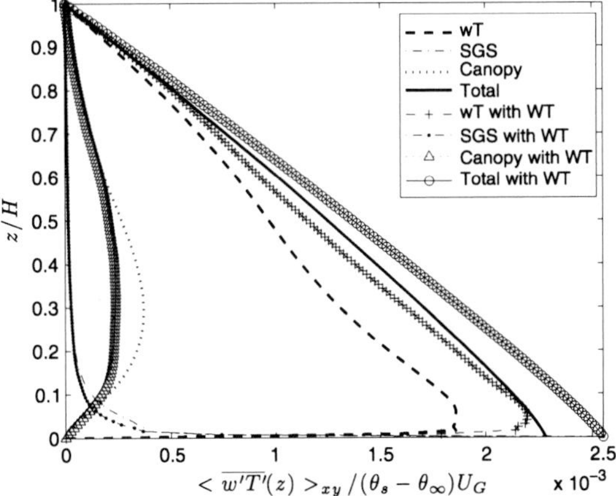

Fig. 2 Vertical profile of the spatially averaged heat flux, $\langle\overline{w'\theta'}\rangle_{xy}$. Symbols show the case with wind turbines. The dashed line shows the Reynolds surface heat flux, the dot-dashed line represents the subgrid-scale heat fluxes and the dotted line represents the diffusive (canopy) heat fluxes. The solid line is the total heat flux, resultant of adding all the three different components. These are normalized with the temperature difference between the surface and the top of the boundary layer times the geostrophic wind.

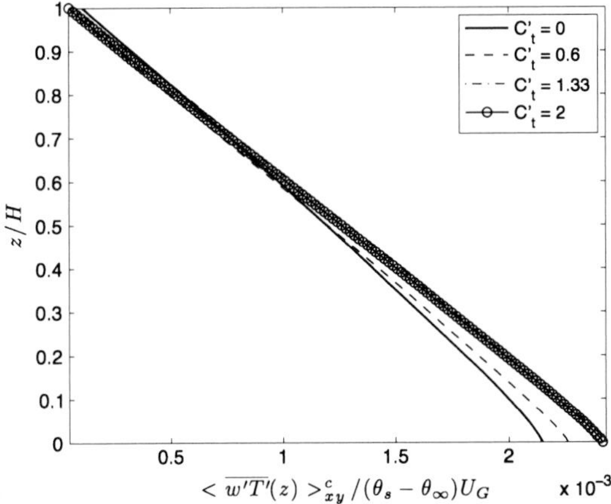

Fig. 3 Heat flux, vertical profiles. Comparison between cases with different thrust coefficients (C_t'). Solid line represents the case where there are no wind turbines ($C_t' = 0$). Dashed line shows the case with the lightest thrust coefficient ($C_t' = 0.6$). The dot-dashed line represents the case with an intermediate loading ($C_t' = 1.33$), and the hollow circle case, represents the case corresponding to the Betz limit ($C_t' = 2$).

4 Conclusions

Results have shown that large wind farms do increase scalar fluxes, as it was earlier foreseen by [11]. This increase is the result of two competing effects: one, a major increase of $u_{*,hi}$ ($\sim 50\%$); two, a decrease due to lower $u_{*,lo}$ near the ground. Also, it has been shown that higher thrust coefficients (C_t') have an increasing effect on the scalar fluxes up to a certain asymptote. This asymptote should be analyzed with more detail. Therefore, we plan on studying different intermediate C_t' values among the ones tested by now, different surface roughnesses (z_0), and wind turbine spacings (S_x, S_y). This will also allow us to better asses the inter-relationship between the increase of scalar fluxes and these relevant wind farm parameters.

Acknowledgements M.C. was supported by (Swiss) SNF 200021-107910/1 Land-atmosphere interaction over complex terrain: large eddy simulation and field experiments. C.M. acknowledges funding from the US National Science Foundation (Projects CBET 0730922 and AGS 1045189).

References

1. Frandsen S., Barthelmie R., Pryor R., Rathmann O., Larsen S., Hojstrup J., Thogersen M.: Analytical modelling of wind speed deficit in large offshore wind farms. Wind. Energy **9**: 39–53, (2006).
2. Calaf M., Meneveau C., Meyers J.: Large Eddy Simulation study of fully developed wind-turbine array boundary layers. Phys. Fluids **22**, 015110, (2010) doi:10.1063/1.3291077.
3. Cal R.B., Lebrón J., Kang H.S., Castillo L., Meneveau C.: Experimental study of the horizontally averaged flow structure in a model wind-turbine array boundary layer. J. Renewable Sustainable Energy **2**, 013106, (2010) doi:10.1063/1.3289735.
4. Barrie D., Kirk-Davidoff D.: Weather response to Management of large wind turbine array. Atmos. Chem. Phys. Discuss. **9**, 2917, (2009).
5. Jimenez A., Crespo A., Migoya E., Garcia J.: Advances in large-eddy simulation of a wind turbine wake. J. of Physics: Conference Series **75**, 012041, (2007).
6. Meyers J., Meneveau C.: Large eddy simulations of large wind-turbine arrays in the atmospheric boundary layer. In Proceedings of the 48th AIAA Aerospace Sciences Meeting, January, Orlando (FL). AIAA paper 2010-827 (2010)
7. Moeng C.-H.: A large-eddy simulation model for the study of planetary boundary-layer turbulence. J. Atmos. Sci. **6**:2311–2330, (1984).
8. Albertson J.D., Parlange M.B.: Surface length-scales and shear stress: implications for land-atmosphere interaction over complex terrain. Water Resour. Res. **35**:2121–2132, (1999a).
9. Germano M., Piomelli U., Moin P., Cabot W.H.: A dynamic subgrid-scale eddy viscosity model. Phys. Fluids A **3**, 1760, (1991).
10. Bou-Zeid E., Meneveau C., Parlange M.B.: A scale dependent Lagrangian dynamic model for large eddy simulation of complex turbulent flows. Physics of Fluids **17**, 025105, (2005).
11. Baidya-Roy S., Pacala S.W., Walko R.L.: Can large wind farms effect local meteorology? Journal of Geophysical Research **109**, (2004).

Coherent Structures in the Flow over Two-Dimensional Dunes

Mohammad Omidyeganeh and Ugo Piomelli

1 Introduction

The fluid flow over rough sand beds in rivers has unique dynamics compared with the flows that occur when the bed is flat. Depending on the flow Reynolds number, the most commonly found river-bed formations are ripples and dunes. Ripples have dimensions much smaller than the river depth, while dunes may reach heights of the order of the depth. Ripples do not affect the dynamics of the whole flow depth whereas dunes influence on the turbulent flow as well as the sediment transport at the whole depth. Dune formation may affect navigation, erosion of bridge piles and other structures, as well as dispersion of contaminants [8].

One feature of the flow over dunes that has attracted significant attention is the variety of very large (with size comparable to the river depth) coherent structures that are observed. Among the large structures are the "boils" [4, 5] (upwelling motions observed at the water surface, usually when a horizontally oriented vortex attaches to the surface); they play an important role in the sediment transport [11, 23], as these eddies lift up sediments and carry them away from the bed.

The uniqueness of the boil structures in flows over dunes has been illustrated in laboratory measurements [4, 5, 9, 14, 15] and field observations [2, 3]. Unfortunately little is known of the generation and evolution of these coherent flow structures. Several studies show that the structures that cause the boils maybe originally loop or horseshoe vortices [9, 14]. In our study we take advantage of the three-dimensional and unsteady information supplied by the LES to investigate, for the first time, the dynamics of the eddies responsible for the boil generation.

Mohammad Omidyeganeh
Department of Mechanical and Materials Engineering, Queen's University, Kingston (Ontario), Canada, e-mail: omidyeganeh@me.queensu.ca

Ugo Piomelli
Department of Mechanical and Materials Engineering, Queen's University, Kingston (Ontario), Canada, e-mail: ugo@me.queensu.ca

H. Kuerten et al. (eds.), *Direct and Large-Eddy Simulation VIII*, ERCOFTAC Series 15, DOI 10.1007/978-94-007-2482-2_39,

Numerical simulations can help to understand the dynamics of these events. Other than numerical simulations using Reynolds-averaged Navier-Stokes equations (RANS) [24], which cannot show turbulent structures, a few large eddy simulations (LES) has been reported [7, 22, 25]. In general, the mean flow and second order statistics were in good agreement with the experiments, but large structures illustrated in the simulations differed greatly between the various simulations, and little attention was paid to free-surface events.

The present work is focused on the visualization and evolution of the boils at the surface. It is shown how these structures are generated close to the bed, convected downstream and raised to the surface. Although the instantaneous measurements could not explain this phenomenon comprehensively, this numerical study analyze it in details.

2 Problem formulation and computational configuration

The velocity field is separated into a resolved (large-scale) and a subgrid (small-scale) field, by a spatial filtering operation [12]. The non-dimensional continuity and Navier-Stokes equations for resolved velocity field are

$$\frac{\partial \overline{u}_i}{\partial x_i}=0; \qquad \frac{\partial \overline{u}_i}{\partial t}+\frac{\partial \overline{u}_i\overline{u}_j}{\partial x_j}=-\frac{\partial \overline{P}}{\partial x_i}-\frac{\partial \tau_{ij}}{\partial x_j}+\frac{1}{Re_b}\nabla^2\overline{u}_i \tag{1}$$

where $Re_b = U_b H_b/\nu$, H_b is the channel average height, and U_b is the average velocity there. Here x_1, x_2 and x_3 are the streamwise, wall-normal and spanwise directions, also referred to as x, y and z. The resolved velocity components in these directions are, respectively, u_1, u_2 and u_3 (or u, v and w); $\tau_{ij} = \overline{u_i u_j} - \overline{u}_i\overline{u}_j$ are the subgrid stresses, which, in the present study, were modeled using an eddy-viscosity assumption

$$\tau_{ij} - \delta_{ij}\tau_{kk}/3 = -2\nu_T\overline{S}_{ij} = -2C\overline{\Delta}^2|\overline{S}|\overline{S}_{ij}, \tag{2}$$

where $\overline{\Delta} = 2\left(\overline{\Delta x\Delta y\Delta z}\right)^{1/3}$ is the filter size, $\overline{S}_{ij} = \left(\partial\overline{u}_i/\partial x_j + \partial\overline{u}_j/\partial x_i\right)/2$ is the resolved strain-rate tensor and $|\overline{S}| = \left(2\overline{S}_{ij}\overline{S}_{ij}\right)^{1/2}$ is its magnitude. The coefficient C was determined using the dynamic model [6] with the Lagrangian averaging technique proposed by Meneveau *et al.* [13], and applied to non-Cartesian geometries by Armenio and Piomelli [1].

The governing differential equations (1) are discretized on a non-staggered grid using a curvilinear finite volume code [20]. The method of Rhie and Chow [18] is used to avoid pressure oscillations. Both convective and diffusive fluxes are approximated by second-order central differences. A second-order semi-implicit fractional-step procedure [10] is used for the temporal discretization. The Crank-Nicolson scheme is used for the wall normal diffusive terms, and the Adams-Bashforth scheme is used for all the other terms. Fourier transforms are used to reduce the three-dimensional Poisson equation into a series of two-dimensional Helmholtz

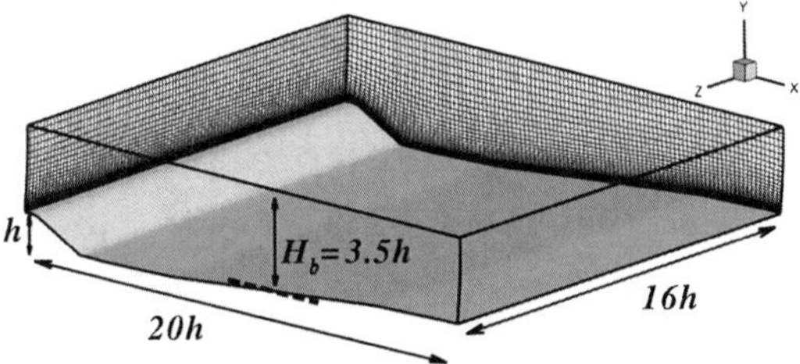

Fig. 1 Sketch of the physical configuration. Every fourth grid line is shown.

equations in wave number space, which are then solved iteratively using the Bi-conjugate Gradient Stabilized (BCGSTAB) method. The code is parallelized using the MPI message-passing library and the domain-decomposition technique, and has been extensively tested for various turbulent flows [16, 17, 20, 21].

The computational configuration (Figure 1) is the same as that used by Stoesser *et al.* [22]. Periodic boundary conditions are used in the streamwise (x) and spanwise (z) directions. At the free surface, the free-slip condition is used (corresponding to an undeformable surface). The Reynolds number is $Re_b = 18,900$. The resolution is the same as that used by Stoesser *et al.* [22] but the domain is twice as wide as that used by previous simulations in order to capture the largest structures in the flow. $416 \times 128 \times 384$ points are used, which results in grid resolution in local wall units, $\Delta s^+ < 19$ (streamwise), $\Delta z^+ < 15$ (spanwise), and $\Delta n^+ < 0.8$ (wall normal). The grid dependency of the first and second order statistics has been studied for grids refined by a factor of 1.3 in all directions; the results showed no difference in the first- and second-order statistics.

3 Results and discussion

We first validated the simulation by comparing statistics with the LES by Stoesser *et al.* [22] and the experiments reported in that paper. The agreement with first- and second-order statistics was very good. The two-point correlations were also examined, and they indicated that the large structures that occur near the surface require a significantly wider domain than that used in previous simulations, resulting in the choice of a $16h$-wide domain.

The large structures at an instant in time are shown in Figure 2. They are visualized as isosurfaces of pressure fluctuations [19], colored by y to highlight their distance from the free surface. Rollers are generated at the dune crest due to the Kelvin-Helmholtz instability. They are convected downstream and interact with the near-wall turbulence, or rise to the surface. The interaction of rollers near the wall with streamwise vortices generated in the developing boundary layer leads to large horseshoe-like structures. Since the flow is highly turbulent, these structures may be distorted, become one-legged or be completely destroyed. The region behind the

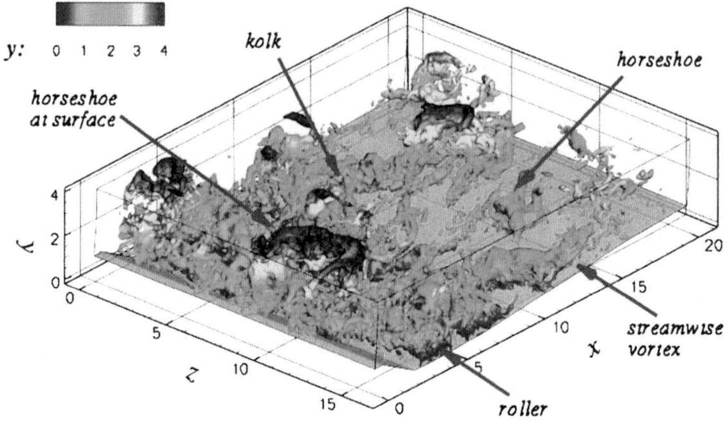

Fig. 2 Instantaneous isosurfaces of pressure fluctuation colored with height.

head (and between the legs) of horseshoe vortices is generally characterized by an intense Q2 event ($u' < 0$ and $v' > 0$), as shown in Figure 3, where the velocity fluctuation vectors in a xy-plane are also shown.

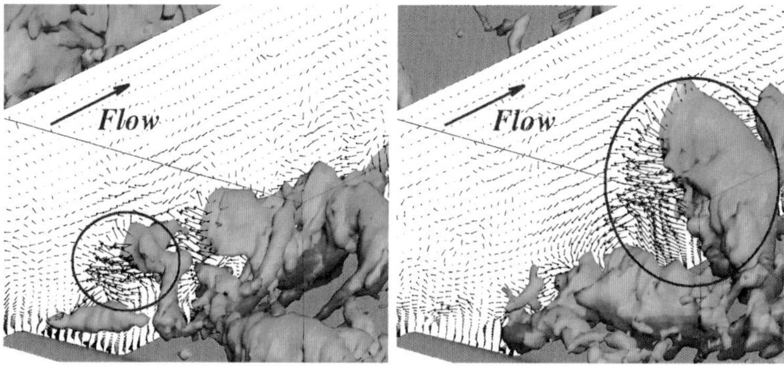

Fig. 3 Evolution of a large horseshoe structure; the time step between snapshots is $6H_b/U_b$.

The structure is elongated and tilted downward as it rises to the surface; eventually its tip touches the surface. Q2 events happen inside the vortex between the legs and Q4 events ($u' > 0$ and $v' < 0$) outside the structure. The vortex is originally vertical (see Figure 3), but is later tilted downward due to the higher mean velocity at the surface (see Figure 4(a)). The degree of inclination of the structure was measured by looking for a plane which passes through the core of the vortex loop, and is about 55° with respect to the xz plane. The signatures of Q2 and Q4 events are also illustrated by the three-dimensional vectors on the inclined plane.

Figure 4(b) shows the vectors on the surface at the instant when the tip of the structure hits the surface. At the center of the boil, the divergence of the surface

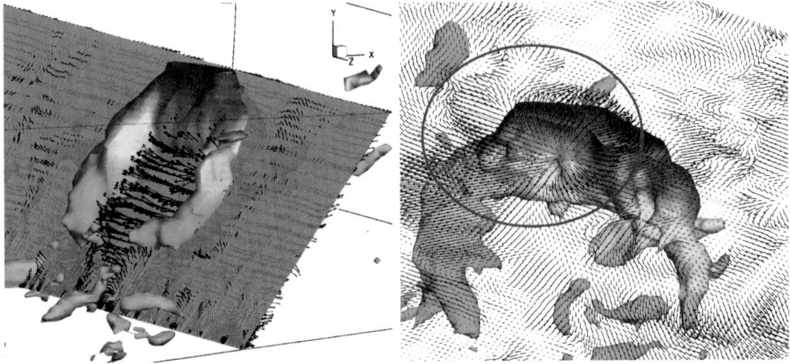

Fig. 4 Visualization of the flow near a large horseshoe structure when it touches the surface (left); boil event on the surface (right).

velocity fluctuations ($\partial u'/\partial x + \partial w'/\partial z$) is locally high, as is the planar turbulent kinetic energy ($u'^2 + w'^2$) on the edge of the boil. These are the signatures of the boils on the surface. Later, the vortex loop is convected downstream, elongated, and distorted. Its legs reach the surface and the upwelling is expanded and weakened.

4 Conclusions

The flow over two-dimensional dunes has unique characteristics among open channel flows. The main distinction between the structures of the flow over dunes and smooth beds is the presence in the former of roller structures that interact with wall turbulence and shape an inclined horseshoe-like vortex that is able to reach the surface and make larger and stronger boils, compared with the common boils found in open channel flows.

We performed for the first time a detailed analysis of the generation and evolution of boils at the surface and showed that the origin of the strong boils in the flow over dunes are the rollers due to Kelvin-Helmholtz instability. The history of these structures shows that they are inclined from their originally vertical orientation, while horseshoe vortices in boundary layer starts from a horizontal orientation. This indicates that these structures are unique to the flow over dunes, in which separation occurs at the crest. It rejects the hypothesis that these structures are originated from the turbulent bursts.

While the boil frequency is fairly low (they were found every $40h/U_b$ time, approximately) their size can be even larger than the flow depth (almost $5 \sim 6h$). Since boils rise to the surface from the bed area, they are known to affect sediment transport significantly. A more quantitative analysis of the contribution of large vortices to mass and momentum transport is presently being carried out.

References

1. Armenio V, Piomelli U (2000) A Lagrangian mixed subgrid-scale model in generalized coordinates. Flow, Turbulence and Combustion 65:51–81
2. Babakaiff SC, Hickin EJ (1996) Coherent Flow Structures in Squamish River Estuary, British Columbia, Canada. In: Ashworth P, Bennett S, Best JL, McLelland S (eds) Coherent Flow Structures in Open Channels, Wiley, New York
3. Best JL, Kostaschuk RA, Villard PV (2001) Quantitative Visualization of Flow Fields Associated with Alluvial Sand Dunes: Results from the Laboratory and Field Using Ultrasonic and Acoustic Doppler Anemometry. Journal of Visualization 4:373–381
4. Best J (2005) Kinematics, topology and significance of dune-related macroturbulence: some observations from the laboratory and field. Spec Publs int Ass Sediment 35:41–60
5. Best J (2005) The fluid dynamics of river dunes: A review and some future research directions. Journal of Geophysical Research, doi: 10.1029/2004JF000218
6. Germano M, Piomelli U, Moin P, Cabot WH (1991) A dynamic subgrid scale eddy viscosity model. Physics of Fluids A 3:1760–1765
7. Grigoriadis DGE, Balaras E, Dimas AA (2009) Large-eddy simulations of unidirectional water flow over dunes. Journal of Geophysical Research, doi: 10.1029/2008JF001014
8. Itakura T, Kishi T (1980) Open channel flow with suspended sediment on sand waves. Proceedings of the Third International Symposium on Stochastic Hydraulics, 599–609
9. Kadota A, Nezu I (1999) Three-dimensional structure of space-time correlation on coherent vortices generated behind dune crest. Journal of Hydraulic Research 37:59–80
10. Kim J, Moin P (1985) Application of a fractional step method to incompressible Navier-Stokes equations. Journal of Computational Physics 59:308–323
11. Kostaschuk RA, Church MA (1993) Macroturbulence generated by dunes: Fraser River, Canada. Sedimentary Geology 85:25–37
12. Leonard A (1974) Energy cascade in large-eddy simulations of turbulent fluid flows. In: Turbulent diffusion in environmental pollution; Proceedings of the Second Symposium 75:237–248
13. Meneveau C, Lund TS, Cabot WH (1996) A Lagrangian dynamic subgrid-scale model of turbulence. Journal of Fluid Mechanics 319:353–385
14. Muller A, Gyr A (1986) On the vortex formation in the mixing layer behind dunes. Journal of Hydraulic Research 24:359–375
15. Nezu I, Nakagawa H (1993) Turbulence in Open-Channel Flows. Balkema
16. Radhakrishnan S, Piomelli U, Keating A, Lopes AS (2006) Reynolds-averaged and large-eddy simulations of turbulent non-equilibrium flows. Journal of Turbulence 7(63):1–30
17. Radhakrishnan S, Piomelli U (2008) Large-eddy simulation of oscillating boundary layers: Model comparison and validation. Journal of Geophysical Research, doi: 10.1029/2007JC004518
18. Rhie C, Chow W (1983) Numerical study of the turbulent flow past an airfoil with trailing edge separation. AIAA Journal 21:1525–1532
19. Robinson SK (1991) Coherent motions in the turbulent boundary layer. Annu. Rev. Fluid Mech. 23:601–639
20. Silva-Lopez A, Palma JMLM (2002) Simulations of isotropic turbulence using a non-orthogonal grid system. Journal of Computational Physics 175:713–738
21. Silva-Lopez A, Piomelli U, Palma JMLM (2006) Large-eddy simulation of the flow in an S-duct. Journal of Turbulence 7(11):1–24
22. Stoesser T, Braun C, Garcia-Villalba M, Rodi W (2008) Turbulence structures in flow over two-dimensional dunes. Journal of Hydraulic Engineering 134(1):42–54
23. Venditti IG, Bennet SJ (2000) Spectral analysis of turbulent flow and suspended sediment transport over dunes. Journal of Geophysical Research 105:22035–22047
24. Yoon JY, Patel VC (1996) Numerical model of turbulent flow over sand dune. Journal of Hydraulic Engineering 122(1):10–18
25. Yue W, Lin CL, Patel VC (2005) Large-eddy simulation of turbulent open-channel flow with free surface simulated by level set method. Physics of Fluids 17:1–12

LES of turbulence around a scoured bridge abutment

F. Bressan, F. Ballio and V. Armenio

1 Introduction

Local scour phenomena around bridge abutments and piers can induce the collapse of hydraulic structures [1]. Understanding of the erosion process is required for the a-priori estimation of the scour-hole geometry. The problem of local erosion is particularly difficult since a complete modeling of the process needs to take into account several phenomena ranging from fluid mechanics to river geomorphology. The turbulence characteristics of the incoming flow field [2], the dynamics of the coherent structures that forms in junction flows [3] and the sediments motion which is characterized by large intermittency [4], are processes that have to be properly considered. The first step of the present research is aimed at the analysis of the coherent structures dynamics around the bridge abutment and how they change in different scour conditions. In fact many authors found out that the lack in understanding these processes is one of the motivation of the incapability of the models to accurately predict the scour-hole geometry and its maximum depth [5]. The second step is aimed at understanding how the coherent structures and their dynamics can influence the scouring process. It is important to know how the fluctuations and the intermittent character of the vortical structures can be involved in the sediment transport and to single out the most important forces that can destabilize the sediments since a clear view of the incipient motion is still missing. This study focuses on the analysis of

F. Bressan
IIHR - Hydroscience & Engineering, University of Iowa, 100 C. Maxwell Stanley Hydraulics Laboratory Iowa City, IA, 52242-1585, USA, e-mail: `filippo-bressan@uiowa.edu`

F. Ballio
Dipartimento di Ingegneria Idraulica, Ambientale, Infrastrutture Viarie, Rilevamento, Politecnico di Milano, Italy, e-mail: `francesco.ballio@polimi.it`

V. Armenio
Dipartimento di Ingegneria CIvile e Ambientale, Università di Trieste, Italy, e-mail: `armenio@dica.units.it`

H. Kuerten et al. (eds.), *Direct and Large-Eddy Simulation VIII*,
ERCOFTAC Series 15, DOI 10.1007/978-94-007-2482-2_40,

the turbulent field around a 45° wing-wall bridge abutment at different phases of the scour phenomenon.

2 Problem formulation

The scour around the bridge abutment is investigated considering three phases of the process that are considered typical in local erosion phenomena. The beginning of the erosion in which the bottom bed is still flat (Flat Bed case, FB) and two advanced configurations: the logarithmic phase in which the evolution of the scour-hole depth goes like the log of time (Logarithmic Scour, LS) and the equilibrium stage in which the process can be considered as quasi-steady (Equilibrium Scour, ES). Due to its own performance in simulating unsteady 3D junction flow, resolved LES is used to compute the flow field while the bathymetry of both the scour-hole configurations is taken from a physical experiment [4]. The bulk Reynolds number $Re_b = \frac{hU_b}{\nu}$, based on the bulk velocity U_b at the inlet section and on the duct height h, is set equal to 7000; correspondingly the friction Reynolds number $U_\tau = \frac{hu_\tau}{\nu}$ is equal to 417. The grid has $256 \times 64 \times 96$ cells along the x-, y-, and z-directions, respectively. In all cases about eight velocity points are placed within the first ten wall units in the wall-normal direction, and the distances between the wall and the first cell face are $\Delta y^+_{min} = \Delta y_{min} \frac{u_\tau}{\nu} = 2$ and $\Delta z^+_{min} = \Delta z_{min} \frac{u_\tau}{\nu} = 2$, respectively, in the two cross sectional directions. This means that the first velocity point off the wall is at $\Delta y^+_1 = \Delta z^+_1 = 1$. Since the bathymetry of the scour-hole is different for each condition analyzed, the grid geometry of the advanced configurations was changed in the vertical direction (z-axis), while the proportion of the cells in the x- and y-directions was kept identical in the three cases. The largest spacing is $\Delta z^+_{max} = 11$ in the flat bed configuration and $\Delta z^+_{max} = 23$ in the equilibrium scour configuration. The grid spacing was always chosen in order to satisfy the requirements of large-eddy simulations for resolving the near-wall viscous sublayer and literature examples indicate that such resolution is adequate for this kind of problem [6]. We use the curvilinear-coordinate LES algorithm developed by [7], with a dynamic mixed SGS model composed of a Smagorinsky part and a scale similar part. More details of the code adopted can be found in [6].

3 Results

The presence of the obstacle creates an adverse pressure gradient which, interacting with the three-dimensional incoming boundary layer, generates a complex system of vortical coherent structures (Figure 1). In the FB configuration the vortex system is composed of a primary vortex (PV), a corner vortex (CV) and a detached shear layer (DSL). The primary vortex PV originates from the interaction between the incoming boundary layer of the bottom bed with the adverse pressure gradients gen-

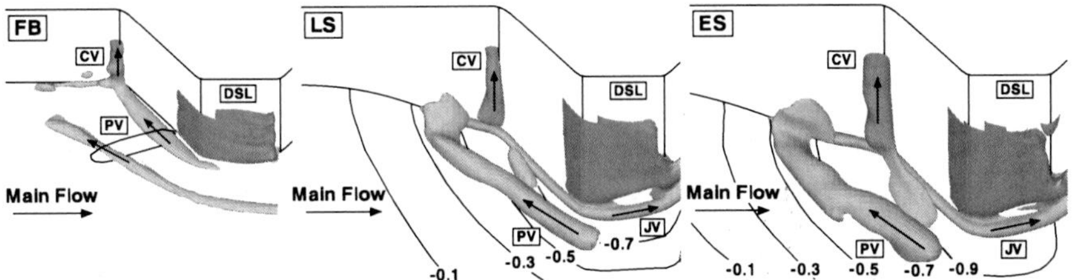

Fig. 1 Vortex system in front of the abutment visualized using the Q criterion. The arrow of the vortical structures indicates the direction of the vorticity and the bathymetry of the scour-hole in the LS and ES cases is shown through isolines.

erated by the obstacle and it is the analogous of the classic horseshoe vortex which forms in case of bridge piers. Since the vortex has a very shallow cross-sectional shape, the parameter Q can only individuate its core and its external part close to the abutment wall. Similarly to the primary vortex PV, the structure CV originates from the interaction between the incoming boundary layer of the vertical duct wall with the adverse pressure gradient of the obstacle. Along the central abutment wall a detached shear layer (DSL) forms as a consequence of the separation induced by the sharp nose edge of the abutment. In the scoured cases (LS and ES) the boundary layer separation is enhanced by the scour-hole edge and the vortex system develops within a larger volume. As a consequence the vortices PV and CV increase in size and they assume a more circular shape. The detached shear layer (DSL) is still present and a new structure called junction vortex (JV) forms in the deeper part of the scour-hole between the abutment walls and the bottom bed. In the ES case the structures are in general larger compared to those of the LS case, still remaining qualitative similar to the latter. Further particulars of the vortex system can be

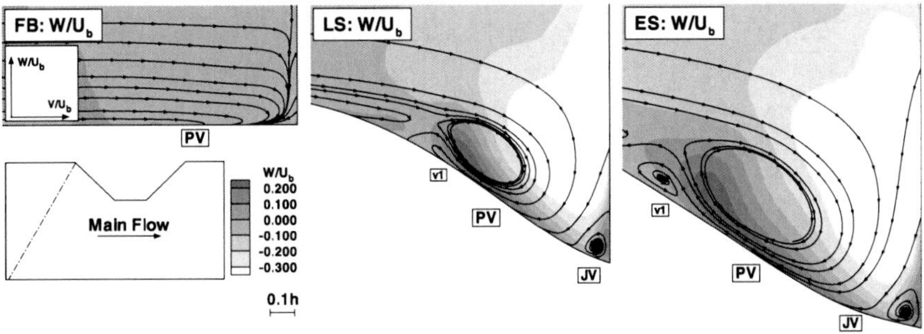

Fig. 2 Non-dimensional vertical velocity W/U_s shown for a vertical section taken from the abutment nose. The formation of the vortices PV and JV in the scoured configurations is highlighted with streamlines.

detected analyzing the vertical velocity W/U_b and the streamlines (Figure 2). The flow field in front of the abutment which is directed towards the bed is commonly called downflow and it shows negative value of the vertical velocity. Its intensity increases by about one order of magnitude in the advanced configurations compared to its value in the FB case and the maximum is detected for the ES case. The larger downflow interacts with the bottom bed and originates the smaller vortex JV as can be seen by the streamlines shown in Figure 2. Once the flow field in front of the abutment reaches the bottom bed, it separates in two directions: part of that turns backwards creating the primary vortex PV and the remaining part bends towards the obstacle and it is pushed upwards forming vortex JV. This structure can be also found in different obstacle geometry [9].

The numerical simulations were able to capture all the coherent structures that are usually found in experimental studies [8] and they were helpful to understand their behavior. A typical feature of many surface mounted obstacles is the bimodal oscillation of these structures: the PV oscillates between two preferential modes that are called *back-flow* mode and *zero-flow* mode [3]. In the *back-flow* mode the vortical structures are pushed backwards by a strong near-wall jet originating from the downflow and the vortex center is located in a position more upstream compared to its time-averaged position. In the *zero-flow* mode the near-wall jet is destroyed and the vortex center shifts downstream. This dynamics can be easily recognized with the probability distribution function (pdf) of the vertical velocity W/U_b shown in Figure 3. In the FB case the thin shape of the pdf indicates that there is a low

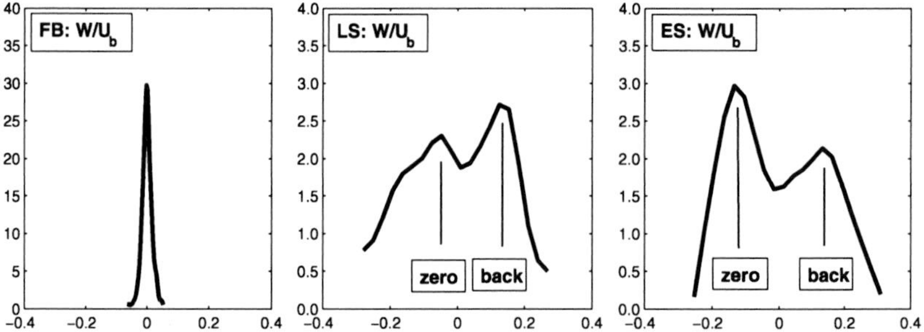

Fig. 3 Pdf of the vertical velocity component W/U_s for a point located in the near-wall region below the primary vortex PV. Mean value has been subtracted.

level of fluctuations and no bimodal dynamics. In the scoured configuration LS and ES the pdf shows a larger variance and two peaks of largest probability that correspond to the preferential modes. The *back-flow* mode is detected when the vertical velocity has a positive value meaning that the structure is moving upwards along the sloping bed. When the structure is going towards the obstacle, and thus is descending along the scour-hole, it generates a negative fluctuation of the vertical velocity (*zero-flow* mode). This oscillation affects directly the Reynolds stresses leading this tensor to have non-zero off-diagonal components (Figure 4). In the FB case the contours of the off-diagonal stress vw/U_b^2 is similar to that of a classical duct flow in which secondary currents are generated in the corner region. In the scoured cases the

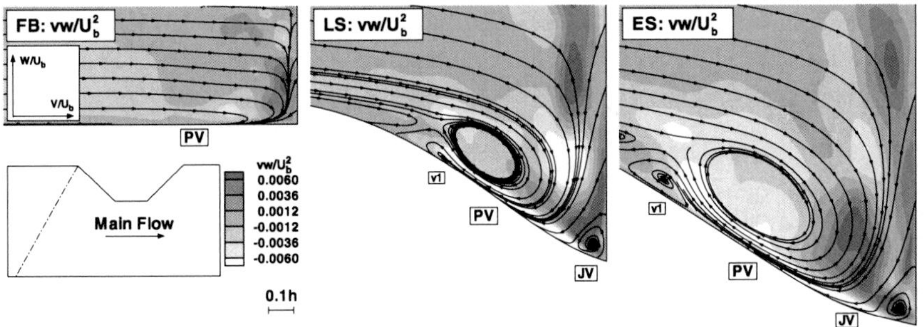

Fig. 4 Off-diagonal Reynolds stress vw/U_b^2 for the three scour configurations.

Reynolds stress is larger and the particular dynamics of the PV generates a negative stress in the near-wall region. This is in fact a consequence of the bimodal dynamics and indicates that the switching of the primary vortex between the two preferential modes induces a strong correlation between the horizontal V/U_b and vertical W/U_b velocity components.

The analysis of the turbulent field has important implications for the understanding of erosion phenomena. The most important contribution to the destabilization of the sediments is known to be the instantaneous bottom shear stress. In order to evaluate the importance of the stress in terms of erosion, the bottom shear stress amplification τ/τ_m is defined as the ratio between the local shear stress and the mean shear stress of a turbulent channel flow in condition of incipient of motion: a value exceeding one indicates incipient motion of sediments. In experimental studies it is reported that sediment transport in the upstream corner region is close to zero at the beginning of the process, then it increases during the logarithmic phase and it is highly intermittent [4]. The probability density function of τ/τ_m in that region

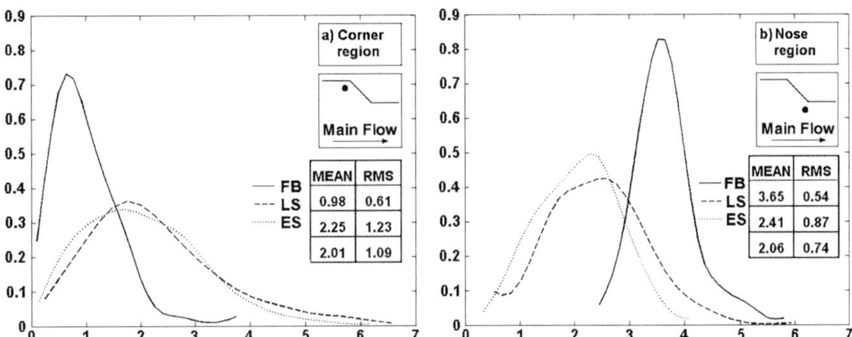

Fig. 5 Probability density function (pdf) of the bottom shear stress amplification τ/τ_m. a) Corner region; b) Nose region. Mean and RMS values are also reported in the legend.

(Figure 5a) shows in fact that in the FB condition the mean value is below the critical conditions. On the contrary when the scour-hole is formed (LS and ES cases) the mean value is large enough to move the sediments. Furthermore the wall-shear stress PDF is affected by heavier right tails indicating that the process is dominated by rare but very intense events of high shear stress. Such events are originated by the bimodal dynamics which, inducing the primary vortex to oscillate parallel to the bottom bed, can originate very large stresses. In the nose region (Figure 5b) the situation is completely different: this is the typical location where the erosion process starts in the experimental studies and once the scour-hole has formed the sediment transport is lower. The pdf in the FB case has a very large mean value and a rather thin shape. In advanced configurations the pdf has a lower mean value but it is still larger than the critical value and besides, coherently with the larger fluctuations of the vortex system, it shows a larger variance.

4 Conclusions

The results obtained indicates that the dynamics of the vortex system changes according to the bathymetry changes. The advanced stages of the process shows bimodal dynamics of the primary vortical structure which is reflected on the Reynolds stress tensor and on its anisotropy. Furthermore this type of fluctuations can induce rare events of large shear stress on the bottom bed and lead the sediment transport to be highly intermittent. In order to accurately simulate local erosion phenomena, the bimodal dynamics has therefore to be properly reproduced.

References

1. Cardoso, A. H., and Bettess, R.: Effects of time and channel geometry on scour at bridge abutments. J. Hydraul. Eng., **125(4)**, 388–399 (1999)
2. Sumer, B. M.: Mathematical modeling of scour: A review. J. Hydraul. Res., **45(6)**, 723–735 (2007)
3. Simpson, R. L.: Junction Flows. Annu. Rev. Fluid Mech., **33**, 415–443 (2001)
4. Radice, A., Porta, G. and Ballio, F.: Local scour at a trapezoidal abutment: sediment motion pattern. J. Hydr. Res., **47(2)**, 250–262 (2009)
5. Ahmed, F., and Rajaratnam, N.: Observations on flow around bridge abutment. J. Eng Mech., **123(1)**, 51–59 (2000)
6. Teruzzi, A., Ballio, F. and Armenio, V.: Turbulent stresses at the bottom surface near an abutment: laboratory-scale numerical experiment. J. Hydraul. Res., **135(2)**, 106–117 (2009)
7. Armenio, V., and Piomelli, U.: Lagrangian mixed subgrid-scale model in generalized coordinates. Flow Turbul. Combust., **65(1)**, 51–81 (2000)
8. Dey, S., and Barbhuiya, A. K.: 3D flow field in a scour hole at a wing-wall abutment. J. Hydraul. Res., **44(1)**, 33–50 (2006)
9. Koken M. and Constantinescu S. G.: An investigation of the flow and scour mechanisms around isolated spur dikes in a shallow open channel. Part 1 and 2. Water Resour. Res., **44** (2008)

Large Eddy Simulation of a neutral and a stratified flow in a plane asymmetric diffuser

F. Roman, S. Sarkar, V. Armenio

1 Introduction

The scope of this paper is to analyze the performance of wall-layer model Large Eddy Simulation (WLES) in the prediction of flow separation under stable stratification. This is a fundamental problem of practical importance both in industrial and in environmental applications. We consider turbulent flow in a plane asymmetric diffuser because this simple configuration is particularly challenging for testing the performance of numerical models to reproduce flow separation. Further, this flow has been previously investigated both experimentally and numerically. A sketch of the geometry is in Fig. 1. Depending on the slope of the diffuser and on the Reynolds number ($Re = ul/\nu$ with u and l characteristic velocity and length scale of the problem, while ν is the kinematic viscosity) separation occurs.

Experiments were carried out by Obi et al. [7] and subsequently by Buice and Eaton [3] and [4]. These experiments were performed at $Re_{bulk} = 9000$ based on the bulk velocity and half channel height measured at the inlet. For the configuration adopted, separation occurs at the first third of the inclined wall. This configuration has been simulated *via* wall-resolving LES by Kaltenbach et al. [9] who obtained a good agreement with the experimental data for the mean velocity, the pressure coefficient and the skin friction. Schluter et al. [11] used different mesh resolutions and different subgrid models, and they found that standard Smagorinsky model gave very bad results.

The performance of Reynolds Averaged Navier Stokes Equations (RANS) models (also using commercial codes) were tested on this problem by Apsley and

F. Roman and V. Armenio
Dipartimento di Ingegneria Civile e Ambientale, Universita' degli studi di Trieste, Piazzale Europa 1, Trieste, Italy, e-mail: froman@units.it,armenio@dica.units.it

S. Sarkar
Department of Mechanical and Aerospace Engineering, University of California San Diego, 9500 Gilman Drive, La Jolla, California, USA, e-mail: sarkar@ucsd.edu

H. Kuerten et al. (eds.), *Direct and Large-Eddy Simulation VIII*, ERCOFTAC Series 15, DOI 10.1007/978-94-007-2482-2_41,

Leschziner [1] and Iaccarino [5]. In general, good results were obtained but strongly dependent on the closure adopted for turbulence; for example, the $k-\varepsilon$ model completely fails in the prediction of separation.

In the present work we use LES in conjunction with wall layer modeling. It will be shown that, even with a very coarse grid, satisfactory results can be achieved. Moreover, we investigate the flow under stratification to see the influence on separation zone.

In section 2 the numerical model is described, in section 3 the results for the neutral case are shown, in section 4 we consider the stratified case. Finally conclusions are presented in section 5.

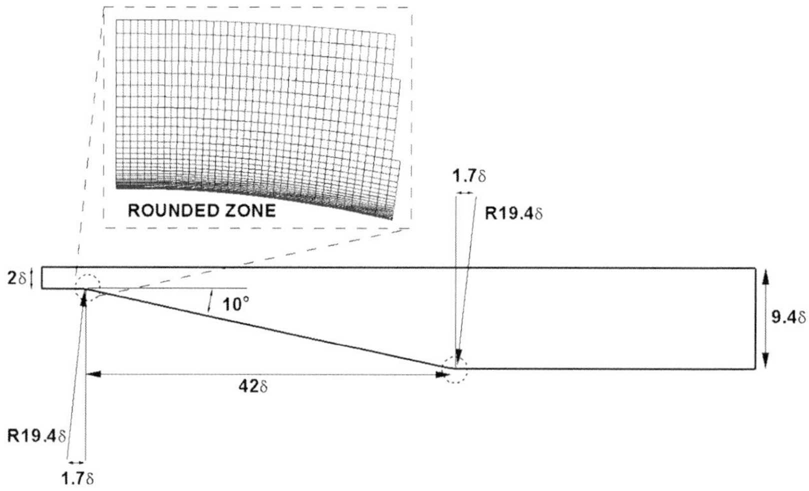

Fig. 1 Diffuser geometry with opening angle of 10 degree. The grid sketch refers to a finer grid than that used in the simulation.

2 The model

We solve the Navier Stokes equations, under the Boussinesq hypothesis. Density variations are considered only in the buoyancy term in the vertical momentum equation. A low pass filter is applied to the equation set, the arising subgrid stresses are modeled through a Dynamic Smagorinsky Model. In the transport equation of the resolved density, the subgrid density fluxes are treated with an eddy diffusivity approach.

We use the same geometry of [3] and [9] with an opening angle of 10 degree. The x,y and z coordinates correspond to the streamwise, wall normal and spanwise direction respectively. The inlet is located at $x/\delta = -5$ (where δ is half channel height at the inlet), the domain end is at $x/\delta = 95$, between $x/\delta = 77$ and the exit plane there is a sponge region. The expansion ratio is $a = \delta_{out}/\delta_{in} = 4.7$. The spanwise length is taken $L_z = 8\delta$. The diffuser geometry is discretized with a coarse grid, especially in the wall normal direction. We use 225x17x65 points respectively in the streamwise, wall normal and spanwise direction. In the streamwise direction the latter 24 points are used for the sponge region. At the inlet plane we use a turbulent field obtained in a pre-run of an equivalent turbulent plane channel flow. At the outlet a convective condition is used. Periodicity is used in the spanwise direction and a wall-layer model is used at the solid walls. The wall model belongs to the class of equilibrium stress models and it is based on the logarithmic law, see [8] for a review.

The numerical code uses finite differences and it is second order accurate in space and time. The time scheme is semi-implicit with Adam-Bashforth for all the terms excepts the diagonal diffusive ones solved with a Crank-Nicolson scheme. The algorithm is based on [13].

3 Neutral case

We first perform the simulation of the diffuser for a neutral case. Based on the friction velocity and the half channel height at the inlet, $Re_\tau = 500$. Fig. 2 shows profiles of the mean velocity and rms fluctuations of a plane channel flow used to provide the turbulent inflow condition to the diffuser. The profile is compared to DNS data from [6] at $Re_\tau = 590$. Considering the coarseness of the wall normal resolution, the results are satisfactory, although the mean velocity profile appears a bit underestimated in the channel core. In fact the centerline velocity $u_c/u_\tau = 20.13$ is lower than an expected value of $u_c/u_\tau = 20.86$, estimated as in [10]. The grid has only 17 points in the wall normal direction, probably too coarse to correctly solve the large energy carrying scales. However the wall model requirements (i.e. $y_1^+ > 30$ with y_1^+ the location of the first computational node out of the wall scaled with ν/u_τ) prevent from using a finer grid at the present value of Re_τ.

Fig. 3 shows the comparison at different locations between the numerical results and the experimental data of [3]. The recirculation region is between $x/\delta = 20.80$ and $x/\delta = 59.80$, while in the reference experiment it is between $x/\delta = 12$ and $x/\delta = 58.40$. The separation point is underestimated, this is due to the very coarse resolution in the wall normal direction as in the streamwise direction. In fact, the separation in the first part of the inclined wall is confined in a thin layer and cannot be properly caught with the grid resolution employed. On the other hand, the reattachment point is quite well predicted although it occurs further downstream. This means that, even with very coarse resolution, reasonable results can be obtained with this geometry.

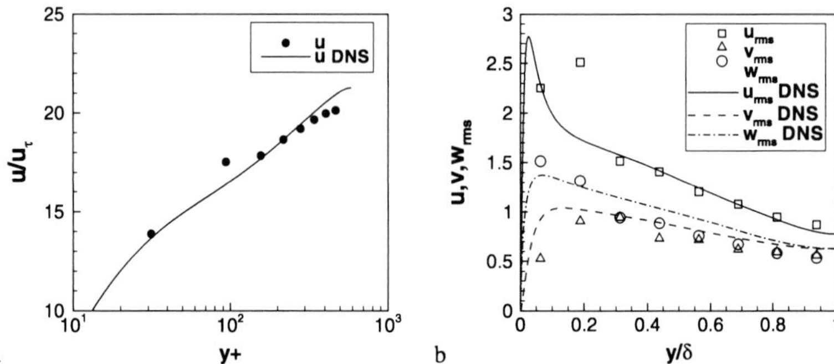

Fig. 2 Neutral Case: a) mean velocity profile at the inflow (circles) compared to DNS data (line) from [6]. b) resolved rms quantities.

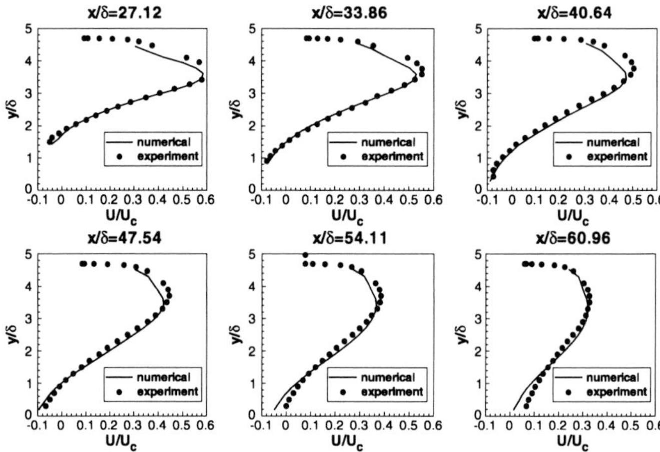

Fig. 3 Neutral Case: comparison at different location of streamwise velocity profile with experimental data from [3].

4 Stratified case

We discuss here the results when the flow is under stable stratification with $Ri_\tau = \Delta\rho g 2\delta/(\rho_0 u_\tau^2) = 19.62$, where $\Delta\rho$ is the imposed constant difference between the density at upper and the bottom boundaries. Fig. 4 shows the comparison between the experimental results in the neutral case and the numerical ones in the stratified case at different locations in streamwise direction. Because of the stratification conditions employed, velocity is higher in the channel core with respect to the neutral case, see for example [2]. The overall behavior in the diffuser seems not affected by stratification and the mean separation zone is shifted downstream. The mean separation bubble occurs between $x/\delta = 20.38$ and $x/\delta = 62.78$ and a comparison with the neutral case can be seen in Fig. 5. A density contour plot is given in Fig. 6; in

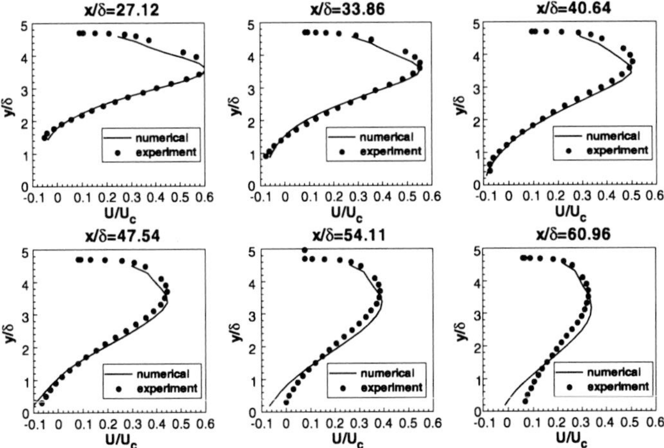

Fig. 4 Stratified Case: comparison at different location of streamwise velocity profile with experimental data from [3].

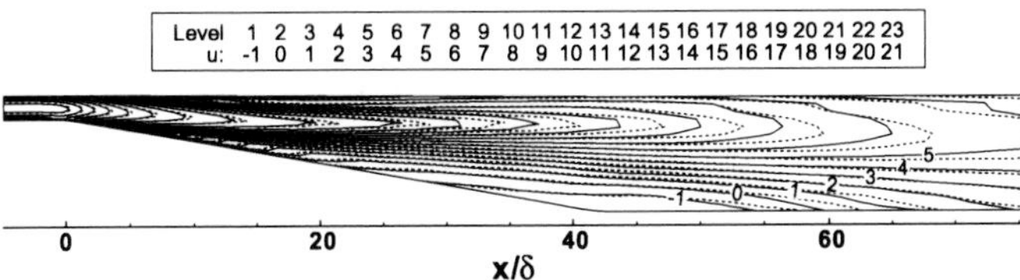

Fig. 5 Contour lines for mean streamwise velocity for neutral (solid lines) and stratified case (dashed lines).

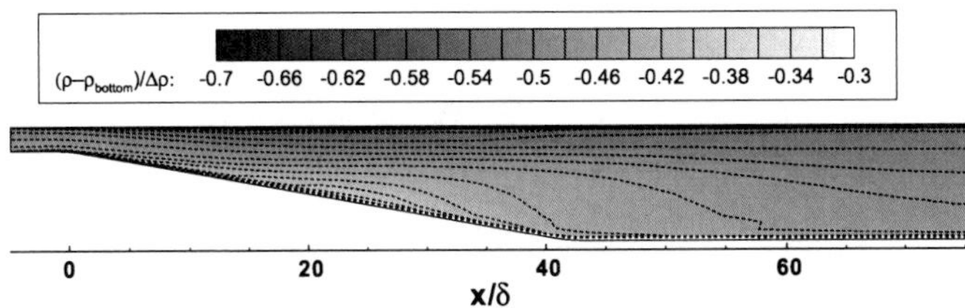

Fig. 6 Stratified Case: contour plot for density.

the middle of the inclined wall, it can be seen that density profile tends to be mixed by the recirculation. In general the separation bubble seems affected more by the increased velocity in the channel owing by buoyancy effect, see [2]. Although no proper comparison is available for this case the results look reasonable.

5 Conclusions

Large eddy simulations with a wall-layer model have been carried out for flow in a plane asymmetric diffuser in both neutral and stratified conditions. The simulations are performed using a very coarse grid especially for the wall normal direction. Results are encouraging, in particular for the neutral case where the availability of data allows proper comparison with experiments. The separation zone is generally well reproduced, in particular the reattachment point is well predicted. For the stratified case we observe a longer separation zone with respect to the neutral case and it seems mostly due to the higher velocity in the channel due to stratification.

References

1. Apsley, D.D., Leschziner, M.A.: Advanced turbulence modelling of separated flow in a diffuser. Flow, Turbulence and Combustion **63**, 81–112 (1999)
2. Armenio, V., Sarkar, S.: An investigation of stably stratified turbulent channel flow using large-eddy simulation. J. Fluid Mech. **459**, 1–42 (2002)
3. Buice, C.U., Eaton, J.K.: Experimental investigation of flow through an asymmetric plane diffuser. CTR Annual research briefs, 243–248 (1996)
4. Buice, C.U., Eaton, J.K.: Experimental investigation of flow through an asymmetric plane diffuser. J. Fluids Eng. **122**, 433–435 (2000)
5. Iaccarino, G.: Prediction of a turbulent separated flow using commercial CFD codes. J. Fluids Eng. **123**, 819–828 (2001)
6. Moser, R.D., Kim, J., Mansour, N.: Direct numerical simulation of turbulent channel flow up to $Re_\tau = 590$. Physics of fluids **11**, 4, 943–945 (1999)
7. Obi, S., Aoki, K., Masuda, S.: Experimental and computational study of turbulent separating flow in an asymmetric diffuser. Ninth Symposium on Turbulent Shear Flows, 305.1–305.4 (1993)
8. Piomelli, U.: Wall-layer models for Large-Eddy simulations. Progress in aerospace sciences **44**, 437–446 (2008)
9. Kaltenbach, H.J., Fatica, M., Mittal, R., Lund, T.S., Moin P.: Study of flow in a planar asymmetric diffuser using large-eddy simulation. J. Fluid Mech. **390**, 151–185 (1999)
10. Nikitin, N.V., Nicoud, F., Wasistho, B., Squires, K.D., Spalart, P.R.: An approach to wall modeling in large-eddy simulations. Phys. Fluids, 2000, **12**(7), 1070-6631
11. Schuster, J.U., Wu, X., Pitsch, H.: Large-Eddy simulation of a separated plane diffuser. 43rd AIAA Aerospace sciences meeting and exhibit Jannuary 10–13, 2005, Reno, NY (2005)
12. Wu, X., Schluter, J., Moin, P., Pitsch, H., Iaccarino, G., Ham, F.: Computational study on the internal layer in a diffuser. J. Fluid Mech. **550**, 391–412 (2006)
13. Zang, Y., Street, R.L., Koseff,J.R.: A non-staggered grid, fractional step method for time-dependent incompressible Navier-Stokes equations in curvilinear coordinates. J. Comp. Physics **114**, 18–33 (1994)

Reynolds Number Influence on the Particle Transport in a Model Estuary

R. Henniger and L. Kleiser

1 Introduction

The details of the transport of riverine sediments to the ocean are not fully understood [3, 7]. Normally the freshwater-particle mixture is lighter than estuarine saltwater such that most river plumes are positively buoyant. Thus, the particles can be transported over relatively large distances with the freshwater current until their settling dominates over the horizontal transport. Generally, two different *settling modes* are known to increase the average particle settling speed significantly [7]: *flocculation* of individual particles forming larger effective aggregates with larger Stokes settling speeds [3] and the settling enhancement due to *turbulence* [7].

In this work we focus on typical laboratory-scale flows (*e.g.* [7]) which already reveal a large range of different length and time scales such that fundamental particle settling mechanisms can be studied. This limitation allows us to perform Direct Numerical Simulations (DNS) in which all relevant scales of the flow are represented accurately. In our numerical model we consider the buoyancy forces arising from the salinity and the particle suspension, the momentum of the freshwater-particle inflow and the turbulent mixing of all phases. Other effects such as flocculation, winds or tides are neglected as in many laboratory-scale experiments. Coriolis forces are not considered due to the small size and time scale of our configuration.

2 Configuration and characteristic numbers

Our model configuration is depicted qualitatively in Fig. 1. Our basins are rectangular boxes with extents $\widetilde{L}_1 \times \widetilde{L}_2 \times \widetilde{L}_3$ (a tilde $\widetilde{(\cdot)}$ denotes dimensional quantities) which are discretized on Cartesian grids with coordinates $\widetilde{x}_1, \widetilde{x}_2, \widetilde{x}_3$. The inlet depth is de-

Institute of Fluid Dynamics, Sonneggstr. 3, 8092 Zurich, Switzerland, e-mail: henniger@ifd.mavt.ethz.ch, kleiser@ifd.mavt.ethz.ch

H. Kuerten et al. (eds.), *Direct and Large-Eddy Simulation VIII*,
ERCOFTAC Series 15, DOI 10.1007/978-94-007-2482-2_42,

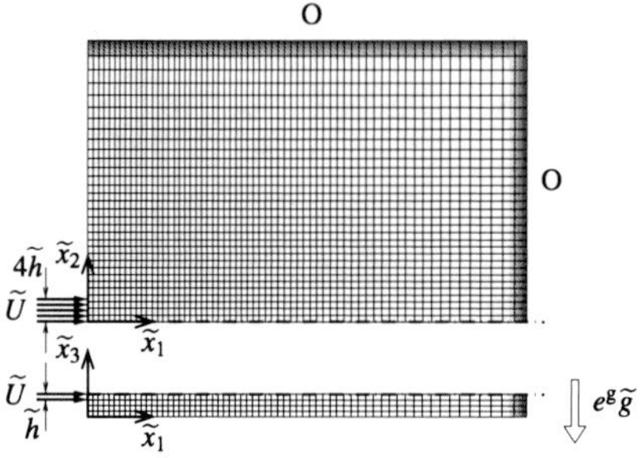

Fig. 1 Simulation setup with strongly thinned out computational grid and salt fringe region (gray). The water inlet is established in a small portion (arrows) and abundant fluid leaves the domain via the outflow boundaries 'O'.

noted as $\widetilde{h}$, the inflow bulk velocity as $\widetilde{U}$ and the gravitational acceleration $\widetilde{g}$ acts in the negative $\widetilde{x}_3$–direction, $e^g \equiv \{0,0,-1\}^T$. To ensure sufficiently large amounts of salinity inside the basins at late times we establish a salinity fringe region at the outflow 'O'. Additionally, we introduce a symmetry plane at $\widetilde{x}_2 = 0$ to save computational effort. We checked experimentally that this measure does not influence the results notably. More details on the configuration are given in [4].

Besides $\widetilde{U}$, $\widetilde{h}$ and $\widetilde{g}$, we use as the reference quantities for our simulations the kinematic viscosity $\widetilde{\nu}$, the densities $\widetilde{\rho}$, $\widetilde{\rho}_{\text{sal}}$, $\widetilde{\rho}_{\text{grain}}$ of the freshwater, saltwater and the particle grains, respectively, the particle diameter $\widetilde{d}$, the maximum particle volume fraction ϕ_V and the suspension diffusivities $\widetilde{D}_{\text{part}}$ and $\widetilde{D}_{\text{sal}}$ ("part" stands for *particle suspension* and "sal" for *salinity*). With the reduced gravitational accelerations $\widetilde{g}'_{\text{part}} \equiv (\widetilde{\rho}_{\text{grain}} - \widetilde{\rho})\widetilde{g}\phi_V/\widetilde{\rho}$ and $\widetilde{g}'_{\text{sal}} \equiv (\widetilde{\rho}_{\text{sal}} - \widetilde{\rho})\widetilde{g}/\widetilde{\rho}$ we define the Reynolds, Schmidt and Richardson numbers as

$$Re \equiv \frac{\widetilde{U}\widetilde{h}}{\widetilde{\nu}}, \quad Sc_k \equiv \frac{\widetilde{\nu}}{\widetilde{D}_k} \quad \text{and} \quad Ri_k \equiv \frac{\widetilde{g}'_k\widetilde{h}}{\widetilde{U}^2}, \quad k = \text{part, sal}, \tag{1}$$

respectively. The particles are characterized by the Stokes number St and the non-dimensional particle Stokes settling velocity $u^{\text{s}}_{\text{part}}$ (in gravity direction e^g),

$$St \equiv \frac{\widetilde{d}^2\widetilde{\rho}_{\text{part}}\widetilde{U}}{18\widetilde{\nu}\widetilde{\rho}\widetilde{h}}, \quad u^{\text{s}}_{\text{part}} \equiv e^g\frac{\widetilde{\rho}_{\text{part}} - \widetilde{\rho}}{\widetilde{\rho}}\frac{\widetilde{d}^2\widetilde{g}}{18\widetilde{\nu}\widetilde{U}}. \tag{2}$$

Generally, we choose similar values for these numbers as in the laboratory experiments of McCool and Parsons [7]. The influence of the Reynolds number on the particle settling is studied in the present work where our largest Reynolds number, $Re = 4\,000$ (*cf.* Table 1), is about at the lower end of these experiments. Furthermore, we set $Ri_{\text{part}} = 0.05$ and $Ri_{\text{sal}} = 0.5$ to obtain a slightly supercritical (hypopycnal) inflow. The non-dimensional particle settling speed is chosen as $|u^{\text{s}}_{\text{part}}| = 0.015$ for which the Stokes number is on the order of 10^{-4} to 10^{-3}. The Schmidt numbers Sc_{part} and Sc_{sal} are in reality at least on the order of hundreds to thousands, however, gravity-driven flows normally depend only weakly on the Schmidt numbers if these are not much lower than unity [2, 8]. This allows us to reduce the Schmidt numbers to $Sc_{\text{sal}} = 1$ and $Sc_{\text{part}} = 2$ to avoid very fine grids resolving steep gradients for $Sc_k \gg 1$.

Table 1 Specifications of the numerical simulations for different Reynolds numbers.

Re	$L_1 \times L_2 \times L_3$	t_{end}	$N_1 \times N_2 \times N_3$	N_t
750	$80 \times 50 \times 4$	1 600	$1153 \times 577 \times 97$	70 000
1 500	$80 \times 50 \times 4$	1 600	$2305 \times 1153 \times 193$	210 000
4 000	$65 \times 40 \times 4$	1 560	$4609 \times 3073 \times 513$	570 000

3 Governing equations

Because the particle Stokes numbers are relatively small, we neglect the influence of the particle inertia which permits an Eulerian approach for the particle suspension, similarly to the salinity. Hence, the non-dimensional transport equations for each of the concentrations c_k can be formulated as

$$\frac{\partial c_k}{\partial t} + ((u + u_k^s) \cdot \nabla) c_k = \frac{1}{Re\, Sc_k} \Delta c_k, \quad k = \text{part, sal}, \tag{3}$$

where u denotes the velocity and t the time ($u_{\text{sal}}^s \equiv 0$). Since all density variations are very small compared to the mean density we can apply the Boussinesq approximation for which the dimensionless incompressible Navier–Stokes equations read

$$\frac{\partial u}{\partial t} + (u \cdot \nabla) u = -\nabla p + \frac{1}{Re} \Delta u + e^g \sum_{k=\text{part, sal}} Ri_k c_k, \tag{4a}$$

$$\nabla \cdot u = 0, \tag{4b}$$

where p denotes the pressure. We use Dirichlet-type boundary conditions for u, whereas advective and Robin-type no-flux boundary conditions are employed for c_k, additionally (*cf.* [4, 8]). We solve these equations numerically with a high-order finite-difference approach in space and explicit Runge–Kutta time integration. More details on the algorithm, its implementation and validation can be found in [5]. The spatial and temporal resolutions of our configurations are listed in Table 1.

4 Particle transport and settling

We add the particles to the freshwater inflow beginning at $t = 250$ where the pure freshwater/saltwater interaction has attained a proper statistically stationary state. The temporal evolution of the suspended particle mass (with Ω as the entire do-

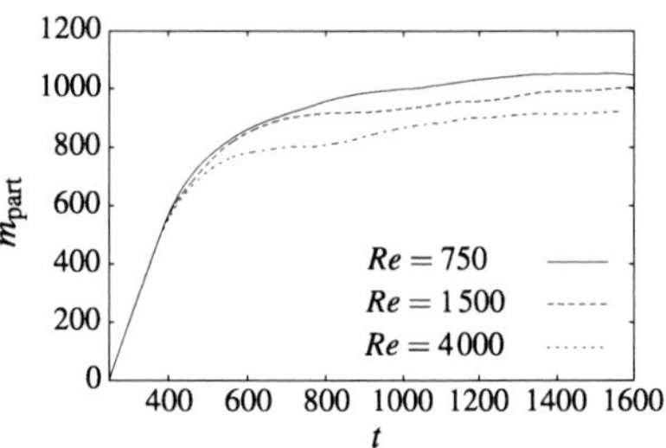

Fig. 2 Particle mass m_{part} over time.

main),

$$m_{\text{part}} \equiv Ri_{\text{part}} \int_{\Omega} c_{\text{part}} \, \mathrm{d}V, \tag{5}$$

is depicted in Fig. 2. The curves level off as soon as the particles touch the ground for the first time ($t \approx 400$) and from $t \approx 600 \ldots 800$ the amounts of suspended particles start to saturate approaching statistically stationary states.

To better understand the settling processes we visualize iso-surfaces and slices through the particle plumes for the initial transient and the statistically stationary states (Fig. 3). In the area close the inflow the particle concentrations are transported more or less passively with the carrier fluid. With growing distance to the inlet their transport speeds decelerate due to the spreading of the freshwater current and inertial forces become less dominant, *i.e.* other effects such as turbulence or particle buoyancy play a more important role. The particles form distinct layers in the vicinity of the freshwater/saltwater interface close to the water surface.

From the same pictures one can observe sheet/finger-like settling convection at early times ($t \approx 300 \ldots 450$) as in the experiments [7]. These structures are aligned with the direction of the surface streamlines (not shown here, *cf.* [4] for more details). For the statistically stationary state the horizontal extents of the particle plumes are about the same as for the transient, however, the plumes are not thin layers in the freshwater/saltwater interface anymore but now fill almost the entire water depth.

5 Average particle settling velocity

To investigate the particle settling more quantitatively we define the *average* particle settling velocity in the gravity direction e^{g},

$$u_{\text{part}}^{\text{s,av}} \equiv \int_{\Omega'} c_{\text{part}} (u + u_{\text{part}}^{\text{s}}) \cdot e^{\text{g}} \, \mathrm{d}V \Big/ \int_{\Omega'} c_{\text{part}} \, \mathrm{d}V. \tag{6}$$

The result of this expression strongly depends on the integration domain Ω'. Since the fastest average particle settling occurs between the near-surface particle plume and the bottom of the basin, we integrate over the entire horizontal domain but vertically only over the layer $1 \leq x_3 \leq 2$ to exclude most of the slowly settling particles

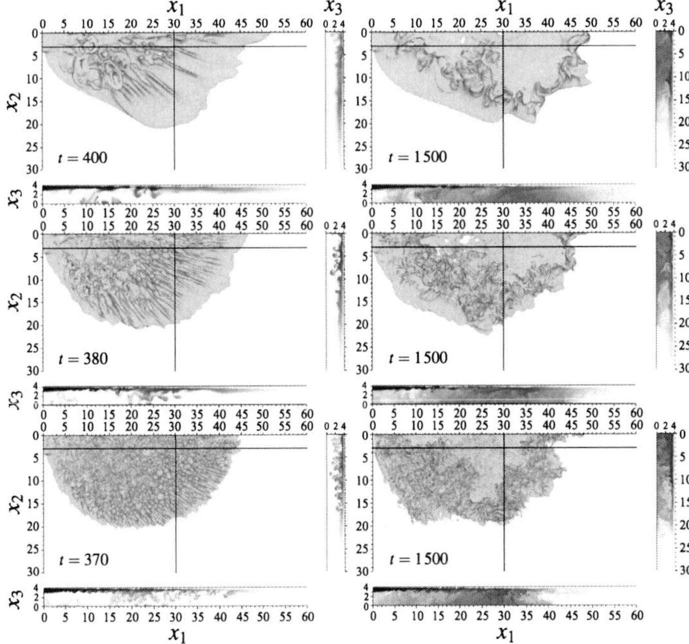

Fig. 3 Particle plumes for different Reynolds numbers and two different times. Iso-surfaces: $c_{\text{part}} = 0.25$, slices: clear fluid (white), maximum particle concentration (black). From top: $Re = 750$, $Re = 1\,500$, $Re = 4\,000$. Left: transient state, right: statistically stationary state.

in the other areas and to demonstrate notable increases of the particle settling speeds compared to the pure Stokes settling velocity $u^{\text{s}}_{\text{part}}$. From Fig. 4 we find for all three simulations that the particle settling velocity is most enhanced around $t \approx 300\ldots450$ where the sheet/finger convection occurs. The relative enhancement of the settling velocity attains up to 700% for the two largest Reynolds numbers and 500% for $Re = 750$. When the statistically stationary states are reached, the average settling velocities are still larger than the Stokes settling velocity, roughly between 10% and 50%. Only for $Re = 750$ we find slightly smaller sustained average settling velocities. These numbers are about in the same order of magnitude as the measurements of McCool and Parsons [7].

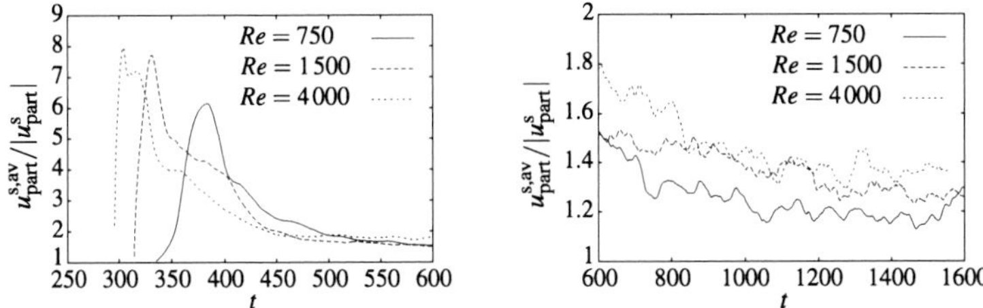

Fig. 4 Relative enhancements of the particle settling velocities over time.

6 Conclusions

Generally, the basic particle settling mechanisms of the experiments [7], *i.e.* sheet and finger settling convection and turbulence-enhanced particle settling, can be observed in our simulations as well. Furthermore, we find average particle settling speed enhancements which are roughly on the same order as in these laboratory measurements. Particularly, we find that only for the smallest Reynolds number investigated, $Re = 750$, the average settling velocities are smaller than for the larger Reynolds numbers, $Re \gtrsim 1\,500$. This suggests that the influence of the Reynolds number on the average particle settling velocity is relatively small beyond $Re \gtrsim 1\,500$.

The settling enhancements of the particle concentrations are caused by the turbulent mixing of the plumes with ambient clear fluid. The ongoing turbulent mixing of the two phases ensures that the particle plumes remain rather heterogeneous which is necessary for an enhanced particle settling [6]. Correspondingly, an extended particle model including particle inertia could increase the heterogeneity and thus increase the average particle settling velocity additionally, *cf.* [1, 6].

Acknowledgements Support for this project has been obtained through the ETH research grant TH-23/05-2. All computations were performed at the Swiss National Supercomputing Centre (CSCS).

References

1. Aliseda, A., Cartellier, A., Hainaux, F., Lasheras, J.C.: Effect of preferential concentration on the settling velocity of heavy particles in homogeneous isotropic turbulence. J. Fluid Mech. **468**, 77–105 (2002)
2. Bonometti, T., Balachandar, S.: Effect of Schmidt number on the structure and propagation of density currents. Theor. Comput. Fluid Dyn. **22**, 341–361 (2008)
3. Geyer, W.R., Hill, P.S., Kineke, G.C.: The transport, transformation and dispersal of sediment by buoyant coastal flows. Cont. Shelf Res. **24**, 927–949 (2004)
4. Henniger, R., Kleiser, L., Meiburg, E.: Direct numerical simulation of a model estuary. In: Sixth Int. Symp. on Turb. Shear Flow Phenom., pp. 1148–1153. Seoul (2009)
5. Henniger, R., Obrist, D., Kleiser, L.: High-order accurate solution of the incompressible Navier–Stokes equations on massively parallel computers. J. Comput. Phys. **229**(10), 3543–3572 (2010)
6. Maxey, M.R., Patel, B.K., Chang, E.J., Wang, L.P.: Simulations of dispersed turbulent multiphase flow. Fluid Dyn. Res. **20**, 143–156 (1997)
7. McCool, W.W., Parsons, J.D.: Sedimentation from buoyant fine-grained suspensions. Cont. Shelf Res. **24**, 1129–1142 (2004)
8. Necker, F., Härtel, C., Kleiser, L., Meiburg, E.: Mixing and dissipation in particle-driven gravity currents. J. Fluid. Mech. **545**, 339–372 (2005)

Dispersal and fallout simulations for urban consequences management

Fernando F. Grinstein, Gopal Patnaik, Adam J. Wachtor
Matt Nelson, Michael Brown, and Randy J. Bos

1 Introduction and Motivation

Hazardous chemical, biological, or radioactive releases from leaks, spills, fires, or blasts, may occur (intentionally or accidentally) in urban environments during warfare or as part of terrorist attacks on military bases or other facilities. The associated contaminant dispersion is complex and semi-chaotic. Urban predictive simulation capabilities can have direct impact in many threat-reduction areas of interest, including, urban sensor placement and threat analysis, contaminant transport (CT) effects on surrounding civilian population (dosages, evacuation, shelter-in-place), education and training of rescue teams and services. Detailed simulations for the various processes involved are in principle possible, but generally not fast. Predicting urban airflow accompanied by CT presents extremely challenging requirements [1, 2, 3].

Crucial technical issues include, simulating turbulent fluid and particulate transport, initial and boundary condition modeling incorporating a consistent stratified urban boundary layer with realistic wind fluctuations, and post-processing of the simulation results for practical consequences management. Relevant fluid dynamic processes to be simulated include, detailed energetic and contaminant sources, complex building vortex shedding and flows in recirculation zones, and modeling of particle distributions, including particulate fallout, as well as deposition, re-suspension and evaporation. Other issues include, modeling building damage effects due to eventual blasts, addressing appropriate regional and atmospheric data reduction,

Fernando Grinstein, Adam Wachtor, Matt Nelson, Michael Brown, and Randy Bos
fgrinstein@lanl.gov; MS F644, Los Alamos National Laboratory, Los Alamos, NM 87545, USA

Gopal Patnaik
patnaik@lcp.nrl.navy.mil; US Naval Research Laboratory, Washington DC 20375, USA

H. Kuerten et al. (eds.), *Direct and Large-Eddy Simulation VIII*,
ERCOFTAC Series 15, DOI 10.1007/978-94-007-2482-2_43,

and, feeding practical output of the complex combined simulation process into "urbanized" fast-response models.

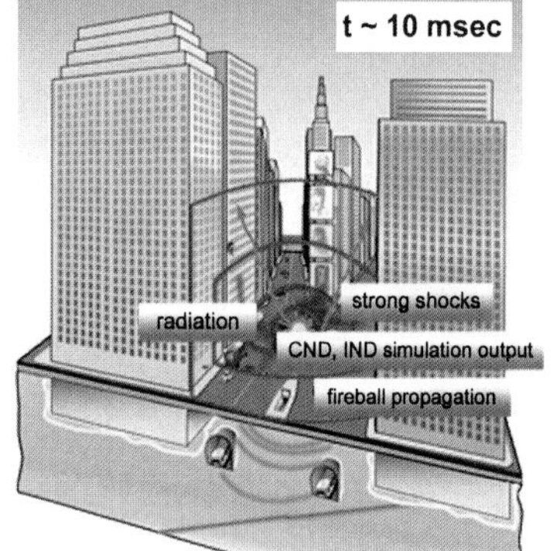

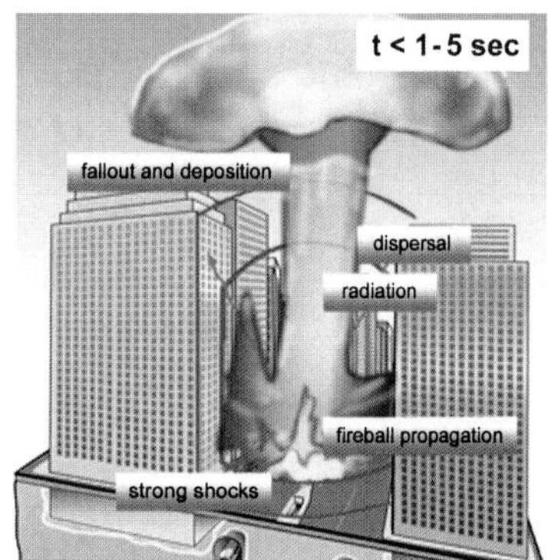

Fig. 1 Relevant early time physical processes associated with a CND or IND explosion.

In this paper we report progress in developing a simulation framework [3] for dispersal and fallout predictions in urban settings (Fig. 1) based on effective linkage of a strong motion hydrodynamics code – capable of simulating detailed energetic and contaminant sources associated with the effects of a conventional or improvised nuclear device (CND or IND) explosion, an implicit large-eddy simulation (ILES) [4] model (FAST3D-CT [2]) – capable of emulating CT due to wind and turbulence fields in the built-up areas, and a fast-running building-aware dispersion model (QUIC) [5].

2 Urban Flow and Dispersal Simulation

Simulating urban flow and dispersion is a problem for time-dependent aerodynamic computational fluid dynamics methods. Unavoidable trade-offs demand choosing between fast (but less accurate) and much slower (but more accurate) models. Relevant time domains can be identified which require appropriate corresponding time-accurate (full physics) simulation codes, involving physical processes that occur in microseconds-to-milliseconds, and seconds-to-one-hour ranges. Target (strong-motion and ILES) codes for these domains are integrated with appropriate mesoscale / atmospheric reduced data. Linking codes between the various time domains allows the results of one code to be used as the initial conditions for the next. The suite of full-physics simulations is used to develop source term, buoyant rise, and flow field parameterizations in urban environments for later use with fast-response high-fidelity analytical model tools.

The CASH strong-motion hydrodynamics code [6] includes appropriate methods for accurately modeling explosions, including state-of-the-art models for high explosive (HE) performance, and for the deformation and failure of other materials. It can model shock wave propagation in the atmosphere and the ground. We envision that the CASH calculation would be run until shock strength and deformation effects are small enough to map to the dispersal code for the remainder of the simulation [3]. Effective use of ILES strategies has been reported for large scale urban flow and dispersal simulations for consequences management [2, 3]. In ILES, the large energy containing structures are resolved whereas the smaller, presumably more isotropic, structures are filtered out and unresolved SGS are emulated with physics capturing finite-volume numerics [4]. Contaminant spread can be effectively simulated based on tracers with suitable advection velocities, sources and sinks. The fallout simulations reported here rely on unsteady buoyant particle advection, parameterized terminal velocities [7], and particle groups selected based on typical relevant available dust particle mass / size distributions [8]. The Quick Urban and Industrial Complex (QUIC) Dispersion Modeling System [5] is a fast response urban dispersion model that runs on a laptop. QUIC is comprised of an empirical-diagnostic building-aware 3D wind field model, a Lagrangian random-walk transport and dispersion model, a pressure solver, and a graphical user interface. QUIC has been extensively evaluated against urban wind tunnel and field experiments (e.g., [9]) and was shown to perform similarly to a suite of CFD models when compared to plume measurements in the NYC Midtown Experiment [10, 11].

Establishing the credibility of the solutions is one of the stumbling blocks of urban simulations. Validation studies with experiments require well-characterized datasets with information content suitable to initiate and evaluate unsteady simulation models as well as the cruder steady-state models. Obtaining full-scale (field) datasets for the inherently complex flows in question is costly and difficult; alternate validation approaches at present (see [2] and chapter 17 in [4]) involve code-to-code comparisons, carrying out detailed comparisons with actual urban field experimental databases as they become available, and comparing urban flow simulations with carefully controlled laboratory-scale wind-tunnel experiments [2, 12].

2.1 Results and Discussion

Detailed urban dispersal studies in typical urban settings based on effective linkage of strong motion CASH AND FAST3D-CT are first addressed in what follows. As noted, in the type of simulations represented by the dispersal of contaminants due to an energetic release of energy in an urban environment, no one code can adequately simulate the full range of physics involved, nor should a user want that to be the case. Codes such as CASH and FAST3D-CT have been specifically developed to simulate very precise ranges of the relevant physics. CASH has the ability to simulate the strong motion regime where shocks are present due to an energetic source (such as a high explosive) but is not able to do the dispersal of contaminants in the atmosphere over a the size of a typical urban setting. On the other hand FAST3D-CT is not able to handle shocks or solid material descriptions. The solution we have proposed is to use results of a strong motion code (e.g., CASH) as detailed initial-condition energetic and contaminant sources to the dispersal code (e.g., FAST3D-CT) at a time when shocks are no longer present or reduced to negligible levels. The calculations then proceed as in a regular dispersal simulation. We have been able to establish such a link using FAST3D-CT to simulate flow over a flat terrain [3]. Ongoing work using CASH energetic and tracer solutions in detailed urban domains as sources for dispersal simulation models will be reported separately.

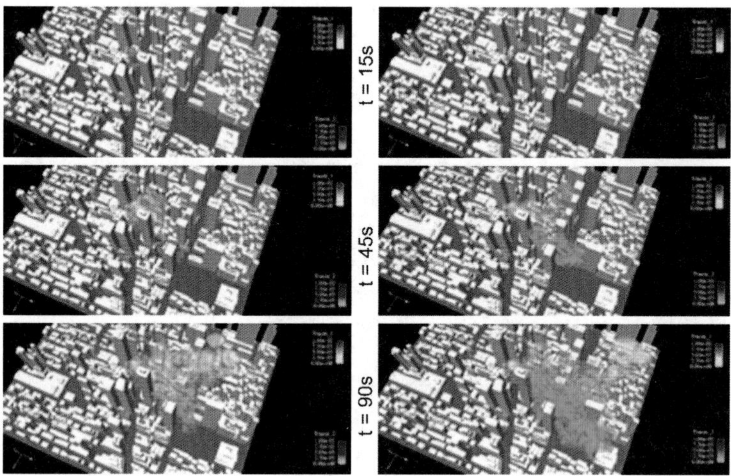

Fig. 2 Dispersal studies on a typical urban setting: effects of release characteristics. Simulations of a 0.25-ton uniform HE energy-pill release with FAST3D-CT. Particle concentration distributions; hot (orange tones) / cold (blue tones) releases at 15m and 90m heights, on the left and right columns, respectively.

The urban dispersal studies described here used "energy-pill" HE releases modeled in terms of initial cubic volumes having 15m side length at a 550 deg-K and 1316 deg-K (approximately equivalent to, 0.25-ton and 1-ton HE yields, respectively). Dispersal is found to be very sensitive to release characteristics, e.g., hot vs. cold (non-buoyant) release, as well as release height (cf. Figs. 2). Particle tracers from FAST3D-CT simulations associated with particle sizes ranging between 7- 200μm, and a HE energy-pill release occurring at 6m height were investigated. Local concentrations and velocity magnitude are found to be very sensitive to urban geometry and significantly more variability is observed at plume edges, due to the presence of less coherent vortical structures.

Comparison of predictions with FAST3D-CT and QUIC codes on effects of particle group specifics (e.g., sizes) were addressed in this context (Figures 3 and 4, for the 1-ton HE case). The simulations were initialized with similar mean flow field and same energy pill, except for QUIC not accounting for the turbulent im-

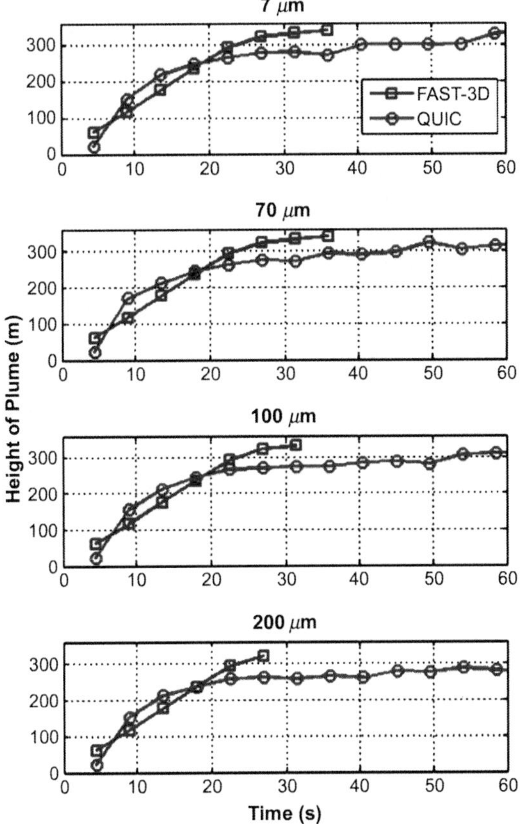

Fig. 3 Plume rise predicted by FAST3D-CT and QUIC for various particulate sizes; FAST3D-CT used geometrically stretched grid above 300m and plotted results do not account for that.

pact of the urban geometry on the (initial) ABL profiles. Figure 3 shows that both codes can give surprisingly similar plume rise results (based on height of peak particle concentration evolution), but actual concentration distributions are qualitatively fairly different (Fig. 4). The comparisons have suggested improvements in the QUIC modeling, specifically, changing the constant temperature assumption in the buoyant cloud model. This will allow for the hotter inner core to rise faster while contaminant will be able to slough off along the edges in the cooler and slower rising outer shell of the buoyant explosive cloud.

Acknowledgements Los Alamos National Laboratory is operated by the Los Alamos National Security, LLC for the U.S. Department of Energy NNSA under Contract No. DE-AC52-06NA25396.

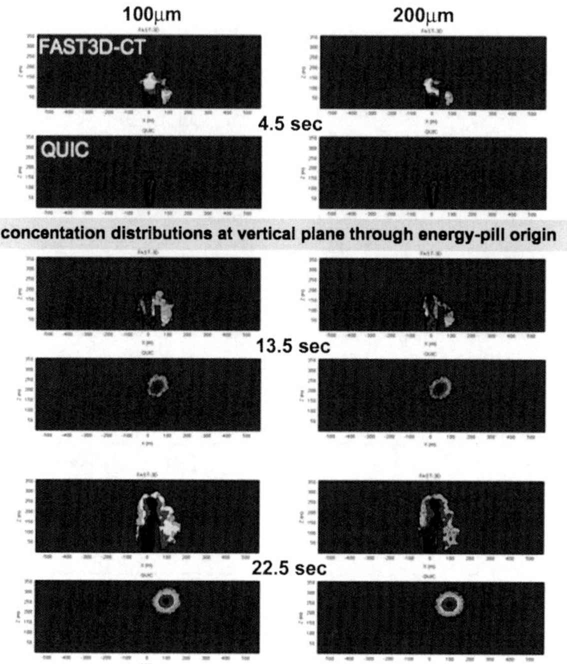

Fig. 4 Comparison of FAST3D-CT and QUIC predictions of particle concentration distributions.

References

1. Britter, R.E., and Hanna, S.R., *Ann. Rev. Fluid Mech.*, **35**, 469–496 (2003).
2. Patnaik, G., Boris, J.P., Young, T.R., Grinstein, F.F., *J. Fluids Eng.*, **129**, 1524–1532 (2007).
3. Grinstein, F.F., Bos, R.J. and Dey, T.N., *ERCOFTAC Bulletin*, **78**, 11–14 (2009).
4. Grinstein, F.F., Margolin, L.G., Rider, W.J., Editors, *Implicit Large Eddy Simulation: Computing Turbulent Fluid Dynamics*, Cambridge University Press (2007).
5. Brown, M., "Urban Dispersion Challenges for Fast Response Modeling", in *Fifth AMS Symposium on the Urban Environment*, LA-UR-04-5129, LANL, Los Alamos, NM (2004).
6. Dey, T.N. and Bos, R.J., "CASH, Version 01," LANL computer code LA-CC-01-053 (2001).
7. Cheng, N.-S., *Powder Technology*, **189** (2009).
8. Pinnick, R.G., Fernandez, G., and Hinds, B.D., *Appl. Optics*, **22**, 95–102 (1983).
9. Singh, B., Hansen, B., Brown, M., Pardyjak, E., *Env. Fluid Mech.*, **8**, 281–312 (2008).
10. Allwine, K.J., Flaherty, J.E., Brown, M., Coirier, W., Hansen, O., Huber, A., Leach, M. and Patnaik, G., "Urban Dispersion Program: Evaluation of six building-resolved urban dispersion models", Official Use Only PNNL-17321 report (2008).
11. Boughton, B.A. and DeLaurentis, J.M., An Integral Model of Plume Rise from High Explosive Detonations, SAND-86-2553C.
12. Lee, M.-Y., Harms, F., Young, T., Leitl, B., and Patnaik, G., Model- and Application-Specific Validation Data for LES-Based Transport and Diffusion Models, 89th AMS Annual Conference, Phoenix AZ (January 2009).

On the Mechanisms of Pollutant Removal from Urban Street Canyons: A Large-Eddy Simulation Approach

Chun-Ho Liu, W.C. Cheng, Tracy N.H. Chung and Colman C.C. Wong

1 Introduction

Urbanization modifies the bottom of the atmospheric boundary layer (ABL) leading to elevated air pollutant concentrations. Over 50% of the world population lives in cities nowadays [1]. Urban air quality is thus a problem of major concern.

The meteorological process in the urban canopy layer (UCL) exhibits a distinct, neighborhood-scale climate full of wakes and recirculations. A street canyon is the generic unit commonly used in urban pollutant transport studies [2]. In urban meteorology, the flows over idealized two-dimensional (2D) street canyons are classified into three characteristic flow regimes, viz., isolated roughness, wake interference, and skimming flow, as functions of the building-height-to-street-width ratio h/b [3]. Similarly, in fluid mechanics, flow over a rough wall with 2D ribs is broadly divided into d-type or k-type, depending on the separation between the ribs [4].

In this paper, large-eddy simulation (LES) is used to examine the pollutant transport in 2D street canyons. We focus on: (1) in the street canyons, (2) at the roof level, and (3) in the UCL so as to address the pollutant removal in urban areas.

2 Mathematical Model and Computational Domain

The LES adopted is modified from the open-source computational fluid dynamics (CFD) code OpenFOAM version 1.6 [5] that consists of the continuity and momentum conservation in isothermal, incompressible flows. The turbulence kinetic energy (TKE) and the Smagorinsky model are used to handle the subgrid-scale (SGS) motions [6]. The pollutant is passive and inert. The Reynolds number Re, which is

---

Chun-Ho Liu, W.C. Cheng, Tracy N.H. Chung and Colman C.C. Wong
Department of Mechanical Engineering, The University of Hong Kong, Pokfulam Road, Hong Kong, e-mail: `liuchunho@graduate.hku.hk`

H. Kuerten et al. (eds.), *Direct and Large-Eddy Simulation VIII*, ERCOFTAC Series 15, DOI 10.1007/978-94-007-2482-2_44,

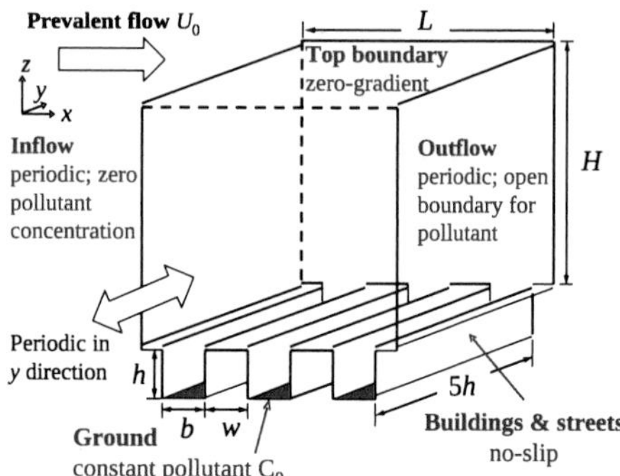

Fig. 1 Computational domain and boundary conditions.

based on the prevailing wind speed U_0 and the building height h, equals 12,000, that is high enough for fully developed turbulent flow in street canyons [7].

Building blocks of width w (= h) and height h are evenly placed at a separation b in the streamwise direction x (Figure 1). The length of the street is equal to the homogeneous spanwise y domain extent (= $5h$). The height of the UCL is H over the street canyons. The prevailing flow, which is normal to the street axis, is driven by the background pressure gradient in the x direction. $\rho u_\tau^2/H$, where ρ is the fluid density and u_τ the friction velocity. The boundary conditions (BCs) of the flow are periodic in the x and y directions. No-slip and zero-gradient BCs are applied, respectively, on the solid boundaries and the domain top. A constant pollutant concentration C_0 is prescribed on the ground of the street canyons. The incoming air is assumed to be free of pollutant while the convective BC is applied at the streamwise outflow to avoid reflection of pollutant back into the computational domain.

The grid spacing is ($0.0216 \leq \Delta x/h \leq 0.0432$, $\Delta y/h = 0.0315$, $0.0216 \leq \Delta z/h \leq 0.0432$) in the street canyons. The horizontal grids in the UCL are the same but $\Delta z/h$ is reduced to 0.0137 to 0.0412 handling the roof-level strong shear. The first element is placed at $z^+ \approx 5$ away from the nearby solid boundary. Another test doubling the spatial resolution was performed but no noticeable difference was observed [8]. Hence, the current LES is fine enough providing consistent numerical results.

3 Results and Discussions

3.1 Inside the Street Canyons

Inside the street canyon, we focus on the spatial behavior of pollutant removal and how it is affected by the flow structure. In the isolated roughness regime $h/b = 1/15$, the prevailing flow in the UCL entrains down to the street level inducing persistent flow reattachment and separation (Figure 2a). The local convective pollutant transfer coefficient Ω (= $\langle \overline{c}\,\overline{w} \rangle / C_0 / U_0$, where $\overline{c}\,\overline{w}$ is the vertical pollutant flux and $\langle \cdot \rangle$ the ensemble average in time and the homogeneous spanwise direction) exhibits a close

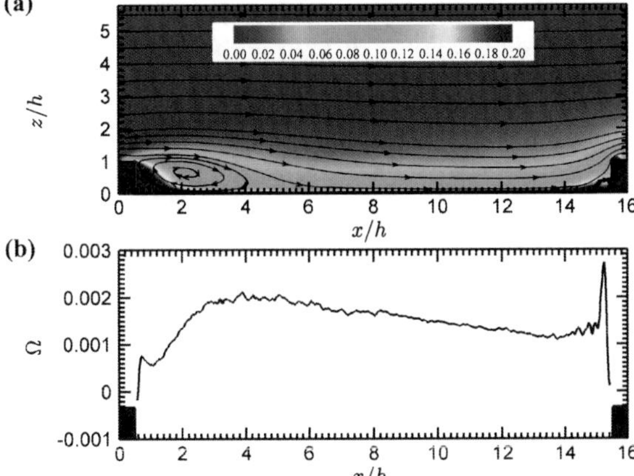

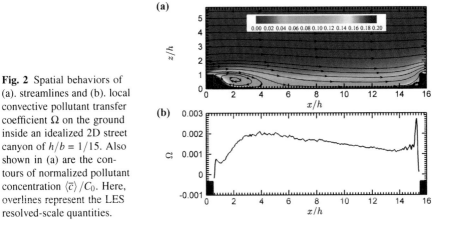

Fig. 2 Spatial behaviors of (a). streamlines and (b). local convective pollutant transfer coefficient Ω on the ground inside an idealized 2D street canyon of $h/b = 1/15$. Also shown in (a) are the contours of normalized pollutant concentration $\langle \bar{c} \rangle / C_0$. Here, overlines represent the LES resolved-scale quantities.

correlation with the recirculating flow (Figure 2b). The flow reattaches on the leeward side initiating the primary clockwise-rotating recirculation. The reattachment point (at $x/h = 5$) falls within the broad maximum of Ω. Another small recirculation is developed near the ground-level windward corner at which the local maximum Ω coincides with the separation point (at $x/h = 15$). These maxima collectively suggest the favorable effects of impingement on local pollutant removal.

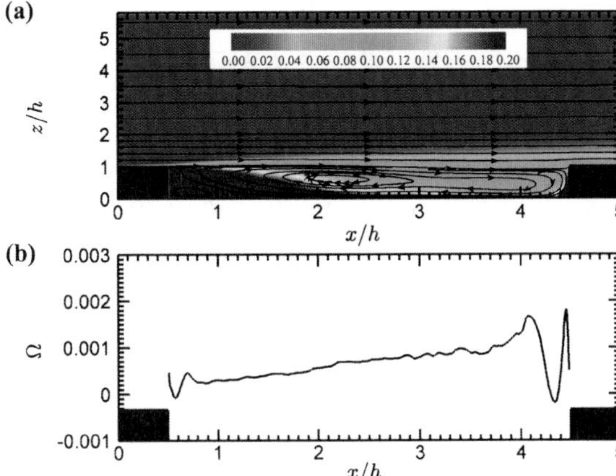

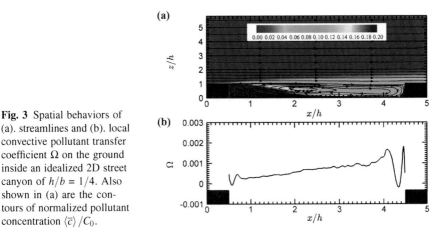

Fig. 3 Spatial behaviors of (a). streamlines and (b). local convective pollutant transfer coefficient Ω on the ground inside an idealized 2D street canyon of $h/b = 1/4$. Also shown in (a) are the contours of normalized pollutant concentration $\langle \bar{c} \rangle / C_0$.

Reducing the street width leads to the skimming flow regime $h/b = 1/4$ (Figure 3a). The reattachment and separation are replaced by a large recirculation in the center core. The recirculation is isolated from the UCL as the prevailing wind no longer entrains down to the ground level. As such, the pollutant removal is mainly governed by the relatively weaker roof-level turbulent vertical pollutant flux. Ω is thus decreased in the skimming flow regime. Besides, the recirculation is in the form of wall jet resulting in a monotonically increasing Ω from the leeward to the windward sides. Neither trough nor peak is observed. A vigorous building down-wash is found close to the facade contributing to the elevated Ω on the windward side.

3.2 Roof Level of the Street Canyons

Table 1 Quadrants of vertical fluxes of momentum $\langle u''w''\rangle$ and pollutant $\langle c''w''\rangle$.

Quadrants	$\langle u''w''\rangle$	$\langle c''w''\rangle$
Outward interactions	$u'' > 0$ and $w'' > 0$	$c'' < 0$ and $w'' > 0$
Ejections	$u'' < 0$ and $w'' > 0$	$c'' > 0$ and $w'' > 0$
Inward interactions	$u'' < 0$ and $w'' < 0$	$c'' > 0$ and $w'' < 0$
Sweeps	$u'' > 0$ and $w'' < 0$	$c'' < 0$ and $w'' < 0$

Along the roof level, conditional sampling is used to examine the nature of the transport processes. Quadrant analysis (Table 1) is applied on five equal-sized segments along the roof level of the street canyon of $h/b = 1$ (Figure 4). The resolved-scale vertical fluxes of momentum $\langle u''w''\rangle$ and pollutant $\langle c''w''\rangle$ are taken into account. The sweeps and ejections, which represent the downward and upward transports, are the dominating events compared with the inward and outward interactions. The scales of the sweeps and ejections are larger on the windward side that are partly attributed to the shear-induced instability at the roof level of the leeward building edge. The small-scale and large-scale transport processes are dominated by the upward ejections and the downward sweeps, respectively, suggesting that the upward transports are influenced by the finite-size street canyons while the downward sweeps are enhanced by the large vortices in the UCL.

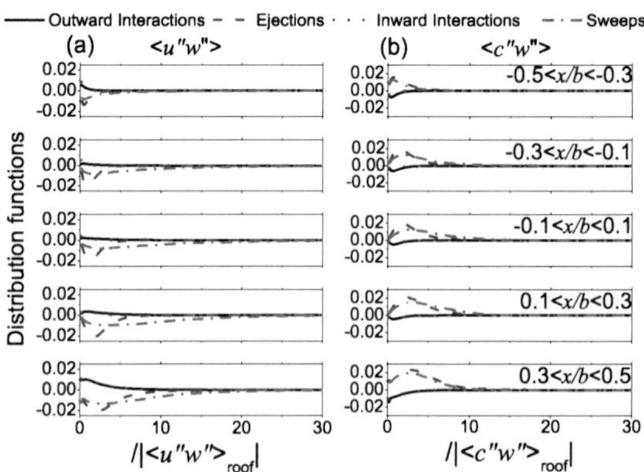

Fig. 4 Conditional sampling of the vertical fluxes of (a). momentum $\langle u''w''\rangle$ and (b). pollutant $\langle c''w''\rangle$ at the roof level of the idealized 2D street canyons of $h/b = 1$.

3.3 Over the Street Canyons

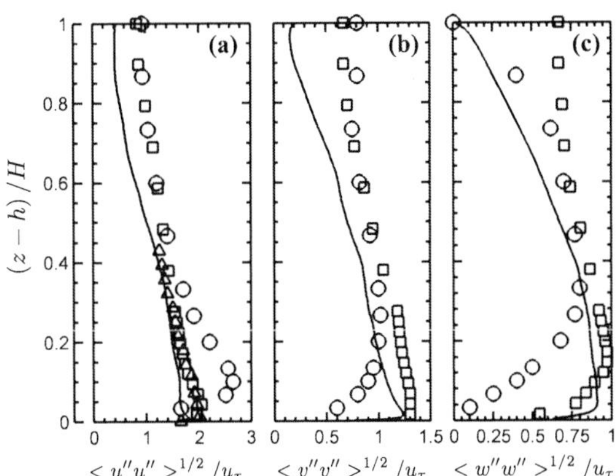

Fig. 5 Vertical profiles of normalized velocity fluctuations: (a). $\langle u''u''\rangle^{1/2}/u_\tau$, (b). $\langle v''v''\rangle^{1/2}/u_\tau$, and (c). $\langle w''w''\rangle^{1/2}/u_\tau$ in the UCL over idealized 2D street canyons of $h = b$. DNS of: ○ flat plate [9], □ 2D ribs of $h/b = 1/8$ [10], and Δ cubical array [11]; and —: current LES. Here, $\phi'' = \overline{\phi} - \langle\overline{\phi}\rangle$.

The turbulence structures in the UCL are key parameters of pollutant dispersion. As shown in Figure 5, the turbulence intensities of the current LES over idealized 2D street canyons are different from those of smooth wall [9], ribs of $h/b = 1/8$ [10], and a cubical array [11]. The maximum horizontal velocity fluctuations are peaked right over the roughness elements while the maximum vertical velocity fluctuation is peaked near the roof at $z = 0.05h$. Hence, the pollutant dispersion is expected to be modified from the conventional Gaussian plume especially along the roof level.

Switching on only the ground-level pollutant source in the first street canyon, the pollutant distribution in the UCL over the idealized 2D street canyons resembles closely the Gaussian-plume shape (Figure 6a). Assuming a Gaussian distribution of pollutant in the vertical direction, the dispersion coefficient σ is increased linearly in the streamwise direction (Figure 6b). Although recirculating flows are found inside the street canyons, no noticeable difference in plume rise p is observed between the street canyons and the building roofs. The rise of the pollutant plume centerline is rather shallow that is less than $2h$ in the LES. Whereas, the current streamwise domain extent (= $24h$) is too short for a converged plume centerline. Similar to the Gaussian plume, the pollutant concentration $\langle\overline{c}\rangle$ inside the street canyons is decreased exponentially in the streamwise direction (Figure 6c). The horizontal profile of the roof-level pollutant concentration in the streamwise direction shows a sharp drop on the windward side of each street canyon, suggesting the vigorous pollutant entrainment from the UCL down to the ground level following the building downwash. The changes in $\langle\overline{c}\rangle$ are diminished with increasing height from the roof level. Hence, the conventional Gaussian plume model should be applied with caution for the calculation of pollutant concentration in the UCL over urban roughness.

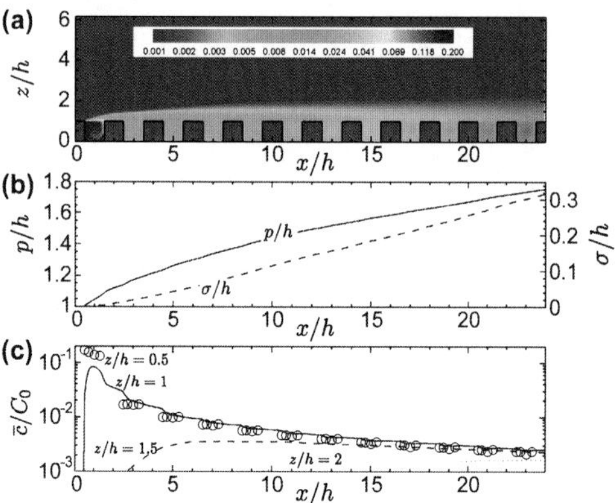

Fig. 6 (a). Contours of normalized pollutant concentration $\langle \overline{c} \rangle / C_0$, (b). plume rise p and dispersion coefficient σ, (c). normalized pollutant concentration profile $\langle \overline{c} \rangle / C_0$ in the urban canopy layer in streamwise direction over idealized 2D street canyons of $h = b$.

4 Conclusions

LESs are performed to examine the mechanism of pollutant removal from idealized 2D street canyons. Inside the street canyon, the flow structures, such as separation and reattachment, are closely coupled to the ground-level pollutant removal. Along the roof level, sweeps and ejections are the dominating events for pollutant transport. In the UCL, the turbulence structures over the street canyons are different from those over other types of urban roughness. As a result, the pollutant distribution over the roof level is slightly different from the conventional Gaussian plume model. Caution should thus be applied in handling pollutant transport problems in urban areas.

References

1. United Nation: World Urbanization Prospects: The 2007 Revision Highlights, United Nations, New York (2008)
2. Britter, R.E. and Hanna, S.R.: Flow and dispersion in urban areas. Annu. Rev. Fluid Mech. **36**, 469–496 (2003)
3. Oke, T.: Street design and urban canopy layer climate. Energy Bldg. **11**, 103–113 (1988)
4. Jiménez, J.: Turbulent flows over rough walls. Annu. Rev. Fluid Mech. **36**, 143–196 (2004)
5. OpenFOAM-1.6: OpenFOAM: The Open Source CFD Toolbox - User Guide, Version 1.6. `http://www.openfoam.com`
6. Chung, T.J.: Computational Fluid Dynamics. Cambridge University Press, Cambridge (2002)
7. Hoydysh, W.G., Griffiths, R.A. and Ogawa, Y.: A scale model study of the dispersion of pollution in street canyons. APCA Paper No. 74-157. In: 67th Annual Meeting of the Air Pollution Control Association, Denver, CO, 9–13 June 1974, (1974)
8. Liu, C.-H. and Wong, C.C.C.: On the pollutant removal, dispersion, and entrainment over two-dimensional idealized street canyons. Q. J. R. Meteorol. Soc., submitted, (2010)
9. Nagaosa, R.: Direct numerical simulation of vortex structures and turbulent scalar transfer across a free surface in a fully developed turbulence. Phys. Fluids. **11**, 1581–1595 (1999)
10. Ashrafian, A., Andersson, H.I. and Manhart, M.: DNS of turbulent flow in a rod-roughened channel. Int. J. Heat Mass Transfer. **25**, 373–383 (2004)
11. Coceal, O., Dobre, A., Thomas, T.G. and Belcher, S.E.: Structure of turbulent flow over regular arrays of cubical roughness. J. Fluid Mech. **589**, 375–409 (2007)

Part V
Compressible Flows and Reactive Flows

Numerical simulations of shock-wave/boundary-layer interaction phenomena

Neil D. Sandham and Emile Touber

1 Introduction

We review recent simulations of shock-induced separation of boundary layers and then consider in detail some properties of the detached shear layer that forms after separation of a turbulent boundary layer at Mach 2.3. Whilst still challenging in terms of numerical methods, due to the simultaneous presence of shock waves and turbulence, both direct numerical simulation and large eddy simulations (LES) are useful tools to investigate fundamental issues of shock-wave/boundary-layer interaction. In particular with LES it is feasible to calculate the long run times and wide domains necessary to study low frequency motions that occur under the reflected shock foot. We show here that a simplified model based on growth rate of the separated shear layer leads to predictions of frequency that are significantly higher than the low-frequency peak seen in LES and experiment.

Direct and large-eddy simulations can be used to investigate physical phenomena in flows with additional complications besides transition and turbulence. In this paper we consider the effect of compressibility, which commonly leads to shock waves whose interaction with turbulence needs to be accurately computed. The shock waves are either imposed on the flow or generated internally in the form of eddy shocklets [11]. High order methods are preferred for turbulent flows due to the reduced number of grid points required for the overall simulation. However, when turbulence is combined with shock waves there is a risk that the additional local dissipation used to capture the shock waves will have a significant effect on the turbulence and, in particular, that the increased dissipation will overwhelm the

Neil D. Sandham
School of Engineering Sciences, University of Southampton, Southampton SO17 1BJ, UK, e-mail: n.sandham@soton.ac.uk

Emile Touber
Department of Aeronautics, Imperial College London, London SW7 2AZ, UK, e-mail: e.touber@imperial.ac.uk

H. Kuerten et al. (eds.), *Direct and Large-Eddy Simulation VIII*, ERCOFTAC Series 15, DOI 10.1007/978-94-007-2482-2_45,

physical dissipation and cause erroneous flow physics. To be useful in practice, the algorithms must not only be stable and accurate, but also scale well in a parallel computing environment. In this paper we will briefly review our numerical approach and then focus on recent applications aimed at understanding some particular phenomena associated with transitional and turbulent flows with shock-wave/boundary-layer interactions.

2 Numerical approach

In previous investigations [7,12] it has been found that a splitting of the Euler terms enhances the numerical stability of high order discretisations applied to the compressible Navier-Stokes equations. We will not go into the details of the 'entropy' splitting here, but the same technique has been used in all of our subsequent work without adjustment of the splitting parameter. In addition to a basic shock-capturing scheme, an artificial compression method was used so that the effect of the additional dissipation was confined to the immediate vicinity of the shock wave. We have since incorporated the Ducros et al. [2] filter that additionally turns off the shock capturing in vortical regions of the flow. In each case the coefficients for the additional filters were fixed using representative test cases and have not been changed subsequently. A useful test case involved an oblique shock wave impinging on a spatially-evolving mixing layer in two dimensions [12]. The problem was deliberately specified with supersonic inflow and outflow to remove boundary effects. Other test cases include compressible turbulent channel flow, which has no eddy shock waves but gives a basic reassurance that the split schemes can work for internal flows where global conservation is important, and a shock-induced separation of a laminar boundary layer.

For LES we have generally found improved behaviour when a sub-grid model is used, even when high-order filtering is present [1]. Since our initial interest was in turbulent spots and other transitional flows, we needed a subgrid model that turns off in laminar flow and does not require any averaging. A suitable method with low computational overhead is the mixed time scale method [3,9]. One coefficient was adapted to a compressible channel flow test case and all subsequent calculations have used the same model constants. For standard boundary layer flow we aim for resolution in the wall-normal direction that is only a factor of two or three away from a DNS. Lower tolerances can be placed in other directions, e.g. within a factor of four to six of DNS for the grid parallel to a wall. The grid generation procedure is generally iterative in nature and consists of running the calculation past the initial flow transient and then looking for regions that are under-resolved and refining the grid as necessary. In other cases we have run extensive grid refinement studies [9].

Large eddy simulation (LES) can be used for more ambitious calculations than are possible with DNS. One of the more challenging applications recently has been to a complete scramjet (supersonic combustion ramjet) intake [5], including transition due to the growth of turbulent spots and various shock-wave/boundary-layer

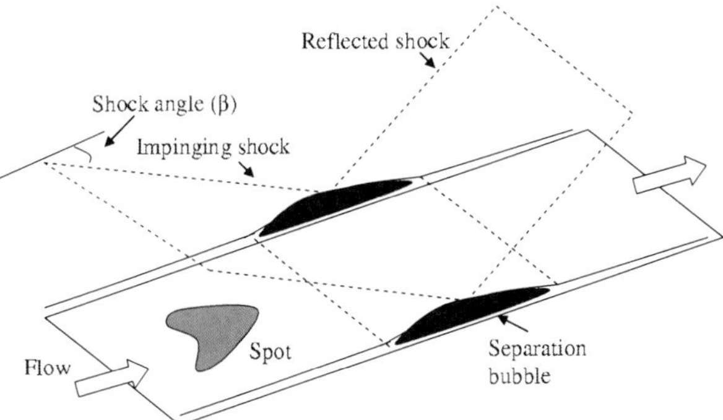

Fig. 1 Sketch of the flow configuration for interaction of a turbulent spot with a shock-induced separation bubble [4].

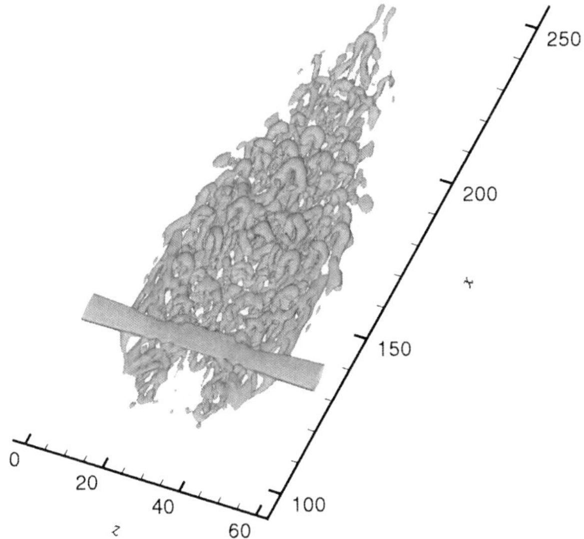

Fig. 2 Surface of constant second invariant, showing a turbulent spot passing through a shock-induced separation bubble [4].

interaction phenomena. A number of separate model studies have also been completed including a case of a turbulent spot passing through a laminar separation bubble induced by an impinging oblique shock wave as sketched on Figure 1 [4]. The turbulent spot was triggered by a local perturbation inside the boundary layer and is approximately the length of the separated flow region when the interaction occurs. Figure 2 shows contours of the second invariant of the velocity gradient tensor during the interaction. The spanwise structure visible in the figure is the recirculation vortex towards the rear of the separation zone. One important observation from the simulation was the enhanced growth of the turbulent spot during the interaction.

This is consistent with a growth mechanism for turbulent spots based on destabilization of the wing tip region [4].

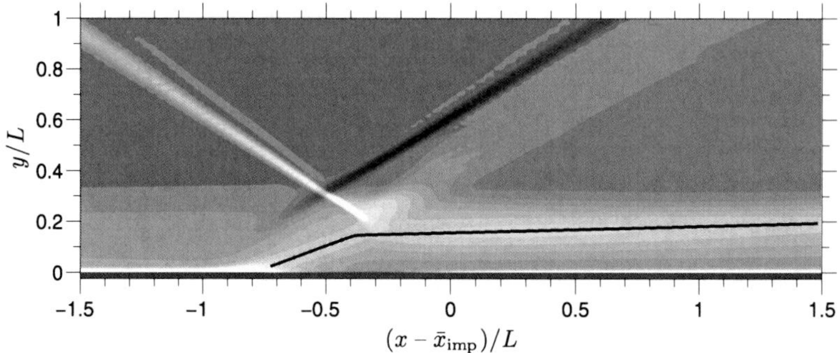

Fig. 3 Plot of $\partial\overline{\rho u}/\partial y$. The additional solid line highlights the center of the detached shear layer. The impinging and reflected shock waves show up as light and dark contours respectively.

3 Turbulent boundary layer separation

Another common test case for experiment and simulation is a turbulent boundary layer subjected to shock impingement. Upstream boundary conditions are critical, as it is important not to trigger spurious effects by inappropriate upstream conditions. In this case a fast implementation of a digital filter approach (given in [8,9] along with other details of the simulations) was found to give a broadband spectrum upstream, with no preferential excitation of particular low frequencies. Figure 3 shows contours of $\partial\overline{\rho u}/\partial y$. The turbulent inflow is given sufficient streamwise distance to reach a fully-developed state before an oblique shock wave is impinged onto it. The shock wave is sufficiently strong to separate the boundary layer, causing a turbulent separation bubble. The reflected shock originates from the separation region and is unsteady. The shock clearly responds to upstream turbulent flow fluctuations at high frequency. Current interest is mainly focused on the strong low-frequency motions that originate near the foot of the reflected shock wave. Large eddy simulations are capable of reproducing the effect, in good agreement with experiments. Figure 4 shows a premultiplied spectrum of wall pressure fluctuations at various streamwise locations near the separation zone. It can be seen how the upstream peak at Strouhal numbers of around 10 is disrupted by the separation, only reforming well downstream of reattachment (at a different Strouhal number since the boundary layer has thickened considerably by then). Close to the separation point we see the development of a strong peak in a very low Strouhal number range, peaking at 0.03 in close agreement with experiments.

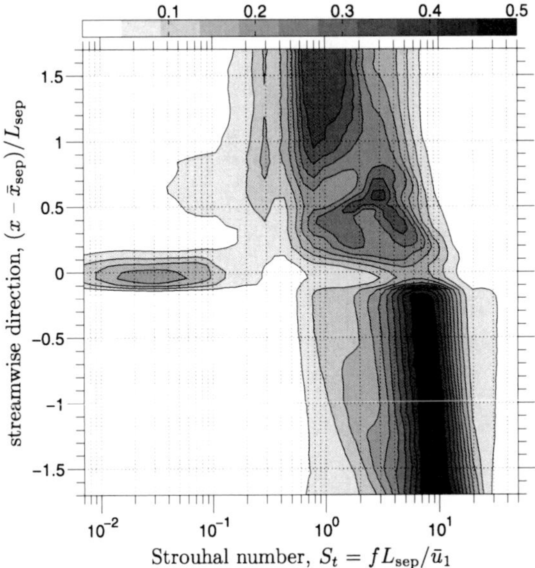

Fig. 4 Premultiplied spectrum of wall pressure fluctuations as a function of streamwise location in the interaction region.

In connection with the low-frequency motion it is of interest to consider the possible role of the separated shear layer. In particular Piponniau et al. [6] have proposed a model for the most energetic Strouhal number ($St = fL/U_1$, where f is the frequency, L is the interaction length, defined as the distance between the inviscid shock impingement point and a linear projection of the reflected shock to the wall, and U_1 is the pre-interaction freestream velocity) as

$$St = \frac{L}{H}\delta'_\omega\left[(1-r)C + \frac{r}{2}\right], \tag{1}$$

where H is the height of the interaction, r is the velocity ratio, $C \approx 0.14$ is a constant and δ'_ω is the vorticity thickness growth rate. The solid line added to figure 3 shows the approximate location of the center of the detached shear layer, which gives $H/L = 0.15$ compared to 0.12 reported in [6]. A local co-ordinate system may be defined following the shear layer centreline, with ξ and ζ as the local tangential and normal components. The time averaged velocity $\overline{u}_\xi(\zeta)$ is shown in figure 5 at several ξ locations. The dots shown on the figure are used to define the upper and lower velocities. The high velocity $\overline{u}^u_\xi$ is defined as the first local maximum of the velocity as one goes in the positive ζ direction. The low velocity $\overline{u}^d_\xi$ is defined as the next inflection point as one goes in the negative ζ direction. If no inflection point is found, the first local minimum is used instead. These two characteristic velocities are shown by large dots on Figure 5 (left frame). They can be used to define a vorticity thickness as $\delta_\omega = (\overline{u}^u_\xi - \overline{u}^d_\xi)/[d\overline{u}_\xi/d\zeta]_{\zeta=0}$ which can then be used as a reference length to scale the velocity profiles, as shown in the right frame of Figure 5 together with an error function for reference. It is clear that the detached shear layers do not conform very well with the standard mixing layer profile, suggesting a large degree of error in applying standard mixing layer correlations to the shock-induced separation problem.

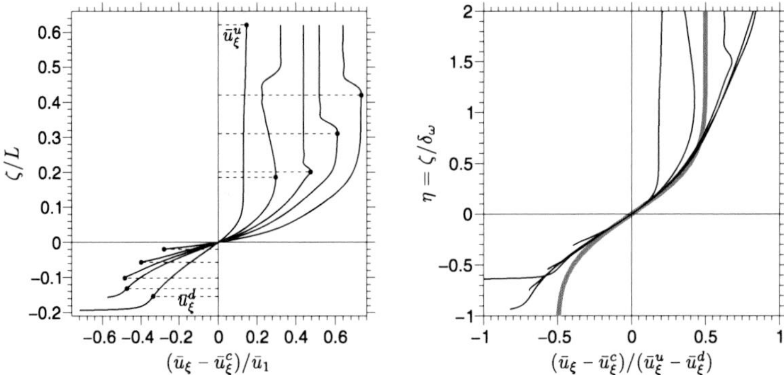

Fig. 5 Velocity profiles at different locations along the detached shear layer. The left hand figure is normalized using the inflow freestream velocity and separation bubble length, whereas the right hand figure is normalized using the vorticity thickness and velocity jump. The thick line shows an error function, for comparison.

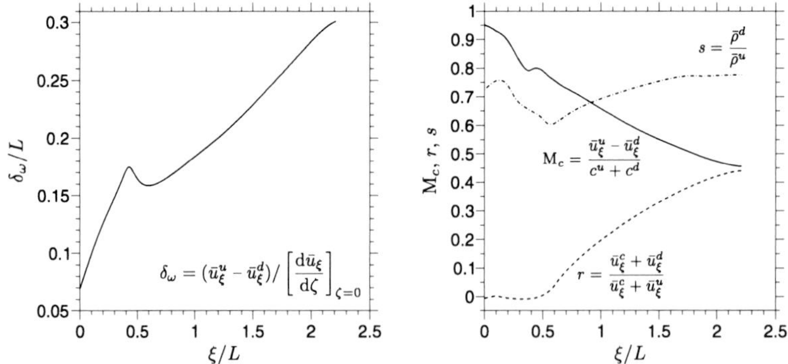

Fig. 6 Properties of the detached shear layer. The left hand figure shows the vorticity thickness. The right hand figure shows the density ratio s, the velocity ratio r and the convective Mach number M_c.

The left frame of Figure 6 shows the development of the vorticity thickness and right frame shows the variation of the local velocity ratio (r), density ratio (s) and convective Mach number (M_c), all of which may be defined once the upper and lower streams are known. It can be seen that the density ratio varies in the range 0.6 to 0.8, the velocity ratio starts at $r = 0$ and shows an increasing trend downstream, reaching $r = 0.4$ one interaction length downstream of the inviscid shock impingement location, and the convective Mach number decreases from 0.95 to 0.5 over the same distance. The relatively high values of M_c suggest strong compressibility effects and reduced growth rates relative to low speed separated flows. Piponniau et al. [6] used

$$\delta'_\omega = 0.08\frac{(1-r)(1+\sqrt{s})}{1+r\sqrt{s}}\Phi(M_c), \tag{2}$$

which leads to $\delta'_\omega \approx 0.05$ at that start of the interaction, compared to $\delta'_\omega \approx 0.3$ at $\xi = 0$ from figure 6. It may therefore be concluded that the simplified mixing layer model is not applicable to the case of a separating turbulent boundary layer and the apparent agreement of the simplified model with experiment shown in [6] is spurious, although the mechanism discussed may be contributing at a higher frequency. In the context of the low-frequency oscillation a more complete analysis based on the unsteady momentum integral equation is reported in [8,10].

4 Conclusion

Direct and large eddy simulations of shock-wave/boundary-layer interaction are helping to unravel the complex physics of these flows. With fully turbulent inflow conditions such interactions are characterized by high-amplitude low-frequency motions that are correctly reproduced in large-eddy simulations with long run times. The mechanisms behind the low frequency part of the spectrum are the subject of current debate. In this contribution we have shown that models based on planar mixing layer characteristics are oversimplified and lead to predictions of the low frequency that are some way from the values found in LES and experiments.

References

1. Almutairi, J.H., Jones, L.E. and Sandham, N.D. (2010) Intermittent bursting of a laminar separation bubble on an airfoil. *AIAA Journal* **48**(2), 414–426.
2. Ducros, F., Ferrand, V., Nicoud, F., Weber, C., Darracq, D., Gacherieu, C. and Poinsot, T. (1999) Large-eddy simulation of the shock turbulence interaction. *J. Comp. Phys.* **152**(2), 517–549.
3. Inagaki, M., Kondoh, T. and Nagano, Y. (2005) A mixed-time-scale SGM model with fixed model parameters for practical LES. *J. Fluids Eng.* **127**, 1–13.
4. Krishnan, L. and Sandham, N.D. (2007) Strong interaction of a turbulent spot with a shock-induced separation bubble. *Phys. Fluids* **19**(1): Art. No. 016102.

5. Krishnan, L., Sandham, N.D. and Steelant, J. (2009) Shock-wave/boundary-layer interactions in a model scramjet intake. *AIAA Journal* **47**(7), 1680–1691.
6. Piponniau, S., Dussauge, J.-P., Debieve, J.-F. and Dupont, P. (2009) A simple model for low-frequency unsteadiness in shock-induced separation. *J. Fluid Mech.* **629**, 87–108.
7. Sandham, N.D., Li, Q. and Yee, H.C. (2002) Entropy splitting for high-order numerical simulation of compressible turbulence. *J. Comp. Phys.* **178**(2), 307–322.
8. Touber, E. (2010) Unsteadiness in shock-wave/boundary-layer interactions. *PhD Thesis*, University of Southampton (available from `http://eprints.soton.ac.uk/161073/`).
9. Touber, E. and Sandham, N.D. (2009) Large-eddy simulation of low-frequency unsteadiness in a turbulent shock-induced separation bubble. *Theo. and Comp. Fluid Dyn.* **23**(2), 79–107.
10. Touber, E. and Sandham, N.D. (2010) Low-order stochastic modelling of low-frequency motions in reflected shock-wave/boundary-layer interactions. *J. Fluid Mech.*, accepted for publication.
11. Vreman, B., Kuerten, H and Guerts, B. (1995) Shocks in direct numerical simulation of the confined three-dimensional mixing layer. *Phys. Fluids* **7**, 2105–2107.
12. Yee, H.C., Sandham, N.D. and Djomehri, M.J. (1999) Low-dissipative high-order shock-capturing methods using characteristic-based filters. *J. Comp. Phys.*, **150**(1), 199–238.

DNS of a canonical compressible nozzle flow

Richard D. Sandberg, Victoria Suponitsky and Neil D. Sandham

1 Introduction

Simulations of subsonic free jets can only compute noise contributions from sources connected with the large-scale structures occurring close to the potential core region and their breakdown to fine-scale turbulence [1]. To account for additional noise sources associated with fine-scale turbulence in the initial shear layers and the interaction of flow with the nozzle lip, the nozzle must be included in the simulation and, moreover, the flow inside the nozzle must be fully turbulent. Previous work including realistic nozzle geometries in the simulation failed to achieve fully turbulent flow at the nozzle exit (e.g. [2]). To overcome the difficulties encountered when using realistic geometries, the problem can be simplified by using a canonical nozzle with well defined turbulent exit conditions. The ultimate goal of this ongoing study is to use a round pipe with sufficient length as a canonical nozzle for direct noise computations of jets. This paper focuses on whether the flow conditions at the pipe exit can be used as well defined turbulent upstream conditions for such calculations. From this perspective the following issues are of interest: (i) the effect of the inflow boundary conditions on the length of the pipe required to obtain fully developed turbulent pipe flow (development length); and (ii) the effect of compressibility on the temperature field. The former is needed to specify the length of the nozzle for full jet calculations, while the latter is related to the correct prescription of the ambient temperature. The spatially developing pipe flow using a turbulent inflow generation was chosen for this study instead of the alternative recycling or precursor simulation techniques because it avoids introducing an undesired artificial recycling frequency, the need to impose a pressure gradient, and minimizes the computational cost and memory requirements.

R. D. Sandberg e-mail: r.d.sandberg@soton.ac.uk
V. Suponitsky, e-mail: v.suponitsky@soton.ac.uk
N. D. Sandham, e-mail: n.sandham@soton.ac.uk
Aerodynamics and Flight Mechanics Research Group, School of Engineering Sciences, University of Southampton, Southampton SO17 1BJ, UK

H. Kuerten et al. (eds.), *Direct and Large-Eddy Simulation VIII*,
ERCOFTAC Series 15, DOI 10.1007/978-94-007-2482-2_46,

2 Methodology and computational details

The compressible Navier–Stokes equations for the conservative variables are solved in cylindrical coordinates using a newly developed finite-difference DNS code. Either a wavenumber-optimized 4th-order accurate compact-difference scheme, or, for simulations with periodic boundary conditions, a 4th-order standard-difference scheme with Carpenter boundary stencils is applied for the spatial discretization in the radial and streamwise directions. A spectral method is used in the azimuthal direction, enabling an axis treatment that exploits parity conditions of individual Fourier modes. Time marching is achieved by an ultra low-storage 4th-order Runge-Kutta scheme. The stability of the code is enhanced by a skew-symmetric splitting of the nonlinear terms. More details about the numerical method can be found in [3].

Three different types of simulations have been carried out: (i) a streamwise periodic pipe, (ii) a spatially developing pipe and (iii) a spatially developing pipe flow exiting into a uniform flow with $M_{coflow} = 0.2$. For validation, the periodic calculations were first carried out at Mach number $M = 0.4$ and Reynolds number $Re = 5000$, based on pipe diameter and bulk velocity. Following [4], the length of the pipe was set to $L_p = 15R$, and 201×72 grid points were used in the streamwise and radial directions, respectively. Grid points were uniformly distributed in the streamwise periodic direction, while a polynomial stretching was used in the radial direction with maximum and minimum grid spacings $\Delta r = 0.025R$ and $\Delta r = 0.00246R$ at the axis ($r = 0$) and wall ($r = R$), respectively. In the spatial simulations, the mean streamwise velocity profile obtained from the periodic pipe calculations is prescribed at the pipe entrance, while mean density and temperature profiles are set to be uniform. Turbulent fluctuations calculated using a digital filter technique [5] are superposed onto the mean flow values. The parameters required for this turbulence inflow technique (integral length scale, etc.) are evaluated from the periodic pipe data. Results indicated that the digital filter parameters significantly affect the development length of the flow. At the outflow boundary non-reflective characteristic boundary conditions were applied in the simulations of the pipe only (no boundary condition was required at the pipe exit in the simulations of the pipe flow exiting into a uniform flow).

All spatial pipe calculations were conducted with 624×68 grid points in the streamwise and radial directions, while the length of the pipe was $L_s = 50R$ (this length was found to be sufficient to achieve well established fully developed turbulent flow close to the pipe exit). In the isolated pipe simulations the grid points were uniformly spaced in the axial direction, whereas in the simulations of the pipe exiting into a uniform flow the grid spacing was equidistant ($\Delta z = 0.084$) up to $z = -1.5$ and then was refined using polynomial stretching towards the pipe exit where $\Delta z = 0.009$. In the radial direction, the maximum and minimum grid spacing were $\Delta r = 0.026$ at the axis and $\Delta r = 0.00258$ at the wall, respectively. Spatial simulations were carried out for Mach numbers $0.64 \leq M \leq 0.84$ and Reynolds number 6700. (The parameters were chosen for future jet noise computations.) In all simulations 64 Fourier modes were employed in the azimuthal direction and isothermal wall boundary conditions were specified.

3 Results

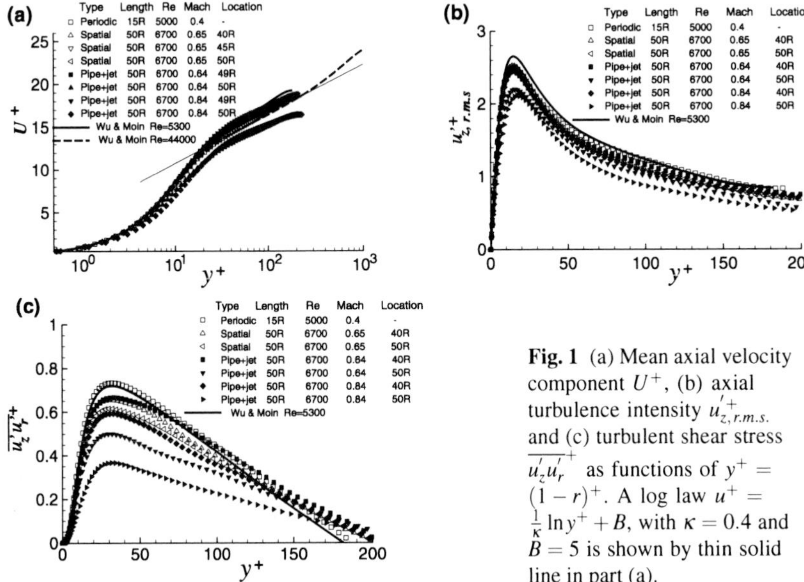

Fig. 1 (a) Mean axial velocity component U^+, (b) axial turbulence intensity $u'^+_{z,r.m.s.}$ and (c) turbulent shear stress $\overline{u'_z u'_r}^+$ as functions of $y^+ = (1-r)^+$. A log law $u^+ = \frac{1}{\kappa}\ln y^+ + B$, with $\kappa = 0.4$ and $B = 5$ is shown by thin solid line in part (a).

Profiles of the mean axial velocity $(U^+ vs. y^+)$[1] for the periodic and spatially developing pipes are shown in Fig. 3(a) along with the recent incompressible results of Wu & Moin [4]. One can see that in the case of isolated pipe simulations (periodic and spatial) current results are in an excellent agreement with each other and also with [4]. In the spatially developing pipe simulation the mean velocity profiles appear to be axially independent for at least $10R$ before the pipe exit. In simulations of the pipe exiting into a uniform flow (denoted as 'Pipe+jet' in the figure and from here on) the shape of the mean velocity profiles is affected by the surrounding flow in the close vicinity of the pipe exit ($50R$), while only one diameter upstream ($49R$) it corresponds to that of fully developed turbulent pipe flow.

Similar results can be observed for the axial intensity profiles plotted in Fig. 3(b): all results, except those at the pipe exit in the 'Pipe+jet' simulations, are in good agreement with [4] and with each other. In the 'Pipe+jet' simulations the intensity profiles are influenced by the expansion into the co-flow further upstream from the pipe exit (up-to about $5R$) compared with the mean velocity profiles. Profiles of the turbulent shear stress (Fig. 1(c)) demonstrate even stronger sensitivity to the simulation details: (i) a significant difference can be observed in the peak amplitudes between the 'Pipe+jet' simulations with $M = 0.64$ and $M = 0.84$ in the vicinity of the pipe exit (in contrast with the profiles of axial intensity), (ii) the difference in

[1] Van Driest scaling was used for Mach numbers $M > 0.4$.

the peak amplitudes becomes less prominent at some distance upstream of the pipe exit (40R), but in all spatial simulations the peak amplitude itself remains somewhat lower than that reported in [4], and (iii) results for the periodic pipe still are in a very good agreement with [4].

Frequency spectra $|\rho\hat{u}_z|$ of the conservative variable ρu_z calculated from the 'Pipe+jet' simulation with $M = 0.84$ and $Re = 6700$ in the vicinity of the pipe exit at two different radial locations are given in Figs. 2(a) and 2(b). First of all one can see that close to the axis (Fig. 2(a)) there is a rapid decrease in energy with the increase in Fourier mode. The situation is different in the region of peak intensities, where the energy of at least five first Fourier modes (from $m = 1$ to 5) are basically the same (Fig. 2(b)). The decrease of the energy with the increase in Fourier mode near the axis can be explained by the different structure of the flow. The turbulent nature of the flow is also evident from this figure showing well resolved spectra up to Strouhal number St_D, based on bulk velocity and pipe-diameter, about 20. It is also worth noting that the amplitudes of sine and cosine parts (shown by solid and broken lines) are the same for all presented Fourier modes.

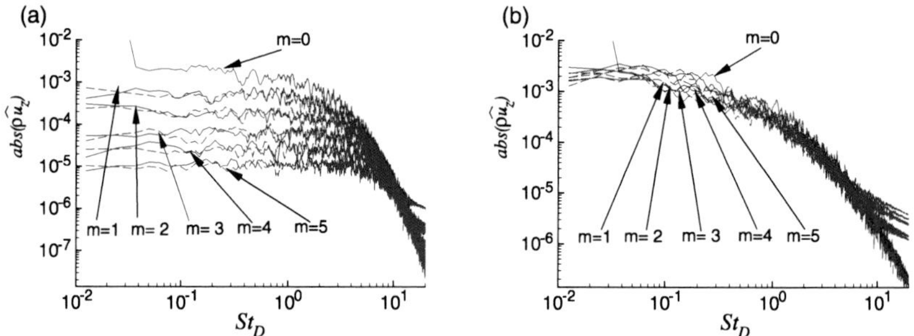

Fig. 2 Spectra $|\rho\hat{u}_z(St_D)|$ of the conservative variable ρu_z obtained for the six first Fourier modes (starting from zero) calculated from the 'Pipe+jet' simulation with $M = 0.84$ and $Re = 6700$ very close to the pipe exit ($z = -0.056R$). For each Fourier mode but zero, the cosine and sine parts are shown by solid and broken lines, respectively. (a) close to the axis at $r/R = 0.1$; (b) in the region of the maximum intensities at $r/R = 0.9$.

The streamwise variation of the mean pressure inside the pipe is given in Fig. 3(a) for the isolated spatial pipes with $M = 0.65$ and 0.72 and for the 'Pipe+jet' simulations with $M = 0.64$ and 0.84. This plot illustrates the development of a pressure gradient inside the pipe (with no pressure gradient being explicitly imposed). The pressure at the pipe exit depends on the Mach number for both the isolated pipe and the 'Pipe+jet' simulations. In the isolated pipe simulations the pressure at the outflow boundary is driven to the normalized pressure of $1/\gamma M^2$ ($\gamma = 1.4$). This pressure is obtained from the equation of state when the normalized density and temperature are set to unity. In the 'Pipe+jet' simulations the pressure at the pipe exit equals that of the surrounding uniform flow (which is also specified as $1/\gamma M^2$). It is worth pointing out that the pressure at the inflow boundary is also given by $1/\gamma M^2$, as normalized temperature and density are prescribed to be unity at the pipe inlet. Consequently, a rapid adjustment of the pressure (and other quantities like density and temperature) occurs in the vicinity of the pipe inlet. Other variations of the inflow boundary conditions, such as allowing pressure/density to drift, or prescribing density/pressure based on an approximate pressure gradient were tested to alleviate the strong inlet gradients. However, these attempts did not lead to any improvement and the current method was retained.

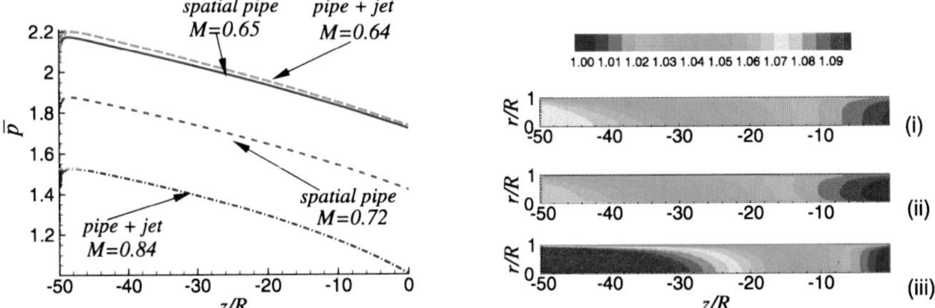

Fig. 3 (a) Axial variation of the mean pressure $\overline{p}$ inside the pipe; (b) Contours of the mean temperature $\overline{T}$ inside the pipe; (i), (ii) and (iii) correspond to the simulations of the isolated spatial pipe with $M = 0.65$,'Pipe+jet' with $M = 0.64$ and 'Pipe+jet' with $M = 0.84$, respectively.

Contour plots of the mean temperature are shown in Fig. 3(b), where (i),(ii) and (iii) correspond to the isolated pipe with $M = 0.65$ and 'Pipe+jet' simulations with $M = 0.64$ and 0.84. The existence of streamwise temperature variations along the pipe can be observed and the variation becomes more pronounced with increasing Mach number. Focusing on the region close to the pipe exit ($-10R < z/R < 0$) we can see that in all cases the temperature variations are insignificant (within $2-3\%$) and near the pipe exit the temperature is approximately one. This knowledge is essential for specifying ambient temperature conditions for isothermal jet simulations.

Finally, instantaneous contours of the azimuthal vorticity component are shown in Fig. 4 for the 'Pipe+jet' simulation with $M = 0.84$ and $Re = 6700$. One can see

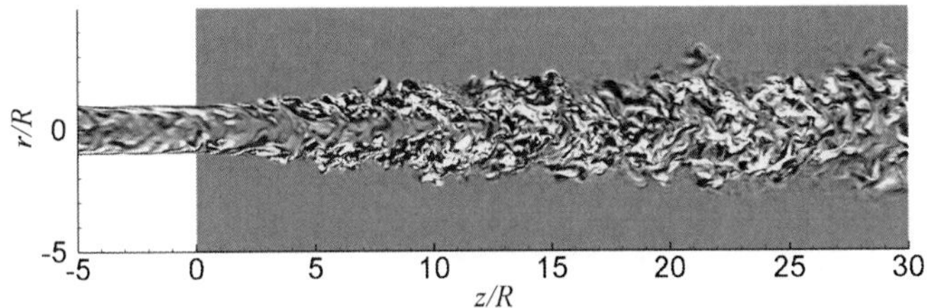

Fig. 4 Instantaneous contours of the azimuthal vorticity component; 'Pipe+jet' simulation with $M = 0.84$ and $M_{coflow} = 0.2$. 20 contour levels are shown from -0.5 to 0.5.

the turbulent nature of the flow near the pipe exit and rapid streamwise development of the turbulent jet. The shear layers appear to be turbulent from the very beginning and the length of the potential core is somewhere between $10R$ and $12R$ which is shorter than that observed in the simulations with laminar inflow conditions.

4 Conclusions

Current results show that fully turbulent inflow conditions for jet calculations are achievable by using a round pipe of a sufficient length as a canonical nozzle. In spatially developing simulations of the isolated pipe the mean and r.m.s properties near the pipe exit are in good agreement with reference data of a fully developed turbulent pipe. In the 'Pipe+jet' simulations the structure of the flow close to the pipe exit is strongly affected by the expansion of the jet into the co-flow, but remains fully turbulent. For the Mach numbers in question temperature variations inside the pipe are insignificant (especially close to the pipe exit) such that no adjustment of the surrounding flow temperature is required for isothermal jet simulations.

Overall it seems that by using this relatively computationally inexpensive canonical nozzle approach, direct jet noise computations which include all noise sources are feasible. Therefore, the next step of this ongoing study is to perform jet simulations in which particular attention will be devoted to the understanding of trailing-edge noise and noise associated with fine-scale turbulence inside the initial shear layers.

Acknowledgements This project was partly supported by EPSRC grant EP/E032028/1 and the Royal Academy of Engineering / EPSRC research fellowship (EP/E504035/1). Computer time for this project was provided by the Cray CoE Science delivery programme and early user time on HECToR phase 2b.

References

1. Freund, J.B. (2001). Noise sources in a low-Reynolds-number turbulent jet at Mach 0.9. *J. Fluid Mech.* **438**, 277–305.
2. Uzun, A. & Hussaini, M.Y. (2009). Simulation of Noise Generation in the Near-Nozzle Region of a Chevron Nozzle Jet. *AIAA J.* **47**(8), 1793–1810.
3. Sandberg, R.D. (2010). Direct numerical simulations of turbulent supersonic axisymmetric wakes. *ERCOFTAC WORKSHOP Direct and Large-Eddy Simulation 8*, Eindhoven, The Netherlands.
4. Wu, X. & Moin, P. (2008). A direct numerical simulation study on the mean velocity characteristics in turbulent pipe flow. *J. Fluid Mech.* **608**, 81–112.
5. Touber, E. & Sandham, N. (2009). Large-eddy simulation of low-frequency unsteadiness in a turbulent shock-induced separation bubble. *Theoretical and Computational Fluid Dynamics*, **23**(2), 79–107.

Direct numerical simulations of turbulent supersonic axisymmetric wakes

Richard D. Sandberg

1 Introduction

Over the last decades, there has been considerable interest in supersonic axisymmetric wakes, or *base flows*. Initially, the motivation of the research was to gain a better understanding of the dynamics of supersonic turbulent flows, and to devise methods for drag reduction. Later, base flows were frequently chosen as a challenging test case for numerical simulations, mainly due to the availability of reliable data from carefully conducted base flow experiments (e.g. Herrin & Dutton [1]), and the fact that a complex flow is generated by a relatively simple geometry, facilitating grid generation. Furthermore, the failure of early RANS calculations to capture some of the characteristic properties of the flow, e.g. a flat base pressure distribution [2], motivated studies employing various RANS and hybrid RANS/LES turbulence models.

Nevertheless, the focus of most numerical investigations has typically been on matching mean flow data from the experiments, rather than attempting to identify and understand the origin and evolution of large-scale vortical structures and their effect on the mean flow. The most likely reason for this is the high Reynolds number of the experiments ($Re_D = 3.3 \times 10^6$, based on the freestream velocity and the base diameter), excluding DNS. Sandberg & Fasel [3] conducted DNS of base flows for Reynolds numbers up to 1×10^5 to identify the fundamental hydrodynamic mechanisms leading to the generation of large structures and studied their effect of on the mean flow, in particular on the base pressure [4]. In all cases a laminar approach boundary layer was specified and simulations were performed using 'symmetric' Fourier transforms in the azimuthal direction, i.e. either cosine or sine transforms were used for symmetric or asymmetric quantities, respectively.

This paper aims at investigating the effect of those two simplifications. DNS were conducted using 'full' Fourier transforms, i.e. using both cosine and sine modes for

---

R. D. Sandberg
Aerodynamics and Flight Mechanics Research Group, School of Engineering Sciences, University of Southampton, Southampton SO17 1BJ, UK, e-mail: sandberg@soton.ac.uk

H. Kuerten et al. (eds.), *Direct and Large-Eddy Simulation VIII*,
ERCOFTAC Series 15, DOI 10.1007/978-94-007-2482-2_47,

each variable, in order to evaluate whether the flow is adequately represented using ‘symmetric’ Fourier transforms. Furthermore, a simulation has been performed with a fully turbulent approach boundary layer to assess whether the global flow and individual azimuthal modes are affected by the change of the inflow condition.

2 Methodology

The compressible Navier–Stokes equations for conservative variables are solved in cylindrical coordinates using a newly developed finite-difference code. For the DNS presented here, a 4th-order central difference scheme with Carpenter boundary stencils is applied for the spatial discretization in the radial and streamwise directions. A spectral method is used in the azimuthal direction, enabling an axis treatment that exploits parity conditions of individual Fourier modes. Time marching is achieved by an ultra low-storage 4th-order Runge–Kutta scheme [5]. The stability of the code is enhanced by a skew-symmetric splitting of the nonlinear terms [6] and an 11 point wave-number optimized filter [7], used after each full Runge–Kutta cycle with a weighting of 0.8 to remove grid-to-grid-point oscillations that might occur using a central scheme to capture shock waves. For the simulation requiring a turbulent inflow, a mean turbulent profile from a precursor simulation is specified at the inlet, superposed with turbulent fluctuations calculated using a digital filter technique [8].

The code has been validated by carrying out simulations of oblique instability waves in supersonic boundary layers, matching linear stability results, and by running low Mach number wakes reproducing incompressible reference data. Also, turbulent pipe flow simulations have been conducted showing good agreement with recently published results [9]. The current DNS at $M = 2.46$ and $Re_D = 1 \times 10^5$ were conducted using 78×10^6 grid points distributed over 512 cores and 177×10^6 points on 2048 cores for the laminar and turbulent approach flow cases, respectively.

3 Results

To study the effect of the approach flow conditions on the global wake behavior, laminar and turbulent approach flow cases were set up, each with a boundary layer thickness at the base corner of $\delta = 0.1R$, similar to the experimental set-up at higher Reynolds number [1].

As shown in figure 1, left, the time-averaged velocity profile resulting at the base corner compares better with the experimental data than the laminar approach flow case, although the profile is not as full as the reference data. To more quantitatively assess the quality of the turbulent boundary layer, Reynolds stress profiles were computed. A direct comparison with the experiments is not meaningful due to the difference in Reynolds number. Therefore, the current case is compared with the incompressible DNS of Spalart [10] because the current $Re_\theta \approx 1{,}500$ is close to the

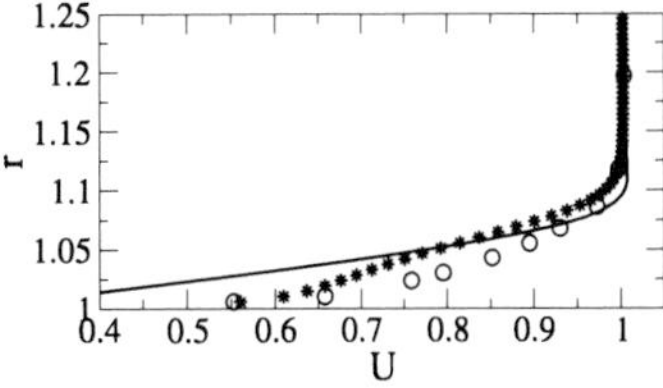

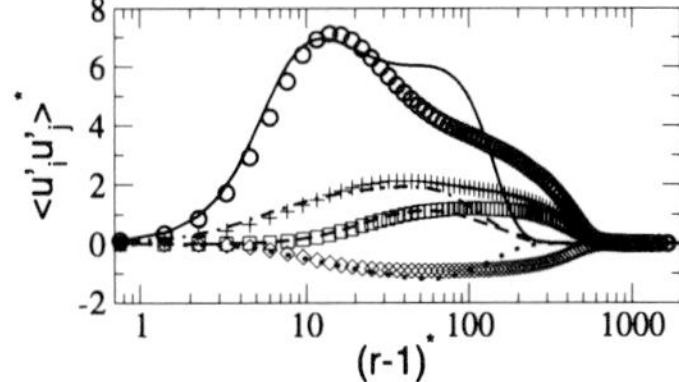

Fig. 1 Approach boundary layer profiles (left); (–) axisymmetric and 3D DNS with laminar inflow profile, (∗) 3D DNS with turbulent inflow profile, (○) experimental data [1]; Reynolds stress tensor components for turbulent approach flow case (right); lines from reference DNS [10], symbols from current DNS; $Re_D = 1 \times 10^5$, $M = 2.46$.

reference data. To compare the data at $M = 2.46$ with the incompressible reference data, a 'semi-local' scaling [11] is used and good agreement in terms of amplitudes and peak locations is found. Nevertheless, despite the similar Re_θ of both cases, the maximum $(r-1)^*$ location at which fluctuations are present in the current DNS is smaller, possibly due to the favorable pressure gradient caused by the corner expansion. Overall, the flow separating at the base corner can be classified as fully turbulent, which was the primary goal of the present study.

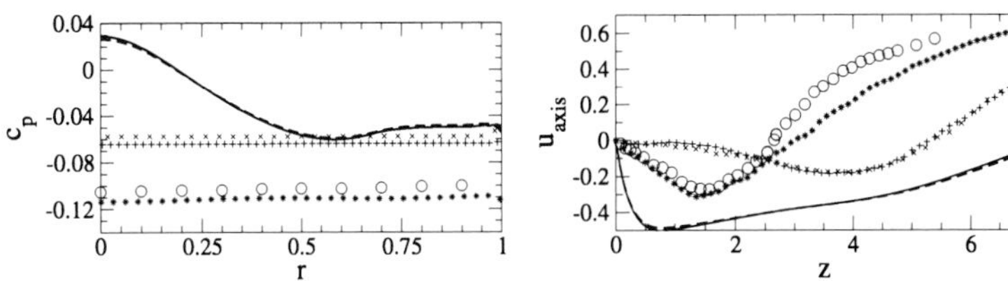

Fig. 2 Base pressure coefficient distribution (left) and streamwise axis velocity (right) from DNS at $Re_D = 1 \times 10^5$, $M = 2.46$; (–) axisymmetric DNS with laminar inflow profile, (- - -) axisymmetric DNS with turbulent inflow profile, (+) 3D DNS with laminar inflow profile, (×) reference DNS with laminar inflow profile [3], (∗) 3D DNS with turbulent inflow profile, (○) experimental data [1].

Time-averaged profiles of base-pressure coefficient and streamwise axis velocity are shown in figure 2. The data from the case with laminar inflow can be seen to agree well with the laminar inflow DNS using 'symmetric' Fourier transforms [4]. The base pressure profile is flat and the amplitude is considerably larger than in the high Reynolds number experiments [1]. The streamwise axis velocity shows a very small wall-normal gradient at the wall and the maximum reverse velocity is considerably smaller in magnitude and located farther downstream than in the experiments. The good agreement with the reference DNS is evidence that the relevant physics can be captured using 'symmetric' Fourier transforms. In the case of a turbulent approach flow, the picture is radically changed. The base pressure coefficient is significantly reduced relative to the laminar inflow case and is now close to the value obtained from the high Reynolds number experiments. The streamwise axis velocity also agrees well with the experimental data in terms of slope at the base, location and magnitude of the maximum reverse flow, and mean recirculation length. However, the developing wake does not show as fast a recovery as the higher Reynolds number, as expected.

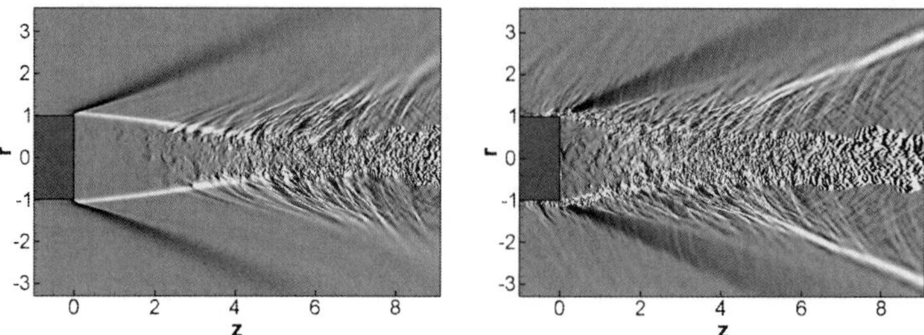

Fig. 3 Contours of streamwise density gradient ($\partial\rho/\partial z$) for DNS with laminar (left) and turbulent (right) approach flow, illustrating the respective flow topologies; $Re_D = 1 \times 10^5$, $M = 2.46$.

For a qualitative view of the differences between the cases with laminar and turbulent inflows, instantaneous contours of the streamwise density gradient are shown in figure 3. In case of a laminar approach flow the initial shear layer is laminar and the onset of transition to turbulence is at $z \approx 2.5$. In contrast, using a turbulent approach flow a fully turbulent initial shear layer can be observed. The increased mixing of the fully turbulent shear layers results in a larger turning angle of the flow at separation, leading to the shorter recirculation region seen in figure 2.

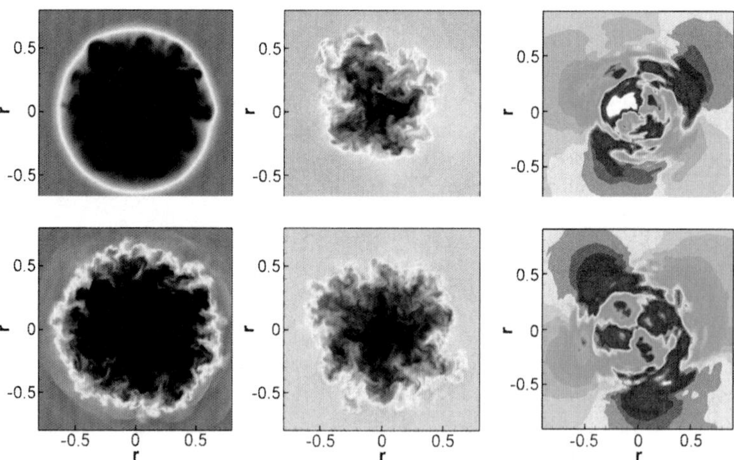

Fig. 4 Endviews of instantaneous contours of local M upstream of recompression (left) and downstream of recompression (middle); time-averaged θ-component of velocity in developing wake (right); DNS with laminar (top) and turbulent (bottom) inflow at $Re_D = 1 \times 10^5$, $M = 2.46$.

Endviews of instantaneous contours of the local Mach number are shown in figure 4. Similar to the case with laminar inflow and 'symmetric' Fourier modes [3, 4], for both cases conducted here approximately 12–14 mushroom-like structures on the inner side of the shear layer can be observed instantaneously, in good agreement with the typical number of 10–14 observed in the experiment. Further downstream, in the developing wake, the 'four-lobe' structure observed at some instances in the experiments and the reference DNS can be vaguely discerned in both cases, although not as clearly. This might be due to the fact that using 'full' Fourier transforms does not invoke any kind of symmetry plane as is the case when using 'symmetric' transforms or can be caused by minute imperfections in experiments. Nevertheless, when looking at the time-averaged azimuthal velocity component, the 'four-lobe' structure in the developing wake appears more clearly, indicating that this structure occurs regularly. As shown previously, this 'four-lobe' structure can be generated by either $m = 2$ or $m = 4$ being the dominant mode in the developing wake. To determine which azimuthal mode is dominant, radial amplitude distributions of selected azimuthal Fourier modes, shown in figure 5, are considered. For both inflow options, the amplitude of the second mode $m = 2$ is considerably larger than that of $m = 4$ in the developing wake. In the current cases, $m = 3$ also is larger in amplitude than $m = 4$, in contrast to the reference DNS with 'symmetric' transforms. $m = 1$ has a considerably larger amplitude when using a laminar approach flow case than for the turbulent approach flow case, suggesting that the turbulent inflow reduces the flapping of the wake compared with the laminar inflow case. Within the recirculation region, $m = 1$ appears to dominate in both cases with the turbulent inflow case showing the largest amplitude.

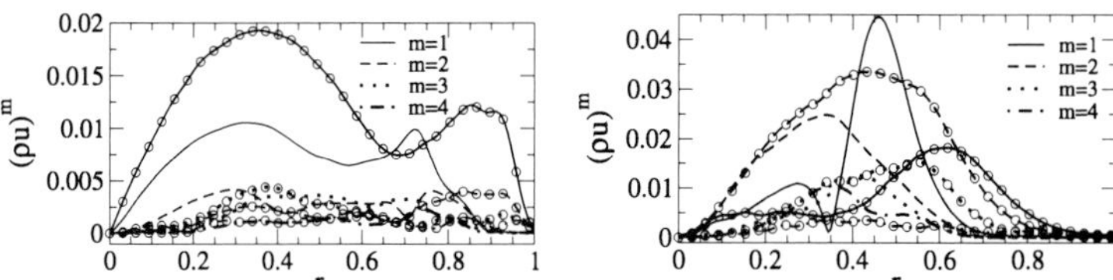

Fig. 5 Radial amplitude distributions of azimuthal Fourier modes of ρu in recirculation region (left) and in developing wake (right); data from DNS using laminar approach flow (lines) and turbulent approach flow (open circles); $Re_D = 1 \times 10^5$, $M = 2.46$.

4 Conclusions

Direct numerical simulations have been conducted of supersonic axisymmetric wakes at $Re_D = 1 \times 10^5$ and $M = 2.46$ with a newly developed Navier–Stokes solver. As opposed to previous DNS studies [3, 4], the current simulations were performed using 'full' Fourier transforms in the azimuthal direction. The case with a laminar inlet condition shows good agreement with the reference data, particularly when looking at mean flow profiles. A 'four-lobe' wake pattern is observed in endviews of time-averaged quantities which appears to be generated by $m = 2$. The strong similarity with the reference case suggests that the relevant physics can be captured using 'symmetric' Fourier transforms, despite their slight underprediction of the significance of $m = 3$. When specifying a turbulent approach flow, mixing is significantly increased in the shear layers, reducing the recirculation length of the wake and decreasing the base-pressure coefficient. Also, the significance of the first helical mode appears considerably reduced compared with the laminar inflow case, leading to a reduced flapping of the developing wake. Although the current case is conducted at a Re_D more than a magnitude lower than the experiments, the mean flow profiles and instantaneous endviews show surprising resemblance with the experimental data, indicating that the most important mechanisms dominating the flow in the experiments can be studied with the DNS presented here.

Acknowledgements Funding for the present study was provided through the Royal Academy of Engineering / EPSRC Research fellowship (EPSRC grant EP/E504035/1) and the simulations were run on the UK high performance computing service HECToR.

References

1. Herrin, J. L. and Dutton, J. C. (1989). Supersonic base flow experiments in the near wake of a cylindrical afterbody. *J. Fluid Mech.*, **199**, 55–88.
2. Sahu, J., Nietubicz, C., and Steger, J. (1985). Navier–Stokes computations of projectile base flow with and without mass injection. *AIAA J.*, **23** (9), 1348–1355.
3. Sandberg, R. D. and Fasel, H. F. (2006). Numerical investigation of transitional supersonic axisymmetric wakes. *J. Fluid Mech.*, **563**, 1–41.
4. Sandberg, R. D. and Fasel, H. F. (2006). Direct numerical simulations of transitional supersonic base flows. *AIAA J.*, **44** (4), 848–858.
5. Kennedy, C., Carpenter, M., and Lewis, R. (2000). Low-storage, explicit Runge–Kutta schemes for the compressible Navier–Stokes equations. *Applied Numerical Mathematics*, **35**, 177–219.
6. Kennedy, C., and Gruber, A. (2008). Reduced aliasing formulations of the convective terms within the Navier–Stokes equations for a compressible fluid. *J. Comp. Phys.*, **227**, 1676–1700.
7. Bogey, C., de Cacqueray, N., and Bailly, C. (2009). A shock-capturing methodology based on adaptative spatial filtering for high-order non-linear computations. *J. Comp. Phys.*, **228**, 1447–1465.
8. Touber, E. and Sandham, N. (2009). Large-eddy simulation of low-frequency unsteadiness in a turbulent shock-induced separation bubble. *Theoretical and Computational Fluid Dynamics*, **23** (2), 79–107.
9. Sandberg, R. D., Suponitsky, V., and Sandham, N. D. (2010). DNS of a canonical nozzle flow. *ERCOFTAC WORKSHOP Direct and Large-Eddy Simulation 8*.
10. Spalart, P. R. (1988). Direct simulation of a turbulent boundary layer up to $Re_\theta = 1410$. *J. Fluid Mech.*, **187**, 61–98.
11. Coleman, G. N., Kim, J., and Moser, R. D. (1995). A numerical study of turbulent supersonic isothermal-wall channel flow. *J. Fluid Mech.*, **305**, 159–183.

DNS of a Variable Density Jet in the Supercritical Thermodynamic State

F. Battista, F. Picano, G. Troiani and C.M. Casciola

1 Introduction

Cryogenic rocket engines, advanced gas turbines and diesel engines are characterized by the injection of a liquid fuel into a high temperature and pressure chamber. Typically the fuel is injected at high enough pressure to be close or above the critical pressure. In these conditions the behavior of the fluid differs strongly from that of a perfect gas. It exhibits large variations of thermodynamic and transport properties also for small temperature changes, with significant effects on mixing and combustion processes. In this context an appropriate numerical simulation should take into account such thermodynamic phenomena via suitable equation of state and transport properties relations.

In literature this issue is typically addressed in several experimental studies of round jets at super-critical pressure see [1, 2, 3]. In sub-critical conditions the mixing/evaporation process leads to the formation of ligaments which break-up in evaporating droplets, while in super-critical conditions the ligaments dissolve in the external environment. In the context of numerical simulations few Large Eddy Simulations (LES) of jet under super-critical conditions have been performed. In particular *Zong et al.* [4] perform a 2*D* large eddy simulation reproducing the thermodynamic conditions of the experimental study provided by *Mayer et al.* [1]. The same conditions are matched in the 3*D* large eddy simulation carried out by *Schmitt et al.* [5]. Moreover an extensive review containing the main results obtained before 2006 is provided in [6].

Aim of this work is to perform a 3*D* Direct Numerical Simulation of a cryogenic jet at weakly supercritical pressure injected in an environment with the same pres-

F. Battista, F. Picano, C.M. Casciola: Dipartimento di meccanica ed aeronautica, "Sapienza" University of Rome, e-mail: `francesco.battista@uniroma1.it`, `francesco.picano@uniroma1.it`, `carlomassimo.casciola@uniroma1.it`

G. Troiani: ENEA C.R. Casaccia, via Anguillarese 301, 00123 Roma Italy, e-mail: `guido.troiani@enea.it`

H. Kuerten et al. (eds.), *Direct and Large-Eddy Simulation VIII*, ERCOFTAC Series 15, DOI 10.1007/978-94-007-2482-2_48,

sure and ambient temperature. Actually the injection and mixing of cryogenic gases occur with characteristic velocities smaller than the local sound speed. To this purpose, an asymptotic Low Mach number formulation of the Navier-Stokes equations coupled with the Van der Waals state equation is derived to perform the simulations.

2 Methodology

The simulations involve a jet injected with a density smaller than the critical one, $\rho_j < \rho_c$, in an environment at weakly super-critical pressure with density $\rho_{ext} > \rho_c$.

In these conditions the perfect gas assumption is not consistent with the thermodynamics of dense gases near the critical point. Here the Van der Waals equation of state (EOS) is used to mimic the supercritical gas,

$$\left(p + a'\rho^2\right)\left(1 - b'\rho\right) = \rho\,\theta \tag{1}$$

where the EOS is expressed in its dimensionless form and p, ρ and θ are the pressure, density and temperature of the gas, respectively. The constants a' and b' depend on the attraction forces between molecules and on their average volume occupied, respectively. Since the Van der Waals constants are function of the critical thermodynamic variables, they are expressed in terms of the ratio between the critical and reference (indicated with the subscript ∞) quantities, $a' = 3(\rho_\infty/\rho_c)(p_c/p_\infty)$ and $b' = (1/3)(\rho_\infty/\rho_c)$. The thermodynamic properties, such as enthalpy ρh, internal

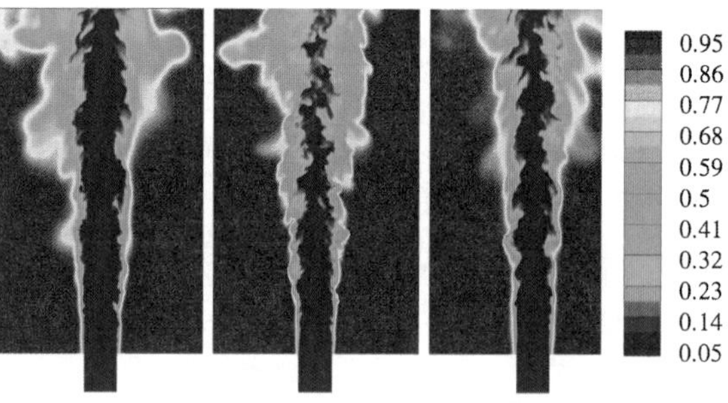

Fig. 1 Snapshot of normalized temperature $(\theta - \theta_{ext})/(\theta_j - \theta_{ext})$ field of the three simulations. From left: real gas with $\rho_j/\rho_{ext} \simeq 10.$, perfect gas with $\rho_j/\rho_{ext} \simeq 4.$, perfect gas with $\rho_j/\rho_{ext} \simeq 10.$

energy ρe and ratio of the constant-pressure to constant-volume specific heats γ, follow by the standard statistical mechanics arguments,

$$\rho e = \frac{1}{\gamma - 1}\rho\theta - a'\rho^2 \quad , \quad \rho h = \rho e + p$$

$$\gamma = \gamma^{pg}\left[\frac{3}{5} + \frac{2}{5}\frac{1}{1 - 2a'\frac{\rho}{\theta}\left(1 - b'\rho\right)^2}\right]$$

where γ^{pg} denotes the specific heat ratio of the perfect gas.

The present derivation of the Low Mach number formulation of the Navier-Stokes equations for a Van der Waals EOS is an extension of that proposed by *Majda & Sethian* [7], for the ideal gas. This approach is particularly attractive in the simulation of cryogenic flows that, despite at small Mach number, cannot be considered strictly incompressible due to the occurrence of strong density gradients [7, 8]. The zeroth-order dimensionless system reads,

$$\frac{\partial \rho_0}{\partial t} + \nabla \cdot (\rho \boldsymbol{u})_0 = 0 \tag{2}$$

$$\frac{\partial (\rho \boldsymbol{u})_0}{\partial t} + \nabla \cdot [(\rho \boldsymbol{u})_0 \boldsymbol{u}_0] = \tfrac{1}{Re} \nabla \cdot \boldsymbol{\tau}_0 - \nabla p_2 + \rho_0 \boldsymbol{f}_0 \tag{3}$$

$$\nabla \cdot \boldsymbol{u}_0 = \frac{1}{p_0} \left[\tfrac{1}{RePr} \nabla \cdot (\mu \nabla \theta)_0 \right] \left[\frac{1}{1 + \frac{(\gamma-2)a'\rho_0^2}{p_0 \gamma} + \frac{2a'b'\rho_0^3}{p_0 \gamma}} \right] \tag{4}$$

$$\theta_0 = \frac{p_0}{\rho_0} \left(1 + a' \frac{\rho_0^2}{p_0} \right) \left(1 - b' \rho_0 \right) \tag{5}$$

where continuity (2) and momentum (3) equations coincide with those of the ideal gas. Energy (4) and EOS (5) equations present instead corrective terms with respect to the real gas behavior. Clearly, the perfect gas formulation is recovered in the limit $a' \to 0$ and $b' \to 0$. The second-order pressure term p_2 provides the dynamic pressure which is the only pressure term present in the incompressible formulation. In the present context this degree of freedom allows a prescribed value of the velocity divergence, eq. (4), as it happens in the original ideal gas formulation [7].

As anticipated our purpose here is to perform a 3*D* Direct Numerical Simulation (DNS) of a turbulent jet based on the above equations. The Eulerian algorithm discretizes the system in a cylindrical domain to address a variable density jet at weakly supercritical conditions. Spatial discretization is based on central second order finite differences in conservative form on a staggered grid while the convective term of scalar equations is dealt with by a bounded central difference scheme to avoid spurious oscillations. Temporal evolution is performed by a low-storage third order Runge-Kutta scheme. Dirichlet (prescribed velocity) conditions are enforced at the inflow, by using a cross-sectional plane of a periodic turbulent pipe flow, obtained by a companion time-evolving DNS. A convective condition is adopted at the outflow, while a traction-free condition is used for the side boundary. More details on the code are available in [9]. Three simulations are performed. The first one reproduces a cold jet of real gas ($S1$) injected in a environment with a pressure weakly supercritical $p/p_c \simeq 1.18$. The density of the environment is one tenth the jet one $\rho_j/\rho_{ext} = 10$, ensuring that $\rho_j < \rho_c < \rho_{ext}$, while the ratio between the jet and environment temperature is $\theta_{ext}/\theta_j \simeq 4.$. The other two simulations $S2$, $S3$ reproduce two perfect gas jets matching either the temperature ratio $\theta_{ext}/\theta_j = \rho_j/\rho_{ext} = 4$ or the density ratio $\theta_{ext}/\theta_j = \rho_j/\rho_{ext} = 10$ of the first one, respectively. All simulations present a Reynolds number based on the jet diameter,

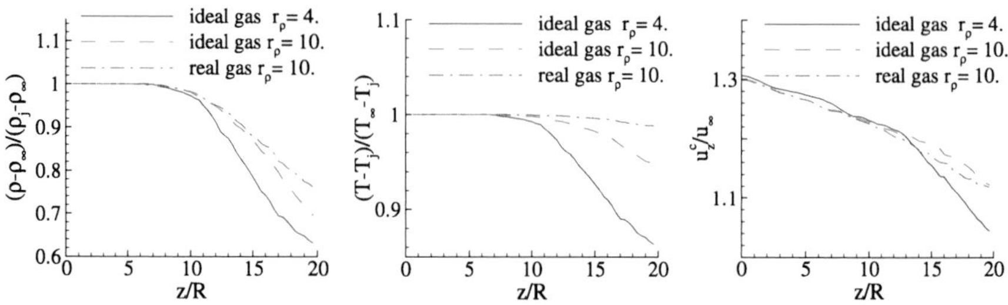

Fig. 2 Mean axial profiles of density (left panel) normalized with the injection and surroundings density $(\rho - \rho_j)/(\rho_{ext} - \rho_j)$, normalized centerline temperature (middle panel) $(\theta - \theta_{ext})/(\theta_j - \theta_{ext})$ and normalized centerline axial velocity u_z^c/u_∞

equal to $Re_D = U_0 D/\nu_\infty = 6000$, with U_0 the bulk velocity. The thermodynamic parameters mimic one of the experiments with the cold jet of nitrogen in nitrogen environment performed by *Mayer et al.* in [1].

The computational domain, $[\phi_{max} \times R_{max} \times Z_{max}] = [2\pi \times 6.2D \times 10D]$ is discretized by $N_\phi \times N_r \times N_z = 128 \times 201 \times 600$ nodes with a stretched mesh in the radial direction to assure a resolved shear layer. The grid size in the jet region is about two/three times the Kolmogorov scale, and is able to accurately capture the strong density gradients which develops in the flow, [10]. After reaching the statistically steady state, about 200 complete fields, separated by $0.125\,D/U_0$, are collected for statistical analysis.

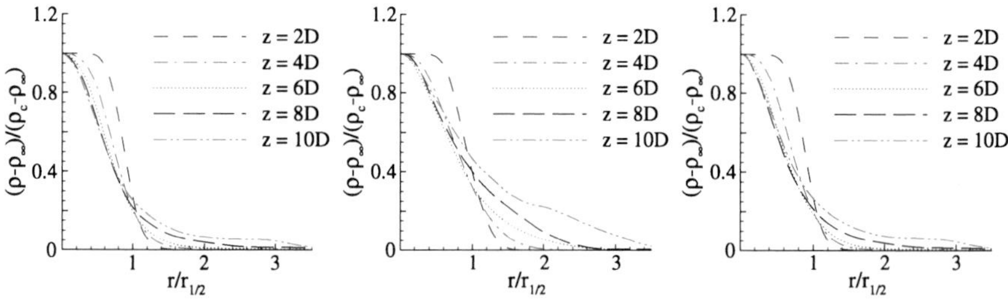

Fig. 3 Mean radial profiles of density normalized with the centerline and surroundings density, $(\rho - \rho_{ext})/(\rho_c - \rho_{ext})$. The radial coordinate is normalized with the radial distance where the averaged velocity is the half of the corresponding centerline value. From left: real gas with $\rho_j/\rho_{ext} \simeq 10.$, perfect gas with $\rho_j/\rho_{ext} \simeq 4.$, perfect gas with $\rho_j/\rho_{ext} \simeq 10.$

In fig. 1 the instantaneous field of normalized temperature is shown for the three simulations. Apparently the density ratio has a larger influence on the jet than the temperature ratio. The core of the jet in the *S*2 case (middle panel) presents in fact a temperature lower that the real gas jet (left panel). Also the perfect gas jet (right panel) with the same density ratio (*S*3) of the real gas flow, has an instantaneous temperature lower though the difference is sensibly smaller than the previous case

$S2$ vs $S1$. The temperature difference in the latter case is probably associated with the thermodynamic properties of the real gas near the critical point.

3 Results

Fig. 2 reports the axial profiles of the Reynolds averaged centerline density (left panel), temperature (middle panel) and velocity (right panel). These mean quantities confirm that the density ratio determines the jet dynamics more than the temperature ratio. The decay of density and normalized temperature is more rapid in configuration $S2$. The behavior in the dense-gas simulation, $S1$, and for the ideal gas at matching density ratio, $S3$, is quite similar although in the far field the ideal gas jet gives a weaker reduction of both density and temperature. In the velocity plots, for $S2$ a curve slope variation occurs at $z \simeq 12D$, while the other jets show a liquid-like behavior. Figs. 3 and 4 provide the mean density and velocity profiles,

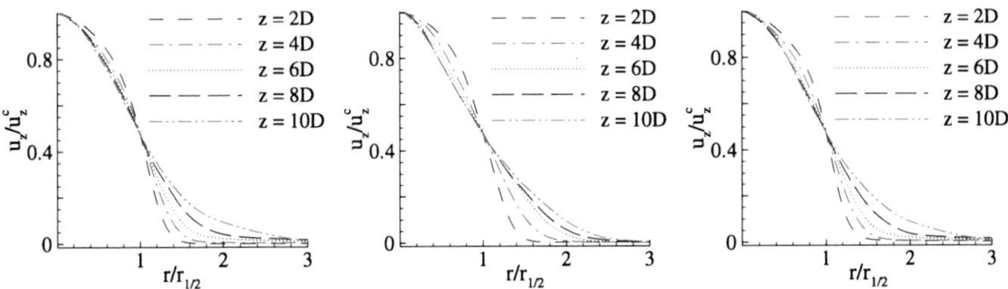

Fig. 4 Mean radial profiles of axial velocity normalized with the centerline mean velocity, u_z/u_z^c. The radial coordinate is normalized with the radial distance where the averaged velocity is the half of the corresponding centerline value. From left: real gas with $\rho_j/\rho_{ext} \simeq 10.$, perfect gas with $\rho_j/\rho_{ext} \simeq 4.$, perfect gas with $\rho_j/\rho_{ext} \simeq 10.$

respectively, as a function of the radial coordinate at five different axial locations. Mean quantities are normalized by their centerline and surroundings values, while the radial coordinate is normalized by the jet half-width $r_{1/2}$, radial distance where the mean axial velocity is the half of the corresponding centerline value. The density profiles, fig. 3, highlight once more that the density ratio plays a crucial role in the jet dynamics. Actually the simulations $S1$ and $S3$ exhibit a similar behavior also of the radial profiles, while the $S2$ differs considerably. In addition, only the density profiles of $S1$ and $S3$ seem to collapse on a unique self-similar curve at small axial distances, $z = 8 - 10D$, presenting noticeable differences for radial distances larger than a couple of $r_{1/2}$. The existence of self-similarity is observed also in the $2D$ LES provided in [4] even at axial distances larger than the present distances. The axial velocity profiles, fig. 4, tend to collapse on a unique self-similar curve in all jets, also for that with the density ratio of 4 ($S2$, middle panel). Nonetheless, also for the velocity appreciable differences emerge at the edge of the jet, more than $r_{1/2}$.

4 Conclusions

A numerical algorithm to deal with Low-Mach number supercritical flows is proposed and used to perform a DNS of real gas turbulent jet in cryogenic conditions. The effect of the real gas EOS is analyzed comparing the results with two simulations of ideal gas matching the density ratio in the first and the temperature ratio in the other. The real gas jet presents a liquid-like behavior showing a density/temperature core less influenced by the surrounding environment, in respect to ideal jets. Hence the mixing in real gas jets appears reduced with respect to ideal conditions. Besides the mean axial velocity and density profiles show a tendency to a self-similar behavior as in the incompressible limit. Nonetheless, differences in the outer region of the jet emerge that could disappear at larger axial distances.

Acknowledgements The work was supported by the Standard HPC-2010 Grant n. std10-284 for providing CPU time and storage at CASPUR High Performance Computing center.

References

1. Mayer, W., Schik, A., Schweitzer, C. & Schaffler, M. (1995). Injection and mixing processes in high pressure LOX/GH2 rocket combustion. Proceeding of AIAA, 16
2. Chehroudi, B., Tally, D. & Coy, E. (2002). Visual characteristics and initial growth rate of round cryogenic jets at sub-critical and super-critical pressures. Physics of Fluids 14:850–861
3. Oschwald, M. (1999). Supercritical nitrogen free jet investigated by spontaneous Raman scattering. Experiments in Fluids 27:497–506
4. Zong, N., Meng, H., Hsieh, S.Y. & Yang, V. (2004). A numerical study of cryogenic fluid injection and mixing under supercritical conditions. Physics of Fluids 16:4248–4261
5. Schmitt, T., Selle, L., Cuenot, B. & Poinsot, T. (2009). Large-Eddy Simulation of transcritical flows. Comptes Rendus Mécanique 337, 6–7:528–538
6. Zong, N. & Yang, V. (2006). Cryogenic fluid jets and mixing layers in trans-critical and supercritical environments. Comb. Science and Technology 178:193–227
7. Majda, A. & Sethian, J. (1985). The derivation and numerical solution of the equations for zero Mach number combustion. Combustion Science and Technology 42, 3:185–205
8. Müller, B. (1998). Low-Mach-number asymptotics of the Navier-Stokes equations. Journal of Engineering Mathematics 34, 1:97–109
9. Picano, F. & Casciola, C.M. (2007). Small-scale isotropy and universality of axisymmetric jets. Physics of Fluids 19:118106
10. Picano, F., Battista, F., Troiani, G. & Casciola, C.M. (2010). Dynamics of PIV seeding particles in turbulent premixed flames. To appear online. Experiments in Fluids. doi:10.1007/s00348-010-0896-y

DNS Study on Control of Turbulent Heat Transfer in Curved Channel

Takashi Uchida, Koji Matsubara, Takahiro Miura and Atsushi Sakurai

1 Introduction

Studies of curved channel flows are clearly important since the flow over curvatures is related to various applications such as turbines, heat exchangers and combustors. Since Wattendorf [1], many works have been made for the flow instability near the concave wall, and it was pointed out that the organized flow called Taylor-Görtler vortex grew due to the centrifugal force.

Moser and Moin [2] directly solved the Navier-Stokes equations for the curved channel flow at the radius ratio of 0.975, and reported various statistics including the budgets of Reynolds stresses. However, they assumed streamwise periodicity, and the growth of the vortex was left ambiguous. Kobayashi et al. [3] experimented on spatial advancement of curved channel flow. However, they did not report turbulence structures due to limitation of experimental technique.

This study is an extension of the work by Matsubara et al. [4]; the spatial advancing direct numerical simulation (DNS) was performed for curved channel flows and the related heat transfer. Purpose of the present study is to reveal effects of a rectangular rib placed in a straight connecter introducing fully developed turbulence into the curved channel. Flow and heat transfer characteristics are discussed for the cases

Takashi Uchida
Graduate School of Science and Technology, Niigata University, Ikarashi 2-nocho 8050, Niigata-shi, 950-2181, Japan, e-mail: f09b127g@mail.cc.niigata-u.ac.jp

Koji Matsubara
Department of Mechanical and Product Engineering, Niigata University, Ikarashi 2-nocho 8050, Niigata-shi, 950-2181, Japan, e-mail: matsu@eng.niigata-u.ac.jp

Takahiro Miura
Graduate School of Science and Technology, Niigata University, Ikarashi 2-nocho 8050, Niigata-shi, 950-2181, Japan, e-mail: f09k008c@mail.cc.niigata-u.ac.jp

Atsushi Sakurai
Department of Mechanical and Product Engineering, Niigata University, Ikarashi 2-nocho 8050, Niigata-shi, 950-2181, Japan, e-mail: sakurai@eng.niigata-u.ac.jp

H. Kuerten et al. (eds.), *Direct and Large-Eddy Simulation VIII*,
ERCOFTAC Series 15, DOI 10.1007/978-94-007-2482-2_49,

where a rib is placed on either an inner or outer wall, and for the case of no rib on the walls.

2 Description of Numerical Procedure

2.1 Configurations

The computational domain and the coordinate system are shown in Figure 1. A rectangular rib is placed on the outer wall or the inner wall of the straight part connecting to the curved part. Fully developed turbulence is introduced through the straight part. The inlet condition is generated by the driver with assuming the streamwise periodicity, and the spatial development of flow and temperature are simulated in the curved domain. For both the driver and the curved domain, periodic condition is applied for the spanwise direction, and walls are treated as non-slip. Thermally, temperature difference is given to the inner walls of the curved domain and the driver against the outer walls, and rib surfaces are heated by the same temperature as the ribbed wall. At the exit of the curved domain, convection outflow condition is used for flow and temperature. In all the simulations, the radius ratio of the curved domain, α, is 0.92, and the cross-section of the rib is $0.25\delta \times 0.20\delta$ where δ means the channel half width. As listed in Table 1, the computational domain has the volume $(3.9+66.7)\delta \times 2\delta \times 7.2\delta$, and $(32+544) \times 61 \times 128$ grid points are distributed to it. Air is assumed as the working fluid, and the Prandtl number, Pr, is prescribed at 0.71. The Reynolds number based on the inlet frictional velocity, $U_{\tau 0}$, and the channel half width, $Re_{\tau 0}$, is set at 150. The resulted mean velocity Reynolds number with the length scale 2δ is 4560, which corresponds to about one fourths of Kobayashi et al.'s experiment [3]

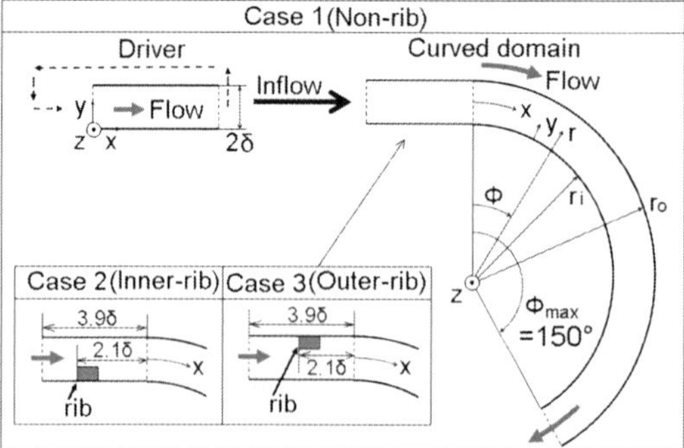

Fig. 1 Computational domain. Fully developed turbulence, generated by the driver with streamwise periodicity, is used as the inlet value of the curved domain.

Table 1 Computational conditions

	$\alpha = r_i/r_o$	N_x	N_y	N_z	L_x/δ	L_y/δ	L_z/δ
Driver	1.00	32	61	128	3.9	2	7.2
Curved domain	0.92	544	61	128	66.7	2	7.2

2.2 Numerical Method

Governing equations used in simulations are the continuity, Navier-Stokes and energy equations for the incompressible Newtonian fluid. Using the curvilinear coordinates (x, y, z), the energy equation can be written as

$$\frac{\partial \Theta^*}{\partial t^*} + \frac{\sigma+1}{y^*+\sigma} U^{++} \frac{\partial \Theta^*}{\partial x^*} + V^{++} \frac{\partial \Theta^*}{\partial y^*} + W^{++} \frac{\partial \Theta^*}{\partial z^*} = \frac{1}{Re_{\tau 0} Pr} \{ (\frac{\sigma+1}{y^*+\sigma})^2 \frac{\partial^2 \Theta^*}{\partial x^{*2}} + \frac{\partial^2 \Theta^*}{\partial y^{*2}} + \frac{1}{y^*+\sigma} \frac{\partial \Theta^*}{\partial y^*} + \frac{\partial^2 \Theta^*}{\partial z^{*2}} \}, \quad (1)$$

where variables are non-dimensionalized by the inlet frictional velocity, the channel half width and the temperature difference, ΔT_w. (x, y, z) corresponds to Cartesian coordinates in the straight parts of the computational domain, and cylindrical coordinates in the curved part. In the curved part, the location of (x, y, z) can be expressed by the angle, ϕ, radial coordinate, r, and the axial coordinate, z. The relationship between (x, y, z) and (ϕ, r, z) can be described by

$$x = \frac{r_i + r_o}{2} \phi, \; y = r - r_i, \; z = z, \quad (2)$$

where r_i and r_o are the curvature radius of the inner wall and that of the outer wall, respectively. Equation (1) corresponds to the Cartesian formula for $\sigma (= 2\alpha/(1-\alpha)) \to +\infty$ $(\alpha \to 1.0)$.

The fractional step method is used for coupling of the continuity and the Navier-Stokes equations. In the time splitting, the second-order Crank-Nicolson scheme is applied for the wall-normal second derivatives, and the second-order Adams-Bashforth method for other terms. As the spatial discretization, fourth-order central difference is used for the convection and diffusion terms. For presentation of statistics, numerical variables are averaged along the spanwise coordinate and the time using a time interval of $40\delta/U_{\tau 0}$.

3 Result

3.1 Nusselt Number

Figure 2 shows distributions of the Nusselt number, Nu. It can be written as

$$Nu = \frac{r_i q_{w_i} + r_o q_{w_o}}{r_i + r_o} \frac{4\delta}{\Delta T_w \lambda}, \tag{3}$$

which is averaged between the inner and outer walls with weighing according to the surface broadness. The weighted averaging procedure accounts for the difference in the extent of walls in each side. In the definition of Nu, q_{w_i} and q_{w_o} are the heat flux on inner and outer walls, respectively, and λ is the fluid thermal conductivity.

In the case of no rib, the Nusselt number slightly increases in the curved channel, and the organized flow is suggested to enhance the heat transfer. In the case of a rib on the inner wall, the Nusselt number is dramatically increased over $0° < \phi < 30°$, and it returns mostly to non-rib values for $\phi > 70°$. In the case of the outer-rib, the heat transfer enhancement around $0° < \phi < 30°$ is not so large as the inner-rib case. However the enhancement continues to appear even far from the rib; $\phi > 70°$.

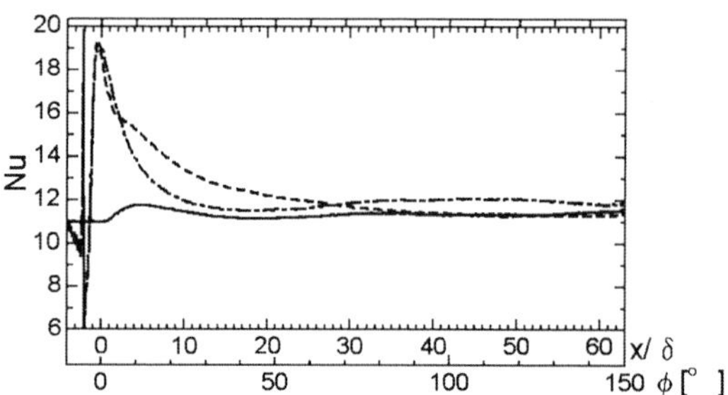

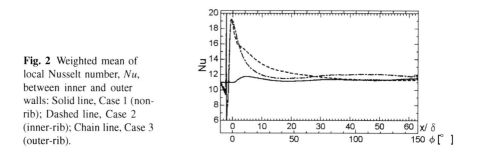

Fig. 2 Weighted mean of local Nusselt number, Nu, between inner and outer walls: Solid line, Case 1 (non-rib); Dashed line, Case 2 (inner-rib); Chain line, Case 3 (outer-rib).

3.2 Mean Velocity Vectors

From the distribution of mean velocity and turbulence intensity, it was suggested that the heat transfer enhancement by the rib was attributed to the mean flow change by the blockage effect and the intensified fluctuation. Since the organized flow of the vortex contained the majority of the turbulent fluctuations (Matsubara et al. [4]), such a flow was thought to have a great impact on increasing heat transfer.

Figure 3 shows time-mean velocity vectors on $y-z$ plane of $\phi = 75°$. In the figure, large-scale vortices are clearly depicted by filtering out the random fluctuations. In the case of the inner-rib, the vortices are weak, and the rib is thought to attenuate the growth of the vortex. In the case of the outer-rib, the vortices are stronger than the non-rib case, and the outer rib is implied to assist the vortex to grow faster.

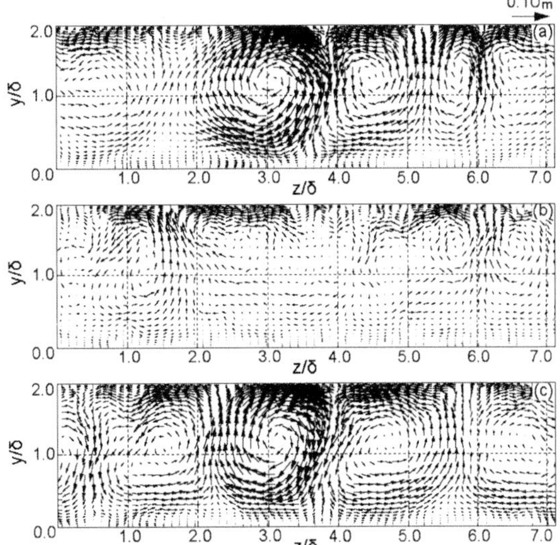

Fig. 3 Mean velocity vectors on $y-z$ plane at $\phi = 75°$. (a) Case 1 (non-rib); (b) Case 2 (inner-rib); (c) Case 3 (outer-rib).

3.3 Power Spectra Density

Figure 4 shows contour lines of the spanwise pre-multiplied power spectra of wall-normal velocity, $k_z E_{vvz}/U_m^2$, at $\phi = 75°$. The figure shows distribution against the

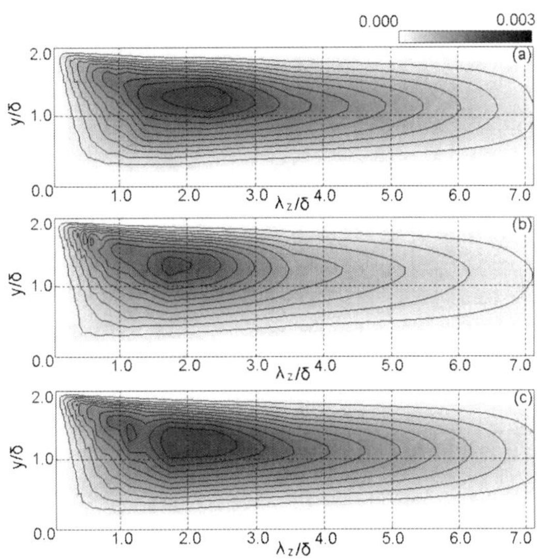

Fig. 4 Contour lines for spanwise pre-multiplied power spectra of wall-normal velocity at $\phi = 75°$, $k_z E_{vvz}/U_m^2$. (a) Case 1 (non-rib); (b) Case 2 (inner-rib); (c) Case 3 (outer-rib).

wall-normal coordinate and the spanwise wave length, λ_z. In the figure, the organized flow is suggested to have a spanwise extent of about 2.0δ, which is equivalent to the full width of the channel. From the figure, it is observed that this organized flow becomes weaker in the inner-rib case but this increases the intensity in the outer-rib case. From this result and the mean velocity, the heat transfer enhancement in the outer-rib case is suggested to occur due to the intensified flow of the organized vortex which can strongly mix the thermal non-uniformity inside the channel through its powerful transport by the wall-normal convection.

4 Conclusion

1. When a rib is attached to the upstream part of a curved channel, heat transfer in the channel is enhanced by the rib effects. Although the heat transfer enhancement vanishes in the latter part of the curved channel for the inner-rib case, heat transfer is continuously good through the channel for the outer-rib case.
2. The time-mean velocity vectors suggested that the organized flow of the large-scale vortex is attenuated by the inner-rib case but is strengthened by the outer-rib case. The pre-multiplied power spectra of the wall-normal velocity showed such trends quantitatively. In the spectra, turbulence energy changed due to attachment of a rib around the wave length mostly equivalent to the scale of a channel width.
3. The changes in the turbulence energy were consistent with the trends of the heat transfer. The heat transfer enhancement in the outer-rib case thus is concluded to occur due to the strengthened fluctuation of the organized flow which can activate the scalar transport normal to the wall.

References

1. Wattendorf, F. L., A study of the effect of curvature on fully developed turbulent flow, Proc. R. Soc., 148, pp. 565–598 (1935).
2. Moser, R. D. and Moin, P., The effects of curvature in wall bounded turbulent flows, J. Fluid Mech., 175, pp. 479–510 (1978).
3. M. Kobayashi, H. Maekawa, Y. Shimizu and K. Uchiyama, An experimental study on a turbulent flow in a two-dimensional curved channel, Transactions of the Japan Society of Mechanical Engineers, Series B, 58-545, pp. 119–126 (1992), in Japanese.
4. K. Matsubara, A. Matsui, T. Miura, K. Kawai and M. Kobayashi, Multi-scale coherent structures of spatially advancing turbulent flows in curved channel, Sixth International Symposium on Turbulent Shear Flow Phenomena, 1, pp. 29-34 (2009).

A Priori Assessment of the Potential of Flamelet Generated Manifolds to Model Lean Turbulent Premixed Hydrogen Combustion

A. Donini, R.J.M. Bastiaans, J.A. van Oijen, M.S. Day and L.P.H. de Goey

1 Introduction

The numerical modeling of combustion systems is a very challenging task. The interaction of turbulence, chemical reactions and thermodynamics in reacting flows is of exceptional complexity. Computing power is too limited to solve practical problems in detail. This problem asks for special treatments in the modeling of flames.

We developed the Flamelet Generated Manifold (FGM) technique [1, 2], which combines advantages of chemistry reduction and flamelet models. The approach is based on the idea that the most important aspects of the internal structure of the flame fronts should be taken into account. In the FGM technique the progress of the flame is generally described by one (or at most a few) progress variable(s) Y, for which a transport equation is solved during run-time. The chemical source term $\dot{Y}$ in the transport equation for Y is derived from the flamelet system. The flow and mixing of elements and enthalpy is described by equations for the enthalpy h and elements Z_j, which are independent of the chemical kinetics. The flamelet system is solved in a pre-processing step for each variable Y, h and Z_j. The corresponding solution for the source term $\dot{Y}$ depends only on Y, h and Z_j which is stored in a data-base. In the CFD, only equations for Y, h and Z_j are solved using the data-base to retrieve all necessary information to update the solution. The data-base is called a manifold. Also, of course, the temperature T and all species variables Y_i can be retrieved from the manifold.

We successfully combined this approach with both DNS and LES for turbulent methane combustion. However combustion of hydrogen rich fuels has become a very important research item. Therefore analysis of DNS with detailed chemistry and comparison with flamelets can give good indications of the predictive powers

A. Donini, R.J.M. Bastiaans, J.A. van Oijen, L.P.H. de Goey
Section Combustion Technology, Department of Mechanical Engineering, Eindhoven University of Technology, P.O. Box 513, 5600 MB Eindhoven, The Netherlands, e-mail: a.donini@tue.nl
M.S. Day
CCSE, Lawrence Berkeley National Laboratory, 1 Cyclotron Road, Berkeley, CA 94720, USA

H. Kuerten et al. (eds.), *Direct and Large-Eddy Simulation VIII*,
ERCOFTAC Series 15, DOI 10.1007/978-94-007-2482-2_50,

of FGM. To this aim we investigated the probability density distributions of the hydrogen source term by processing the DNS data of Aspden et al [3], in which detailed chemistry is used, compared with a set of perturbed flamelet solutions [4].

2 Description of the numerical method

We will investigate the probability density distribution by processing the data of Aspden et al [3]. These simulations are based on a low Mach number numerical formulation of the reacting flow equations. A mixture-averaged model for species diffusion is used and the transport coefficients, thermodynamic relationships and hydrogen kinetics are obtained from the GRI-Mech 2.11 model [5] with carbon species removed. The integration algorithm has local grid refinement and is second-order accurate in space and time. The method is presented and tested in [6].

The configuration, schematically represented in Figure 1(a), consists of a rectangular domain with a 3 cm square cross-section, and a height of 9 cm, oriented such that the flame propagates downward. The freely propagating flame is initialized at the top, while the cold fuel mixture ($T = 298$ K, H2-air, $\phi = 0.31$ and 0.37) was specified at the square bottom boundary. The simulations were performed with a base grid of $128\times128\times384$, using two levels of factor-of-two refinement that dynamically tracked the region of high reactivity as the flame evolved into a quasi-steady configuration. The effective resolution at the flame surface was maintained at 58.6 microns.

In the flamelet regime of unstable or turbulent combustion laminar, though perturbed, flames are assumed to exist. Reference flamelet solutions (see [4] for a complete description of flamelet equations) were computed with the same mechanism using the CHEM1D code [7]. By applying stretch and/or changing equivalence ratio changes due to local perturbations are taken into account.

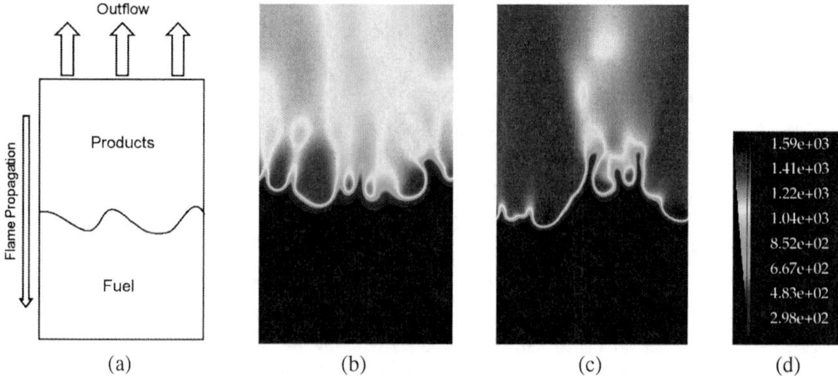

Fig. 1 (a) - Schematic representation of the domain. (b) - Vertical slice of the domain showing the temperature field at $\phi = 0.31$. (c) - Temperature field at $\phi = 0.37$. (d) - Temperature palette, [K].

3 Results

Figure 1(b) and 1(c) show the temperature field in a vertical section of the simulation. We clearly see how the flat-laminar flame is distorted and broken by thermo-diffusive instabilities. Preferential diffusion leads hydrogen to spread into hot regions, which will therefore burn more intensely than the flat laminar flame. However, looking at the temperature and mass fraction fields closely, it is still possible to identify perturbed laminar structures. It appears that velocity gradients near the flame front have a strong influence, leading us then to identify flame stretch as a cause of local fluctuations in consumption rate from the laminar unperturbed flamelet value. The analysis of joint probability density distributions (*JPDF*s) of the

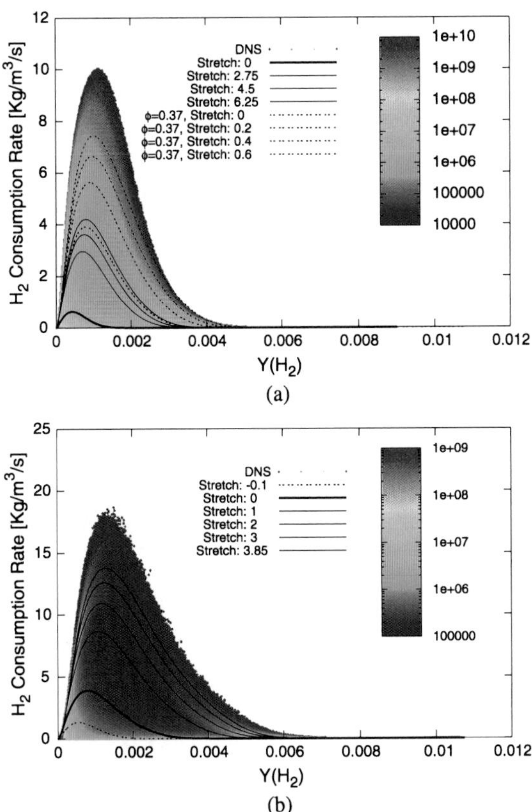

Fig. 2 Joint probability density distributions of the mass fraction of hydrogen against its consumption rate, comparison with one-dimensional simulations at different stretches. Cases at $\phi = 0.31$ (a) and $\phi = 0.37$ (b).

local amounts of fuel versus its consumption rate is a way to compare the DNS data with the results from flamelet calculations. In Figure 2 we can see the *JPDF*s for the two equivalence ratio cases. In an enlarged view (not shown) we can see how in both cases the probability shows high probability zones near the origin and in the zone of high mass fraction and very low consumption rate. These values are characteristic respectively of zones in the burnt region and in the unburnt, but both far from the flame active region. More interesting are the central regions, where we can clearly see a "band" of more probable trend that have a noteworthy similarity with a laminar flamelet. This interesting feature suggests us a direct comparison with the laminar data. Therefore a set of one-dimensional steady laminar simulations were computed using the CHEM1D code. The simulations included a set of unstretched flames across a range of inlet fuel mixtures, and a set of steady 1D stretched flames at the two mixtures, 0.31 and 0.37.

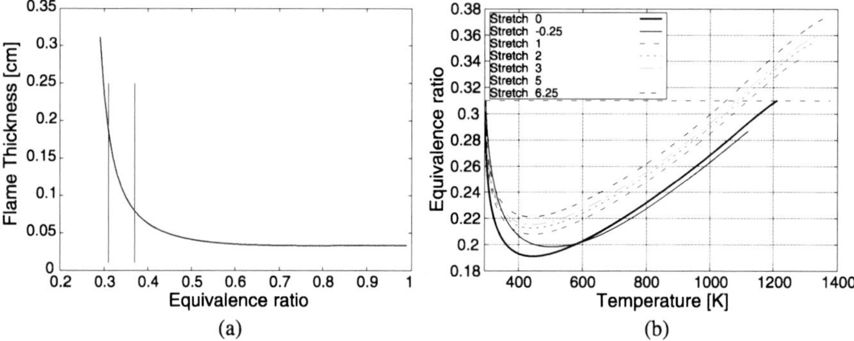

Fig. 3 Results of flamelet calculations: (a) - Laminar flame thickness as function of the equivalence ratio. (b) - Laminar local equivalence ratio as function of temperature through the flame, every slope is a different rate of stretch (inlet $\phi = 0.31$).

In Figure 2 we examine the direct comparison of DNS and perturbed one-dimensional flamelets at different stretch rates (curvature seems to affect the results in a much less effective magnitude). Here the stretch is non-dimensionalized by the flat laminar values of flame thickness δ_L ($\delta_L = 0.182$ and 0.078 cm, respectively for $\phi = 0.31$ and $\phi = 0.37$) and flame speed s_L ($s_L = 4.91$ and 15.64 cm/s, respectively). We can see how a good fit of the high-probability areas is achieved for relative high stretch rates for $\phi = 0.37$, Figure 2(b). However, an important difference that we notice in the case of $\phi = 0.31$ is that not all the DNS data is covered by the stretched laminar data. The observation of the values of stretch necessary to fill the high probability DNS data is also noteworthy, indeed we discover that much higher values are shown for the case at $\phi = 0.31$. Both these effects are caused by preferential diffusion, which generates zones with locally modified equivalence ratio, where hydrogen lean flames behavior is significantly affected.

Regions of negative stretch are present in the important cusp regions. It appears that only with limited negative stretch stable stationary solutions can be obtained by solving the flamelet equations. Larger values of the (negative) stretch lead to instationary flamelets; the flame extinguishes. This is an important region of the *JPDF* since there are a relative large number of observations. Indeed these zones

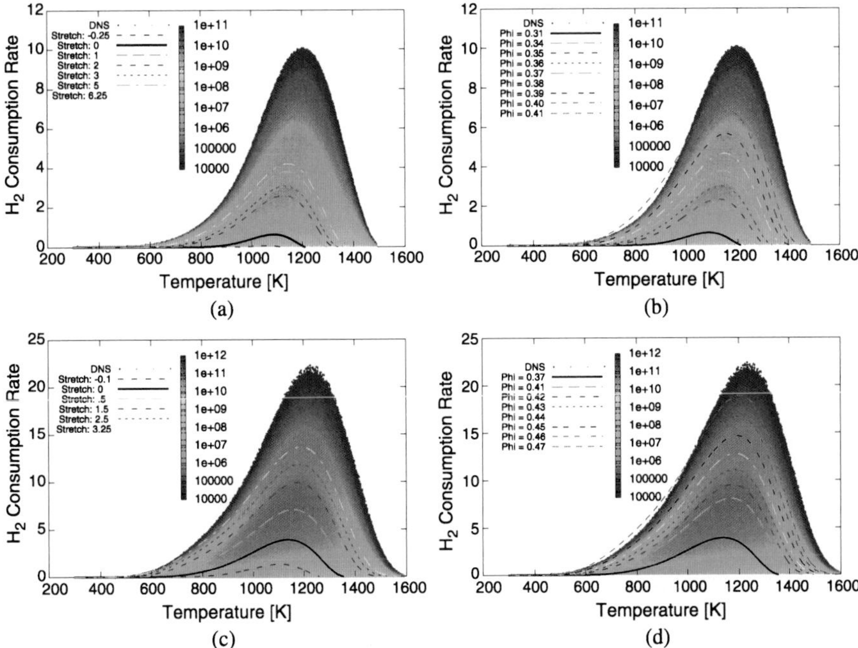

Fig. 4 Joint probability density distributions of temperature against hydrogen consumption rate. Cases at ϕ = 0.31 (a and b) and ϕ = 0.37 (c and d). In the left plots the *JPDF*s are in comparison with one-dimensional simulations at different stretch rate, while in the right comparison with different equivalence ratios.

can probably not be captured by steady flamelets. However it is demonstrated that the statistics of the unstable flame propagation are still well captured by stationary flamelets [8].

Additionally, looking at the flame thickness as function of equivalence ratio, Figure 3(a), we see that the basic flame at the lower equivalence ratio, ϕ = 0.31, is much thicker than the situation for ϕ = 0.37. This allows a more intense turbulence-chemistry interaction.

The evaluation of the local equivalence ratio along the laminar simulations is also of significant importance, defining it as $\phi_{loc} = s \cdot (Z_H/Z_O)$, where s is the mass stoichiometric ratio (in this case $s = 7.937$), and Z_j is the elemental mass fraction of element j. From Figure 3(b) we can first recognize the effect of preferential diffusion, that leads to a lower equivalence ratio through the flame, but more importantly we can see how stretch locally increases the value of equivalence ratio, leading to wide super-adiabatic zones. This important effect indeed gives rise to an improved burning intensity in those zones that are affected from a higher hydrogen concentration, and in fact in Figure 2(a) we can verify that laminar flames with increased equivalence ratio can also cover the DNS data.

To have a better assessment of these important zones, the evaluation of *JPDF*s of hydrogen consumption rate against temperature is useful, as shown in Figure 4.

We can clearly see from Figure 4(c) that in the case of $\phi = 0.37$ the important super-adiabatic zone is well covered by the laminar stretched simulations. Conversely, at $\phi = 0.31$ (Figure 4(a)), the perturbed laminar data is not able to cover properly the DNS results, and only higher (and unrealistic) values of stretch would be able to do it. This leads us to believe that the preferential diffusion effect in this case is so high to strongly modify the local equivalence ratio. Therefore, a curious feature is shown in Figure 4(b), where the DNS data is compared with one-dimensional simulations at increasing inlet equivalence ratio. We can notice how a perfect fit is achieved in the most important super-adiabatic region just simulating the strong variation of local equivalence ratio due to preferential diffusion. This can also be noticed in the case at $\phi = 0.37$ (Figure 4(d)). Therefore a final observation is that the DNS behavior can be covered either by flamelets with higher equivalence ratio than the inlet, or by strongly stretched flamelets.

4 Conclusions

In the present contribution, we describe the analysis for a future extension of FGM for the modeling of premixed flames with DNS including non-unit Lewis numbers to predict preferential diffusion effects. With this purpose we analyze the data of a DNS of lean premixed hydrogen combustion, investigating the probability density function of the hydrogen consumption rate. It is shown the one-dimensional results with increasing stretch rate or equivalence ratio can mimic in the correct way most of the DNS data. The essential extinguishing behavior at negative stretch in cusp regions has to be studied further.

References

1. van Oijen, J.A. & de Goey, L.P.H. (2000). Modelling of premixed laminar flames using flamelet-generated manifolds. *Combust. Sci. Technol.* **161**, 113–131.
2. van Oijen, J.A., Bastiaans, R.J.M. & de Goey, L.P.H. (2007). Low-Dimensional Manifolds in Direct Numerical Simulations of Premixed Turbulent Flames. *Proc. Combust. Inst.*, **31**, 1377–1384.
3. Aspden, A.J., Day, M.S. & Bell, J.B. (2010). Characterization of Low Lewis Number Flames. *Proceedings of the Combustion Institute.* **33**.
4. de Goey, L.P.H. & ten Thije Boonkkamp, J.H.M. (1999). A Flamelet Description of Premixed Laminar Flames and the Relation with Flame Stretch. *Combust. Flame.* **119**(3), 253–271.
5. Bowman, C.T., et al., GRI-Mech 2.11. `http://www.me.berkeley.edu/gri_mech`.
6. Day, M.S. & Bell, J.B. (2000). Numerical Simulation of Laminar Reacting Flows with Complex Chemistry. *Combust. Theory Modelling.* **4**, 535–556.
7. CHEM1D. `http://www.combustion.tue.nl/chem1d`.
8. Bastiaans, R.J.M. & Vreman, A.W. (2010). Numerical simulation of instabilities in lean premixed hydrogen combustion. *Submitted to Int. J. Numer. Method. H.*

Numerical Analysis of a Swirl Stabilized Premixed Combustor with the Flamelet Generated Manifold approach

T. Cardoso de Souza, R.J.M. Bastiaans, B.J. Geurts and L.P.H. De Goey

1 Introduction

In this paper the effectiveness of LES for modeling premixed methane combustion will be investigated in the context of gas turbine modeling. The required reduction of the chemistry is provided by the flamelet generated manifold (FGM) approach of van Oijen [1]. For turbulence-chemistry interactions an algebraic model is used to calculate variations which are used to invoke a pre-assumed pdf, Vreman et al. [2]. The algebraic model has a tunable parameter.

Academic cases have been researched with this model so far and now we assess the potential of the model for turbine combustors, specially the role of the algebraic model, which depends on the appropriate choice related with the tunable parameter. One of the main objectives is to predict the turbulent flame brush at the right location. Since the approach is consistent with a RANS approach, also these kind of simulations (Realizable $k-\varepsilon$) are presented. It follows that the location of the flame brush is predicted too much upstream. This is consistent with a too low value of the parameter, which was also suggested in [2].

The paper starts with a short description of the method and physical problem. Then the manifold is discussed following a presentation of RANS and LES results. The paper ends with conclusions.

2 Combustion Modeling

The flamelet generated manifold (FGM) technique [1] makes use of correlations of species to reduce the number of equations, in the sense that the combustion process

Eindhoven University of Technology, P.O. Box 513, 5600 MB Eindhoven, The Netherlands, e-mail: t.cardoso.de.souza@tue.nl
B.J. Geurts: Twente University, P.O Box 217, 7500AE Enschede, The Netherlands

H. Kuerten et al. (eds.), *Direct and Large-Eddy Simulation VIII*,
ERCOFTAC Series 15, DOI 10.1007/978-94-007-2482-2_51,

is described in terms of a few control variables. The strength of the FGM reduction technique is that the number of independent control variables, starting with a single reaction progress variable c, can be increased for a better description of the chemical phenomena.

At the present lean condition, it is assumed that the progress variable can be sufficiently described by carbon dioxide, CO_2. To check this assumption, thermo-chemistry variables were plotted as function of CO_2 and indeed these variables can be described without ambiguity using just carbon dioxide as the reaction progress variable. Due to the limited space, these plots will not be shown here.

In the FGM technique the real source term can be used, instead of its definition based on empirical correlations as in [3]. The flamelet generated manifold uses a physical laminar source term directly retrieved from a laminar flamelet database generated from a one-dimensional laminar flame calculation. More information about the FGM procedure can be found in van Oijen [1].

If the purpose is to understand the physical aspects of reacting flows in high swirl flow conditions, and therefore combustion occurring in a turbulent flow field, the interactions between chemistry and turbulence must be considered properly. In this approach the flamelet approach combined with a pre-assumed PDF closure method for the chemical source term will be used in order to describe the turbulence-chemistry interactions. For this we will use a commonly used β-PDF function. In a second step, the laminar manifold will be integrated in terms of this function. Now the variables are tabulated as a function of the mean reaction progress variable $\tilde{c}$ and its respective variance $var(c)$. Therefore a transport equation for $\tilde{c}$ and an algebraic equation for $var(c)$ are taken into account, as in Vreman et al. [2]. Here the progress variable is defined as the scaled mass fraction of carbon dioxide $c \sim Y_{CO_2}$ and the variance is taken as

$$var(c) = \frac{a^2 \Delta_k^2}{12} \left(\frac{\partial \tilde{c}}{\partial x_k} \right)^2 \tag{1}$$

An estimation related with the computational time between the current method and cold flow simulations without a lookup, gives an expectation of comparable computing times. This since only one additional transport equation will be solved.

In the FGM approach, chemistry is tabulated *a priori* aside from the flow solution, and therefore a fast look-up procedure to retrieve the thermo-chemical variables from the integrated database, which structured nature is known, is possible.

3 Numerical procedure

The code used here is FLUENT 12.0 which is a commercial finite-volume code. To simulate the flow field for the reacting flow case, the turbulence models used were the Reynolds-averaged Navier-Stokes, RANS (Realizable $k-\varepsilon$) and LES, with sub-grid terms modeled using the dynamic Smagorinsky model. The combustor used in the present numerical investigation is referred to as SimVal, as developed by Stakey and Yip [4]. It consists of a slot swirler, a nozzle section and a combustion section that is encased by a 180 mm diam quartz tube followed by an exhaust section. Further details can be found in [4]. Most of the numerical data obtained were collected using a mass flow rate fixed at 0.045 kg/s, which corresponds to a bulk flow velocity in the nozzle section of approximately 40 m/s.

The three-dimensional computational domain encompassed the region from the exit of the swirl plate to 9 cm into the exhaust section and was comprised of approximately 1.305×10^6 hexahedral cells in a multi block structured grid.

The boundary conditions were considered through the same conditions described in [3], thus the individual slots of the swirl plate at the entrance of the domain were modeled as velocity inlets with specified axial and tangential velocity corresponding to a 30° flow angle, an inlet temperature $T_{inlet} = 530K$ and $c = 0$ for the scalar (progress variable).

The fluid used was a fully premixed mixture (CH_4/air) at an equivalence ratio $\phi = 0.6$, and the kinematic viscosity was assumed to be constant and based on the preheat temperature.

The time-step used for the LES simulations was 5μs, and the total time of simulation 145.3 ms., which corresponds to approximately 36 flow-through times. The discretization scheme used was a second-order bounded central differencing scheme in space and a second-order backward differencing scheme for the temporal discretization. The scalar equations were discretized with a second-order upwind scheme. We have taken a to be equal to 1.

4 Results

In order to generate the flamelet database for the present case, in a first step the governing laminar free premixed flame is calculated using the package CHEM1D, developed at Eindhoven Technology University. For this case complex chemistry and transport are applied. For the present work, the laminar flame speed, S_L was calculated using the GRI3.0 mechanism which contains 325 elementary reactions between 53 species. It was found that $S_L = 44.5$ cm/s.

In a second step, turbulence-chemistry interactions are taken into account by integrating the laminar database using the β-PDF function. This results in values of the thermo-chemical variables tabulated as a function of c and $var(c)$, e.g., the source term $\dot{c} = \dot{c}(\tilde{c}, var(c))$. The referred steps are shown in Fig. 1 and for both steps, the manifold generation process and its integration in terms of β-PDF, the routines used were developed by Vreman et al. [2].

In order to see the capability related with FGM to simulate combustion in gas turbines, some first results for the turbulent reactive flow field modeled using *RANS* (Realizable $k - \varepsilon$) and FGM as the combustion model were obtained. Considering that the flame front can be related with intermediate species, such as OH, results for the contours of OH mass fractions and also for the axial velocity along the cutting plane through the center of combustor are shown in Fig. 2.

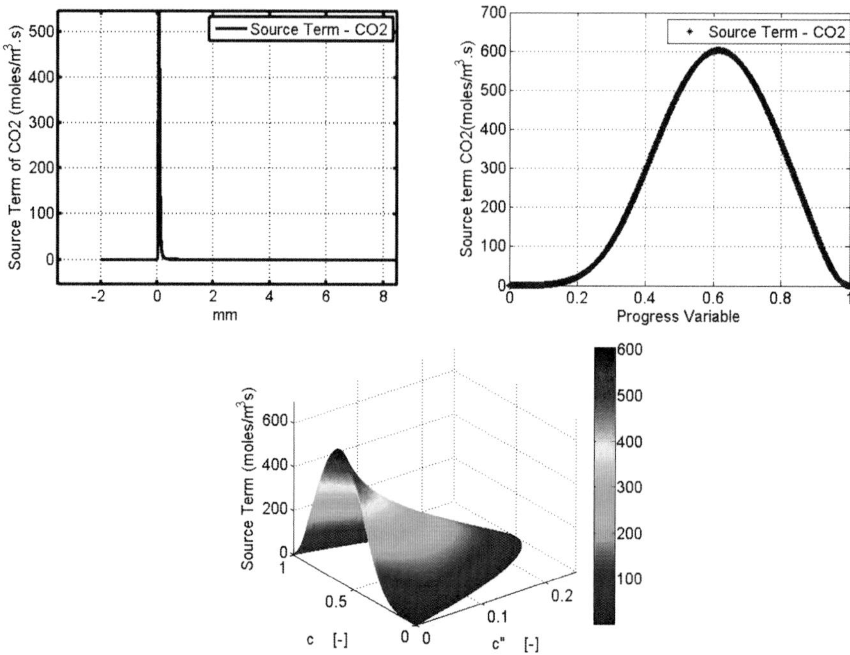

Fig. 1 Upper left: Source term of CO_2 at pressure of 1 bar as a function of the physical space, x. Upper right: Source term of CO_2 as a function of c, $1D$ manifold for the laminar flame (FGM). Lower figure: $2D$ turbulent manifold, source term of CO_2 as a function of c and $var(c)$.

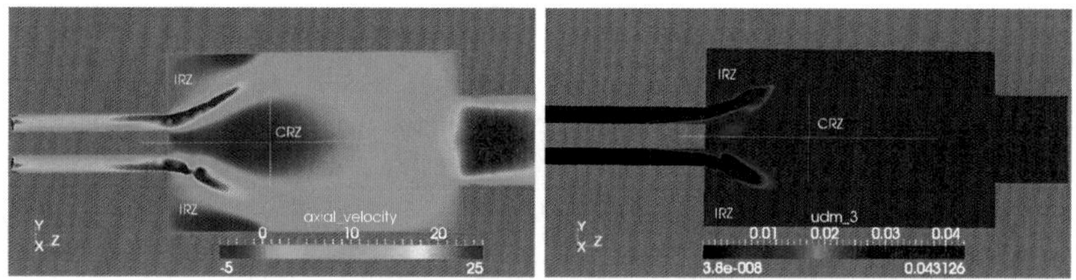

Fig. 2 Results from RANS/FGM. Left: Axial velocity contours. Right: OH contours.

A central recirculation zone (CRZ) and the inner recirculation zone (IRZ) are observed as expected and indicated in the figure. Also, the flame location and shape can be directly observed from the OH contours, and are in agreement with the observed flame shape [3]. Nevertheless, these experimental results show that combustion is located a little further downstream of the combustor dump (entrance) plane.

Considering the intrinsically dynamic nature of the process, LES simulations were also performed in order to verify the effects of unsteadiness and large turbulent scales on the flame front. According to the ratio of flame thickness to the LES filter width, which is of the order of unity, it is possible to conclude that subgrid wrinkling is not important here [2]. Here the flame thickness, $\delta_F \approx 0.6$ mm, is based on the maximum temperature gradient value computed from the previous 1D laminar flamelet calculations. The LES filter width, $\Delta_k \approx 0.75$ mm, is determined from an average of the grid resolution at the combustor dump plane.

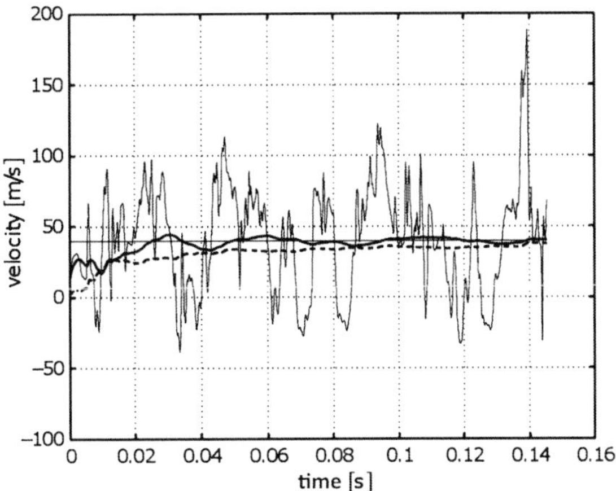

Fig. 3 Axial velocity at a single point at the dump plane, thin line: basic signal, straight drawn thin line: mean axial velocity at the inlet for reference (40 m/s), thick drawn line: running mean, thick dashed line: running rms.

Time series for the axial velocity and for the scalar were monitored in a single point at the combustor dump plane, and the respective cumulative temporal mean and rms of the axial velocity are shown in Fig. 3. It can be visually observed that the results are almost converged with respect to the mean and rms of the axial velocity at the dump plane, being 40.5 and 37 m/s respectively at the end. By analyzing the autocorrelation function we can determine the (Taylor) micro scale to be 0.0015 s and the integral scale to be 0.039 s, which coincides with 97 and 3.7 occurrences in the total simulation time. Note that these values are evaluated from a time series of the LES results. Thus the meaning of these numbers requires careful interpretation. The time history for the progress variable (not shown) shows the unsteady nature related with the mixture alternating between unburnt and burnt. This means that the stabilization of the mean flame comprises the dump plane. In the experiments this is found to be located more downstream. Clearly the present description does not suffice completely.

The effect of the unsteadiness in particular related to the shear layer in the flame region is visualized in Fig. 4. Here it can also be observed that the combustion takes place in a region comprising the dump plane.

The algebraic model for the variance $var(c)$ strongly depends on the parameter a in Eq. (1). In fact the optimum value for the referred parameter is not the same for grids with different filter width, Δ_k [2]. Defining the mean source term in terms of a presumed β-PDF model shows that low values for $var(c)$ implies in large values for the source term, and as can be seen in Fig. 1(c) this could be the reason related with the existence of flame also in the upstream of the combustor dump plane.

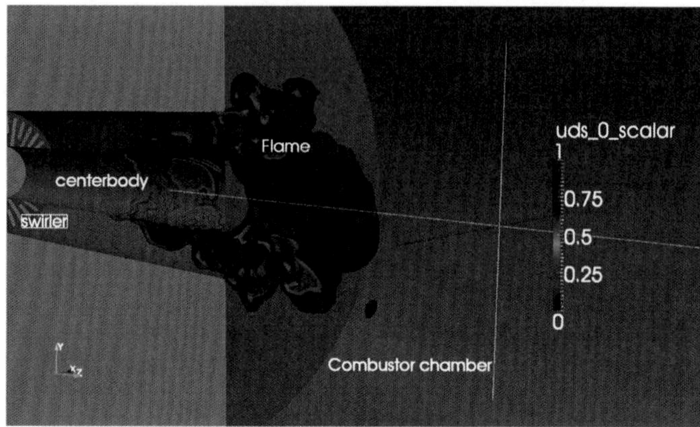

Fig. 4 Cutting plane of a 3*D* instantaneous flame front showing isosurfaces of reaction progress variable, c ($UDS_{scalars}$). The flame zone location and the combustor components are shown.

5 Conclusion

Results concerned with the LES and RANS simulations were obtained for the reactive flow in a gas turbine burner considering the use of FGM as the combustion model.

As an overall conclusion, the simulations performed with FGM shows that through an appropriate look-up procedure, combustion features in gas turbine conditions can be reproduced with a reasonable computational effort. Clearly LES gives much more insight compared to RANS. However, they both give a consistent burning region that is a little too much upstream. To improve the results an evaluation on the parameter of the variance model should be performed. Values approaching $a = 2$ are suggested for comparable cases, [2]. Using a transport equation or a dynamic procedure would also be potential options. In the future non-ideal premixing and heat-loss can be taken into account.

References

1. Van Oijen, J.A.: Flamelet-generated manifolds: development and application to premixed laminar flames. PhD thesis, Eindhoven: University of Technology Eindhoven, (2002)
2. Vreman, A.W., van Oijen, J.A., de Goey, L.P.H., Bastiaans, R.J.M.: Subgrid scale modeling in large-eddy simulation of turbulent combustion using premixed flamelet chemistry. Flow, Turbulence and Combustion, **82**, No. 4, 511–535, (2009)
3. Strakey, P.A., Woodruff, S.D., Williams, T.C., Schefer, R.W.: OH-Planar Fluorescence Measurements of Pressurized, Hydrogen Premixed Flames in the SimVal Combustor. AIAA Journal, **46**, No. 7, 1604–1613 (2008)
4. Strakey, P.A., Yip, M.J.: Experimental and Numerical Investigation of a Swirl Stabilized Premixed Combustor Under Cold-Flow Conditions. J.Fluids Engineering. **129**, 942–953 (2007)

Direct Numerical Simulation of highly turbulent premixed flames burning methane

Gordon Fru, Gábor Janiga and Dominique Thévenin

1 Introduction

The last century has witnessed soaring gas prices, deteriorating air quality and alarming global climate changes. In recent years, increasing concerns have been raised with respect to the environmental impacts of energy consumption via the combustion of fossil fuels, for instance in stationary power generation and transportation, emitting greenhouse gases and air pollutants. As a result, governments now set more and more stringent standards. Hence, it is essential to understand and improve combustion processes, in order to reduce fuel consumption and pollutant emissions as much as possible.

As an additional source of information combined with experimental and theoretical studies, numerical simulation, in particular, Direct Numerical Simulations (DNS) provides an efficient way of analyzing combustion phenomena of gaseous mixtures. DNS is attractive because it is – usually – affordable and – always – informative, and offers the most accurate description of turbulent reacting flows when applicable [1].

The DNS approach consists in solving exactly all the physical space and time scales embedded in the representative system of equations, without any approximate model, making it one of the most powerful tools to predict and simulate complex flows. To simulate realistic systems burning relevant fuels and subjected to large Reynolds numbers, the number of computational cells needed can be overwhelming [2]. Fortunately enough, the rapid growth of computational capabilities presents significant opportunities for combustion DNS. Progress in computational power allow for increased grid size, larger number of time steps, ensemble averaging [3] and a more complete temporal development of the turbulent flame with high statistical reliability. Harnessing computational power to simulate systems with higher

G. Fru, G. Janiga and D. Thévenin,
Lab. of Fluid Dynamics and Technical Flows, Otto-von-Guericke-Universität Magdeburg, Universitätsplatz 2, D-39106 Magdeburg, Germany, e-mail: atanga@ovgu.de

H. Kuerten et al. (eds.), *Direct and Large-Eddy Simulation VIII*, ERCOFTAC Series 15, DOI 10.1007/978-94-007-2482-2_52,

chemical complexity at higher Reynolds numbers is equally important in order to investigate practically relevant fuels. In this manner, turbulent combustion processes can be considered under highly realistic conditions using DNS, providing for the ever increasing need for detailed information on the turbulent flame structure and the general physical understanding of the involved, fully coupled physicochemical phenomena.

Two main challenges have been addressed here: increasing the chemical complexity (considering methane flames) while increasing simultaneously the turbulence intensity. In the following section, the configuration and numerical methods are described. A discussion of the obtained simulation results is given in Sect. 3 before concluding the paper.

2 Considered configuration

A stoichiometric premixed methane-air flame is considered in a high-intensity turbulent environment using a 16 species, 25-step chemical kinetics scheme, providing accurate results for lean up to stoichiometric conditions [4, 5]. The initial mixture composition is $Y_{CH_4} = 0.05517$ and $Y_{O_2} = 0.22$ at $T = 300$ K on the fresh gas side, with an appropriate nitrogen complement.

One of the main objectives in this paper is to carry out a parametric study to investigate the effect of turbulence intensity. In order to increase in a flexible and robust manner the turbulent Reynolds number Re_t on fine-grain parallel systems, a hybrid of two turbulence generators based on Digital Filtering (DF) [6] and Random Noise Diffusion (RND) [7] has been successfully derived, implemented and parallelized on massively parallel computers.

Considering in particular the DF technique [7], a 1D series r_{m_x} of random data (with $\overline{r_{m_x}} = 0$) is convoluted with a linear, digital and non-recursive filter such that

$$u_{m_x} = \sum_{n=-N_x}^{N_x} b_n r_{m_x+n}; \qquad b_k = \widetilde{b}_k \Big/ \left(\sum_{j=-N_x}^{N_x} \widetilde{b}_j^2 \right)^{1/2}; \qquad \widetilde{b}_k = \exp\left(-\frac{\pi k^2}{2n_x} \right) \quad (1)$$

where b_k is the filter coefficient. Assuming that Δx is the homogeneous grid spacing, the desired length scale is computed as $l_x = n_x \Delta x$. The half width of the filter kernel is specified by N_x. This relation is extended to 3D by convolution of three one-dimensional filters: $b_{ijk} = b_i b_j b_k$. With this approach it becomes possible to implement efficiently a fully parallel generation of turbulent fields on massively parallel machines, unlike the complex inverse FFT-based algorithm used in previous versions of the employed code [8]. In this manner, much higher values of Re_t can be achieved.

Twelve numerical experiments at increasing turbulent intensities, u' (all within the Thin Reaction-Zone regime [9] – see Fig. 1(a)) have been realized. The u', Re_t and eddy turn-over time (τ) are given in Table 1. The characteristic mixture viscosity, $\nu = 1.56\ 10^{-5}$ m^2/s, the integral length scale $l_t = 0.48$ cm, problem configura-

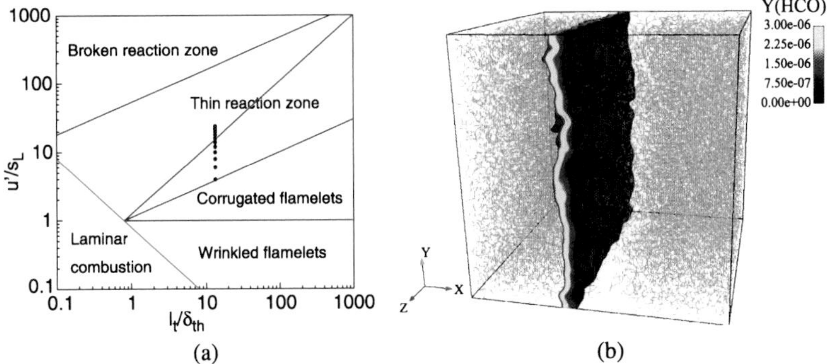

Fig. 1 (a) Modified combustion diagram of Peters [9] with positions of current DNS; (b) Computational domain showing instantaneous HCO iso-contour surrounded by vortex cores

tion, domain size $L = 2.0$ cm (Fig. 1(b)), number of grid points per coordinate direction $N = 801$ (hence a uniform spatial resolution of 25 μm), mixture equivalence ratio $\Phi = 1.0$ and initial flame profiles are identical for all computations. Stable

Case	1	2	3	4	5	6	7	8	9	10	11	12
u' (m/s)	0.50	1.00	2.00	3.00	4.00	5.00	6.00	7.00	8.00	9.00	10.0	12.0
τ (ms)	9.55	4.77	2.39	1.59	1.19	0.95	0.80	0.68	0.60	0.53	0.48	0.40
Re_t	152	305	610	916	1221	1527	1832	2137	2443	2748	3054	3665

Table 1 Initial turbulence parameters

profiles from a laminar 1D computation are linearly extended in the remaining directions at $t = 1.0$ ms and superimposed with a pseudo turbulent field computed with the method described above.

The DNS code employed is the massively parallel, finite-difference, fully three-dimensional and detailed physicochemical flame solver *parcomb* [8, 10]. Derivatives are computed using centered explicit schemes of order six, relying on the skew-symmetric scheme [11] for the convective terms. Time integration is with an explicitly 4th-order Runge-Kutta scheme. Boundary conditions are treated with the help of the extended Navier-Stokes Characteristic Boundary Condition (NSCBC) [12] technique. Transport coefficients and chemical kinetics are treated following methods similar to those used in the standard packages CHEMKIN, TRANSPORT and EGLIB. The problem is solved using 512 cores on the *IBM HPCx Power5* system (EPCC) and require typically a week of user waiting time per Case. For the results presented bellow, the data is analysed at $t = 2\tau$, leading to equilibrium conditions between turbulence and chemistry in time-decaying turbulence.

3 Results and discussion

Highly turbulent conditions (Re_t up to 3 665) have been accessed. All results have been analyzed using the in-house Matlab-based library *Anaflame* [13]. The instantaneous flame structure is exemplified in Fig. 2, where the instantaneous progress variable based on temperature $c = (T - T_u)/(T_b - T_u) = 0.5$ (T_u and T_b are the fresh and burnt gas temperatures) is plotted at the same non-dimensional time for two different turbulent intensities ($\mathrm{Re}_t = 305$ & 3 665). The initially planar flame is wrinkled by the superimposed turbulent field soon after initialization. Previously, DNS easily accessed mild turbulence conditions such as the one shown in Fig. 2(a) where the flame is only slightly contorted. Comparing the twelve cases, it is observed that the amount of wrinkling increases strongly from Case 1 to 6 and tends to saturate afterwards. This is an indication that Re_t should exceed noticeably 1 000 in order to reach realistic conditions, depending of course on the application. For higher values of Re_t, considerable structural modifications are observed, in particular flame–flame interactions, leading to pinch off as evident in the higher Re_t snapshot in Fig. 2(b). For highly turbulent cases, pinch off and mutual annihilation of flame surface due to interactions are found to be a dominant mechanism limiting the flame surface area generated by wrinkling due to turbulence.

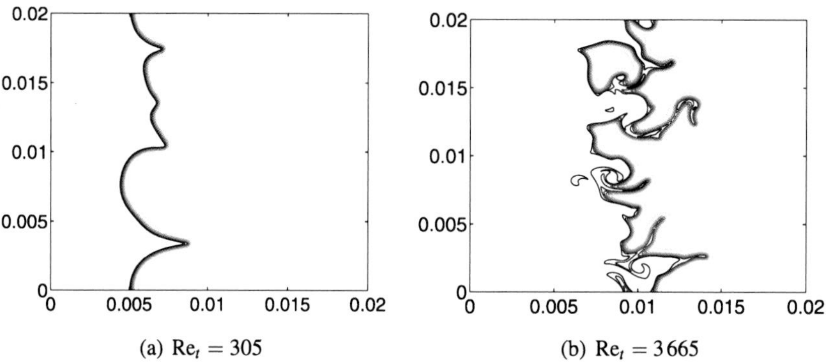

Fig. 2 Instantaneous iso-surface of the temperature field for different Re_t

Another interesting observation is the overall displacement of the turbulent flame front within the computational domain with time. The conventional behavior of a laminar premixed flame front, steadily migrating towards the fresh gas mixture is observed only under low Reynolds number conditions (see Fig. 2(a)). This propagation is not visible at realistic Re_t values, such as seen in Fig. 2(b), at least for the short time scales considered here.

In the literature, both experiments and numerical simulations report contradictory claims on the impact of turbulent stirring on flame thickness, highlighting the need for further studies. DNS data is hereby analyzed to determine if the flame thickness

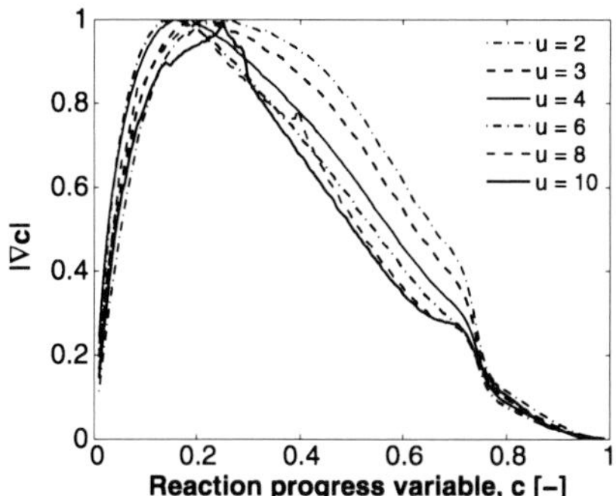

Fig. 3 Conditional mean of $|\nabla c|$ for different cases, increasing Re_t

actually increases or decreases with increasing turbulent intensity. The reciprocal of the magnitude of the flame progress variable gradient, $|\nabla c|^{-1}$, provides a measure of the flame thickness when averaged over fixed intervals of c. The normalized conditional mean of $|\nabla c|$ is plotted in Fig. 3 for six different turbulent intensities (Cases 3, 4, 5, 7, 9 & 11). The obtained profiles decrease with turbulent intensity. Hence, flame thickening is observed. Within the reaction zone ($0.3 \leq c \leq 0.7$), a comparison of the last three Cases in Fig. 3 shows a relatively slow rate of decrease compared to the first three cases. Between Cases 10 and 12, the increase becomes almost negligible. This might indicate that any further increase in turbulence intensity beyond a threshold level does not result in thicker flames. Our observations are in agreement with those by Chen et al. [14] in a parametric study considering a slot-burner Bunsen flame with lean premixed combustion of methane.

The influence of turbulence on the flame structure can be further quantified by probability density functions (PDF). For instance, the local tangential and normal strain rates extracted along the flame front (defined as the isosurface of the reaction progress variable $c = 0.5$) for different Re_t are shown in Fig. 4. The peak PDF of the local tangential strain rate is highly flattened when increasing turbulence intensity. The local normal strain rate shows two peaks under mild turbulent conditions, merging into a single peak at higher Re_t. Large modifications are also observed for instance in the PDF of the mean flame curvature. These characteristic quantities control in particular extinction limits and are therefore vital for modeling purposes.

4 Concluding remarks

A parametric study of the impact of turbulence intensity on a premixed stoichiometric methane–air flame has been performed using detailed physicochemical models. Highly turbulent conditions (Re_t up to 3 665) have been accessed. For realistic cases, where Re_t exceeds considerably 1 000, considerable structural modifications such as flame–flame interactions, pinch off and local re-ignition fronts become prominent. On average, the flame responds to the increasing turbulence intensity by getting

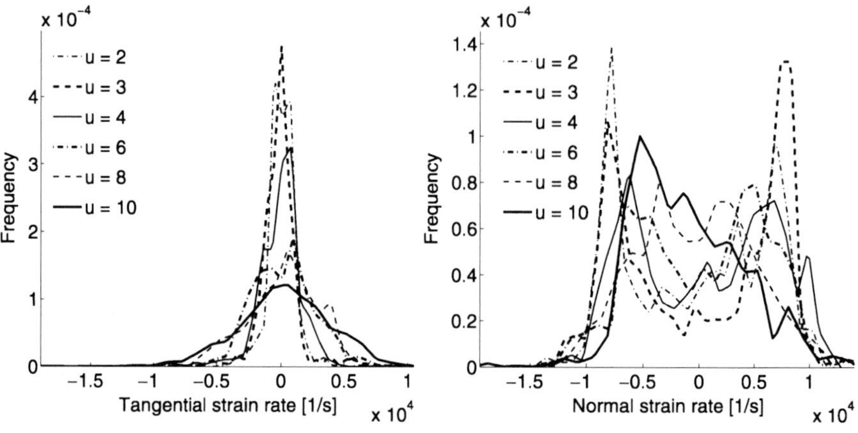

Fig. 4 PDF of tangential and normal strain rates for different Cases, increasing Re_t

thicker. All the observations point to the fact that analyzing DNS results at high Reynolds numbers is essential to obtain realistic information for modeling purposes.

Acknowledgements Access to various supercomputers was granted by the European Union through the research project *DNS-HiRe* of the DEISA Extreme Computing Initiative (DECI).

References

1. T. Poinsot and D. Veynante. *Theoretical and numerical combustion*. R. T. Edwards, second edition, 2005.
2. P. Moin and K. Mahesh. *Annu. Rev. Fluid Mech.*, 30:539–578, 1998.
3. H. Shalaby and D. Thévenin. *Flow Turb. Comb.*, 84(3):357–367, 2010.
4. M. D. Smooke and V. Giovangigli. *Formulation of the premixed and nonpremixed test problems*, volume 384 of *Lecture Notes in Physics*, pages 1–28. Springer-Verlag, 1991.
5. S. James and F. A. Jaberi. *Combust. Flame*, 123:465–487, 2000.
6. M. Klein, A. Sadiki, and J. Janicka. *J. Comput. Phys.*, 186:652–665, 2003.
7. A. Kempf, M. Klein, and J. Janicka. *Flow Turb. Comb.*, 74:67–84, 2005.
8. R. Hilbert and D. Thévenin. *Combust. Flame*, 138:175–187, 2004.
9. N. Peters. *Turbulent combustion*. Cambridge University Press, 2000.
10. A. Laverdant. Technical Report RT 2/13635 DEFA, The French Aerospace Lab., ONERA, 2008.
11. A. Honein and P. Moin. *J. Comput. Phys.*, 201:11–19, 2004.
12. T. Poinsot and S. Lele. *J. Comput. Phys.*, 101(1):104–129, 1992.
13. C. Zistl, R. Hilbert, G. Janiga, and D. Thévenin. *Comput. Vis. Sci*, 12:383–395, 2009.
14. J. H. Chen, A. Choudhary, B. de Supinski, M. DeVries, E. R. Hawkes, S. Klasky, W. K. Liao, K. L. Ma, J. Mellor-Crummey, N. Podhorszki, R. Sankaran, S. Shende, and C. S. Yoo. *Comput. Sci. Discov.*, 2:31, 2009.

A New Subgrid Breakup Model for LES of Spray Mixing and Combustion

S. Srinivasan, E.O. Kozaka and S. Menon

1 Introduction

Characterizing the inflow conditions, especially of droplet distribution at the computational inlet is one of the challenging aspects for a successful multiphase Large Eddy Simulation (LES). Here, we investigate spray modeling by simulating the experiments of acetone spray mixing by Chen et al. [1]. In the experiments, the turbulent spray is generated by a shear driven nebulizer and is carried through a $7L/D$ circular pipe into a mixing chamber [see Fig. 1(a)]. Two separate simulation campaigns are performed. In the first set of simulations, only the internal-flow within the injector is studied using breakup modeling. These breakup simulations are used to assess a new multi-scale breakup procedure by comparing the predicted droplet profiles at the injector exit plane with data. Experimental data at the injector exit plane is shown to be dilute and therefore, in the next set of simulations, this dilute spray exiting into the mixing chamber is simulated using different inflow profiles. Comparison with data using two different subgrid mixing models is used to highlight key features of the far field development of the spray and the gaseous flow.

2 Formulation

The LES formulation is cast in a Lagrangian-Eulerian framework for modeling respectively, the dispersed and the carrier phases. Subgrid closure in the momentum and energy equations are performed using a dynamic eddy viscosity ν_t computed using the subgrid kinetic energy, *ksgs* for which a transport equation is solved. Species closure is performed using either, the conventional gradient diffusion approximation, or, alternatively, using the subgrid Linear Eddy Model (LEM). Full two-way

---

Georgia Institute of Technology, Atlanta, Georgia 30332, e-mail: srikant.srinivasan@ae.gatech.edu, kozaka@gatech.edu, suresh.menon@ae.gatech.edu

H. Kuerten et al. (eds.), *Direct and Large-Eddy Simulation VIII*, ERCOFTAC Series 15, DOI 10.1007/978-94-007-2482-2_53,

liquid-gas coupling is facilitated through source terms for mass, momentum, energy, species and *ksgs*. Full details of the multiphase LES and LEMLES equations and the closures are described elsewhere [2], and therefore, avoided here for brevity.

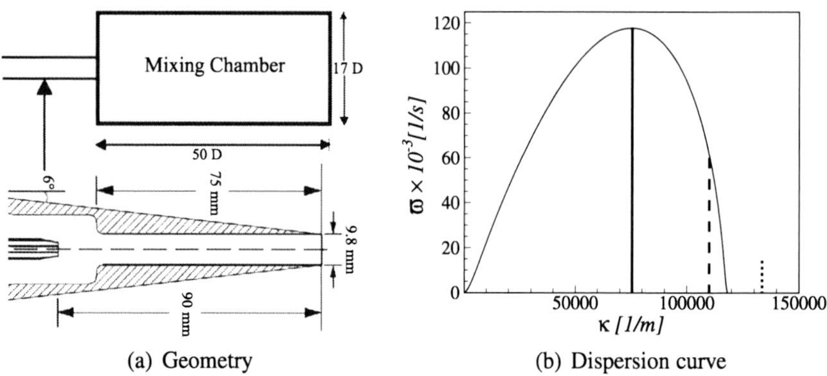

Fig. 1 *(a)* Mixing chamber and the detailed injector geometry [1] and *(b)* The most unstable mode and two other modes introduced in the simulations are shown.

The breakup of the liquid jet is modeled using the linear stability analysis by Reitz [3]. The wave growth rate as a function of the wave number for a given Weber number, $We = \rho_g u_s^2 d/\sigma$ is obtained from this analysis. Here, ρ_g, u_s, d and σ represent gas density, slip velocity, droplet diameter and liquid surface tension, respectively. Only the fastest growing mode given by the maxima of the dispersion curve shown in Fig. 1(b) is usually used to compute the droplet stripping rate, the child particle sizes and velocities. Recent experimental studies, however, show that the final droplet distribution may be a collection of multiple distributions formed by different mechanisms [4]. In another study [5] of liquid jets in cross-flow, a correlation between the boundary layer turbulence and spray formation was suggested. Based on these observations, a multi-scale breakup model is proposed in this paper.

The wave-growth rate profile as a function of wavenumbers in Fig. 1(b) is the classical dispersion relation [3]. The primary breakup model (henceforth, referred to as "baseline") incorporates only the mode with the fastest growth rate, and as a result a nominally monodisperse distribution of child-particles is produced. This contradicts experimental observations of wider range of particle-sizes. The finer drop-size distribution of the experiments is attributed in part to the interaction of turbulent scales with the liquid jet that are not included in the stability model. Sensitivity studies undertaken to study the effect of additional modes reveal the dependence of droplet size distribution on the location of the chosen wave-number relative to the maximum unstable mode. In particular, it is observed that wavenumbers smaller than the most unstable mode has little effect, whereas higher wavenumbers (and hence, smaller droplets) result in a distribution that captures the middle-range of the experimentally measured droplet sizes. To recover the smallest end of the experimental drop-size distribution however, inclusion of modes outside the dispersion curve is found to be necessary. The physical source of these smaller droplet modes is currently being investigated. For example, Fig. 1(b) shows three modes used in the reported results: the most unstable mode, and modes that are 20% and 40% of the most unstable mode. This breakup approach is henceforth referred to as the "multi-mode" approach.

3 Results and Discussion

The details of the LES solver are given elsewhere [2]. The simulated geometry for the breakup studies is a cuboidal domain that axially spans the region between the nebulizer-exit plane and a location $3D$ downstream of the injector exit-plane ($D = 9.8mm$, is the injector diameter). The injector grid has $90,000$ grid points with a minimum grid spacing of $1mm$. The experimental test case HFS [1] is considered here with droplets of size $0.5mm$ injected at $1m/s$ and $300K$ along with a co-flow of air at $60m/s$ and $300K$. The algorithm for computing the child-particle sizes and velocities is similar to the original model [3].

Results from both the baseline and the multimode breakup cases are shown in Fig. 2 plotted at a location near the injector exit where data is also available. The mean diameters are defined as $d_{ij} = (\sum_{k=1}^{n} g_k {d_k}^i / \sum_{k=1}^{n} g_k {d_k}^j)^{1/(i-j)}$ computed from a histogram with bins of mean diameter, d_k and probabilities, g_k. The multimodal breakup approach produces a droplet size distribution [Fig. 2(a)] in which the mean diameter, d_{10} and the Sauter mean diameter, d_{32} are distinctly separated in agreement with data. The baseline model on the other hand, produces a nominally monodisperse distribution significantly over-predicting the data. Droplet velocities show a size dependence in Fig. 2(b) that is absent in the baseline results. The multimode breakup model also predicts significant radial droplet dispersion with droplets concentrated close to the injector-axis as seen in Fig. 2(c).

The injection plane for the dilute simulations is at a distance $2.5D$ upstream of the injector exit. Around 2.5 million grid-points with clustering in the shear-layers and with a minimum grid-size, $\overline{\Delta} = 0.2mm$ is used. The turbulent Reynolds number range is $90-140$ and the Kolmogorov length scale, η is in the $10\mu m - 50\mu m$ range. For LEMLES, the subgrid field is resolved with 24 subgrid cells so that the scalar dissipation scale (the Batchelor scale), $\Lambda_b = \eta/\sqrt{Sc}$ is resolved. Unity Lewis number assumptions are used with the Schmidt number, $Sc = 0.72$. Droplets are tracked using a parcel approach with 2 particles per parcel and a parcel-cutoff radius of $10nm$ to yield approximately 75,000 parcels tracked in the domain.

The profiles for the distribution of droplet number in the inflow plane are shown in Fig. 2(c). Note that this information is not available from the experiments and while the breakup simulation shows qualitative agreement further study is still needed to simulate the full rig. Instead, two extreme cases of droplet distribution are investigated here to see the sensitivity of the far field evolution to the inflow droplet distribution. Case $GC1$ uses the gradient closure $-\widetilde{u_i'Y_k'} = \nu_t/Sc_t \partial\widetilde{Y}_k/\partial x_i$

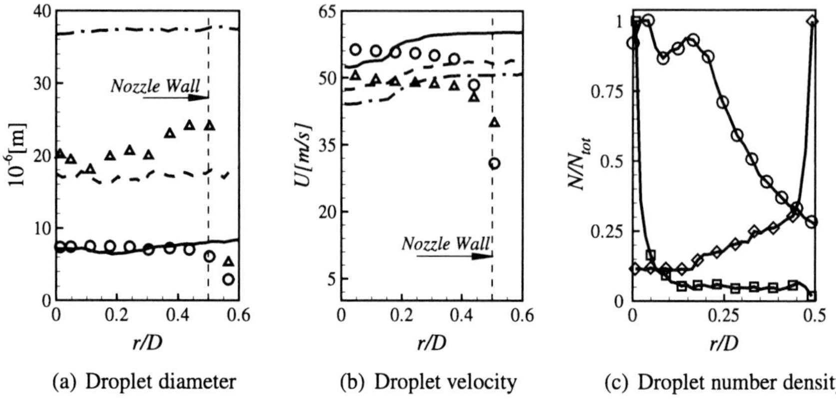

(a) Droplet diameter (b) Droplet velocity (c) Droplet number density

Fig. 2 Prediction of droplet diameter and velocity at the injector exit plane using the multimode model. (a) Droplet diameter, ○: experimental d_{10}, △: experimental d_{32}, solid line: multimode d_{10}, dashed line: multimode d_{32}, dash-dotted line: baseline breakup d_{10}, dotted line: baseline breakup d_{32} (b) Droplet velocity, experimental, ○: $d < 5\mu m$, △: $30\mu m < d < 40\mu m$, multi-mode, dashed line: $d < 5\mu m$, dash-dotted line: $30\mu m < d < 40\mu m$, baseline breakup, solid line: $30\mu m < d < 40\mu m$ (c) Number density, ○: Multimode, □: $GC1$, ◇: $GC2$.

and a radial number-density distribution of droplets according to a $1/r$ profile. This profile is the result of an initialization procedure based on the randomization of the particle positions over (r,θ), where θ is the azimuthal coordinate of the injected droplet. While this method produces a uniform number distribution per unit radius when integrated over θ (at every r), the occupancy per unit area is non-uniform with a singularity at the axis. Case $GC2$ also uses gradient closure but with a different initialization procedure in which the randomization procedure is based on Cartesian coordinates, (x,y) conditioned on $x^2+y^2 \leq D^2/4$. While this produces a uniform distribution per unit area for a quiescent flow field, velocity gradients associated with the boundary layer results in particle accumulation close to the walls. Case *LEM* uses LEM subgrid scalar transport with the Cartesian initialization for droplets. The number density distribution for these simulations are shown in Fig. 2(c). It is seen that the underlying gas-field is largely unaffected since all cases operate in the dilute regime. Thus, the scaling behavior associated with a single-phase turbulent jet is recovered, with the length of the potential core accurately predicted to be $5D$. Also, the peak value of the centerline axial *rms* velocity and its location are also predicted accurately (not shown).

Cases $GC1$ and $GC2$ represent two extreme cases to assess the impact of inlet number distribution. As shown in Fig. 3(a), redistribution of droplet concentration occurs beyond the potential core region, $x/D > 5$. The net result is the convergence of the droplet distribution at downstream locations despite the disparate inflow profiles. This is attributed to Stokes number effects shown in Fig. 3(b), in which the heavier droplets persist in the core, whereas the lighter droplets, being sensitive to the turbulent flow-field, disperse over a greater radial expanse. Here, the Stokes

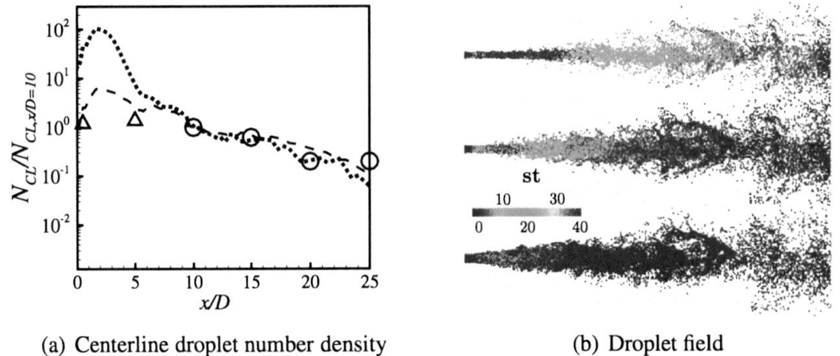

(a) Centerline droplet number density (b) Droplet field

Fig. 3 The decay rate of drop number density is plotted along the centerline in (a) for *GC*1: dotted lines, *GC*2: dashed lines and data: symbols. Triangular symbols indicate erroneous measurements [1]. In (b), droplets are colored by their local *St*, and shown separately for - top: $20\mu m \leq D < 40\mu m$, middle: $20\mu m \leq D < 10\mu m$, bottom: $10\mu m \leq D < 1\mu m$

number, *St* is defined as $St = t_d/t_\eta$ where, t_d is the droplet response time and t_η is the flow response time [1]. With the total number of droplets a function of mass flow rate and the class-based distribution constrained by mean diameters (d_{10} and d_{32}), the centerline convergence is therefore, the result of preferential Stokes number based migration of droplets relative to the centerline. The d_{32} distributions for all cases compare favorably with data at all locations as shown in Fig. 4(a). At around $x/D = 5$, located close to the end of the potential core, d_{32} peaks at off-centered locations due to the residual footprints of the off-centered d_{32} distributions imposed at the injection plane. The trend of decreasing centerline diameter with increasing downstream distance is due to the progressive vaporization of the smaller droplets.

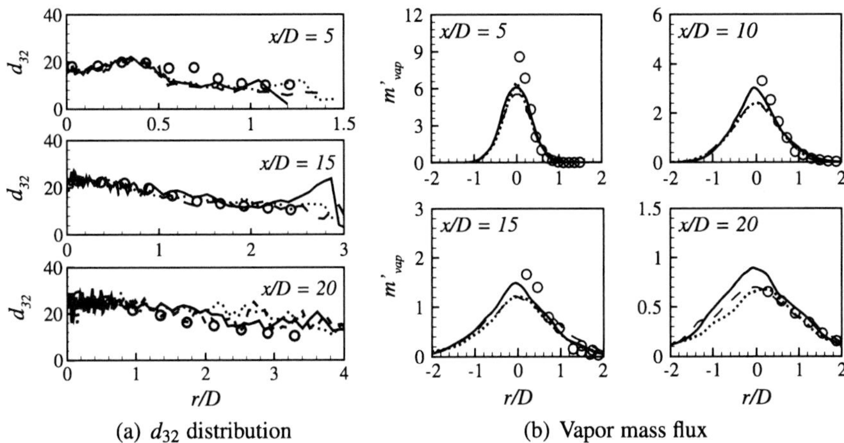

(a) d_{32} distribution (b) Vapor mass flux

Fig. 4 The radial distribution of d_{32} and vapor mass flux at various axial locations. Dotted line: *GC*1, dashed line: *GC*2 and solid line: *LEM*

The acetone vapor mass flux distribution is shown in Fig. 4(b). Except for the centerline at $x/D = 5$, a good agreement is obtained at all locations for all cases. The under-prediction at $(x/D, r) = (5, 0)$ could be due to the experimental uncertainties in the region, $x/D < 5$ [1]. Compared with the *GC* cases, the profiles from *LEM* are relatively sharper and more consistent with data. This can partly be attributed to the subgrid resolution of the Λ_b. In gradient closure if Λ_b has to be resolved a resolution more than an order of magnitude higher would be needed.

4 Conclusion

An approach for LES of spray mixing is discussed here with a specific focus on the initial spray formation due to the breakup process in a highly turbulent shear flow. A new multi-mode breakup model is proposed based on past observations that high shear turbulence can modify the primary breakup process. The most unstable mode and additional modes are used to generate child-particles during breakup. It is shown that this approach captures the variations in droplet sizes and velocities seen in the experiments. The droplet distribution at the exit of the injector assembly is dilute for this case. Therefore, a dilute inflow into the mixing chamber is simulated using different initial conditions and it is shown that this approach produces distributions that are largely independent of the imposed droplet profiles on account of redistribution of droplets in the turbulent far field. Differences in the potential core are however significant, and can significantly affect the flow-field globally in a reacting scenario. Finally, LES using LEM and gradient diffusion closures are compared and reveal the importance of a proper subgrid scalar mixing model even in dilute spray mixing flows. The use of the modified breakup model to eliminate this uncertainty will be the focus of a future work on reacting flows.

References

1. Chen, Y.-C., Stårner, S.H., Masri, A.R. (2006) A detailed experimental investigation of well-defined, turbulent evaporating spray jets of acetone. Int. J. Multiphase Flow. 32:389–412
2. Patel, N., Menon, S. (2008) Simulation of spray-turbulence-flame interactions in a lean direct injection combustor. Combust. Flame. 153:228–257
3. Reitz, R.D. (1987) Modeling atomization processes in high-pressure vaporizing sprays. Atomisation Spray Tech. 3:309–337
4. Marmottant, P., Villermaux, E. (2004) On spray formation. J Fluid Mech. 498:73–111
5. Arienti, M., Soteriou, M.C. (2009) Time resolved proper orthogonal decomposition of liquid jet dynamics. Phys. Fluid. 21, doi:10.1063/1.3263165

LES Modeling of a Turbulent Lifted Flame in a Vitiated Co-flow Using an Unsteady Flamelet/Progress Variable Formulation

Matthias Ihme and Yee Chee See

1 Introduction

In this work, an unsteady flamelet/progress variable (UFPV) model is applied in large-eddy simulation of a lifted methane/air flame in a vitiated co-flow. In this burner configuration, the flame is stabilized by autoignition. This ignition mode is of particular relevance to a number of practical applications, including furnaces, internal combustion engines, and flame stabilization in augmentors.

Autoignition of a reactant mixture is typically initiated in localized regions of low scalar dissipation rate having a mixture composition that favors short ignition times. Since the prediction of autoignition events, however, is strongly dependent on the structure of the surrounding turbulent reacting flow field, combustion models are required that are able to provide an accurate characterization of the spatio-temporal flow field. Although LES techniques have been shown to provide improved predictions for turbulent mixing processes, these localized ignition kernels are computationally not resolved. Therefore, subgrid-scale closure models are required to characterize effects of unresolved scales and ignition kinetics. Another computational challenge arises from the transient evolution of these localized ignition events. Since such ignition events are only inadequately represented by steady-state flamelet models, it is therefore necessary to utilize an unsteady combustion model. To this end, an unsteady flamelet/progress variable (UFPV) model has been developed [1]. In the following, the mathematical model describing the UFPV formulation and the presumed PDF closure is summarized. The experimental configuration and compu-

Matthias Ihme
Department of Aerospace Engineering, University of Michigan, 1320 Beal Avenue, Ann Arbor, MI 48109-2140, USA, e-mail: mihme@umich.edu

Yee Chee See
Department of Aerospace Engineering, University of Michigan, 1320 Beal Avenue, Ann Arbor, MI 48109-2140, USA, e-mail: seeyc@umich.edu

H. Kuerten et al. (eds.), *Direct and Large-Eddy Simulation VIII*,
ERCOFTAC Series 15, DOI 10.1007/978-94-007-2482-2_54,

tational setup are discussed in Sec. 3, and computational results are presented in Sec. 4. The paper finishes with conclusions.

2 Mathematical Model

For the prediction of autoignition in lifted flames, an unsteady flamelet/progress variable model in combination with a low-Mach number variable density formulation is employed [1]. In this approach, the Favre-filtered equations describing the conservation of mass and momentum are solved, and information about density and thermodynamic properties are obtained from the UFPV combustion model. In this combustion model, all thermochemical quantities are obtained from the solution of the unsteady flamelet equations [2],

$$\partial_t \boldsymbol{\phi} - \frac{\chi_Z}{2} \partial_Z^2 \boldsymbol{\phi} = \dot{\boldsymbol{\omega}} , \tag{1}$$

which are solved prior to the LES calculation. In Eq. (1), $\dot{\boldsymbol{\omega}}$ corresponds to the source term of all species and temperature, which are collectively denoted by the vector $\boldsymbol{\phi} = (\boldsymbol{Y}, T)^T$, Z is the mixture fraction, and $\chi_Z = \chi_{Z,\mathrm{st}} F(Z)$ is the scalar dissipation rate of the mixture fraction. The solution of Eq. (1) is tabulated in a chemistry library in terms of three scalar quantities, namely the mixture fraction, the reaction progress parameter Λ, and the stoichiometric scalar dissipation rate $\chi_{Z,\mathrm{st}}$:

$$\boldsymbol{\psi} = \boldsymbol{\psi}(Z, \Lambda, \chi_{Z,\mathrm{st}}) \ . \tag{2}$$

where $\boldsymbol{\psi} = (\boldsymbol{\phi}, \dot{\boldsymbol{\omega}}, \nu, \alpha, \ldots)^T$ denotes the vector of all thermochemical quantities. This mixture-fraction-independent parameter Λ is related to the reaction progress variable C, and is defined so that each flamelet can be uniquely identified. The reaction progress variable is defined from a linear combination of reaction products as $C = Y_{\mathrm{CO}} + Y_{\mathrm{CO_2}} + Y_{\mathrm{H_2O}} + Y_{\mathrm{H_2}}$. The reaction progress parameter, having a unique value for each flamelet, corresponds to C evaluated at stoichiometric condition [3].

For the LES prediction of turbulent reacting flows, the state relation (2) must be formulated for Favre-filtered quantities. In the following, Favre-filtered thermochemical quantities are computed from Eq. (2) by employing a presumed joint PDF for mixture fraction, reaction progress parameter, and stoichiometric scalar dissipation rate, $\widetilde{P}(Z, \Lambda, \chi_{Z,\mathrm{st}})$. A beta PDF is used to model the mixture fraction distribution, and the distribution of $\chi_{Z,\mathrm{st}}$ is modeled by a delta function [4]. A so-called statistically most-likely distribution (SMLD) [5, 6, 7] is employed for Λ, which can be written as

$$P_{\mathrm{SML},j}(\Lambda) = Q(\Lambda) \exp\left\{ \sum_{i=0}^{j} a_i \Lambda^i \right\} , \tag{3}$$

where $j = 2$ denotes the number of enforced moments, and $Q(\Lambda)$ is the so-called *a priori* PDF [7], accounting for bias in composition space. The coefficients a_i, appearing in Eq. (3), are determined by enforcing the first two moments of the reaction progress variable so that the Favre-filtered thermochemical quantities can be expressed in terms of $\widetilde{C}$ and $\widetilde{C''^2}$. With this, the chemistry table can be written as:

$$\widetilde{\psi} = \widetilde{\psi}(\widetilde{Z}, \widetilde{Z''^2}, \widetilde{C}, \widetilde{C''^2}, \widetilde{\chi}_{Z,\mathrm{st}}) \,. \tag{4}$$

In the following, Eq. (4) is used to provide information about all thermochemical quantities in the governing equations. In addition to the solution of the Navier-Stokes equations, four additional transport equations for the first two moments of mixture fraction and progress variable are required to close the system of equations in the UFPV model. Details on the modeling of these equations are discussed in Ref. [1].

3 Experimental Configuration and Computational Setup

The experiment used for the validation of the autoignition model corresponds to the vitiated co-flow burner, which was experimentally studied by Cabra *et al.* [8]. The experimental setup consists of a central fuel pipe with a diameter of $D_{\mathrm{ref}} = 4.57$ mm, through which a methane/air mixture at a temperature of 320 K is supplied. The jet exit velocity is $U_{\mathrm{ref}} = 100$ m/s. The Reynolds number based on the fuel nozzle diameter, exit velocity, and kinematic viscosity of the fuel mixture is 24200, and the value of the stoichiometric mixture fraction is $Z_{\mathrm{st}} = 0.177$. The co-flow consists of reaction products from a premixed hydrogen/air combustion. The co-flow has a diameter of 210 mm and is surrounded by an exit collar to prevent entrainment of ambient air into the flame.

The Favre-filtered governing equations are solved in cylindrical coordinates [9]. The geometry is non-dimensionalized by the jet nozzle diameter D_{ref} and the computational domain is $90\,D_{\mathrm{ref}} \times 30\,D_{\mathrm{ref}} \times 2\pi$ in axial, radial, and circumferential direction, respectively. The axial direction is discretized with 256 grid points and 150 grid points are used in radial direction. The circumferential direction is equally spaced and uses 64 points, resulting in a total number of approximately 2.5 million grid points. More details on the computational setup can be found in Ref. [1].

4 Results

The instantaneous and averaged temperature fields obtained from the UFPV model are shown in Fig. 1. The solid line in these figures corresponds to the isocontour of the stoichiometric mixture fraction. From the instantaneous temperature field, obtained from the LES calculation, it can be seen that up to approximately $25\,D_{\mathrm{ref}}$ downstream of the jet exit, fuel and oxidizer mix without significant heat release. Following this inert mixing zone, a transition region between $30 \leq x/D_{\mathrm{ref}} \leq 50$ is apparent, in which the temperature increases; however, some intermittent pockets with low temperature are evident. Beyond a distance of 50 nozzle diameters above the jet exit the flame is continuously burning, and some entrainment of fluid from the co-flow into the flame core can be observed from the instantaneous flow field results.

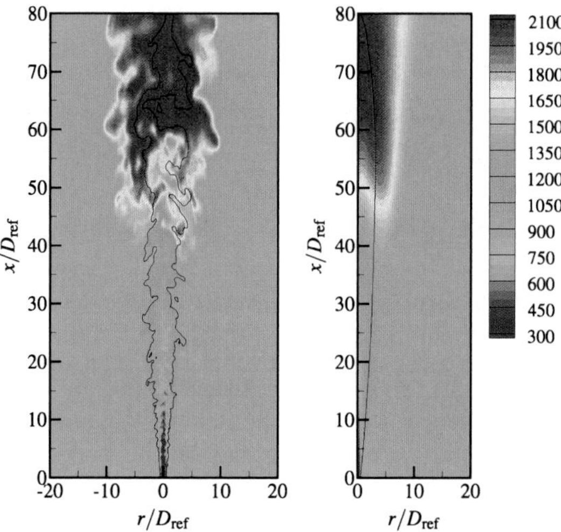

Fig. 1 Instantaneous (left) and averaged (right) temperature fields obtained from the UFPV model. The solid line shows the location of the stoichiometric mixture fraction, $Z_{st} = 0.177$.

Favre-averaged results for mixture fraction and temperature along the jet centerline are shown in Fig. 4. Apart from the slight overprediction in the transition region for $25 \leq x/D_{ref} \leq 60$ the prediction of the mean mixture fraction from the UFPV model is in good agreement with the experimental data.

Radial profiles for species mass fractions of H_2O, CO_2, and CO are compared with experimental data in Fig. 5. The first three measurement locations, shown in this figure, correspond to locations in the transition region, in which ignition takes place. Radial profiles for the water mass fraction are shown in the top row. The simulations accurately predict the shape and peak location of the water formation, and only some discrepancies that are largely confined to the centerline region are evident. A similar good agreement is obtained for the CO_2 mass fraction (middle row), which shows that the carbon dioxide formation increases with increasing downstream location. A comparison of the carbon monoxide mass fraction is shown in the bottom row. In this context it is noted that Raman-measurements are used for model comparisons, since the experimentally reported LIF-data exhibit inconsistencies with the remaining measurements.

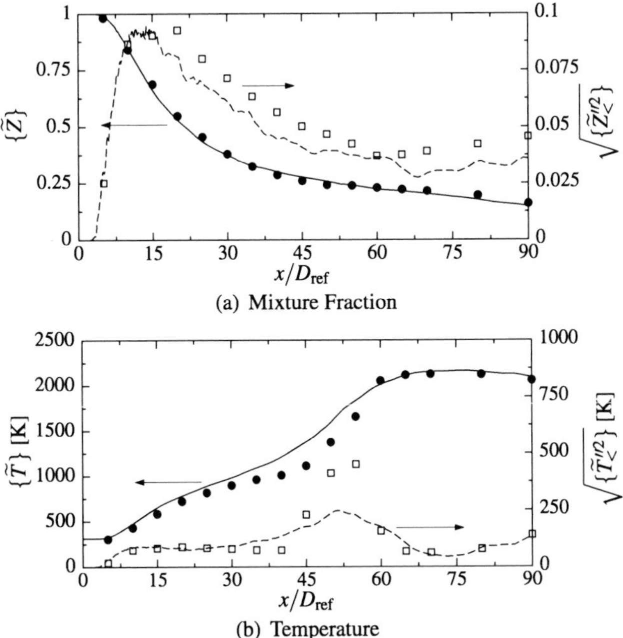

Fig. 2 Comparison of measured (symbols) and calculated (lines) mean statistics of mixture fraction and temperature along the centerline for the Cabra-flame.

5 Conclusions

An unsteady flamelet/progress variable model has been applied to the prediction of autoignition in a lifted flame. The model is an extension to the steady flamelet/progress variable approach, and employs an unsteady flamelet model to describe the evolution of all thermochemical quantities during the flame ignition process. In the UFPV model, all thermochemical quantities are parameterized by mixture fraction, reaction progress parameter, and stoichiometric scalar dissipation rate. The particular advantage of this model over previously developed unsteady flamelet formulations is that in the UFPV model the flamelet time is replaced by physical quantities, which lead to significant simplifications in the computation and parameterization of the thermodynamic state space. A presumed PDF closure model is employed to evaluate Favre-averaged thermochemical quantities. For this a beta-distribution is used for the mixture fraction, a statistically most-likely distribution is employed for the reaction progress parameter, and the distribution of the stoichiometric scalar dissipation rate is modeled by a Dirac delta function. The UFPV model was applied to LES of a lifted flame in a vitiated co-flow. Simulation results are compared with experimental data and good agreement between measurements and predictions is obtained.

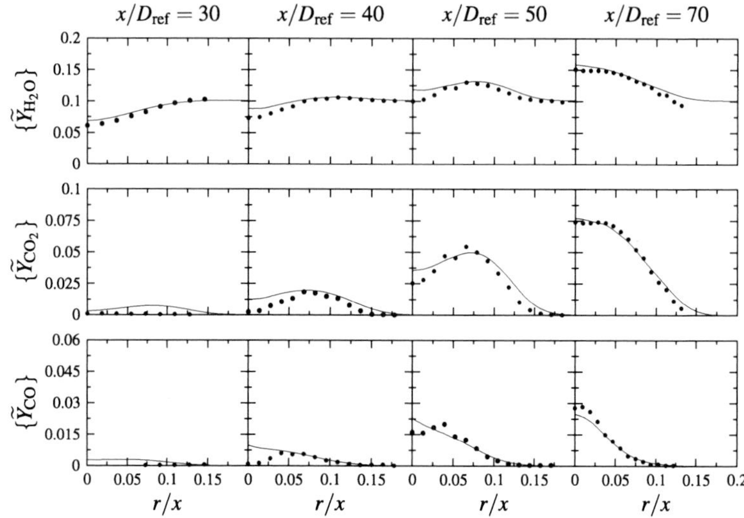

Fig. 3 Comparison of radial profiles for species mass fractions of H_2O, CO_2, and CO. Symbols denote experimental data, and simulation results are shown by lines.

Acknowledgements The authors gratefully acknowledge financial support through the Office of Naval Research under contract N00014-10-1-0561 with Dr. H. Scott Coombe as technical monitor.

References

1. M. Ihme and Y. C. See. Prediction of autoignition in a lifted methane/air flame using an unsteady flamelet/progress variable model. *Combust. Flame*, 157:1850–1862, 2010.
2. N. Peters. *Turbulent Combustion*. Cambridge University Press, Cambridge, 2000.
3. M. Ihme and H. Pitsch. Prediction of extinction and reignition in non-premixed turbulent flames using a flamelet/progress variable model. 1. A priori study and presumed PDF closure. *Combust. Flame*, 155:70–89, 2008.
4. H. Pitsch and M. Ihme. An unsteady/flamelet progress variable method for LES of nonpremixed turbulent combustion. *AIAA Paper 2005-557*, 2005.
5. E. T. Jaynes. Information theory and statistical mechanics. *Phys. Rev.*, 106(4):620–630, 1957.
6. C. H. Shannon. A mathematical theory of communication. *Bell System Tech. J.*, 27(3):379–423, 1948.
7. S. B. Pope. A rational method of determining probability distributions in turbulent reacting flows. *J. Non-Equilib. Thermodyn.*, 4:309–320, 1979.
8. R. Cabra, J.-Y. Chen, R. W. Dibble, A. N. Karpetis, and R. S. Barlow. Lifted methane-air jet flames in a vitiated coflow. *Combust. Flame*, 143:491–506, 2005. Data available from `http://www.me.berkeley.edu/cal/vcb/data/VCMAData.html`.
9. C. D. Pierce and P. Moin. Progress-variable approach for large-eddy simulation of non-premixed turbulent combustion. *J. Fluid Mech.*, 504:73–97, 2004.

Direct Numerical Simulation of a Turbulent Reacting Wall-Jet

Zeinab Pouransari, Geert Brethouwer and Arne V. Johansson

1 Introduction

The turbulent wall-jet includes a number of interesting fluid mechanics phenomena with close resemblance to many mixing and combustion applications. During the last decades, both DNS [1, 2], and LES [3] have been used to study the turbulent wall-jet. Ahlman et al. (2009) performed DNS of nonisothermal turbulent wall jets. Earlier in 2007, Ahlman et al. investigated turbulent statistics and mixing of a passive scalar for an isothermal case by means of DNS. The first three-dimensional DNS of a reacting turbulent flow was performed by Riley et al. (1986) who simulated a single reaction of two scalars, without heat release, for a mixing layer. Recently, Knaus et al. (2009) studied the effect of heat release in non-premixed reacting shear layers [4].

In the present investigation, DNS is used to study a simple reaction in a plane turbulent wall-jet. The flow is compressible and involves a single step reaction between an oxidizer and a fuel species. At the inlet, fuel and oxidizer enter the domain in a non-premixed manner. The reaction is temperature independent and does not release heat. Since the flow is uncoupled from the reaction, the influence of turbulent mixing on the reaction can be studied in the absence of temperature effects. Statistics of the downstream development and reaction of the oxidizer and fuel are studied and compared to the statistics of a passive scalar in the non-reacting wall-jet, [1].

2 Governing Equations

The conservation equations of total mass, momentum and energy read

Linné Flow Centre, KTH Mechanics, SE-100 44 Stockholm, Sweden, e-mail: zeinab@mech.kth.se

H. Kuerten et al. (eds.), *Direct and Large-Eddy Simulation VIII*, ERCOFTAC Series 15, DOI 10.1007/978-94-007-2482-2_55,

$$\frac{\partial \rho}{\partial t}+\frac{\partial \rho u_j}{\partial x_j}=0, \qquad \frac{\partial \rho u_i}{\partial t}+\frac{\partial \rho u_i u_j}{\partial x_j}=-\frac{\partial p}{\partial x_i}+\frac{\partial \tau_{ij}}{\partial x_j},$$

$$\frac{\partial \rho E}{\partial t}+\frac{\partial \rho E u_j}{\partial x_j}=-\frac{\partial q_j}{\partial x_j}+\frac{\partial (u_i(\tau_{ij}-p\delta_{ij}))}{\partial x_j}.$$

Here ρ is the total mass density, u_i are the velocity components, p is the pressure and $E=e+\frac{1}{2}u_iu_i$ is the total energy. The heat fluxes are approximated by Fourier's law according to $q_i=-\lambda\frac{\partial T}{\partial x_i}$, where λ is the coefficient of thermal conductivity and T is the temperature. The viscous stress tensor is defined as $\tau_{ij}=\mu\left(\frac{\partial u_i}{\partial x_j}+\frac{\partial u_j}{\partial x_i}\right)-\mu\frac{2}{3}\frac{\partial u_k}{\partial x_k}\delta_{ij}$ where μ is the dynamic viscosity. The fluid is assumed to be calorically perfect and obeys the ideal gas law.

2.1 Species equations

The wall-jet reaction is modeled by a simple irreversible reaction where oxidant species O and a fuel species F react to form a product P, which is described as $aO+bF \longrightarrow cP$. Stoichiometric coefficients of one are used for all species and the molecular weights are also assumed to be equal for the reactants and hence the mass reaction rates of the species are $\dot{\omega}_o=\dot{\omega}_f=\dot{\omega}=-\frac{1}{2}\dot{\omega}_p$. The reaction mass rate is formulated as $\dot{\omega}=k_r\rho^2\theta_o\theta_f$, where k_r is the reaction rate coefficient. No heat release is used in the simulation and only the oxidizer and fuel mass fractions are computed since the product does not enter in the reaction rate expression. The product mass fraction can be recovered from the reactant mass fractions and the total density. Apart from the reactants, a conserved scalar equation is also solved in order to compare the reacting and non-reacting statistics. Conservation of the reacting species masses are governed by

$$\frac{\partial \rho \theta_i}{\partial t}+\frac{\partial}{\partial x_j}(\rho\theta_i u_j)=\frac{\partial}{\partial x_j}\left(\rho\mathscr{D}\frac{\partial \theta_i}{\partial x_j}\right)-\dot{\omega},$$

where $\theta_i=\theta_o$ or θ_f is the mass fractions of the oxidizer and fuel. A diffusion coefficient $\mathscr{D}$, equal for all scalars, is used.

3 Numerical Method

The governing equations are solved by a fully compressible Navier-Stokes solver, with sixth order compact finite difference for spatial discretization and third order Runge-Kutta for temporal integration. A schematic of the configuration is shown in

Fig. 1, where h is the inlet jet height. At the top, there is a small inflow to compensate for the jet entrainment. The simulation box size in terms of h and the numerical grid specification is listed in table 1. Reynolds, Schmidt and Mach numbers at the jet

Table 1 Simulation cases, h is the inlet jet height. L_i and N_i are the domain size and number of grid points in the i-direction. Reynolds number is defined as $Re = U_j h/\nu_j$, j is used to denote properties at the inlet jet center. And the Schmidt number is $Sc = \nu/\mathscr{D}$.

Case	$L_x \times L_y \times L_z$	$N_x \times N_y \times N_z$	Re	Sc
Isothermal	$47 \times 18 \times 4.8$	$384 \times 192 \times 128$	2000	0.72
Reacting	$35 \times 17 \times 3.6$	$320 \times 192 \times 128$	2000	0.72

inlet are $Re = 2000$, $Sc = 0.72$ and $M = 0.5$, respectively; and a constant coflow of 10% of the inlet jet velocity is used. The reaction time scale is finite and of the same order as the mixing time scale. A Damköhler number of Da = 3, in terms of inlet properties, is used. At the inlet, the fuel is added in the jet and the oxidizer is added in the coflow. The inlet oxidizer mass fraction is half of that of the fuel.

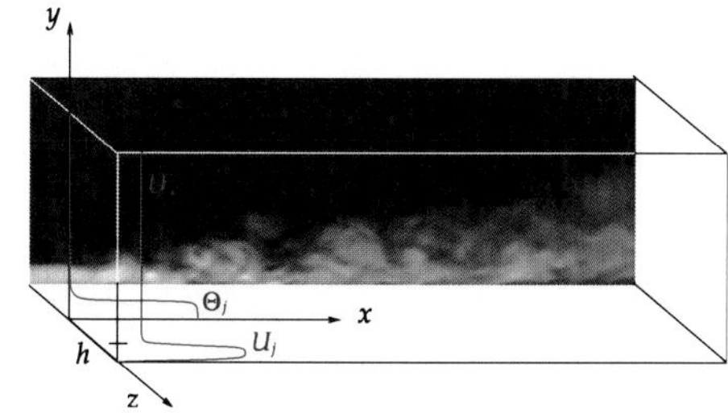

Fig. 1 Schematic of the wall-jet computational domain.

4 Reacting Wall-Jet Results

4.1 Verification with non-reacting wall-jet

Turbulent statistics from the reacting simulations are verified with the non-reacting wall-jet [1]. Inlet normalized mean streamwise velocity, friction velocity and the half-width profiles are demonstrated in Figs. 2(a),(b). The development of u_τ closely follows that of the non-reacting results whereas the half-width of the jet, $y_{1/2}$ is slightly smaller in the turbulent region. Cross stream profiles of the conserved scalar fluctuation intensities and wall-normal fluxes are presented, (Figs. 2(c),(d)), using outer layer normalization; Both compare well with the non-reacting wall-jet results.

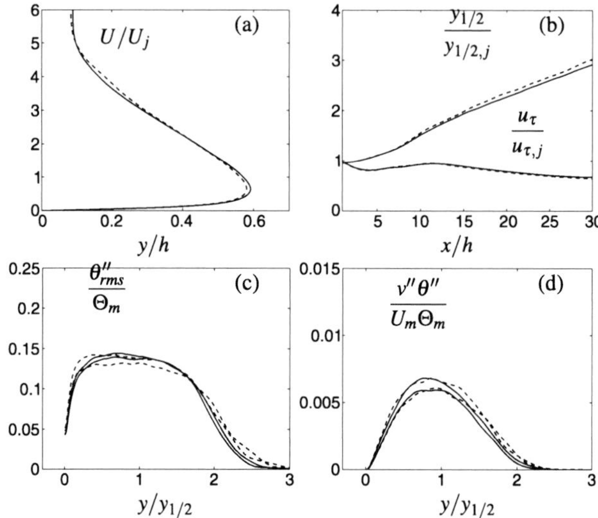

Fig. 2 Comparison of reacting jet (solid) to non-reacting jet (dashed) results. (a) Inlet normalized mean streamwise velocity (b) friction velocity and the half-width (c) cross stream profiles of the conserved scalar fluctuation intensities (d) wall-normal fluxes of the conserved scalar.

4.2 Reacting wall-jet simulation

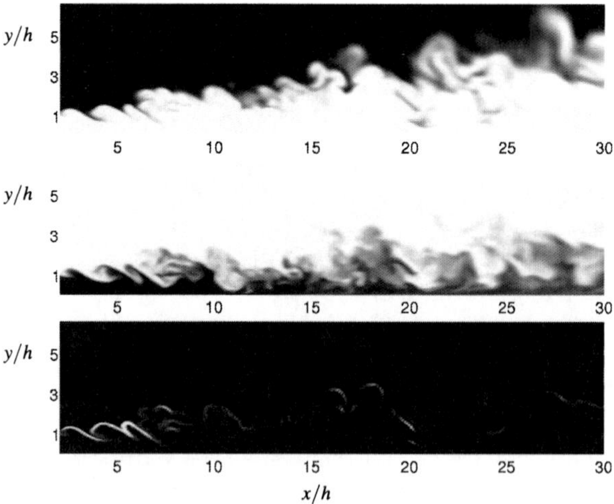

Fig. 3 Snapshots of oxidizer concentration (top), fuel concentration (middle) and reaction rate (bottom).

The reaction starts immediately after the inlet and continues afterward. The snapshots of the species and the reaction term show that the reaction mainly occurs in the upper shear layer in thin sheet like structures; see Fig. 3. However, reactions also take place near the wall, but to a lesser extent.

4.3 Reactants statistics

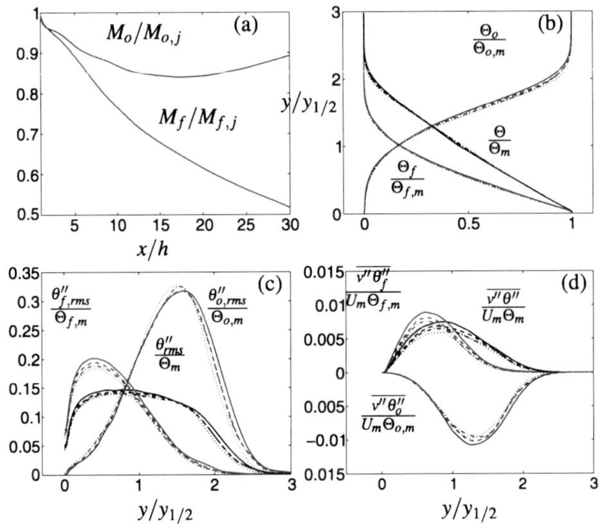

Fig. 4 Statistics of oxidizer (blue), fuel (red) and passive (black) scalars at several downstream positions. Solid: $x/h = 27$, dashed: $x/h = 25$, dash-dotted: $x/h = 23$, dotted: $x/h = 21$ (a) reactants mass flux (b) cross-stream profiles of the reacting scalars (c) fluctuation of the reacting scalars (d) wall-normal fluxes of the reacting scalars.

The downstream development of the fuel mass flux, (Fig. 4(a)), shows that about 30% of the fuel is consumed prior to $x/h = 15$; and in the remaining turbulent part, the fuel consumption is approximately linear. The oxidizer mass flux increases after $x/h = 17$ as a result of the oxidizer influx at the top. As the cross stream profiles of the scalar concentrations confirm, Fig. 4(b), all non-reacting and reacting scalar profiles collapse when using outer scaling.

Considering Figs. 4(c) and (d), the fluctuation of the reacting scalars have started decreasing around the half-width position, where the reaction is very intense. The location of the maximum fluctuation intensities of the reactants are close to the corresponding maximum mean gradients locations. The maximum fluctuation values are also higher than that of the conserved scalar as a result of the sharper gradients of the reactant concentrations.

4.4 Probability density functions (PDF's)

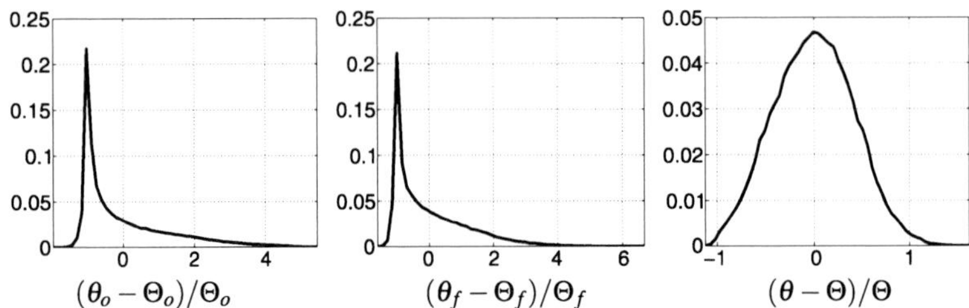

Fig. 5 Probability density functions of fuel, oxidizer & passive scalar at $y/y_{1/2} = 1$. Downstream position is $x/h = 20$, for plots shown here. Minimum and maximum concentrations in each plot matches the borders of plot and zero on x-axis points to the mean concentration.

It is of importance to investigate the progression of the reaction; with this aim in mind, we study the pdf's of fuel, oxidizer and a passive scalar at several downstream positions at different distances to the wall.

The pdf's of the fuel concentration have long exponential tails in the outer shear layer, which is an evidence for the existence of high intermittency. Close to the wall, the distributions of fuel and passive scalar are almost identical, but as it is observed in Figs. 5, in the vicinity of the half-width of the jet, the passive scalar concentration pdf relax to a near-Gaussian distribution. Worth to notice is that the shape of the fuel and oxidizer pdf curves are most similar at the half-width planes at any particular downstream position.

Acknowledgements Computer time provided by Swedish National Infrastructure for Computing (SNIC) is gratefully acknowledged. The financial support from the Swedish National Research program of the Centre for Combustion Science and Technology (CECOST) is very much acknowledged.

References

1. D. Ahlman, G. Brethouwer & A.V. Johansson: Direct numerical simulation of a plane turbulent wall-jet including scalar mixing. Phys. Fluids **19**, 065102 (2007).
2. D. Ahlman, G. Velter, G. Brethouwer & A.V. Johansson: Direct numerical simulation of non-isothermal turbulent wall-jets. Phys. Fluids **21**, 035101 (2009).
3. A. Dejoan & M.A. Leschziner: Large eddy simulation of a plane turbulent wall jet. Phys. Fluids **17**, 025102 (2005).
4. R. Knaus & C. Pantano: On the effect of heat release in turbulence spectra of non-premixed reacting shear layers. J. Fluid Mech. **17**, 025102 (2009).

Novel Developments in Subgrid-Scale Modeling for Space Plasma. Weakly compressible Turbulence in the Local Interstellar Medium

A.A. Chernyshov, K.V. Karelsky, and A.S. Petrosyan

1 Introduction

In this article, recent progress in the development of large eddy simulation (LES) for the study of compressible magnetohydrodynamic (MHD) turbulence of space plasma [1, 2] is given (for LES applications in hydrodynamics of neutral gas see [3] and for incompressible MHD turbulence see [4]). The proper subgrid-scale (SGS) model relevant for space and astrophysical turbulence is selected based on this development, and the efficiency of LES for the investigation of the local interstellar gas turbulence is shown. In particular, the developed LES allows to explain observed data about the Kolmogorov-like spectrum of density fluctuations on the basis of three-dimensional modeling and, thus, to confirm a hypothesis that the weakly compressible regime of MHD turbulence in the local interstellar medium is fulfilled and density fluctuation are a passive scalar.

A still unexplained observation is that density fluctuations in the local interstellar medium exhibit a Kolmogorov-like spectrum over an extraordinary range of scales with a power close to $-5/3$. In spite of the compressibility and the presence of a magnetic field in the local interstellar medium, density fluctuations nevertheless admit a Kolmogorov-like power law, an ambiguity that is not yet completely resolved by any fluid/kinetic theory or computer simulations. The fact that the Kolmogorov-like turbulent spectrum stems from purely incompressible fluid theories of hydrodynamics, and magnetohydrodynamics offers the simplest possible turbulence description in an isotropic and statistically homogeneous fluid. Yet, because the observed density fluctuations in the interstellar medium possess a weak degree of compression, the direct application of such simplistic turbulence models to understand the density spectrum is not completely evident. Besides, the local interstellar gas is not a purely incompressible medium and can have many instabilities due to gradients in the fluid velocity, density, magnetic field, where incompressibility is certainly not a

Theoretical section, Space Research Institute of the Russian Academy of Sciences, Profsoyuznaya 84/32, 117997, Moscow, Russia, e-mail: apetrosy@iki.rssi.ru

H. Kuerten et al. (eds.), *Direct and Large-Eddy Simulation VIII*,
ERCOFTAC Series 15, DOI 10.1007/978-94-007-2482-2_56,

good assumption. Using the advantages of LES, the nontrivial regime of compressible magnetohydrodynamic turbulence of space plasma when initially supersonic fluctuations become weakly compressible is studied. Such mixing regime of the local interstellar medium is important for interpretation of scintillation measurement data that are detected with great sensitivity by the Very Long Baseline Interferometer (VLBI).

2 Subgrid-scale models

To obtain the MHD equations governing the motion of the filtered (that is resolved) eddies, the large scales are separated from the small. Applying the Favre-filtering operation, we can rewrite the MHD equations for compressible fluid flow. The overbar denotes the ordinary filter and the tilde the Favre filter. The following notations are used: ρ is the density; u_j the velocity in the direction x_j; B_j is the magnetic field in the direction x_j; $\sigma_{ij} = 2\mu S_{ij} - \frac{2}{3}\mu S_{kk}\delta_{ij}$ is the viscous stress tensor; $S_{ij} = 1/2\,(\partial u_i/\partial x_j + \partial u_j/\partial x_i)$ is the strain rate tensor; μ is the coefficient of molecular viscosity; η is the coefficient of magnetic diffusivity. It is assumed that the relation between density and pressure is polytropic and has the following form: $p = \rho^\gamma$, where γ is a polytropic index and it is supposed that $\gamma = 5/3$.

The effect of the small-scale turbulence on the filtered part of MHD equations is defined by the following subgrid terms: $\tau^u_{ij} = \bar{\rho}\,(\widetilde{u_i u_j} - \tilde{u}_i\tilde{u}_j) - \frac{1}{M_a^2}(\overline{B_i B_j} - \bar{B}_i\bar{B}_j)$ and $\tau^b_{ij} = (\overline{u_i B_j} - \tilde{u}_i\bar{B}_j) - (\overline{B_i u_j} - \bar{B}_i\tilde{u}_j)$.

We consider five SGS models and three of them (Smagorinsky's model for MHD, Kolmogorov's model for MHD-turbulence, model based on cross-helicity) use the eddy-viscosity assumption to simulate the diffusive transport and dissipation of kinetic and magnetic energy. One of them (scale-similar model) is based on the assumption that the most active SGSs are those closer to the cutoff, and that the scales with which they interact are those immediately above the cutoff wave number. Also, the last SGS closure considered in the present work is the mixed model for MHD turbulence. The differences between SGS models for magnetic energy were shown to decrease with decreasing magnetic Reynolds number and all models discussed above demonstrate good agreement with DNS results at small values of Re_m. The effect of subgrid-scale closures increases with magnetic Reynolds number for modeling of compressible MHD turbulence, but the rate of dissipation of the magnetic energy decreases with increasing Re_m. The best results are shown for the Smagorinsky, the Kolmogorov and the cross-helicity models for the evolution of the magnetic energy [1]. The same behavior was observed for the cross-helicity: the influence of subgrid-scale parameterizations increases with Re_m. For kinetic energy a larger discrepancy of LES results was observed with decreasing Re_m using various SGS closures. The scale-similarity model shows the worst results, however, the other SGS closures increase calculation accuracy. For time dynamics of both magnetic turbulent intensities and kinetic turbulent intensities, the impact of SGS parameterizations on the results increases with Re_m.

The Mach number M_s exerts essential influence on the results. The discrepancy between DNS and LES results for kinetic energy increases with M_s. The Smagorinsky model and the cross-helicity model yield the best agreement with DNS for various Mach numbers. The deviations in results for magnetic energy, on the contrary, decrease with increasing M_s. It should be noted that the magnetic energy reaches a stationary level more rapidly with reducing M_s. The Smagorinsky model shows the best results for cross-helicity both for high and low Mach numbers. The turbulent intensities obtained by LES agree better with DNS results with increasing Mach number.

On the whole, the best results are obtained by the Smagorinsky model for MHD and the model based on cross-helicity [1]. The scale-similarity model does not provide sufficient dissipation of both kinetic and magnetic energy, and it is necessary to use this SGS closure only together with an eddy viscosity model (for example, with the Smagorinsky model), that provided a basis idea for mixed model.

3 LES implications for turbulence in the local interstellar medium

There is growing interest in observations and explanation of the spectrum of the density fluctuations in the interstellar medium. These fluctuations are responsible for radio wave scattering in the interstellar medium and cause interstellar scintillation fluctuations in amplitude and phase of radio waves. A Kolmogorov-like $k^{-5/3}$ spectrum of density fluctuations has been observed for a wide range of scales in the local interstellar medium (from an outer scale of a few parsecs to scales of about 200 km).

The sound speed excited by the small-scale turbulent motion is defined as: $\breve{c}_s = \sqrt{\gamma}\rho^{(\gamma-1)/2}$ and the sonic turbulent Mach number: $\breve{M}_s = \frac{\sqrt{<|u|^2>}}{\breve{c}_s}$. The fluctuating Alfvén speed is defined as: $\breve{u}_a = \frac{\breve{B}}{\sqrt{4\pi\breve{\rho}}}$. Consequently, the turbulent magnetic Mach number is $\breve{M}_a = \frac{\sqrt{<|u|^2>}}{\breve{u}_a}$. Here $u_{rms} = \sqrt{<|u|^2>}$ is the root-mean-square velocity. The turbulent Reynolds numbers and the turbulent plasma beta $\breve{\beta}$ are defined correspondingly. Note that the temporal evolution of these local quantities is important for understanding the fluctuation characteristics of turbulent MHD flows in the local interstellar medium.

For the study of three-dimensional compressible MHD turbulence in the interstellar medium we use the potential of LES [1, 2] developed above. We use the Smagorinsky model for compressible MHD for the SGS parametrization that showed accurate results for a range of similarity numbers [1].

The initial isotropic turbulent spectrum was chosen for the kinetic and magnetic energies in Fourier space to be close to k^{-2} with random amplitudes and phases in all three directions. The choice of such spectrum as initial condition is due to velocity perturbations with an initial power spectrum in Fourier space similar to that of developed turbulence. This is equivalent to start with developed turbulence. In addition, the choice of such spectrum is physically motivated by its rapid convergence as the inertial range spectrum of the energy yields a k^{-3} spectrum through a forward cascade mechanism. Moreover, any single discontinuous shock wave also has such a power spectrum as that of a step function. Taking the Fourier transform of many shocks does not change this power law. Initial conditions for the velocity and the magnetic field have been obtained in physical space using inverse Fourier transform.

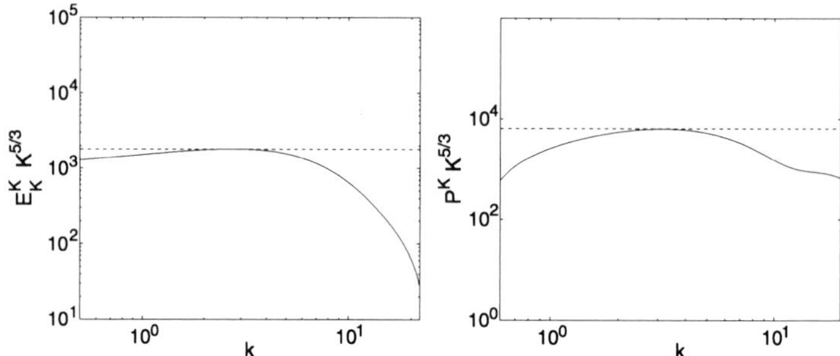

Fig. 1 Normalized and smoothed spectrum of kinetic energy E_k^k(left) and density fluctuations P^k(right), multiplied by $k^{5/3}$. Notice that the spectrum is close to $\sim k^{-3}$ in a forward cascade regime of decaying turbulence. However, there is well-defined inertial Kolmogorov-like range of $k^{-5/3}$ that confirms observation data.

We perform three-dimensional numerical simulations of compressible MHD turbulence for the study of interstellar turbulence. A fourth-order accurate numerical code for the MHD equations in conservative form is used. Periodic boundary conditions for all three dimensions are applied. A uniform mesh with 64^3 grid cells is used for the LES. The simulation domain is a cube with dimensions of $\pi \times \pi \times \pi$. The initial hydrodynamic turbulent Reynolds number is $Re \approx 2000$ and the magnetic Reynolds number is $Re_m \approx 200$. The value of Re is higher than that of Re_m because effects of ambipolar diffusion in the interstellar medium leads to an increased magnetic diffusion and, consequently, to a decrease of Re_m. Other parameters used for modeling of the interstellar medium are: Alfvenic Mach number and sonic Mach number $M_a \approx M_s \approx 2.2$, polytropic index $\gamma = 5/3$ and time step $dt = 0.3 \cdot 10^{-3}$. Different resolution and initial conditions do not qualitatively change the numerical results. The model constants in the SGS closures are determined by means of the dynamic procedure for a test filter with width twice as large as the base filter.

Compressible MHD turbulence in the local interstellar medium is studied using LES for turbulence modeling and subsequent numerical solution of the system of resolved magnetohydrodynamic equations. Notwithstanding the fact that supersonic flows with a high value of large-scale Mach numbers are characterized in interstellar medium, nevertheless, there are subsonic fluctuations of weakly compressible components of the interstellar medium. These weakly compressible subsonic fluctuations are responsible for the emergence of a Kolmogorov-type spectrum in interstellar turbulence which is observed in experimental data. It is shown that density fluctuations are a passive scalar in a velocity field in weakly compressible magnetohydrodynamic turbulence and demonstrate a Kolmogorov-like spectrum in a dissipative range of the energy cascade (Fig. 1). The spectral indices of density fluctuations and kinetic energy are almost coincident. Notice that the range with Kolmogorov-like spectrum exists the same as kinetic energy spectrum, with the same wave numbers $2 \leq k \leq 5$. On the whole, the density fluctuation spectrum demonstrates similar behavior in Fourier space, as kinetic energy spectrum. Consequently, we infer that density fluctuations are a passive scalar in weakly compressible subsonic turbulent flow. Furthermore, theoretical models of turbulence support that any physical characteristic of flow, that propagates passively in large-scale or ambient velocity component of the background turbulence, demonstrates similar spectrum.

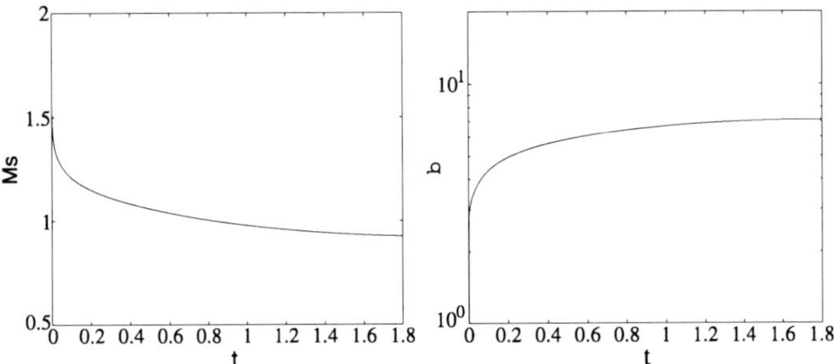

Fig. 2 Decay of turbulent small-scale Mach number $\breve{M}_s$ with time (left). A transition from a supersonic $\breve{M}_s > 1$ to a subsonic $\breve{M}_s < 1$ can be observed. Time dynamics of the turbulent plasma beta $\breve{\beta}$ in compressible MHD turbulence (right). The MHD plasma is strongly magnetized initially and then, as the turbulence evolves, the plasma becomes less magnetized.

It is shown in Fig. 2 (left), that the turbulent sonic Mach number decreases significantly from a supersonic turbulent regime ($\breve{M}_s > 1$), where the medium is strongly compressible, to a subsonic value of Mach number ($\breve{M}_s < 1$), describing weakly compressible flow. This fact indicates that the turbulent cascades associated with the nonlinear interactions in combination with the dissipative effects at the small scales predominantly cause the supersonic MHD plasma fluctuations to damp strongly leaving primarily subsonic fluctuations in the MHD fluid.

In the interstellar medium, the transition of MHD turbulent flow from a strongly compressible to a weakly compressible state not only transforms the characteristic supersonic motion into subsonic motion, but also attenuates plasma magnetization, which is shown in Fig. 2 (right) because plasma beta $\breve{\beta}$ increases with time, thus, the role of magnetic energy decreases in comparison with plasma pressure. Fig. 2 (right) demonstrates that the thermal pressure does not exceed the magnetic energy (that is $\breve{\beta} \leq 1$) in the initial time interval in fully compressible magnetohydrodynamic flow. Plasma particles coupled to the magnetic field lines are expelled from their gyro orbits due to the increase of plasma pressure role in comparison with magnetic energy. This fact leads ultimately to a reduced plasma magnetization and hence plasma fluctuations, and transit into $\breve{\beta} > 1$ regime and subsonic weakly compressible flow. The turbulent plasma beta $\breve{\beta}$ can be written down as: $\breve{\beta} \simeq \frac{8\pi\breve{p}}{\breve{B}^2} \sim \frac{\breve{c}_s^2}{\breve{u}_a^2} \sim \frac{\breve{M}_a^2}{\breve{M}_s^2}$. Since MHD plasma evolves to $\breve{\beta} > 1$ regime, from it follows that Alfvenic turbulent small-scale Mach number $\breve{M}_a$ decreases. Monotonic decrease of $\breve{M}_s$ corresponds to a higher value of $\breve{\beta}$, that is, MHD flow becomes increasingly weakly compressible. Moreover, it follows from the time dynamics of compressible magnetohydrodynamic flow, the magnetosonic fluctuations weaken faster than the Alfvenic ones (Alfvenic modes nevertheless decay due to dissipation too). Besides, the gradual increase of the turbulent plasma beta $\breve{\beta}$ leads to change of speed of turbulent cascade in subsonic regime of the compressible MHD flow. The high plasma beta $\breve{\beta}$ state implies that the shear Alfvenic modes propagate more slowly than sound waves. When magnetized compressible plasma decreases and the turbulent plasma pressure evolves to exceed the turbulent magnetic energy, the perturbations are essentially non-magnetized, that is, the situation is hydrodynamic-like.

Besides, the anisotropy of turbulent flow is considered and it is demonstrated that large-scale flow shows anisotropic properties while small-scale structures are isotropic. Numerical simulation shows various behaviour of the velocity components in spectral cascade at the lower wave number and almost lack of distinctions for large Fourier modes (not illustrated). This indicates that larger scales show a greater anisotropization. Anisotropy and symmetry breakdown are caused first of all by the magnetic field at low value of the plasma beta β when the role of the magnetic field is substantial. Anisotropic cascades are observed to be due to propagating compressible acoustic modes that hinder spectral transfer in the local Fourier space at high value of β when the role of the magnetic field is little. These modes in compressible MHD turbulence could be excited either by a large-scale or ambient velocity component of the background hydrodynamic turbulence. Mixing properties of compressible turbulence in the local interstellar medium predicted by LES method are important to understand effects of radio waves propagation and their scattering in observations data.

Acknowledgements This work has been partly supported by Russian Foundation for Basic Research project 08-08-00687-a. A.Ch. gratefully acknowledges the financial support from Dynasty Foundation.

References

1. A.A. Chernyshov, K.V. Karelsky, A.S. Petrosyan, Physics of Fluids **19**(5), 055106 (2007)
2. A.A. Chernyshov, K.V. Karelsky, A.S. Petrosyan, Physics of Fluids **20**(8), 085106 (2008)
3. E. Garnier, N. Adams, P. Sagaut, *Large Eddy Simulation for Compressible Flows* (Springer Science+Business Media B.V., Netherlands, 2009)
4. W.C. Müller, D. Carati, Phys. Plasmas **9**(3), 824 (2002)

Part VI
Rayleigh-Bénard Flow

Numerical simulations of rotating Rayleigh-Bénard convection

Richard J.A.M. Stevens, Herman J.H. Clercx, and Detlef Lohse

1 Introduction

The Rayleigh-Bénard (RB) system is relevant to astro- and geophysical phenomena, including convection in the ocean, the Earth's outer core, and the outer layer of the Sun. The dimensionless heat transfer (the Nusselt number Nu) in the system depends on the Rayleigh number $Ra = \beta g \Delta L^3/(\nu\kappa)$ and the Prandtl number $Pr = \nu/\kappa$. Here, β is the thermal expansion coefficient, g the gravitational acceleration, Δ the temperature difference between the bottom and top, and ν and κ the kinematic viscosity and the thermal diffusivity, respectively. The rotation rate H is used in the form of the Rossby number $Ro = (\beta g \Delta/L)/(2H)$. The key question is: How does the heat transfer depend on rotation and the other two control parameters: $Nu(Ra, Pr, Ro)$? Here we will answer this question by giving a summary of our results presented in [1, 2, 3].

We present both experimental measurements [1] (fig. 1) and results from direct numerical simulation (DNS) (fig. 2). They cover different but overlapping parameter ranges and thus complement each other. The convection apparatus that was used in the experiments is described in detail as the "medium sample" in Ref. [4]. All measurements were made at constant imposed Δ and Ω, and fluid properties were evaluated at $T_m = (T_t + T_b)/2$ [1].

Richard Stevens
Dept. of Applied Physics, University of Twente, Enschede, The Netherlands, e-mail: r.j.a.m.stevens@utwente.nl

Herman Clercx
Dept. of Applied Physics, Eindhoven University, The Netherlands; Dept. of Mathematics, Twente University, The Netherlands, e-mail: h.j.h.clercx@tue.nl

Detlef Lohse
Dept. of Applied Physics, University of Twente, Enschede, The Netherlands, e-mail: d.lohse@utwente.nl

H. Kuerten et al. (eds.), *Direct and Large-Eddy Simulation VIII*,
ERCOFTAC Series 15, DOI 10.1007/978-94-007-2482-2_57,

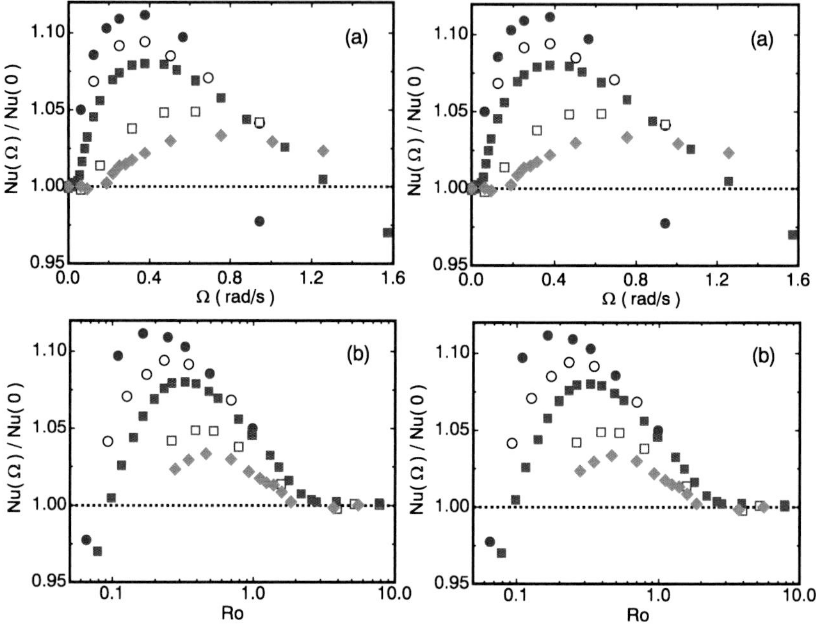

Fig. 1 The ratio of the Nusselt number $Nu(\Omega)$ in the presence of rotation to $Nu(\Omega = 0)$ for $Pr = 4.38$ ($T_m = 40.00$°C). (a): Results as a function of the rotation rate in rad/sec. (b): The same results as a function of the Rossby number Ro on a logarithmic scale. Solid circles: $Ra = 5.6 \times 10^8$ ($\Delta = 1.00$ K). Open circles: $Ra = 1.2 \times 10^9$ ($\Delta = 2.00$ K). Solid squares: $Ra = 2.2 \times 10^9$ ($\Delta = 4.00$ K). Open squares: $Ra = 8.9 \times 10^9$ ($\Delta = 16.00$ K). Solid diamonds: $Ra = 1.8 \times 10^{10}$ ($\Delta = 32.00$ K). Here and in fig. 2 experimental uncertainties are typically no larger than the size of the symbols. Figure adapted from [1].

2 Numerical Method

In the DNS we solved the three-dimensional Navier-Stokes equations within the Boussinesq approximation in a three dimensional cylindrical domain

$$\frac{D\mathbf{u}}{Dt} = -\nabla P + \left(\frac{Pr}{Ra}\right)^{1/2} \nabla^2 \mathbf{u} + \theta \hat{z} - \frac{1}{Ro} \hat{z} \times \mathbf{u}, \tag{1}$$

$$\frac{D\theta}{Dt} = \frac{1}{(PrRa)^{1/2}} \nabla^2 \theta, \tag{2}$$

with $\nabla \cdot \mathbf{u} = 0$. Here $\hat{z}$ is the unit vector pointing in the opposite direction to gravity, $\mathbf{u}$ the velocity vector, and θ the non-dimensional temperature, $0 \leq \theta \leq 1$. Finally, P is the reduced pressure (separated from its hydrostatic contribution, but containing the centripetal contributions): $P = p - r^2/(8Ro^2)$, with r the distance to the rotation axis. The equations have been made non-dimensional by using, next to L and Δ, the free-fall velocity $U = \sqrt{\beta g \Delta L}$. The resolution is sufficient to represent the small scales both inside the bulk of turbulence and in the boundary layers (BLs) (where the grid-point density has been enhanced) for the parameters employed here. Nu is calculated as in ref. [5] and its statistical convergence has been verified [1, 2].

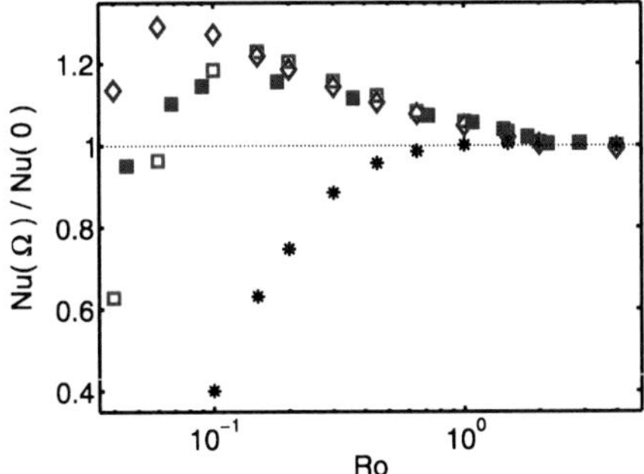

Fig. 2 The ratio $Nu(\Omega)/Nu(\Omega = 0)$ as function of Ro on a logarithmic scale. Blue solid squares: $Ra = 1 \times 10^9$ and $Pr = 6.4$ (DNS) [6]. Open squares: $Ra = 1 \times 10^8$ and $Pr = 6.4$ (DNS). Open diamonds: $Ra = 1 \times 10^8$ and $Pr = 20$ (DNS). Stars: $Ra = 1 \times 10^8$ and $Pr = 0.7$ (DNS). Figure adapted from [1]. Additional experimental and DNS data for $Ra = 2.73 \times 10^8$ and $Pr = 6.26$ can be found in figure 3 of ref. [1].

3 Results

In [1] we showed that the heat-flux enhancement can be as large as 30% and depends strongly on Pr and Ra (figs. 1 and 4). The increased heat transfer is due to Ekman pumping; i.e. due to the rotation, rising or falling plumes of hot or cold fluid are stretched into vertical vortices that suck fluid out of the thermal BLs adjacent to the bottom and top plates (fig. 3b). For $Pr = 6.4$ thermal structures are basically confined to these vortices, whereas for $Pr = 0.7$ they are much shorter and broadened, because the larger thermal diffusion, which makes the Ekman pumping ineffective. This results in a decreasing heat transfer at lower Pr (fig. 3a). Along the same line of reasoning one can explain the reduced effect of Ekman pumping at higher Ra, see figure 1, because with increasing Ra the turbulence is enhanced which leads to a larger eddy thermal diffusivity, promoting a homogeneous mean temperature in the bulk.

In [3] we found that there is an optimal Pr number for the heat transfer enhancement at a fixed Ro, see figure 4. The observation that there is a Pr number for which the heat transfer enhancement is largest suggests that there should be at least two competing effects, which strongly depend on the Pr number, that control the effect of Ekman pumping. One wonders why Ekman pumping becomes less effective for higher Pr numbers. Clearly, it must have a different origin than the reduced effect of

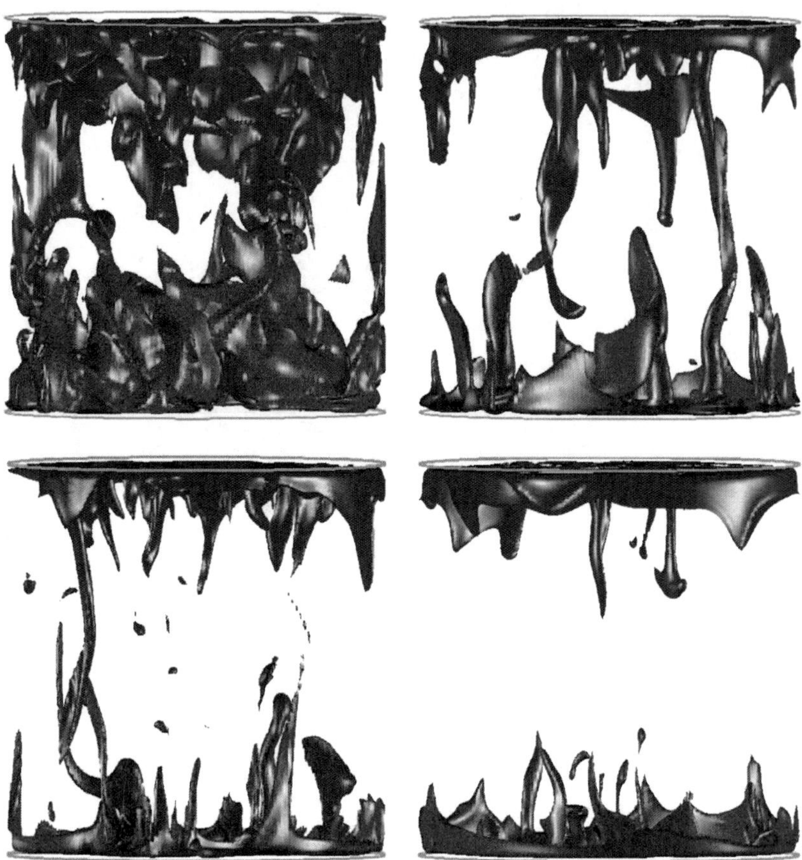

Fig. 3 3D visualization of the temperature isosurfaces in the cylindrical sample of $\Gamma = 1$ at 0.65Δ (originating from bottom) and 0.35Δ (originating from top), respectively, for $Pr = 0.7$ (left upper plot) and $Pr = 6.4$ (right upper plot), $Pr = 20$ (left lower plot), and $Pr = 55$ (right lower plot) for $Ra = 10^8$ and $Ro = 0.30$. The snapshots were taken in the respective statistically stationary regimes. Figure adapted from [3].

Ekman pumping at lower Pr. Indeed, there is an important difference between the high and the low Pr number regime, namely the relation between the thickness of the thermal and the kinetic boundary layers (BL). For the low Pr number regime the kinetic BL is thinner than the thermal BL and therefore the fluid that is sucked into the vertical vortices is very hot. When the Pr number is too low this heat will spread out in the middle of the cell due to the large thermal diffusivity. For somewhat higher Pr number the fluid that is sucked out of the thermal BL is still sufficiently hot and due to the smaller thermal diffusivity the heat can travel very far from the plate in the vertical vortices. In this way Ekman pumping can increase the heat transfer for moderate Pr. In the high Pr number regime the kinetic BL is much thicker than the thermal BL. Therefore, the Ekman vortices forming in the bulk do not reach the thermal BL and hence the temperature of the fluid that enters the vertical vortices is much lower. This is schematically shown in figure 5.

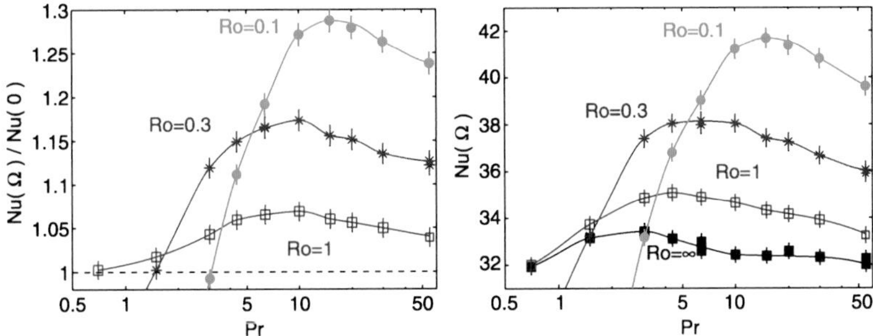

Fig. 4 The heat transfer as function of *Pr* on a logarithmic scale for $Ra = 1 \times 10^8$. Solid squares, open squares, stars, and solid circles indicate the results for $Ro = \infty$, $Ro = 1.0$, $Ro = 0.3$, and $Ro = 0.1$, respectively. Figure adapted from [3].

An investigation of the temperature isosurfaces [1, 3] revealed long vertical vortices as suggested by Ekman-pumping at $Pr = 6.4$, while these structures are much shorter and broadened for the $Pr = 0.7$ case due to the larger thermal diffusivity at lower *Pr* [1]. In figure 3 we show the temperature isosurfaces at $Ro = 0.30$ for several *Pr* to identify the difference between the high and moderate *Pr* number cases. The figure reveals that the vertical transport of hot (cold) fluid away from the bottom (top) plate through the vertical vortex tubes is strongly reduced in the high-Pr number regime compared to the case with $Pr = 6.4$. This is illustrated in figure 3 where the threshold for the temperature isosurfaces is taken constant for all cases. Indeed, a closer investigation shows that the vortical structures for the higher *Pr* are approximately as long as for the $Pr = 6.4$ case. This confirms the view that the temperature of the fluid that is sucked into the Ekman vortices decreases with increasing *Pr*.

4 Conclusions

To summarize, we studied the *Ra* and *Pr* number dependence of the heat transport enhancement in rotating RB convection [1, 2, 3]. We showed that at a fixed *Ro* number there is a *Pr* number for which the heat transfer enhancement reaches a maximum. This is because Ekman pumping, which is responsible for the heat transfer enhancement, becomes less effective when the *Pr* number becomes too small or too large. At small *Pr* numbers the effect of Ekman pumping is limited due to the large thermal diffusivity due to which the heat that is sucked out of the BLs rapidly spreads out in the bulk. At high *Pr* numbers the temperature of the fluid that is sucked into the vertical vortices near the bottom plate is much lower, because the thermal BL is much thinner than the kinetic BL, and therefore the effect of Ekman

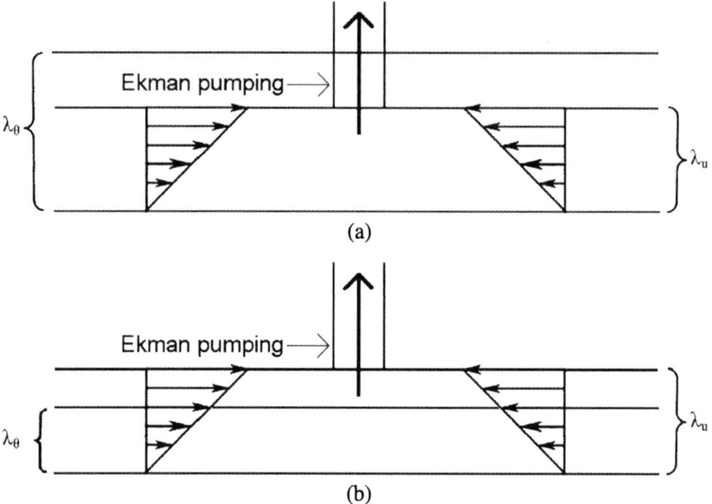

Fig. 5 a) Sketch for the low *Pr* number regime where the Ekman vortices reach the thermal BL. b) Sketch for the high *Pr* number regime where the Ekman vortices do not reach the thermal BL. Figure adapted from [3].

pumping is lower. Furthermore, heat transport enhancement also decreases with increasing *Ra*, because of a larger eddy thermal diffusivity.

Acknowledgements We thank R. Verzicco for providing us with the numerical code and Jin-Qiang Zhong and Guenter Ahlers for many inspiring discussions. The work is supported by the Foundation for Fundamental Research on Matter (FOM) and the National Computing Facilities (NCF), both sponsored by NWO. The numerical simulations were performed on the Huygens cluster of SARA in Amsterdam.

References

1. J.-Q. Zhong, R. J. A. M. Stevens, H. J. H. Clercx, R. Verzicco, D. Lohse, and G. Ahlers, *Prandtl-, Rayleigh-, and Rossby-number dependence of heat transport in turbulent rotating Rayleigh-Bénard convection*, Phys. Rev. Lett. **102**, 044502 (2009).
2. R. J. A. M. Stevens, J.-Q. Zhong, H. J. H. Clercx, G. Ahlers, and D. Lohse, *Transitions between turbulent states in rotating Rayleigh-Bénard convection*, Phys. Rev. Lett. **103**, 024503 (2009).
3. R. J. A. M. Stevens, H. J. H. Clercx, and D. Lohse, *Optimal Prandtl number for heat transfer in rotating Rayleigh-Bénard convection*, New J. Phys. **12**, 075005 (2010).
4. E. Brown, D. Funfschilling, A. Nikolaenko, and G. Ahlers, *Heat transport by turbulent Rayleigh-Bénard convection: Effect of finite top- and bottom conductivity*, Phys. Fluids **17**, 075108 (2005).
5. R. J. A. M. Stevens, R. Verzicco, and D. Lohse, *Radial boundary layer structure and Nusselt number in Rayleigh-Bénard convection*, J. Fluid. Mech. **643**, 495 (2010).
6. R. P. J. Kunnen, H. J. H. Clercx, and B. J. Geurts, *Breakdown of large-scale circulation in turbulent rotating convection*, Europhys. Lett. **84**, 24001 (2008).

Direct Numerical Simulation and Lagrangian Particle Tracking in turbulent Rayleigh Bénard convection

H.J.H. Clercx, V. Lavezzo and F. Toschi

1 Introduction

Over the past years, turbulent convection has been the subject of extensive studies (see e.g. [1, 5, 6]), which attempted to determine the main flow features and the contribution of different parameters to the heat transfer in various geometries, but only few of them focused on Lagrangian statistics. Lagrangian tracking can put some light on the local properties of the flow by gathering information on the temperature and velocity fields along the particle trajectory [7, 10]. This, in particular, has direct relevance for many industrial and environmental applications where the fluid heat transfer is modified by the presence and deposition of particles on the walls (e.g. nuclear power plants, petrochemical multiphase reactors, cooling systems for electronic devices, pollutant dispersion in the atmospheric boundary layer, aerosol deposition etc.).

However, a detailed description of the interaction between a large number of particles and a turbulent flow in a complex configuration such as a Rayleigh Bénard cell, is a difficult task, even with the computational resources today available. Therefore, one has to rely on different approaches to model, as accurately as possible, the statistical effect of the turbulent structures on particle behavior.

In recent years, the lattice Boltzmann method (LBM) has been developed into an alternative and promising numerical scheme for simulating fluid flows with or without a dispersed phase. Unlike the conventional numerical schemes based on discretizations of macroscopic continuum equations, LBM is based on simplified kinetic models that incorporate the essential physics of mesoscopic scales, so that

H.J.H. Clercx · V. Lavezzo · F. Toschi
Department of Applied Physics, Eindhoven University of Technology, 5600 MB Eindhoven, The Netherlands

F. Toschi
Department of Mathematics and Computer Science and International Collaboration for Turbulence Research, Eindhoven University of Technology, 5600 MB Eindhoven, The Netherlands

H. Kuerten et al. (eds.), *Direct and Large-Eddy Simulation VIII*, ERCOFTAC Series 15, DOI 10.1007/978-94-007-2482-2_58,

the macroscopic averaged properties obey the desired macroscopic equations (see e.g. the review articles [2, 4]). The LBM approach is particularly suited to describe complex fluids or geometries and provides many advantages including an easy implementation of boundary conditions and good efficiency on massively parallel machines. For these reasons the LBM method can be considered a powerful tool to achieve a good statistical accuracy while retaining a low computational cost and algorithm complexity.

A Lattice Boltzmann method coupled with Lagrangian particle tracking has been used in this work to test the influence of the grid resolution on the accurate description of a turbulent convective flow, with particular attention to the region close to the walls. To this aim, analysis of the effects of hot plumes formation on the resuspension of tracer particles has been performed. By following in time particles initially released on a horizontal plane close to the bottom wall, a correlation between particle temperature and the fluid structures responsible for their suspension has been found.

2 Description of the numerical methods

The Lattice Boltzmann method is a well established tool described in detail in many previous publications (see e.g. [9] and [2]), so here only the main features characterizing this technique are reported. The simplest lattice Boltzmann equation, under the BGK simplification, can be written as follows:

$$f_i(x+\Delta t c_i, t+\Delta t) - f_i(x,t) = -\frac{\Delta t[f_i(x,t) - f_i^{eq}(x,t)]}{\tau} + F_i \Delta t \tag{1}$$

where $f_i(x,t) = f_i(x, v = c_i, t), i = 1, \dots, n$ is the probability of finding a particle at lattice site x at time t, moving along the lattice direction defined by the discrete speed c_i during a time step Δt. The left-hand side of the equation represents the so-called streaming term, which corresponds to the evolution in time of the probability function f. The RHS describes the collision via a simple relaxation towards local equilibrium f_i^{eq} (a local Maxwellian distribution expanded to second order in the fluid speed) in a time lapse τ. This relaxation time fixes the fluid kinematic viscosity as $\nu = c_s^2(\tau - 1/2)$, where c_s is the speed of sound of the lattice fluid, ($c_s^2 = 1/3$ in the present work). Finally, F_i represents the effects of external forces. The set of discrete speeds must be chosen such that rotational symmetry is fulfilled. Once this symmetry is secured, the fluid density $\rho = \sum_i f_i$, and speed $u = \sum_i f_i c_i/\rho$ can be shown to evolve according to the (quasi-incompressible) Navier-Stokes equations of fluid-dynamics. In this paper, we will refer to the nineteen speed D3Q19 model shown in Figure 1a. The temperature, modeled as a scalar $T = \sum_i g_i$, provides an external forcing via a buoyancy term, which is added in the equation of the evolution of the fluid population.

The reference geometry, shown schematically in Figure 1b, is a Cartesian slab bounded by two horizontal walls having dimensions equal to the size of the computational grid employed for the calculations. The grid resolution has been varied from 64 to 256 in the streamwise (x) and spanwise (y) directions and from 32 to 128 in the wall normal (z) direction, in such a way that the aspect ratio Γ was maintained fixed and equal to 2. In all cases considered, the grid is equispaced in all directions. Periodic boundary conditions are imposed on both the velocity and the temperature field in the two homogeneous directions. No-slip boundary is enforced at the walls for the velocity field whereas constant temperature is considered for the energy. The no-slip condition is implemented using a mid-grid bounce back method which assures a second-order accuracy. With this approach the wall is situated between two nodes. The constant temperature, on the contrary, is imposed on the first and the last lattice nodes, so a small slip temperature is appearing at the wall. This boundary condition, although simplified, has been considered accurate enough for the purposes of the present study, as the error introduced is comparable for all the employed grid resolutions.

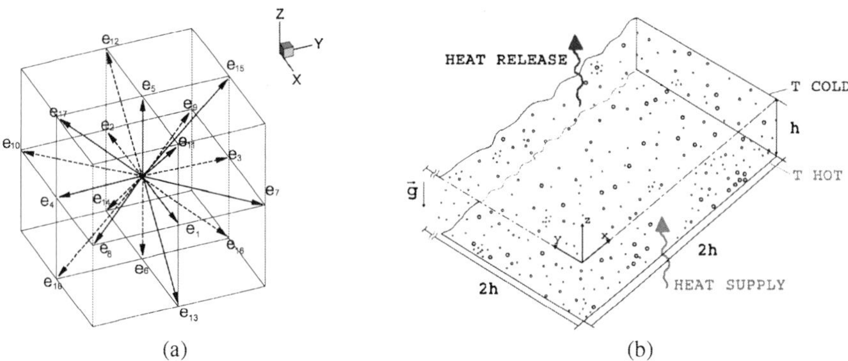

Fig. 1 (a) Scheme of the D3Q19 model for the velocities in the LBM, (b) Sketch of the computational domain.

Parameters characterizing the temperature field are the Prandtl number which was set equal to 1 and the Rayleigh number maintained fixed at $5 \cdot 10^6$. The Rayleigh number is defined as $Ra = (\alpha g \Delta T H^3)/\nu\kappa$ where g is the gravitational acceleration, α the coefficient of thermal expansion, ΔT the temperature difference between the two walls, κ the thermal diffusivity and ν is the fluid kinematic viscosity. The Prandtl number is defined as the ratio of the kinematic viscosity to the thermal diffusivity of the fluid as $Pr = \nu/\kappa$. As by definition in LBM the grid resolution corresponds to the size of the computational domain, to maintain the same Ra number for different grid sizes i.e. for different values of H, the value of αg and κ have been varied.

A Lagrangian approach has been used to track a swarm of 160.000 tracer particles dispersed in the domain. Particles are initially released at random positions on a plane placed at 0.5 lattice unit from the bottom wall. This particular configuration has been chosen to capture the effects of the plumes on particle resuspension at the early stages of the simulation, when hot plumes start to form at the bottom plate and particles are still non-homogeneously distributed in the domain. As particles have zero inertia, their equation of motion reduces to the simple kinematic relationship $dx/dt = u$, where u is the fluid velocity at the particle position and their temperature is equal to the fluid one calculated at the particle position. A linear interpolation scheme is used to obtain both the fluid velocity and temperature at the point of the particle.

3 Results

In turbulent convection problems, it is important to resolve, as accurately as possible, the region close to the wall where the velocity and temperature boundary layers are formed. The Lattice Boltzmann method, in its simplest implementation, is based on a regular grid, so using a grid too coarse can be considered similar to an non-homogeneous Implicit Large Eddy Simulation. To have a better insight on the influence of the grid resolution on average statistics, three grids have been tested. Instantaneous results are shown in Figure 2, where snapshots of temperature isocontours are visualized. Figures are taken at the same instant with a grid refinement increasing from left to right. Strong discrepancies are clearly visible, with a temperature field becoming more coarse grained as the resolution is decreased. This is in accordance with Figure 3a where the energy spectra calculated at the same instant i.e. after about five large circulation turn over times are shown. As the resolution increases the spectrum shifts towards higher frequencies. This suggests an increase in the capability of correctly capturing the effects of the smaller scales and a development towards a more intermittent and turbulent flow.

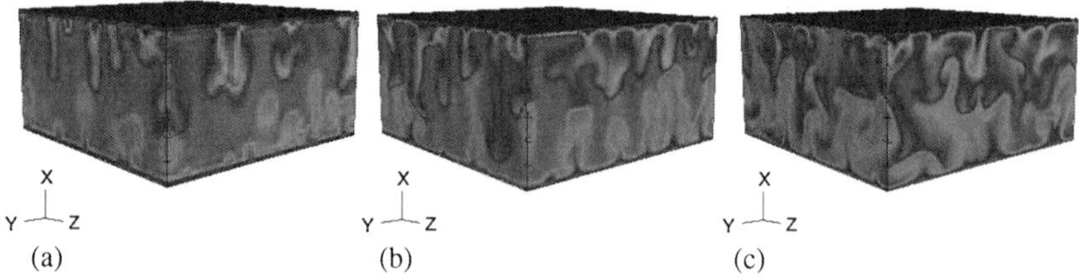

Fig. 2 Instantaneous temperature isocontours taken at the same instant using different grids (a) $64 \times 64 \times 32$, (b) $128 \times 128 \times 64$ and (c) $256 \times 256 \times 128$.

The Nusselt number varying with time, shown in Figure 3c, seems to confirm this behaviour with a Nu increasing with increasing grid resolution after the peak value at a time equal to about five large circulation turn over times, but the available statistics are not enough to fully disclose the Nu dependence on the grid resolution and further investigations are still necessary. The value of the Nusselt number has been calculated within the Oberbeck-Boussinesq approximation for incompressible flow (see e.g. [1]) as:

$$Nu = \frac{1}{H}\int_0^H \frac{< u_z T >_{A,t} - \kappa\partial_z < T >_{A,t}}{\kappa\Delta T/H} = 1 + \frac{H}{\kappa\Delta T} < u_z T >_{V,t} \tag{2}$$

where u_z is the fluid velocity in the wall normal direction, T the temperature and $< \cdot >_{\cdot,t}$ denotes the average over time and over a plane A or the volume V. The mean temperature profile, depicted in Figure 3b, exhibits no strong variation among the different grids, as averaged macroscopic quantities are less sensitive to the grid resolution. The computational grid has a direct impact also on the trajectory of particles dispersed in the flow as visible in Figure 4a, where the mean square displacement of particle position varying in time is shown. The statistic has been calculated with reference to the particle initial position and the time is normalized by the large circulation time scale. A coarser grid seems to under-predict single particle dispersion causing, with its implicit filtering action, particles to be confined in a smaller region of the domain.

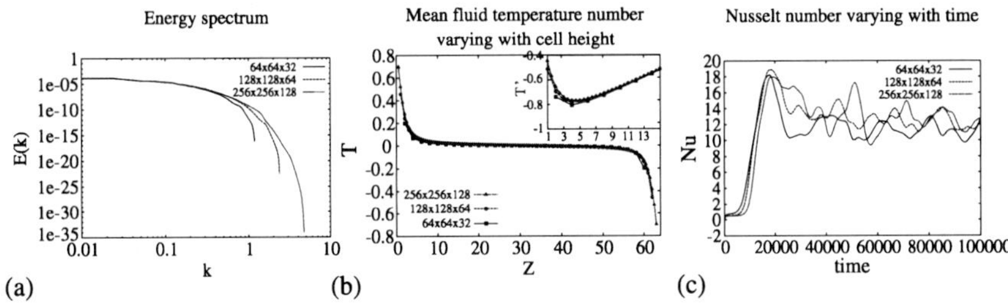

Fig. 3 (a) Instantaneous fluid energy spectra, (b) mean fluid temperature varying with the cell height z and (c) mean Nusselt number varying with time. Comparison between different grid resolutions: ···△··· high, - -○- - -medium and —□— low.

The reason for this can be found in the close relationship between the small scale thermal plumes developing at the wall and particles. The sheetlike plumes close to the wall are responsible, in their morphological evolution, for the formation of mushroomlike plumes in a region confined to few δ_{th} (where δ_{th} is the boundary layer thickness) [8, 6]. These three dimensional plumes are accountable, depending on their temperature, for the upward or downward particle transport. Changing the grid resolution modifies the statistical presence of sheetlike plumes and consequently affects the overall dispersion of particles in the cell. In Figure 4b an instantaneous snapshot of particles colored with their temperature is shown. It is possible to notice that cold regions of fluid (light gray) are surrounded by warm (darker) regions that correspond to the sheetlike structures. Tracer particles are "compressed" at the edges of these systems, and bursts of particles are starting to rise due to the formation of mushroom like type of structures, thus confirming the importance of these plumes in particle suspension.

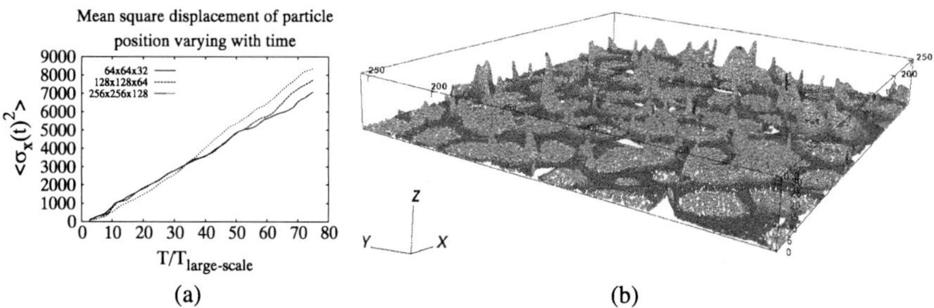

Fig. 4 (a) Mean square displacement of particle position varying in time obtained with three different grid resolutions. (b) Snapshot of particles released in the flow on the more refined grid.

4 Conclusions

A Lattice Boltzmann method coupled with Lagrangian particle tracking has been employed to study the effect of under resolved plumes on tracer particles in turbulent convection. By following in time particles initially released on a horizontal plane close to the bottom wall, it has been possible to show the importance of the mushroom like type of structures in particle resuspension. A test on the effect of the grid resolution on the ability to capture averaged statistics, but also on particle behavior has been discussed. Increasing the grid resolution leads to a better description of the small scales of the flow which are more effective in particle dispersion. Indeed, particles released in a flow obtained with the most refined grid are traveling longer distances with respect to those in the other two resolutions.

References

1. Ahlers, G., Grossman, S. and Lohse, D. (2009) Heat transfer and large scale dynamics in turbulent Rayleigh-Bènard convection. *Rev. Mod. Phys.*, **81**, 503–537.
2. Benzi, R., Succi, S. and Vergassola, M. (1992) The Lattice Boltzmann equation: Theory and applications, *Phys. Rep.* **222**, 145–197.
3. Calzavarini, E., Toschi, F. and Tripiccione, R. (2002) Evidences of Bolgiano-Obhukhov scaling in three dimensional Rayleigh-Bénard convection. *Phys. Rev. E*, **66** 016304.
4. Chen, S. and Doolen, G.D. (1998) Lattice Boltzmann method for fluid flows. *Annu. Rev. Fluid Mech.* **30**, 329–364.
5. Kunnen, R.P.J., Clercx, H.J.H., Geurts, B.J., van Bokhoven, L.J.A., Akkermans, R.A.D. and Verzicco, R. (2008) Numerical and experimental investigation of structure-function scaling in turbulent Rayleigh-Bénard convection. *Phys. Rev. E* **77**, 016302 1–13.
6. Lohse, D. and Xia, K. (2010) Small-Scale Properties of Turbulent Rayleigh-Bénard Convection. *Annu. Rev. Fluid. Mech.*
7. Schumacher, J. (2009) Lagrangian studies in convective turbulence. *Phys. Rev. E* **79**, 056301.
8. Shishkina O. and Wagner C. (2008) Analysis of sheet-like plumes in turbulent Rayleigh-Bénard convection. *J. Fluid Mech.* **599**, 383–404.
9. Succi, S. (2001) The Lattice Boltzmann Equation for Fluid Dynamics and Beyond. *Oxford Science Publications, Oxford.*
10. van Aartrijk, M. and Clercx, H.J.H. (2008) Preferential concentration of heavy particles in stably stratified turbulence. *Phys. Rev. Lett.* **100**, 254501 1–4.

Turbulent convection in a Rayleigh-Bénard cell with solid horizontal plates of finite conductivity

T. Czarnota and C. Wagner

1 Introduction

Turbulent Rayleigh-Bénard convection (RBC) is one of the classical problems of fluid dynamics. In spite of the great effort made in the past to understand the complex physical mechanisms in this type of flow, there are still many open questions which must be answered. In order to understand the effect of conductive horizontal plates on the flow structure in a Rayleigh-Bénard cell the differences between finite and infinite conductivity of the horizontal plates were the aim of many experimental and numerical studies, see [1, 4, 10]. Actually, the horizontal plates are assumed to be isothermal in most numerical simulations while they are conductive in experiments. Even more, the finite conductivity of the plates is one explanation for discrepancies between results obtained in experiments and numerical simulations. A fixed temperature boundary condition corresponds to infinite thermal conductivity of the plates while an imposed heat flux models poorly conducting plates. The effect of these two boundary conditions was recently compared by [10]. Their studies indicated that the heat transfer is suppressed for $Ra > 10^9$ in simulations with plates of finite conductivity whereas for lower Ra the two cases leed to similar flows which agree with observations by [4]. They concluded that below $Ra \approx 10^{10}$ the plate conductivity plays no significant role in the $Nu \sim Ra^{2/7}$ scaling. Nevertheless, the results of [10] as well as those of [4] refer to a cell with infinitely thin plates while the finite thickness of the plates is considered in our simulations. In the present study the scaling exponent β in the Nusselt-Rayleigh relation, i.e. $Nu \sim Ra^{\beta}$ obtained for the cell with highly conducting aluminium plates is similar as those obtained by [5], who studied RBC in the same convection cell and for the same Pr and similar Ra. However, the scaling exponent obtained for the cell with poorly conducting plexiglas plates is found to be slightly lower than those obtained by [5].

DLR Institute for Aerodynamics and Flow Technology, Göttingen, Germany
e-mail: tomasz.czarnota@dlr.de, claus.wagner@dlr.de

H. Kuerten et al. (eds.), *Direct and Large-Eddy Simulation VIII*,
ERCOFTAC Series 15, DOI 10.1007/978-94-007-2482-2_59,

2 Computational setup

The direct numerical simulations are conducted in a rectangular domain with aspect ratio $\Gamma = \hat{W}/\hat{H} = 1$, where $\hat{H}$ is the height and $\hat{W}$ the width of the enclosed fluid. The lateral walls are adiabatic and no-slip and impermeability conditions are applied to all solid walls. The outer sides of the top and bottom solid plates are isothermal with non-dimensional temperatures $T_{top} = -0.5$ and $T_{bot} = +0.5$, respectively. The temperature at the interfaces depends on the solution of the equation (5). Fig. 1 illustrates the geometry and boundary conditions used for the 'solid-fluid case'. The dimensionless thickness of the heating and cooling plates equals $h_s = h_h = h_c = 0.065$. Recently RBC in this cell has been investigated in [5, 6] by assuming the heating and cooling plates to be isothermal and infinitely thin: h_h and $h_c \to 0$ (the 'fluid case').

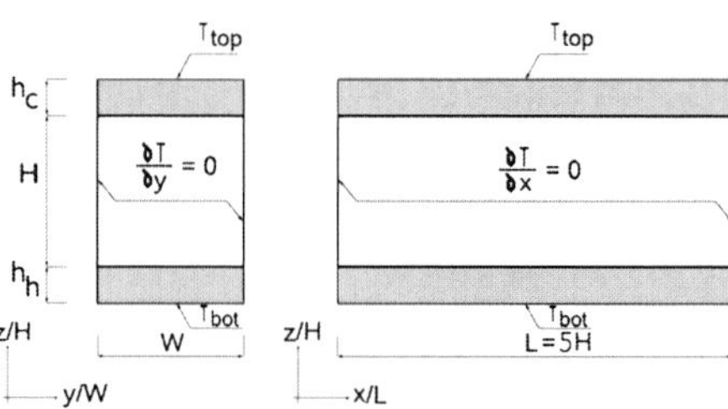

Fig. 1 Geometry of the convection cell with temperature boundary conditions.

The considered governing equations are the incompressible Navier-Stokes equations derived under the assumption of the Boussinesq approximation. Their non-dimensionalization is carried out using $x_i = \hat{x}_i/\hat{W}$, $u_i = \hat{u}_i/(\hat{\alpha}\hat{g}\hat{\Delta T}\hat{H})^{1/2}$, $T = (\hat{T} - \hat{T}_0)/\hat{\Delta T}$, $p = \hat{p}/(\hat{\rho}\hat{\alpha}\hat{g}\hat{H}\hat{\Delta T})$ and $t = \hat{t}(\hat{\alpha}\hat{g}\hat{H}\hat{\Delta T})^{1/2}/\hat{W}$, where $\hat{\alpha}$ is the thermal expansion coefficient and $\hat{\rho}$ is the density. $\hat{g}$ represents the gravitational acceleration which acts in vertical z-direction. $\hat{\Delta T}$ denotes the temperature difference between the outer sides of the heating and cooling plates and $\hat{H}$ is the height of the fluid layer. We denote dimensional quantities with $\hat{}$ and dimensionless without. Finally, the dimensionless form of the governing equations is given by eq. (1)–(2).

$$\frac{\partial u_j}{\partial x_j} = 0, \qquad \frac{\partial u_i}{\partial t} + u_j \frac{\partial u_i}{\partial x_j} + \frac{\partial p}{\partial x_i} = \nu \frac{\partial^2 u_i}{\partial x_j^2} + T\delta_{3i} \tag{1}$$

$$\frac{\partial T}{\partial t} + u_j \frac{\partial T}{\partial x_j} = \kappa_{fluid} \frac{\partial^2 T}{\partial x_j^2}. \tag{2}$$

Here, $u_i (i = x, y, z)$ are the velocity components in i-direction, T and p represent the temperature and pressure, respectively and δ_{ij} is a Kronecker symbol. The non-dimensional kinematic viscosity ν and thermal diffusivity κ_{fluid} are defined by (3) and (4), respectively.

$$\nu = (Pr/\Gamma^3 Ra)^{1/2} \tag{3}$$

$$\kappa_{fluid} = (1/\Gamma^3 Ra Pr)^{1/2}. \tag{4}$$

The present boundary conditions are such that heat conduction through the solid plates needs to be computed, which is done by solving conduction equation

$$\frac{\partial T}{\partial t} = \sqrt{\frac{1}{\Gamma^3 Ra Pr}} \frac{\hat{\kappa}_{solid}}{\hat{\kappa}_{fluid}} \frac{\partial^2 T}{\partial x_j^2}. \tag{5}$$

In eq. (5) $\hat{\kappa}_{solid}$ and $\hat{\kappa}_{fluid}$ are dimensional thermal diffusivities of the solids and the fluid, respectively. In the present study we simulate Rayleigh-Bénard convection for two different thermal boundary conditions at the horizontal plates. One is called 'aluminium-fluid case' and the other one 'plexiglas-fluid case' assuming that the solid plates are made out of aluminium and plexiglas, respectively. Considering aluminium the thermal diffusivity $\hat{\kappa}_{solid}$ equals $8.418 \times 10^{-5} m^2/s$, while for plexiglas $\hat{\kappa}_{solid} = 7.49 \times 10^{-8} m^2/s$ and for air $\hat{\kappa}_{fluid} = 2.216 \times 10^{-5} m^2/s$. The dimensionless parameters, Prandtl and Rayleigh numbers, are defined by (6) and (7), respectively.

$$Pr = \hat{\nu}/\hat{\kappa} \tag{6}$$

$$Ra = \hat{\alpha}\hat{g}\hat{H}^3 \hat{\Delta T}/\hat{\nu}\hat{\kappa}_{fluid}. \tag{7}$$

In order to investigate the role of the non-uniform, transient solid-fluid boundary conditions on the flow field the results obtained for the 'fluid case' and the 'solid-fluid cases' are compared for the same effective Rayleigh number Ra_{eff} and thus, the same effective dimensionless temperatures T_{eff} as defined in eq. (8) and (9), respectively.

$$Ra_{eff} = \hat{\alpha}\hat{g}\hat{H}^3 \Delta\hat{T}_{eff}/\hat{\nu}\hat{\kappa}_{fluid} \tag{8}$$

$$T_{eff} = T\Delta T/\Delta T_{eff} \tag{9}$$

where ΔT_{eff} denotes the temperature difference between top and bottom solid-fluid interface which is not known a priori. Thus, the Ra_{eff} and T_{eff} used for comparison of the considered cases is calculated in the post-processing. The method used for the direct numerical simulations is based on the volume balance procedure by [7] and the second order accurate explicit Euler-Leapfrog time discretization scheme. Spatial derivaties and cell face velocities are approximated by piecewise integrated fourth-order accurate polynomials, where the velocity components are stored on staggered grids as described in more detail in [8]. In order to sufficiently resolve the boundary layers the grid points are clustered in the vicinity of the walls using a hyperbolic tangential and the minimum number of nodes in the thermal boundary layer is estimated for all considered cases with the criterion by [9] which provides the needed grid points to sufficiently resolve a Prandtl-Blasius type boundary layer. This approach is valid for Ra numbers below 10^9 and leads to at least 10 grid points which are needed to resolve the thermal boundary layer for the Ra number considered in this study.

3 Results

An important quantitative result is the behavior of the Nusselt number as a function of the Rayleigh number. In order to investigate whether the $Nu \sim Ra$ follows the 2/7-scaling found by [2] and later proposed by [3] the heat transfer is calculated from eq. (10). The mean Nusselt number represents the heat flux which is time- and area-averaged in seven horizontal cross-sections S_z (top and bottom interface, half of the cell, single and double thickness of the thermal boundary layer). The error of the Nu number, estimated by varying the length of the time-averaging is less then 0.5% for each considered case.

$$Nu = \frac{\langle u_1 T \rangle_{t,S_z} - \kappa_{fluid} \langle \frac{\partial T}{\partial x_1} \rangle_{t,S_z}}{\kappa_{fluid} \Delta T / H}. \tag{10}$$

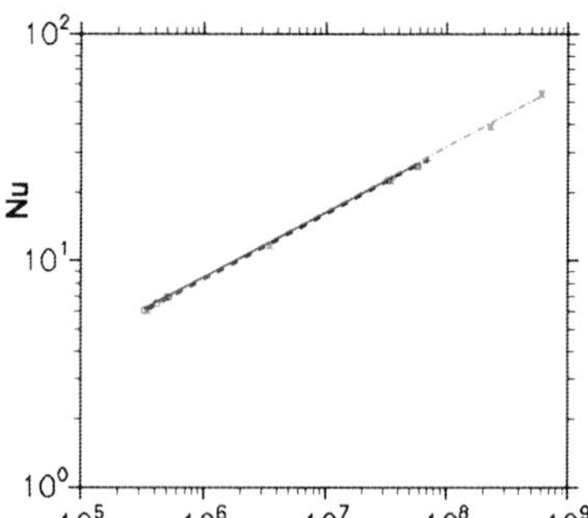

Fig. 2 Nusselt number Nu versus Rayleigh number Ra; (x) 'fluid case', (- · -) $Nu \sim Ra^{0.286}$; (+) 'aluminium-fluid case', (- -) $Nu \sim Ra^{0.285}$; (o) 'plexiglas-fluid case', (—) $Nu \sim Ra^{0.283}$.

Figure (2) reflects the $Nu \sim Ra$ dependence obtained for the 'fluid case' and the 'solid-fluid cases'. The β exponent obtained for highly conducting aluminium plates equals 0.265 which is similar to these obtained by [4, 5, 10]. This behavior of the $Nu \sim Ra$ relation holds because the temperature at the solid-fluid interfaces coincides with that applied for the 'fluid case'. A fit of five Ra data points generated for the 'plexiglas-fluid case' yields $Nu \approx 0.17 \times Ra^{0.283}$, whereas [5] obtained the Ra exponent $\beta = 0.286$ for $3.5x10^5 < Ra < 6.0x10^8$. From our results, we conclude that the deviation from the 2/7-scaling in case of poorly conducting plates is due to the fact that the solid-fluid boundary conditions are non-uniform. Changes of the time-averaged temperature gradient at the bottom interface are illustrated in figure (3). In fact, the heat transferred from the solid-fluid interfaces into the air is directly proportional to the temperature gradients at these interfaces. The present analysis reveals that with increasing Ra numbers, the conditions at the solid-fluid interfaces better reflect constant heat flux than constant temperature which was also observed by [4, 10]. Additionally, poorly conducting plates reveal smaller and more homogeneous temperature gradients at the interfaces than infinitely thin plates, which leads to slightly lower Nu. The thermal plumes interacting with the poorly conducting horizontal plates modify the plate temperature such that the temperature difference across the fluid layer is smaller than in the highly conducting case. In order to maintain the same effective Ra number, the temperature difference between outer sides of the top and bottom plates is fixed higher for the 'plexiglas-fluid' case what is visible in figure (4). Further, the comparison of the mean temperature profiles presented in figure (4) shows that, non-uniform conditions at the interfaces for the 'plexiglas-fluid' case have no influence on the mean temperature field.

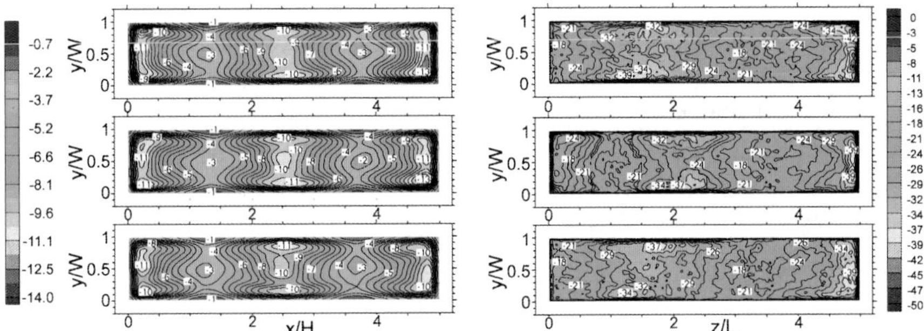

Fig. 3 Distribution of time-averaged temperature gradients at the bottom interface; left figures: $Ra_{eff} = 3.5 \times 10^5$; right figures: $Ra_{eff} = 5.8 \times 10^7$; 'fluid case' (upper figures), 'aluminium-fluid case' (middle figures), 'plexiglas-fluid case' (lower figures).

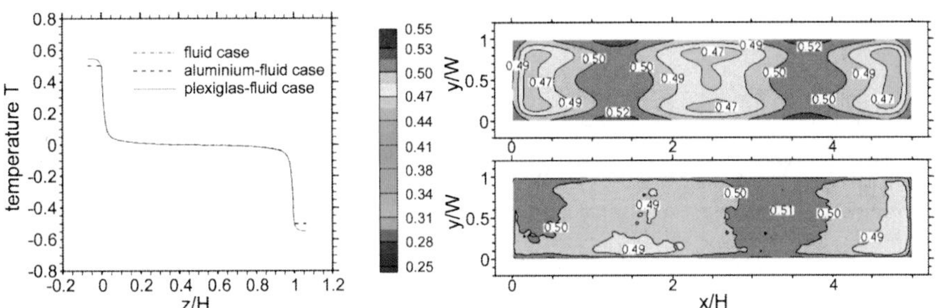

Fig. 4 Left figure: the mean temperature profile for $Ra_{eff} = 5.8 \times 10^7$; right figures: time-averaged temperature distribution at the bottom interface, upper/lower: $Ra_{eff} = 3.5 \times 10^5/Ra_{eff} = 5.8 \times 10^7$.

4 Conclusions

Many numerical studies of Rayleigh-Bénard convection performed in the past use fixed temperature or fixed flux boundary conditions of infinitely thin plates. In the present study we assume fixed temperature boundary conditions at the outer side of the solid plates with finite thickness. Using such boundary conditions allows to analyse the influence of non-uniform, transient solid-fluid boundary conditions on the flow field and the heat transfer. Analyzing the results of the conducted direct numerical simulations it is found that taking into account the finite thickness and conductivity of the top and bottom plates might slightly change the exponent of the Nusselt-Rayleigh relation. In particular, the Ra exponent obtained for the 'plexiglas-fluid case' for $3.27 \times 10^5 < Ra < 5.8 \times 10^7$ equals 0.283. However, the scaling exponent obtained for the 'aluminium-fluid case' is similar as those obtained by [4, 5, 10]. It is shown that these differences in Nusselt-Rayleigh relation depends on the physical properties of the boundaries, and hence they appear due to non-uniform boundary conditions at the solid-fluid interfaces. Further, it is found that with increasing Ra numbers, the conditions at the solid-fluid interfaces better reflect constant heat flux than constant temperature which is in a good agreement with the results obtained by [4, 10]. Additionally, poorly conducting plates reveal smaller and more homogeneous temperature gradients at the interfaces than infinitely thin plates, which leads to slightly lower Nu. Finally, it is observed that the non-uniform boundary conditions at the interfaces for poorly conducting plates have almost no influence on the mean temperature field.

References

1. Brown, E., Nikolaenko, A., Funfschilling, D. and Ahlers, G. (2005), Heat transport in turbulent Rayleigh-Bénard convection: Effect of finite top- and bottom-plate conductivity, Phys. Fluids, Vol. 17, 075108.
2. Castaing, B., Gunarante, G., Heslot, F., Kadanoff, L., Libchaber, A., Thomae, S., Wu, X.Z., Zaleski, S. and Zanetti, G. (1989), Scaling of hard thermal turbulence in Rayleigh-Bénard convection, J. Fluid Mech., Vol. 204, pp. 1–30.
3. Grossmann, S. and Lohse, D. (2000), Scaling in thermal convection: A unifying theory, J. Fluid Mech., Vol. 407, pp. 27–56.
4. Johnston, H. and Doering, C.R. (2009), Comparison of Thermal Convection between Conditions of Constant Temperature and Constant Flux, PRL 102, 064501.
5. Kaczorowski, M. and Wagner, C. (2007), Direct Numerical Simulation of Turbulent Convection in a Rectangular Rayleigh-Bénard Cell, 5th International Symposium on Turbulence and Shear flow Phenomena, Vol. 2, pp. 499–504.
6. Kaczorowski, M. and Wagner, C. (2009), Analysis of the thermal plumes in turbulent Rayleigh-Bénard convection based on well-resolved numerical simulations, J. Fluid Mech., Vol. 618, pp. 89–112.
7. Schumann, U., Grötzbach, G. and Kleiser, L. (1979) Direct numerical simulations of turbulence, In Prediction Methods for Turbulent Flows number, VKI-lecture series 1979 2. Von Kármán Institute for Fluid Dynamics.
8. Shishkina, O. and Wagner, C. (2007), Boundary and interior layers in turbulent thermal convection in cylindrical containers, Int. J. Sci. Comp. Math., Vol. 1(2/3/4), pp. 360–373.
9. Shishkina, O., Stevens, R.J.A.M., Grossmann, S. and Lohse, D. (2010), Boundary layer structure in turbulent thermal convection and its consequences for the required numerical resolution, New J. Phys., Vol. 12, 075022.
10. Verzicco, R. and Sreenivasan, K.R. (2008), A comparison of turbulent thermal convection between conditions of constant temperature and constant heat flux, J. Fluid Mech., Vol. 595, pp. 203–219.

Non-Oberbeck-Boussinesq effects in three-dimensional Rayleigh-Bénard convection

Susanne Horn, Olga Shishkina and Claus Wagner

1 Introduction

To study the classical problem of Rayleigh-Bénard convection, i.e. a fluid layer confined between a heating-plate at the bottom and a cooling-plate at the top, a common assumption is that all material properties are temperature independent, except for the density ρ within the buoyancy part, that changes like

$$\rho(T) = \rho_0 (1 - \alpha \cdot (T - T_0)), \tag{1}$$

with a constant isobaric expansion coefficient α. In combination with the condition of an incompressible fluid this is the so-called Oberbeck-Boussinesq (OB) approximation [2, 5].

However, as illustrated in Fig. 1, obviously, for water neither the isobaric expansion coefficient α nor the kinematic viscosity ν, thermal diffusivity κ and the heat conductivity Λ can be considered to be constant with temperature, if the temperature difference ΔT between the top and bottom plate becomes too large. In particular, this also means, if one wants to reach high Rayleigh numbers (see eq. (6)), by simply increasing ΔT, the fluid properties become non-uniform and it is necessary to consider non-Oberbeck-Boussinesq (NOB) effects.

A more rigorous deduction was given by Gray & Giorgini [4]. Their theoretical investigations show that the region of validity of the OB approximation in the case of the operating fluid water is very restrictive. That is, for the here considered mean temperature of $T_m = 40°C$, the temperature difference ΔT must be less than $0.3K$

Susanne Horn
DLR - Institute of Aerodynamics and Flow Technology, Göttingen (Germany)
e-mail: Susanne.Horn@dlr.de

Olga Shishkina
e-mail: Olga.Shishkina@dlr.de

Claus Wagner
e-mail: Claus.Wagner@dlr.de

H. Kuerten et al. (eds.), *Direct and Large-Eddy Simulation VIII*, ERCOFTAC Series 15, DOI 10.1007/978-94-007-2482-2_60,

in order to guarantee a residual error of at the most 10%. However, experiments revealed that certain NOB effects can compensate each other and still for several quantities, e.g. the Nusselt number Nu, the same values as under OB conditions can be obtained [1].

We aim to clarify and quantify the actual influence of temperature dependent material properties in turbulent Rayleigh-Bénard convection in water by means of three-dimensional Direct Numerical Simulations (DNS).

2 Numerical methodology

Firstly, to employ OB assumptions, we make use of a parallel version of the finite volume code described by Shishkina & Wagner [8, 9]. Secondly, for the purpose of investigating NOB effects, the code has been advanced by taking temperature dependent material properties into account, discussed in more detail below.

2.1 Non-Oberbeck-Boussinesq effects

The viscosity $X = \nu$, thermal conductivity $X = \Lambda$ and the density in the buoyancy term $X = \rho$ are described by polynomials up to cubic order,

$$\frac{X - X_m}{X_m} = \sum_i a_i (T - T_m)^i, \tag{2}$$

where the coefficients a_i are adopted from Ahlers et al. [1], while the density ρ, except within the buoyancy term, and the isobaric specific heat capacity c_p are set constant to their values at the arithmetic mean temperature T_m. This approach is accurate for water (see Fig. 1) and accounts for the major relevant non-Oberbeck-Boussinesq effects.

Hence, the equations to be solved for obtaining the flow characteristics are the continuity equation for incompressible fluids (3), the Navier-Stokes equation (4) and the heat transfer equation (5), including the aforementioned material functions:

$$\partial_i u_i = 0, \tag{3}$$

$$\rho_m(\partial_t u_i + u_j \partial_j u_i) + \partial_i p = \partial_i\big(\nu \rho_m(\partial_j u_i + \partial_i u_j)\big) + (\rho_m - \rho) g \delta_{i3}, \tag{4}$$

$$\rho_m c_{p,m}(\partial_t T + u_i \partial_i T) = \partial_i(\Lambda \partial_i T). \tag{5}$$

Here u_i are the velocity components, g is the gravitational acceleration, δ_{ij} is the Kronecker symbol, where the third component is pointing in opposite direction of the buoyany force, and ∂_t and ∂_i denote the temporal and spatial partial derivatives, respectively.

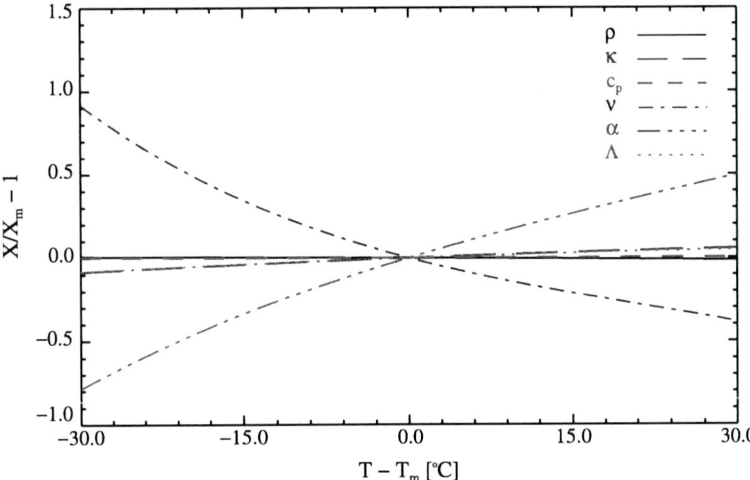

Fig. 1 Relative deviations of water properties X from their values X_m at a mean temperature of $T_m = 40°$, according to Ahlers et al. [1]; black solid line, density ρ; green dashed line, thermal diffusivity κ; orange short dashed line, specific heat capacity c_p; purple dashed dotted line, kinematic viscosity ν; blue dashed triple-dotted line, expansion coefficient α; pink dotted line, thermal conductivity Λ.

2.2 Computational setup

We perform our simulations of Rayleigh-Bénard convection in a cylindrical domain and apply a conservative 4th order finite volume method based on the Chorin ansatz [3] and on a direct solution of the resulting Poisson equation.

The control parameters for the simulations are essentially the Rayleigh and the Prandtl number defined at the mean temperature, as well as the aspect ratio,

$$Ra = \frac{\alpha_m g \Delta T H^3}{\kappa_m \nu_m}, \quad Pr = \frac{\nu_m}{\kappa_m}, \quad \Gamma = \frac{D}{H}, \tag{6}$$

with D being the diameter and H the height of the cylinder. At the moment we restrict ourselves to Rayleigh-Bénard cells with an aspect ratio of one.

As boundary conditions, we impose that the lateral walls are adiabatic and the top and bottom plate are isothermal with a fixed dimensionless temperature of $\hat{T}_t = \frac{T_t - T_m}{\Delta T} = -0.5$ and $\hat{T}_b = \frac{T_b - T_m}{\Delta T} = 0.5$, respectively. We apply impermeability and no-slip conditions for the velocity and a 2π-periodicity in ϕ-direction.

For our performed DNS we use staggered computational meshes consisting of $N_z \times N_\phi \times N_r = 384 \times 512 \times 192$ nodes. These nodes are distributed equidistantly in ϕ-direction and non-equidistantly in z- and r-direction. They are clustered in the boundary layers to fulfil the requirements of a DNS [6], that is for $Pr > 1$ the global mesh size is everywhere

$$h^{global} \leq \frac{H}{Ra^{1/4}(Nu-1)^{1/4}} \tag{7}$$

and to ensure that also the NOB simulations are properly resolved twice as many nodes N as required in the OB case are placed in the thermal and viscous boundary layer, λ_θ and λ_v,

$$N(\lambda_\theta) > 2 \cdot \left(0.69 Nu^{1/2}\right), \qquad N(\lambda_v) > 2 \cdot \left(0.70 Nu^{1/2} Pr^{1/3}\right). \tag{8}$$

Consequently, for our simulations of water with $Nu^{OB} = 32.5$ [7], $Pr = 4.38$ and $Ra = 10^8$ we placed at least 12 nodes in the viscous boundary layers.

3 Results

We investigated three different NOB cases with $\Delta T = 20K, 40K$ and $60K$, each with a mean temperature of $T_m = 40°C$, and compared them to the OB case.

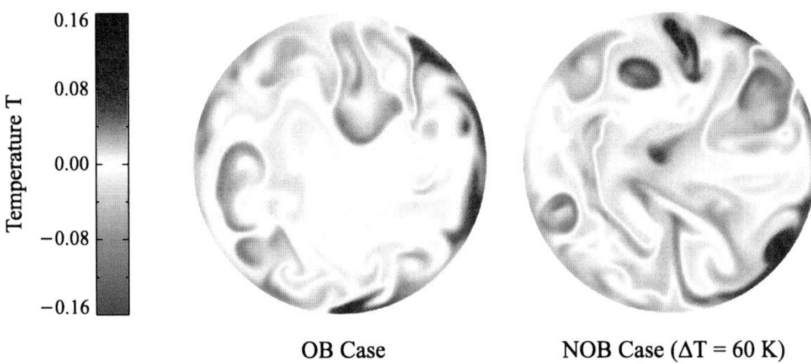

Fig. 2 Cross sections of the dimensionless temperature field for the working fluid water with $Pr = 4.38$ and $Ra = 10^8$ through the centre of the cylinder. A bigger area of the temperature in the NOB case is above the arithmetic mean temperature T_m (red), while in the OB case most of the area is the actual T_m and the rest is equally cold and warm. The exact values of the center temperature for all 3 different NOB cases can be found in table 1.

One of the best known NOB phenomena is the deviation of the temperature in the centre of the cell T_c and the arithmetic mean temperature $T_m = \frac{T_t + T_b}{2}$. In our simulations we found an increasing centre temperature with increasing NOBness, i.e. increasing ΔT, the actual values are given in table 1. To visualize this feature a slice through the center of the cylinder is presented in Fig. 2 and the temperature field for the OB case (2a) and the NOB case with $\Delta T = 60K$ (2b), respectively, are shown.

This, of course, also affects the temperature probability functions (see Fig. 3) and the mean temperature profiles (see Fig. 4).

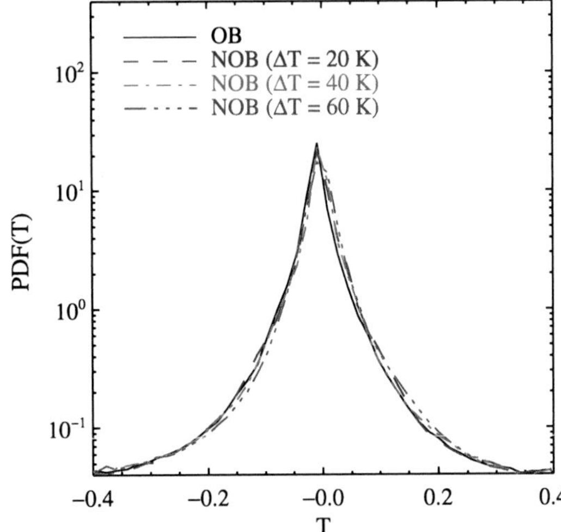

Fig. 3 Instantaneous normalized probability density function (PDF) of the dimensionless temperature over the whole volume; comparison of three different NOB cases with the OB case. All NOB cases show a shift to the right, i.e. the warm tail.

Table 1 Centre temperatures

Case	T_c
OB	40.00°C
NOB ($T_m = 40°\text{C}, \Delta T = 20K$)	40.08°C
NOB ($T_m = 40°\text{C}, \Delta T = 40K$)	40.11°C
NOB ($T_m = 40°\text{C}, \Delta T = 60K$)	40.28°C

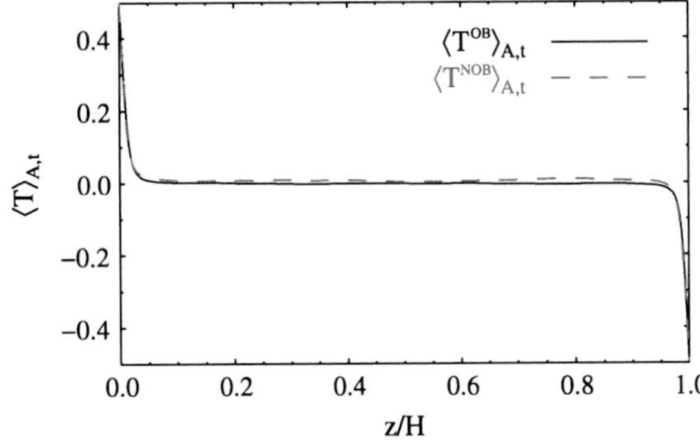

Fig. 4 Mean temperature profiles averaged in time and in the r-ϕ-plane A for water ($Pr = 4.38$, $Ra = 10^8$) in the OB case (solid black line) and in the representative NOB case with $\Delta T = 60K$ and $T_m = 40°C$ (blue dashed line). In the NOB case the enhanced centre temperature T_c and becomes visible.

4 Conclusions and outlook

In the presented study, three different NOB simulations were compared to an OB simulation of Rayleigh-Bénard convection of water in a cylindrical container of unity aspect ratio. For this purpose an already existing code was advanced by implementing the most relevant temperature dependencies of the material properties. We obtained modified mean temperature profiles and PDFs and an increase of the centre temperature T_c with the temperature difference ΔT, i.e. NOBness. In order to validate our findings, they will be checked against experimental data, e.g. from Ahlers et al. [1], and two-dimensional simulations by Sugiyama et al. [10] and compared to the three-dimensional OB simulations with the same Rayleigh and Prandtl numbers by Shishkina & Thess [7]. An objective for future work is to investigate other fluids and higher Rayleigh numbers, since it is very likely that NOB effects will be more pronounced then.

Acknowledgements The authors acknowledge support by the *Deutsche Forschungsgemeinschaft (DFG)* under grant SH405/2-1. Furthermore, the authors would also like to thank the *Leibniz-Rechenzentrum (LRZ)* in Garching for providing computational resources on the national supercomputer HLRB-II under grant pr47he.

References

1. Ahlers, G., Brown, E., Fontenele Araujo, F., Funfschilling, D. Grossmann, S. Lohse, D.: Non-Oberbeck-Boussinesq effects in strongly turbulent Rayleigh-Bénard convection. J. Fluid Mech. **569**, 409–445 (2006)
2. Boussinesq, J. V.: Theorie Analytique de la Chaleur 2. Gauthier-Villars (1903)
3. Chorin, A. J.: A Numerical Method for Solving Incompressible Viscous Flow Problems. Journal of Computational Physics **2**, 12–26 (1967)
4. Gray, D. D., Giorgini, A.: The validity of the Boussinesq approximation for liquids and gases. Int. J. Heat Mass Transfer **19**, 545–551 (1976)
5. Oberbeck, A.: Ueber die Wärmeleitung der Flüssigkeiten bei Berücksichtigung der Strömungen infolge von Temperaturdifferenzen. Annalen der Physik **243**, 271–292 (1879)
6. Shishkina, O., Stevens, R. J. A. M., Grossmann, S., Lohse, D.: Boundary layer structure in turbulent thermal convection and its consequences for the requiered numerical resolution. New J. Physics **12**, 075022 (2010)
7. Shishkina, O., Thess, A.: Mean temperature profiles in turbulent Rayleigh-Bénard convection of water. J. Fluid Mech. **633**, 449–460 (2009)
8. Shishkina, O., Wagner, C.: A fourth order finite volume scheme for turbulent flow simulations in cylindrical domains. Computers & Fluids **36**, 484–497 (2007)
9. Shishkina, O., Wagner, C.: Boundary and interior layers in turbulent thermal convection in cylindrical containers. Int. J. Computing Science and Mathematics **1**, 360–373 (2007)
10. Sugiyama K., Calzavarini, E., Grossmann, S., Lohse, D.: Flow organization in two-dimensional non-Oberbeck-Boussinesq Rayleigh-Bénard convection in water. J. Fluid Mech. **637**, 105–135 (2009)

Analysis of the large-scale circulation and the boundary layers in turbulent Rayleigh-Bénard convection

M. Kaczorowski, O. Shishkina, A. Shishkin, C. Wagner and K.-Q. Xia

1 Introduction

Rayleigh-Bénard convection (RBC) is a phenomenon occurring in fluids heated and cooled by a wall from below and above, respectively. The vertical heat transfer through the fluid is primarily defined by the Rayleigh number $\mathscr{R}a = \alpha g H^3 \Delta T/(\nu\kappa)$, the Prandtl number $\mathscr{P}r = \nu/\kappa$ and the aspect ratio $\Gamma = W/H$ of the convection cell, where the fluid is characterized through the thermal expansion coefficient α, the kinematic viscosity ν and the thermal diffusivity κ. The geometry is characterized by the height H and the width W of the cell and the temperature difference between the horizontal plates is ΔT.

The large scale circulation (LSC) is an important feature for the understanding of the heat transfer mechanisms in turbulent convection. Recently the $\mathscr{R}a$-dependency of the LSC has been investigated experimentally by Sun & Xia [16]. Their experimental findings obtained in a cylindrical geometry with $\Gamma = 1$ and $\mathscr{P}r = 4.3$ suggest that the LSC is $\mathscr{R}a$-dependent in terms of its Reynolds number $\mathscr{R}e_\ell = U_\ell\,\ell/\nu$ based on its mean velocity U_ℓ and its path length ℓ. As pointed out by Ahlers [1] a $\mathscr{R}a$-dependency of the path length of the LSC would introduce another parameter in the Grossmann & Lohse (GL) theory (see [4]), since the scale ℓ would no-longer be constant as assumed by the theory. We therefore investigate the scaling of the LSC in the cube numerically, hence extending the analysis of [16] to a different geometry, while keeping $\mathscr{P}r$ the same.

However, the LSC not only shows a $\mathscr{R}a$-dependency, but also causes a different scaling of the viscous boundary layers at the horizontal plates (p) and the side-walls (sw), see [8, 12]. Additionally Qiu & Tong [10] find that velocities at the side-walls and the horizontal plates are different. The ratio of the plate and side-wall velocities and viscous boundary layer (BL) thicknesses was theoretically investigated by

The Chinese University of Hong Kong - Dept of Physics, Hong Kong, China and DLR Göttingen - Institute of Aerodynamics and Flow Technology, Göttingen, Germany, e-mail: matthias@phy.cuhk.edu.hk

H. Kuerten et al. (eds.), *Direct and Large-Eddy Simulation VIII*, ERCOFTAC Series 15, DOI 10.1007/978-94-007-2482-2_61,

Grossmann & Lohse [3]. Assuming a *plate filling* (PF) and a *laterally restricted* (LR) flow, they respectively derive the following relations for the ratio of the BL thickness: $\delta_{sw}/\delta_p = f_{LR}$ or $\delta_{sw}/\delta_p = f_{PF}$, where

$$f_{LR} = \Gamma^{-1} \quad \text{and} \tag{1}$$

$$f_{PF} = 4a\Gamma^{-5/2}\,\mathscr{Re}_p^{-1/2}, \tag{2}$$

with $a \approx 0.5$ (see [9]) and $\mathscr{Re}_p = U_p H/\nu$.

2 Governing equations and numerical method

In the present analysis we assume the applicability of the Boussinesq approximation, and hence an incompressible fluid. Substitution of $x_i = \hat{x}_i/\hat{H}$, $u_i = \hat{u}_i/(\hat{\alpha}\hat{g}\Delta T\hat{H})^{1/2}$, $T = (\hat{T} - \hat{T}_0)/\Delta T$, $p = \hat{p}/(\hat{\rho}\hat{\alpha}\hat{g}\hat{H}\Delta T)$ and $t = \hat{t}(\hat{\alpha}\hat{g}\Delta T\hat{H})^{1/2}/\hat{H}$ into the governing equations, where $\hat{\cdot}$ denotes dimensional values, yields the following set of non-dimensional equations

$$\begin{aligned} \partial\mathbf{u}/\partial t + \mathbf{u}\cdot\nabla\mathbf{u} + \nabla p &= \mathscr{Ra}^{-1/2}\mathscr{Pr}^{1/2}\Delta\mathbf{u} + T\mathbf{e}_z, \\ \partial T/\partial t + \mathbf{u}\cdot\nabla T &= \mathscr{Ra}^{-1/2}\mathscr{Pr}^{-1/2}\Delta T, \\ \nabla\cdot\mathbf{u} &= 0. \end{aligned} \tag{3}$$

Here $\mathbf{u}$ is the velocity vector, T and p represent the temperature and pressure, respectively. The gravity vector is acting in the vertical direction, *i.e.* the $-\mathbf{e}_z$ direction. Equations (3) are discretized employing the finite volume method on Cartesian, non-equidistant meshes in all three coordinate directions. For details on the discretization and the solver the reader might refer to [13] and [6, 7], respectively.

The geometry of the cell is defined by a square horizontal cross-section and an aspect ratio $\Gamma \approx 1$. We note that the aspect ratio of our geometry is identical with the aspect ratio of the cylindrical cells chosen by Shishkina & Thess [15] and Funfschilling *et al.* [2]. We employ adiabatic boundary conditions in all horizontal directions and isothermal boundary conditions on the top and bottom walls, respectively. No-slip and impermeability conditions are applied to all walls. The fluid under investigation is characterized by $\mathscr{Pr} = 4.38$, *i.e.* similar to water. After initializing the flow field the simulations have been run for at least $t \approx 140$ time units, corresponding to at least 4 LSC turn-over times, before statistical data was collected. For $\mathscr{Ra} \leq 3\times10^7$ $178\times178\times194$ grid points are used to resolve the 2 horizontal and the vertical direction and $194\times194\times222$ grid points are used for $\mathscr{Ra} \geq 5\times10^7$. Hence, the resolution requirements proposed by Shishkina *et al.* [14] is satisfied for $\mathscr{Ra} \lessapprox 3\times10^8$. For $\mathscr{Ra} \leq 10^9$ the Grötzbach criterion [5] is fulfilled.

3 Results

Global flow structure

The heat flux, namely the Nusselt number $\mathscr{N}u = (\mathscr{R}a\mathscr{P}r)^{1/2}u_zT - \partial T/\partial z$, of the present simulations are compared with DNS data [15] and experimental data for a cube and a cylinder with the same $\mathscr{P}r$ and Γ [12, 11]. Both, numerical and experimental data show that there are only negligible differences in $\mathscr{N}u$ between the two geometries. Comparing the mean $\mathscr{N}u$ calculated from the above definition with calculations based on the viscous and thermal dissipation rates yields a maximum error in $\mathscr{N}u$ of $\pm 1\%$.

The streamlines in figure 1 (*a*) calculated from the wall shear stresses illustrate the mean flow structure on the bottom plate of the cube for $\mathscr{R}a = 8 \times 10^8$. It reveals that the mean flow is oriented in one of the diagonal directions of the cell, showing a plate filling behavior, cf. [3], with complex flow structures in the vicinity of the side-walls. A visualization of the flow in both diagonal planes is provided in figure 1 (*b*), where the formation of the LSC and secondary flow structures can be observed.

Characterization of the LSC

A PDF of the flow direction φ in location p ($z \approx \delta_u$) is shown in figure 2. It can be seen that the mean flow is oriented in the diagonal direction. The strength fluctuations, *i.e.* φ_{rma}, strongly depends on $\mathscr{R}a$ and can reach $\varphi_{rms} > 40°$ for $\mathscr{R}a = 10^7$ and reduce to $\varphi_{rms} < 20°$ for $\mathscr{R}a = 10^9$.

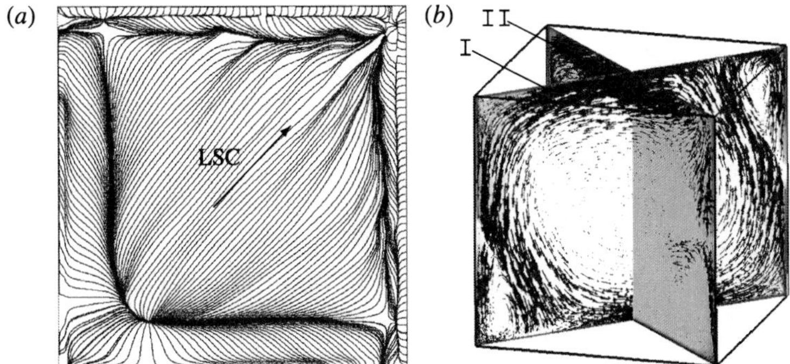

Fig. 1 Mean flow for $\mathscr{R}a = 8 \times 10^8$: (*a*) Streamlines of the wall shear stress (wind stress) on the bottom plate of the convection cell illustrating the 'plate filling' (cf. [3]) behavior of the flow and the complex flow structure near the side walls. (*b*) Temperature distribution (color range from blue to red for $-0.05 \leq T \leq 0.05$) and 3D velocity vectors in the diagonal planes of the cube. The LSC is observed in plane I and secondary flow structures in plane II.

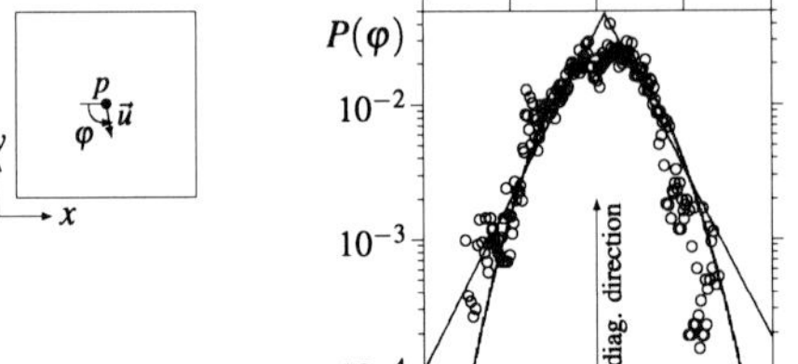

Fig. 2 PDF of the flow direction φ from a point measurement in p ($\circ$) for $\mathscr{Ra} = 8 \times 10^8$. A Gaussian (thick line) and a Laplacian (thin line) distribution are given for reference.

In order to analyse the LSC, we characterize it by the maximum in-plane velocities in any radial direction around the center of the plane. Figure 3 (a) shows the points in the LSC-plane determined through this method, illustrating the $\mathscr{Ra}$-dependence of the LSC's shape. The local velocity U_{LSC} of the LSC is plotted in figure 3 (b) as a function of its angular position β, which is zero in the middle of the left wall and defined in the counter clock-wise direction, *i.e.* the flow direction. It is observed that the velocities are highest close to the walls, where the flow is almost parallel to the horizontal plates or the side walls. Integration of the path length ℓ of the LSC and averaging the local velocities to $U_\ell = \langle U_{LSC\ell} \rangle$, allows to calculate the Reynolds number of the LSC using its mean properties: $\mathscr{Re}_\ell = U_\ell \, \ell / \nu$. Experimentally the LSC Reynolds number is typically measured using a local point measurement, *e.g.* in p. In Figure 4 (a) the Reynolds number scaling based on $\mathscr{Re}_\ell$ is therefore compared with $\mathscr{Re}_p = U_p H / \nu$, where U_p is the velocity at the edge of the viscous BL in the center of the bottom plate. Note that $U_p = U_{LSC}(\beta = 90°)$. It can be seen that both methods exhibit a slightly different $\mathscr{Re}$-$\mathscr{Ra}$ scaling: $\mathscr{Re}_\ell \sim \mathscr{Ra}^{0.62 \pm 0.01}$ and $\mathscr{Re}_p \sim \mathscr{Ra}^{0.66 \pm 0.02}$; for the sake of improved accuracy the least-squares fit was carried out for the compensated data. The scaling $\mathscr{Re} \sim \mathscr{Nu}^\alpha$ is analyzed in figure 4 (b). Theoretical predictions based on Prandtl-Blasius theory suggest $\alpha = 2$, which is confirmed by our results based on the mean quantities of the LSC.

Boundary layer scaling

In order to test the validity of (1) and (2) the ratio of the BL thicknesses (based on the slope method) in the central position of the horizontal plates, denoted as p, and the side-walls at mid-height, denoted as sw, is analyzed. Since $\mathscr{Re}_p \sim \mathscr{Ra}^{0.66}$, it follows from (2) that in the case of plate filling flow $\delta_{sw}/\delta_p = f_{PF} \sim \mathscr{Ra}^{-0.33}$ and for laterally restricted flow $\delta_{sw}/\delta_p = f_{LR} = const$. Figure 5 reveals that $\delta_{sw}/\delta_p \approx 1$ and close to the prediction of f_{LR}.

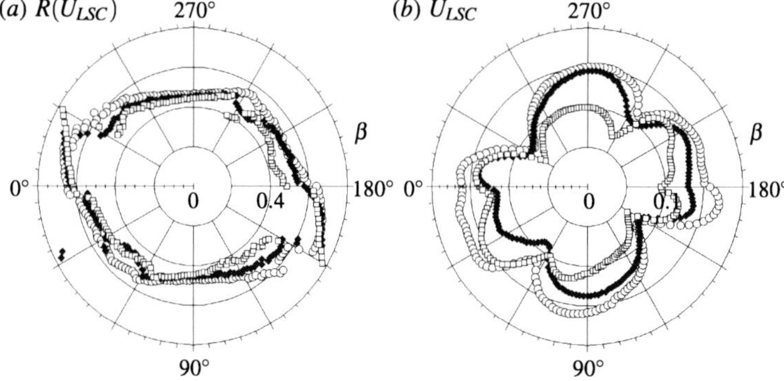

Fig. 3 Properties of the the LSC: (a) Position of the characteristic velocities U_ℓ in the main diagonal I and (b) U_ℓ as a function of its angular position β for $\mathscr{R}a = 3 \times 10^7$ (□), $\mathscr{R}a = 3 \times 10^8$ (♦) and $\mathscr{R}a = 10^9$ (○).

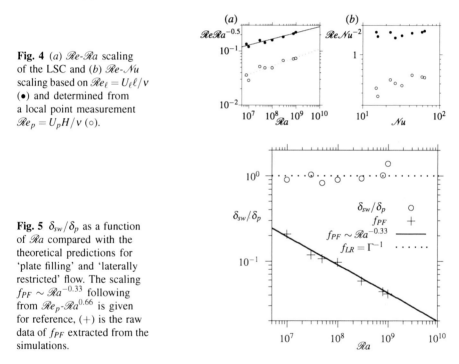

Fig. 4 (a) $\mathscr{R}e$-$\mathscr{R}a$ scaling of the LSC and (b) $\mathscr{R}e$-$\mathscr{N}u$ scaling based on $\mathscr{R}e_\ell = U_\ell \ell / \nu$ (•) and determined from a local point measurement $\mathscr{R}e_p = U_p H / \nu$ (○).

Fig. 5 δ_{sw}/δ_p as a function of $\mathscr{R}a$ compared with the theoretical predictions for 'plate filling' and 'laterally restricted' flow. The scaling $f_{PF} \sim \mathscr{R}a^{-0.33}$ following from $\mathscr{R}e_p$-$\mathscr{R}a^{0.66}$ is given for reference, (+) is the raw data of f_{PF} extracted from the simulations.

4 Conclusions

It is found that the velocities of the LSC along its path are not constant and are functions of $\mathscr{R}a$. Comparison of the scalings of $\mathscr{R}e$-$\mathscr{R}a$ of the LSC obtained through a point measurement at p (as typically done in experiments) and averaged along the LSC path show different scalings: $\mathscr{R}e_\ell \sim \mathscr{R}a^{0.62\pm0.01}$ and $\mathscr{R}e_p \sim \mathscr{R}a^{0.66\pm0.02}$. The Reynolds-Nusselt scaling is found to be close to the theoretical prediction

$\mathscr{Re} \sim \mathscr{Nu}^2$ based on Prandtl-Blasius theory. Despite the observation that the flow (visualized on the bottom plate) looks 'plate filling', analysis of the $\mathscr{Ra}$-scaling of δ_{sw}/δ_p reveals that the flow behaves in a laterally restricted way.

Acknowledgements MK acknowledges the EU Science and Technology Fellowship China for funding this work and the Chinese University of Hong Kong's Information and Technology Service Centre for providing the computational resources for this study.

References

1. G. Ahlers, S. Grossmann, and D. Lohse. Heat transfer & large-scale dynamics in turbulent Rayleigh-Bénard convection. *Rev. Mod. Phys.*, 81(2):503–537, 2009.
2. D. Funfschilling, E. Brown, A. Nikolaenko, and G. Ahlers. Heat transport by turbulent Rayleigh–Bénard convection in cylindrical samples with aspect ratio one and larger. *J. Fluid Mech.*, 536:145–154, 2005.
3. S. Grossmann and D. Lohse. On geometry effects in Rayleigh-Bénard convection. *J. Fluid Mech.*, 486:105–114, 2003.
4. S. Grossmann and D. Lohse. Fluctuations in turbulent Rayleigh-Bénard convection: The role of plumes. *Phys. Fluids*, 16(12):4462–4472, 2004.
5. G. Grötzbach. Spatial resolution requirements for direct numerical simulation of Rayleigh-Bénard convection. *J. Comp. Phys.*, 49:241–264, 1983.
6. M. Kaczorowski, A. Shishkin, O. Shishkina, and C. Wagner. volume 96 of *Notes on Numerical Fluid Mechanics and Multidisciplinary Design*, pages 381–388. Springer, 2008.
7. M. Kaczorowski and C. Wagner. Analysis of the thermal plumes in turbulent Rayleigh-Bénard convection based on well-resolved numerical simulations. *J. Fluid Mech.*, 618:89–112, 2009.
8. S. Lam, X.-D. Shang, S.-Q. Zhou, and K.-Q. Xia. Prandtl-number dependence of the viscous boundary layer and the Reynolds-number in Rayleigh-Bénard convection. *Phys. Rev. E*, 65:066306, 2002.
9. L. D. Landau and E. M. Lifshitz. *Fluid Mechanics*. Pergamon Press, Oxford, 1987.
10. X.-L. Qiu and P. Tong. Large-scale velocity structures in turbulent thermal convection. *Phys. Rev. E*, 64:036304(13), 2001.
11. X.-L. Qiu and K.-Q. Xia. Spatial structure of the viscous boundary layer in turbulent convection. *Phys. Rev. E*, 58(5):5816(5), 1998.
12. X.-L. Qiu and K.-Q. Xia. Viscous boundary layers at the sidewall of a convection cell. *Phys. Rev. E*, 58(1):486(6), 1998.
13. O. Shishkina, A. Shishkin, and C. Wagner. Simulation of turbulent thermal convection in complicated domains. *J. Comput. Appl. Math.*, 226:336–344, 2009.
14. O. Shishkina, R. J. A. M. Stevens, S. Grossmann, and D. Lohse. Boundary layer structure in turbulent thermal convection and consequenses for the required numerical resolution. *N. J. Phys.*, 075022, 2010.
15. O. Shishkina and A. Thess. Mean temperature profiles in turbulent Rayleigh-Bénard convection of water. *J. Fluid Mech.*, 633:449–460, 2009.
16. C. Sun and K.-Q. Xia. Scaling of the Reynolds number in turbulent thermal convection. *Phys. Rev. E*, 72:067302(6), 2005.

On DNS and LES of natural convection of wall-confined flows: Rayleigh-Bénard convection

I. Rodríguez, O. Lehmkuhl, R. Borrell and C.D. Pérez-Segarra

1 Introduction

Turbulent natural convection of a fluid inside an enclosure heated from below (Rayleigh-Bénard convection), has been object of many theoretical and experimental investigations [1, 2]. Over the past decades numerical simulations have become a powerful tool for providing extensive data in turbulence structures and flow dynamics, but flow statistics for DNS at relative high Ra numbers are still limited by insufficient time integration [3]. In this sense, LES can be an attractive alternative for the resolution of natural convection problems at high Ra numbers. As LES models the smallest scales of the fluid their results are not only dependent on the grid resolution and the spatial and temporal discretization but also on the choice of the appropriate subgrid scale stress (SGS) model for describing the flow behavior. There are scarce long-term first and second order statistics results in literature for comparing LES results. Furthermore, time integration for most of the statistical data available does not guarantee their independence with the LSC reversals. Thus, the objective of this work is twofold: i) to provide useful long-term accurate statistical data by means of DNS of a cylindrical enclosure of aspect ratio ($\Gamma = D/H$) 0.5 at $Ra = 2 \times 10^9$ and $Pr = 0.7$ and, ii) to assess the behavior of different LES models by direct comparison with our DNS results.

2 Description of numerical method

The governing equations are discretized on a collocated unstructured grid arrangement, by means of second-order spectro-consistent schemes [4]. Such discretization

---

I. Rodríguez, C.D. Pérez-Segarra: Heat and Mass Transfer Technological Center (CTTC), Polytechnical University of Catalonia (UPC), e-mail: ivette@cttc.upc.edu
O. Lehmkuhl, R. Borrell: TermoFluids S.L., Spain

H. Kuerten et al. (eds.), *Direct and Large-Eddy Simulation VIII*,
ERCOFTAC Series 15, DOI 10.1007/978-94-007-2482-2_62,

preserves the symmetry properties of the continuous differential operators, and ensure both, stability and conservation of the global kinetic-energy balance on any grid. Energy transport is also discretized by means of a spectro-consistent scheme. An explicit third-order Gear-like scheme [5] based on a fractional step method is used for time-advancement algorithm, except for the pressure gradient where a first-order backward Euler scheme is used. LES studies have been performed using different SGS models: i) the Smagorinsky (SMG), ii) the dynamic eddy viscosity model (DEV), iii) the wall-adapting local-eddy viscosity (WALE) [6] and iv) the variational multiscale method (VMS) [7]. In order to analyse the influence of the SGS models, a coarse DNS (no-model) has also been computed on the same grids where LES have been used.

3 Results

The numerical computations have been performed at a Rayleigh number ($Ra = g\beta\Delta T H^3/\nu\alpha$) $Ra = 2\times10^9$ and a Prandtl number ($Pr = \nu/\alpha$) of $Pr = 0.7$ in a cylindrical cavity of aspect ratio $\Gamma = 0.5$. For DNS, the mesh has been constructed considering that the smallest flow scales must be well solved. The smallest scales at the hot and cold walls are imposed by viscous and thermal boundary layers while grid size at the bulk must be lesser than Kolmogorov scale. According to Grötzbach estimates [1], the thickness of the thermal boundary layer is $\delta_\theta = H/2Nu$ while Kolmogorov length scale for $Pr \leq 1$ is of the order of $\eta/H \leq \pi\sqrt{Pr}/(NuRa)^{1/4}$, where H is the height of the cylinder. The estimates for δ_θ and η have been used to select an optimal grid for DNS, with grid points clustered near the walls. Using Niemela [2] scaling relation ($Nu = 0.124Ra^{0.309}$), an a priori estimate of Kolmogorov scale yielded $\eta/H = 4.787\times10^{-3}$ and the wall boundary layer thickness was on the order of $\delta_\theta/H = 5.39\times10^{-3}$. Since both constrains must be attained, the computational mesh has 20591×128planes (~ 2.63M CV) with 8 grid points within the boundary layers. This should be enough to satisfy recent criteria about the number of grid points within the boundary layer [8]. LES and the coarse DNS have been performed in a set of refined grids of 4321×32planes (~ 138272 CV), 8756×32planes (~ 138272 CV) and 8756×64planes(~ 138272 CV). Due to the conservative formulation it is expected to achieve good stability properties even with the coarse grid. However, a minimum of 3 grid points within the boundary layer has been imposed in all cases.

3.1 DNS results

In order to obtain first and second order statistics for comparison with LES results it is important to guarantee that these results are statistically well converged. To do so, we have located numerical probes at different positions of the cavity where flow dynamics have different characteristic time. Then measurements are taken simultaneously at all the azimuthal planes of the computational domain. These probes have provided instantaneous data of the velocity components and temperature over more than 1600 time units (TU). Here reference time (t_0) has been defined as a function of the free-fall velocity ($U_0 = \sqrt{g\beta \Delta T H}$) and the height of the domain. Figure 1(a) shows the sampled data for the temperature from two of these locations. The inspection of these instantaneous data shows that the flow seems to be alternating its circulation with a periodic behavior.

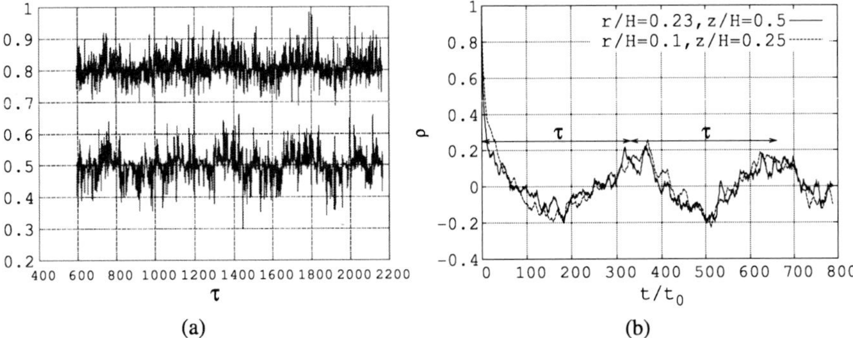

Fig. 1 (a) Time series of temperature fluctuations at two different locations of the cavity. bottom: ($r/H = 0.23$, $z/H = 0.5$); top: ($r/H = 0.1$, $z/H = 0.25$). For clarity, top time series are displaced. (b) Temperature fluctuation autocorrelation function at two locations in the cavity

In order to analyse this periodic motion, we have calculated autocorrelations from the time series of the DNS data, for the different probes. Figure 1(b) shows the autocorrelation for temperature fluctuations for the probes shown in fig. 1(a).

In the figure, a well-defined periodic oscillation of the temperature fluctuations can be observed. These oscillations, which have a very-long time period, have been also observed in the vertical velocity fluctuations (not shown here) and it seems to be related to oscillations in the large-scale flow motion in the sense that it is alternating directions. Further investigation in this periodic behavior are carried out calculating the histogram of the velocity at $r/H = 0.25$, $z/H = 0.95$ (not shown), which exhibits a bimodal shape with two probable peaks at opposite directions. This behavior in the velocity histogram corroborates the above observations that the large scale circulation alternates directions with certain periodicity. This periodic behavior has been reported before by Niemela et al. [2] for an aspect ratio $\Gamma = 1$ and $Ra = 6 \times 10^{13}$ and by Verdoold et al. [9] in an aspect ration $\Gamma = 4$ cavity. As in previous observations, a large-scale flow survives to this alternating circulation, but its mean value is very small if it is compared to the value of the small-scale fluctuations.

The existence of this periodic oscillation with a very low frequency should affect converged statistics. According to the results obtained, the dynamics of the large scale circulation presents a periodic behavior that occurs with a large temporal scale of about 331 TU. Thus, in order to ensure that statistics of DNS data are adequate and that all transients effects are washed out, the averaging of the first and second order statistics has been carried out for more than $1600t_0$ ($t_0 = H/U_0$) which corresponds with almost 5 cycles of the periodic behavior observed.

3.2 LES results

Results of the LES computations carried out with the coarse mesh are compared with DNS results in Figure 2. In the figure, the vertical distribution of the axial velocity and temperature along the cylinder axis is shown. As can be seen, small differences are observed in the temperature distribution for all the models except the SMG model. However, larger differences are obtained in the axial velocity distribution. In all cases, WALE and DEV model are the ones with a better behavior. Thus, for the sake of brevity, results with the finer grids are given with these models only.

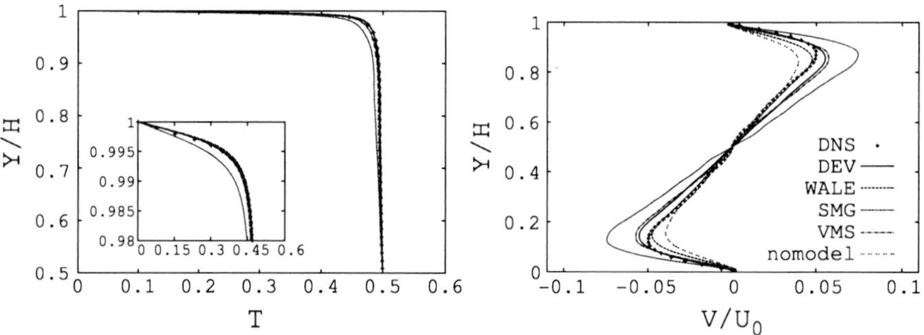

Fig. 2 Comparison of vertical profiles at the axis of the cylinder with the coarse grid (4321× 32 planes). (a) temperature (b) axial velocity $\overline{v_z/U}$

Figure 3 shows the vertical distributions of the root mean square time-averaged fluctuations of the temperature and radial velocities for the SGS models and also for the coarse DNS for the grids 8756× 32 and 8756× 64. As can be seen, with the medium sized grid the DEV model is capable of predict very well the peak in the stresses for both temperature and axial velocity fluctuations. Although WALE model overpredicts slightly the peak in the stresses, its behavior in the core of the cavity is very similar to the DEV model.

The average Nusselt number at the horizontal walls have been calculated and is given in table 1. In the table it is shown that even for the coarse grid, there is quite good agreement with the reference DNS data. In fact, differences in the prediction of the Nusselt number for the coarse grid without any SGS model is quite good, being less than 5% in all cases. These good results can be attributed to the use of a conservative formulation, which preserves the kinetic-energy balance, together with the large integration time used, ensuring temporal converged statistics. However, similar to second order statistics, for the prediction of the maximum and minimum peaks in the heat flux, a SGS model and the largest mesh have been necessary. In this case, the Dynamic Smagorinsky predicts quite well these quantities for the medium and large grids. It is important to highlight that the performance of WALE model is also

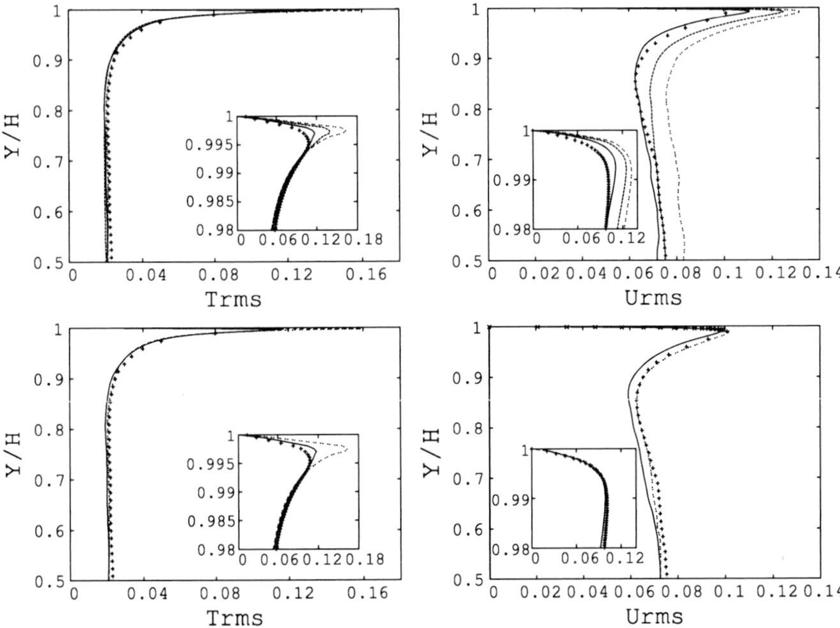

Fig. 3 Comparison of vertical profiles at the axis of the cylinder. (top) temperature fluctuations and and radial velocity fluctuations with the 8756× 32 planes grid. (bottom) temperature fluctuations and radial velocity fluctuations with the 8756× 64 planes grid. For legend see fig. 2

quite good, if we take into consideration that in terms of CPU time, WALE model is faster than Dynamic Smagorinsky. Thus, this feature can be of special interest if the resolution of industrial complex flows at high Rayleigh numbers are required.

Grid	Model	$\overline{Nu}$	$\varepsilon[\%]$	Nu_{max}	$\varepsilon[\%]$	Nu_{min}	$\varepsilon[\%]$
20591×128	DNS	84.89	-	94.8	-	49.07	-
4321×32	no model	85.73	0.99	112.39	18.56	33.15	32.47
4321×32	WALE	81.02	4.50	111.00	17.08	36.18	26.26
4321×32	DEV	81.56	3.92	107.28	13.16	35.2	28.26
8756×32	no model	84.61	0.32	116.17	22.54	36.37	25.88
8756×32	WALE	82.60	2.69	106.69	12.54	36.63	25.35
8756×32	DEV	81.84	3.59	104.65	10.39	35.79	27.06
8756×64	no model	84.35	0.63	103.69	9.37	42.96	12.45
8756×64	WALE	-	-	-	-	-	-
8756×64	DEV	84.37	0.61	100.55	6.06	43.08	12.2

Table 1 Comparison of the average, maximum and minimum Nusselt numbers between LES and DNS results for the different grids solved.

4 Conclusions

DNS and LES of the Rayleigh-Benard convection at $Ra = 2 \times 10^9$ in a cylindrical cavity of aspect ratio $\Gamma = 0.5$ have been carried out using a symmetry preserving formulation. The analysis of the DNS data for the velocity and temperature fluctuations has shown that LSC oscillates with a periodic behavior of $t_0 = 331$TU, which would affect the time convergence of the statistical data. Thus, in order to obtain converged statistical data for comparison with turbulence models, it is required at least a few cycles of integration. The DNS data presented in this work have been integrated for more than 1600 TU. Comparison with the different SGS models show that both WALE and Dynamic Smagorinsky (DEV) models predicts quite well the DNS data even for the coarse grid, although results obtained with the DEV model are in better agreement. However, WALE model would to be a good alternative to the Dynamic Smagorinsky model specially if we consider the CPU time required. The comparison of the average Nusselt number shows that this quantity is quite well predicted even for the coarse grid and without model. These good results can be attributed to the use of a conservative formulation, which preserves the kinetic-energy balance, together with the large integration time used, ensuring temporal converged statistics. Thus, for practical applications. with a minimum number of grid points within the boundary layer it is possible to reproduce the results of the DNS for the average Nusselt number with a 5% of difference but with the saving in time and computational resources.

Acknowledgements This work has been by financially supported by the Ministerio de Educación y Ciencia, Secretaría de Estado de Universidades e Investigación, Spain (ref. ENE2009-07689) and by the Collaboration Project between Universidad Politècnica de Catalunya and Termo Fluids S.L. (ref. C06650).

References

1. G. Grötzbach. Spatial resolution requirements for direct numerical simulation of the Rayleigh-Bénard convection. *Journal of Computational Physics*, 49:241–264, 1983.
2. J.J. Niemela, L. Skrbek, K.R. Sreenivasan, and R.J Donnelly. Turbulent convection at very high Rayleigh numbers. *Nature*, 404, 2000.
3. G. Amati, K. Koal, F. Massaioli, K.R. Sreenivasan, and R. Verzicco. Turbulent thermal convection at high Rayleigh numbers for a Boussinesq fluid of constant Prandtl number. *Physics of Fluids*, 17(121701), 2005.
4. R. W. C. P. Verstappen and A. E. P. Veldman. Symmetry-Preserving Discretization of Turbulent Flow. *Journal of Computational Physics*, 187:343–368, May 2003.
5. G.M. Fishpool and M.A. Leschziner. Stability bounds for explicit fractional-step schemes for the Navier-Stokes equations at high Reynolds number. *Computers and Fluids*, 38:1289–1298, 2009.
6. F. Nicoud and F. Fucros. Subgrid-scale stress modelling based on the square of the velocity gradient tensor. *Flow, Turbulence and Combustion*, 62:183–200, 1999.
7. T.J.R. Hughes, L. Mazzei, and K.E. Jansen. Large eddy simulation and the variational multiscale method. *Computing and Visualization in Science*, 3:47–59, 2000.
8. R.J.A.M Stevens, R. Verzicco, and D. Lohse. Radial boundary layer structure and nusselt number in Rayleigh-Benard convection. *Journal of Fluids Mechanics*, 649:495–507, 2010.
9. J. Verdoold, M. J. Tummers, and K. Hanjalic. Oscillating large-scale circulation in turbulent Rayleigh-Bénard convection. *Physical Review E*, 73(056304), 2006.

Part VII
Industrial Applications

The use of Direct Numerical Simulations for solving industrial flow problems

Claus Wagner, Andrei Shishkin and Olga Shishkina

1 Introduction

At the end of the last decade it was shown that predictions by means of Direct Numerical Simulations (DNS) agree well with experimental results obtained with Laser Doppler Anemometry and Particle Image Velocimetry (see for example Eggels et al. [3]) if weakly turbulent flows, i.e. low Reynolds numbers, are considered. In spite of the widely accepted merit of DNS for fundamental flow studies until now the technique could not shake off the prejudice that it is of little use for solving industrial flow problems. The reason might be that the required computational resources increase with approximately the third power of the Reynolds number and most of the industrially relevant flows, and in particular aircraft or vehicle aerodynamics, are characterized by very high Reynolds numbers. In this regard Spalart [10] estimated in the year 1999, that it will take until 2080 for DNS to be applicable to such flows. However, in the last years we performed a number DNS-studies which are relevant for various industrial branches. The common objective of these incompressible flow simulations was to produce a reliable and comprehensive flow data base for the validation and improvement of corresponding Reynolds-averaged Navier-Stokes simulations (RANS). The latter rely on turbulence models which are known to perform well for simple shear flows but not in general.

2 Industrially relevant DNS

It is generally accepted that DNS results play an import role in the enhancement and calibration of statistical and subgrid scale turbulence models needed to provide

Claus Wagner, Andrei Shishkin and Olga Shishkina
Department of Fluid Systems, Institute of Aerodynamics and Flow Technology, German Aerospace Center (DLR), Bunsenstr. 10, 37073 Göttingen, Germany, e-mail: claus.wagner@dlr.de

H. Kuerten et al. (eds.), *Direct and Large-Eddy Simulation VIII*, ERCOFTAC Series 15, DOI 10.1007/978-94-007-2482-2_63,

closure of Reynolds-averaged and filtered transport equations solved in RANS, Detached Eddy Simulation (DES) and Large-Eddy Simulation (LES). DNS results also play an important role in identifying resolution requirements for thermal and kinetic turbulent boundary layers. However, the most promising development is that DNS is more and more applied to conduct simulations of industrially relevant flows. In the following, results of such DNS are presented and discussed to demonstrate the use of DNS for validation of industrially used Computational Fluid Dynamics (CFD) methods and the required turbulence models.

2.1 LES of the flow around airfoils at high angle of attack

In the past aircraft manufacturers interest in improving the aerodynamic performance of their aircraft was the drive for new developments in Computational Fluid Mechanics (CFD). Lately also the prediction of wind turbine aerodynamics is receiving increasing interest due to the economical impact of the associated wind energy industry. In spite of the undertaken efforts, one of the remaining and most challenging problems is the numerical prediction of flows around airfoils or blades at high or changing angles of attack. The latter is associated with the so-called dynamic stall leading to varying and high aerodynamic loads on the wings or blades. However, the accurate and reliable prediction of such massively separated flows is extremely difficult by means of RANS.

One of the reasons is that RANS-based methods are not sufficiently susceptible to unsteady flow phenomena. In this respect, DNS and LES have proven to be a more promising alternative.

One major effort for pushing the LES technique to application was the European LESFOIL project which ended in 2001 and had the aim to position LES as an alternative to RANS. In this project the Aerospatiale-A airfoil was selected as a common configuration for benchmarking and comparison. Further, the Reynolds number $Re = 2.1 \times 10^6$ and angle of attack $\alpha = 13.3^\circ$ were considered. The results of this project are documented in the final report [2].

With the aim of using a method with spectral accuracy we performed DNS and LES of the flow around a FX-79-W151 airfoil segment (chord length c) using the spectral/hp-element method (SEM) by Karniadakis and co-workers [5] for a Reynolds number based on the undisturbed flow and the chord length of up to $Re = 10^5$ and an angle of attack of 12° using the unstructured grid presented in Fig. 1 (left). The needed computational resources of the SEM exceed that of standard finite-volume methods for structured grids by approximately two orders of magnitude as shown in Shishkin and Wagner [8]. Anyhow the SEM delivers accurate results of the turbulent flow separation on the suction side of an airfoil which is of use for validation of RANS- and DES-results.

The size of the computational domain used in the DNS and LES is $25c \times 20c \times 1c$ in streamwise, vertical and spanwise directions, respectively. In view of the high resolution needed in the vicinity of the airfoil segment up to 50% of the mesh elements are spent in a thin layer around the airfoil segment as shown in Fig. 1 (left). Further, the wake region is resolved with approximately 30 % of the mesh elements and the size of the elements decreases with increasing distance from the trailing edge. Additionally, in the vicinity of the outflow boundary vortical flow structures are dissipated in a sponge zone before they leave the computational domain. The hybrid structured/unstructured mesh consists of 2116 elements and provides $\approx 5.5 \cdot 10^6$ degrees of freedom (i.e. globally independent modes) for the polynomial order $P = 9$ for each of the 64 Fourier planes. The computational resources required to conduct the LES for a Reynolds number of $Re = 5 \cdot 10^4$ with the "2D+Fourier" SEM version (the memory usage is 20 GByte for $5.5 \cdot 10^6$ degrees of freedoms; 5.5 CPU seconds are needed for one time step on 32 AMD Opteron 1.7GHz processors) are much lower than those of the full 3D-SEM considered in [8].

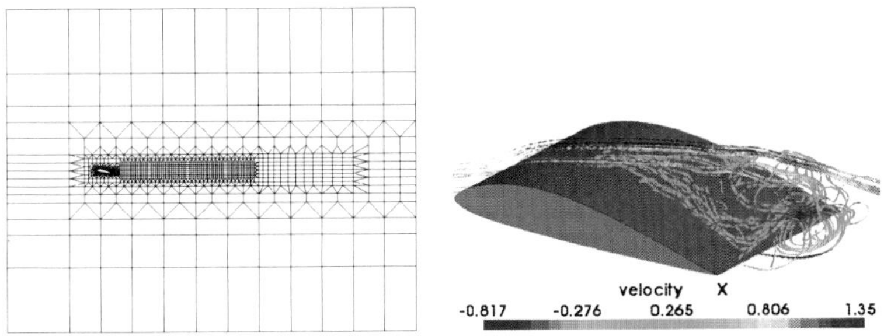

Fig. 1 Computational mesh with $5.5 \cdot 10^6$ degrees of freedoms for solving the incompressible Navier-Stokes equations (left) and perspective view of the airfoil geometry and streamlines over the suction side predicted with LES for $Re = 5 \cdot 10^4$.

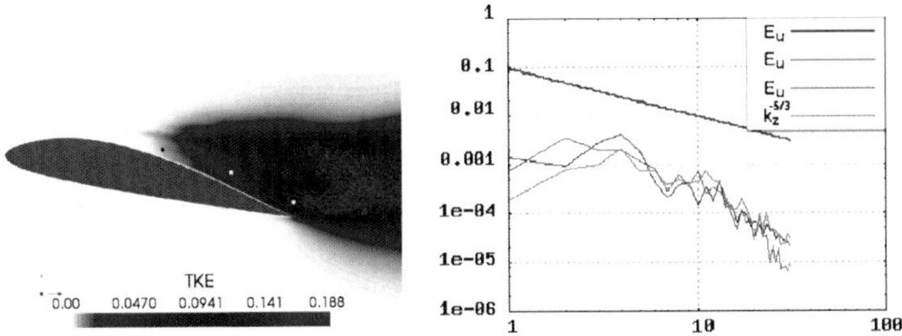

Fig. 2 With LES computed turbulent kinetic energy distribution (left) and energy spectra of the three velocity components taken at the first marked position on the suction side of the airfoil for $Re = 5 \cdot 10^4$.

Fig. 1 reflects the airfoil segment of depth c and the predicted streamlines of the velocity distributions in the recirculation zone as obtained in LES for $Re = 5 \cdot 10^4$ and Fig. 2 (left) the computed turbulent kinetic energy distribution. The first square symbols in Fig. 2 (left) from the left located in the separation zone upstream the trailing edge indicate the position where the also presented energy spectrum was evaluated. The latter reflects a decay of more than two orders of magnitude within the separation zone. The highest wave number of the energy spectra resulting from a fully resolved DNS defines the analogue of the Kolmogorov scale which is known to decrease with increasing Reynolds number. This leads to an increase of computational resources required for DNS which scales approximately with the third power of the Reynolds number. For an Airbus A320 with a mean chord length of the wings of 3 m, a wing span of 34.1 m, ratio of wing span and mean chord length of 9.5, wing area of 122.6 m^2, body length of 37.57 m, body diameter of 3.96 m, body area of approx. 450 m^2 and at a cruising speed of 233.33 m/s this leads to $Re = 5.0 \cdot 10^7$. In order to compute the flow by DNS or well-resolved LES a computational mesh of $\approx 10^{20}$ degrees of freedom (grid points) and 10^{15} cores (10^{14} GByte RAM) of todays processor technology are needed. It is obvious that this is out of reach for some time to come and confirms Spalart's [10] message.

2.2 DNS of the turbulent flow in a pipe with orifices

In many technical applications pipe systems are used to distribute the airflow. A widely used control device is the orifice installed for example in the aircraft's air distribution system, where it controls the volume rate of air delivered to the cabin. If the turbulent flow interacts with the orifice broadband noise is generated which can affect the comfort of passengers in vehicles and aircraft. Until now the computational costs for accurate numerical predictions of the generation and propagation of broadband noise are huge. Therefore, the industrially preferred approach is to conduct RANS that provides the mean flow and the turbulent kinetic energy distribution, which are used to model the broadband noise sources.

To develop and validate the above discussed hybrid approach an arrangement of two consecutive orifices was investigated numerically and experimentally as a representative configuration for a part of the aircrafts air distribution system. The investigated configuration consist of a 500 mm long inflow section followed by a foam covered 30 mm wide and 15 mm high orifice which serves as a turbulator. The distance of 300 mm to the second 2 mm wide and 15 mm high orifice defines the test section. Finally there is a 500 mm long outlet section. For this configuration RANS ($k - \omega$ turbulence model) were performed for mean flow velocities between 5 and 25 m/s corresponding to Reynolds numbers based on the mean velocity and the diameter of a straight pipe section of $3 \cdot 10^5 < Re < 1.6 \cdot 10^6$. Figure 3 (a) illustrates the RANS results obtained for $Re = 6 \cdot 10^5$ (10m/s). Not shown is the predicted mean streamwise velocity which increases in the vicinity of the turbulator and the orifice and reaches its maximum values close to the edge of turbulator and orifices, respectively. Furthermore, a large flow separation region associated with large tke-values develops downstream the turbulator and downstream the orifice as shown in Figure 3(a). However, the absolute tke-maximum is found upstream the orifice's leading edge. This is in contradiction to results of also conducted flow field

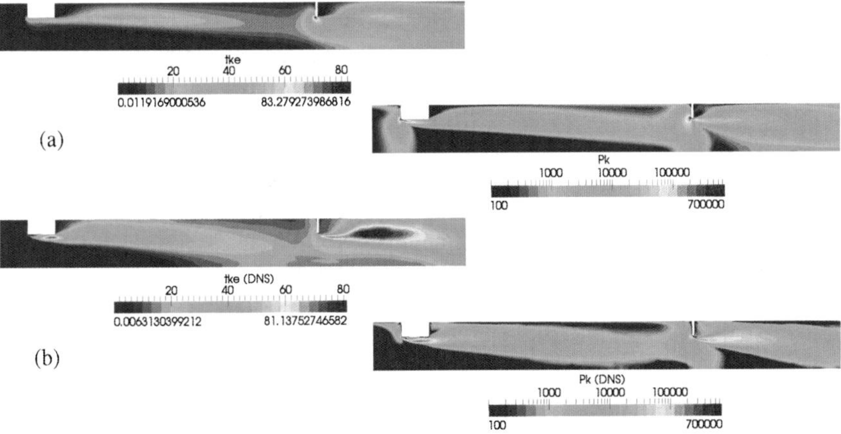

Fig. 3 Distributions of the turbulent kinetic energy (upper-left) and the tke-production P_k (upper-right)in a pipe with two orifices predicted in RANS (a) and distributions of the turbulent kinetic energy (lower-left) and the tke-production P_k (lower-right) in a pipe with two orifices predicted in DNS (b).

measurements. To clarify this we conducted a DNS for $Re = 6 \cdot 10^5$ and the distance of the test section of 300 mm with a grid consisting of $2.7 \cdot 10^7$ points. Although the DNS predicts a mean velocity field which agrees well with that of the RANS (not shown) the tke-maximum presented in Fig. 3 (b) is not located upstream the second orifice. A detailed analysis of the terms in tke-transport equation revealed that the production terms P_k used for the RANS ($k - \omega$ model) and that calculated with the DNS-data differ significantly as shown in Fig. 3 right, (a) and (b). Further the modeled P_k maximum was also located upstream the orifice. Subsequent limiting of the production term of the $k - \omega$-model improved the agreement between DNS and RANS considerably.

2.3 DNS of the turbulent Czochralski flow

The most common way to grow single silicon crystals is the so called Czochralski crystal growth process denoting a hot and non-transparent silicon melt at melting temperature of 1485 K held by a rotating crucible as illustrated in Fig. 4. Approximately 90% of the worldwide demand for single silicon crystal needed by the semiconductor industry is produced with the Czochralski method. By experience it is known that large temperature fluctuations within the melt lead to an increase of the defect concentration in the single silicon crystal and therefore to a larger amount of degraded silicon crystals.

Since it is almost impossible to measure flow velocities and temperatures within the hot silicon melt, DNS offers the unique possibility to study the flow in all detail and to understand the involved transport mechanisms. This is a prerequisite for the development of an efficient method which can damp the large temperature fluctuations in the vicinity of the crystal. Therefore, DNS was conducted for crucible diameters up to 0.34 m and temperature differences between the heated crucible and the crystal which is at melting temperature of silicon (see Wagner and Friedrich [11]) as high as 90.7 K using a finite-volume method with staggered cylindrical grids with up to $129 \times 128 \times 172$ points in vertical, circumferential and radial directions, respectively. The corresponding Grashof numbers $Gr = \beta g R_c^3 \Delta T / \nu^2 < 1.57 \cdot 10^9$ and Marangoni numbers $Ma = c_{tc} R_c \Delta T / \gamma < 2.82 \cdot 10^4$, where β denotes the thermal expansion coefficient, ν the kinematic viscosity, g the gravitational acceleration, ΔT the difference between the temperatures at the crucible and the crystal, c_{tc} the thermocapillary coefficient, γ the thermal diffusivity and R_c the radius of the crucible. Further, the crucible rotates with up to 10 rpm in opposite direction to the rotation (20 rpm) of the crystal. Fig. 4 (central) reflects isolines of the instantaneous radial velocity component and Fig. 4 (right) the respective isotherms as predicted in DNS for $Gr = 10^8$ and $Ma = 3.6 \cdot 10^4$. From these simulations it was concluded that the large temperature fluctuations in the vicinity of the crystal edge are generated by local disturbances of the balance between centrifugal forces, buoyancy forces and surface tension at the edge of the crystal.

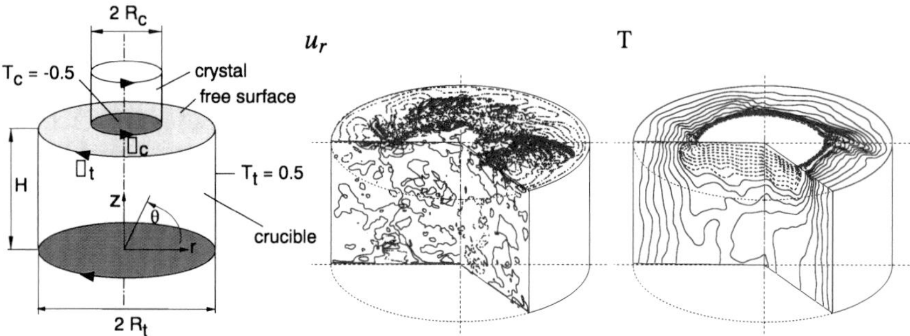

Fig. 4 Computational domain and boundary conditions used for DNS of the turbulent Czochralski flow (left) and predicted isolines of an instantaneous field of the radial velocity component (central) and isotherms (right) for $Gr = 10^8$ and $Ma = 3.6 \cdot 10^4$.

2.4 DNS of turbulent mixed convection in a room

Confined mixed convection is of essential importance for a variety of practical applications like e.g. air-supply in offices or residential buildings, the cooling of microelectronic devices, and the air-conditioning of vehicle and aircraft cabins. The numerical prediction of such flows by means of CFD allows for an optimization of the cabin design with respect to the thermal comfort of passengers, while excessive prototyping can be prevented. One key issue during this task is to select an appropriate turbulence model. Systematic comparative studies performed so far reveal large deviations between computations of mixed convection obtained with different turbulence models (see [1] and [4]). More recently, Lin et al. [6] performed RANS and LES of such flows to study the dispersion of airborne pathogen in aircraft cabins. In Lin et al. [7] they also favorably compared instantaneous velocity components and energy spectra obtained in LES of the flow in a generic cabin model with stereoscopic PIV measurements at five different locations. In order to provide a reliable

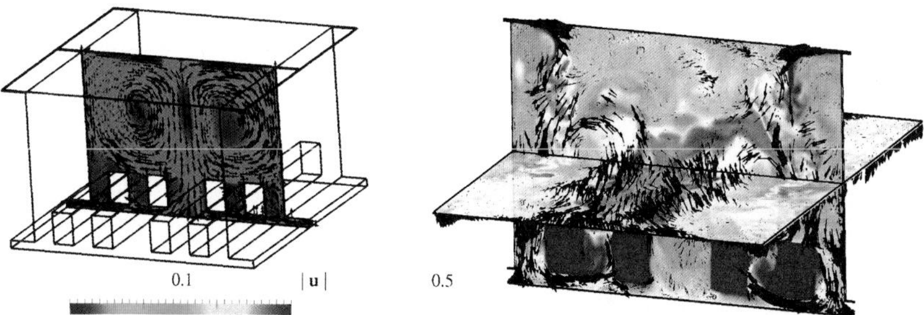

Fig. 5 In RANS of forced convection generated mean velocity field in a central cross section of the 5m deep, 3m high and 4m wide generic room with air inlets/outlets at the top/bottom of the side wall and heated obstacles inside (left). Instantaneous temperature distributions $-0.5 < T < 0.5$ with superimposed velocity vectors predicted by means of DNS in the generic room for $Gr = 2.14 \cdot 10^8$ and $Re = 2.37 \cdot 10^4$ (right).

validation data base and thus to replace expensive validation experiments, we conduct DNS of mixed convection in generic room or cabin using the finite-volume method described in Shishkina et al [9]. The computational domain illustrated in Fig. 5 (left) is 5m deep, 3m high and 4m wide and is equipped with heated obstacles and 4m wide air inlet and outlet channels located at the top and bottom of the side walls, respectively. Fig. 5 (left) shows a RANS flow field in the central cross section computed for the forced convection case ($Gr = 0$) with a mean velocity at the inlet channels of $u_{inlet} = 0.23$ cm/s. For comparison Fig. 5 (right) reflects instantaneous temperature fields and superimposed velocity vectors computed in DNS of mixed convection for the same inlet velocity $0.23cm/s$ and a temperature difference of $\Delta T = 0.11$ K. Here ΔT is the temperature difference between the heated obstacles and the colder inlet flow. The dimensionless parameters in this DNS are $Gr = \beta g H^3 \Delta T / \nu^2 = 2.14 \cdot 10^8$ and $Re = H u_{inlet} / \nu = 2.37 \cdot 10^4$ with H the height of the room. Currently we are performing a DNS for $Gr = 0.913 \cdot 10^{10}$ and $Re = 1.54 \cdot 10^5$ corresponding to an inlet velocity 0.8 m/s and a temperature difference $\Delta T = 4.8$ K on staggered grids with $1536 \times 1024 \times 936$ points using the finite-volume method described in [9]

3 Conclusions

It was shown that Direct Numerical Simulations of the aerodynamics of complete aircraft for cruising speed Reynolds numbers will be impossible in the foreseeable future. Anyhow there are many important industrial flow problems for which the DNS technique can provide the data needed to validate RANS and DES computations. It has been shown, that DNS can be applied for solving turbulent pipe flows of technical relevance, the turbulent flow in Czochralski crystal growth configurations and turbulent mixed convection in rooms with realistic dimensions. Although DNS asks for extreme computational resources the obtained details of the turbulent flows make it possible to improve turbulence models needed by industry.

References

1. Costa JJ, Oliveira LA and Blay D (1999) Test of several versions for the $k-\omega$ type turbulence modelling of internal mixed convection flows. Int. Jour. Heat and Mass Transfer 42.
2. Davidson L, Cokljat D, Fröhlich J et al (2003) (eds) LESFOIL Large Eddy Simulation of Flow Around a High Lift Airfoil, Notes in Numerical Fluid Mechanics, vol. 83, Springer.
3. Eggels JGM, Unger F, Weiss MH et al (1994) Fully developed turbulent pipe flow: a comparison between direct numerical simulation and experiment. J Fluid Mech. 268:175–209
4. Günther G, Bosbach J, Pennecot J et al (2006) Experimental and numerical simulations of idealized aircraft cabin flows. Aerospace Science and Technology. 10:563–573.
5. Karniadakis GE and Sherwin S (1999) Spectral/HP Element Methods for CFD. Oxford University Press.
6. Lin C-H, Horstmann RH, Ahlers MF et al (2005) Numerical Simulation of Airflow and Airborne Pathogen Transport in Aircraft Cabins Part I: Numerical Simulation of the Flow Field. In: Proc. of the ASHRAE winter meeting 2005, 755–763.
7. Lin C-H, Wu TT, Horstman RH et al (2006) Comparison of large eddy simulation predictions with particle image velocimetry data for airflow in a generic cabin model. HVAC&R Research, 12(3c):935–951.
8. Shishkin A, Wagner C (2007) Direct Numerical Simulation of a Turbulent Flow Using a Spectral/hp Element Method. In: Notes on Numerical Fluid Mechanics and Multidisziplinary Design, 92, Springer Publisher, 405–412.
9. Shishkina O, Shishkin A and Wagner C (2008) Simulation of turbulent thermal convection in complicated domains. J Comput and Appl Maths, doi:10.1016/j.cam.2008.08.008.
10. Spalart PR (1999) Strategies for turbulence modeling and simulations. In: Rodi W and Laurence D (eds) Engineering Turbulence Modelling and Experiments-4, Elsevier Science Ltd., 3–17
11. Wagner C and Friedrich R (2004) Direct Numerical Simulation of momentum and heat transport in idealized Czochralski crystal growth configuration. IJHFF 25:431–443.

High-order direct and large eddy simulations of turbulent flows in rotating cavities

Serre E.

1 Introduction

The simulation of rotating cavities flows is a major issue in computational fluid dynamics and engineering applications such as disk drives used for digital disk storage in computers, automotive disk brakes, and especially in turbomachinery (see a review in [8]).

Centrifugal and Coriolis forces produce a secondary flow in the meridian plane composed of two thin boundary-layers along the disks separated by a non-viscous geostrophic core where the axial gradient of pressure nearly equilibrates the Coriolis force. That produces adjacent coupled flow regions that are radically different in terms of the flow properties and the thickness scales involving very challenging simulations. Very thin non-parallel boundary layers governing the flow stability develop along the stationary and the rotating disks, called thereinafter Bödewadt and Ekman layer, respectively.

As a consequence, besides its primary concern to industrial applications, the rotating disks problem has also proved a fruitful means of investigating transition to turbulence and the effects of mean flow three-dimensionality on the turbulence and its structure [6, 11]. Indeed, these boundary layers are three-dimensional from their inception and constitute one of the few non-trivial three-dimensional cases whose Navier-Stokes solutions exist.

Linear stability analysis has revealed that such boundary layer is subject to two generic types of instability. An inviscid instability, due to the inflectional nature of the velocity profile, is labeled type I, whereas type II is due to the combined action of viscous and Coriolis effects (see the review by [2]). The matter of the transition scenario is currently much debated around the idea that a global instability might take place and lead to transition to turbulence. A new light on the problem was given by Lingwood's discovery [5] that such boundary layer underwent transition

Serre, M2P2 UMR 6181, CNRS - Universite Aix-Marseille, France, e-mail: Eric.Serre@L3m.univ-mrs.fr

H. Kuerten et al. (eds.), *Direct and Large-Eddy Simulation VIII*,
ERCOFTAC Series 15, DOI 10.1007/978-94-007-2482-2_64,

from convective to absolute behavior at a local Reynolds number (dimensionless radius, r/δ, $\delta = (\nu/\Omega)^{1/2}$)) (507) close to the onset of turbulence (513). But to this day, if further studies have confirmed these local linear stability results, no general agreement exists concerning their outcome in terms of global behavior. Pier's theoretical work [9], showing the possible existence of a global nonlinear elephant mode, together with Davies and Carpenter's numerical results [3], demonstrating linear global stability, would imply a subcritical global bifurcation, which has been demonstrated in [14]. This behavior would then be the outcome of the competition between the stabilizing non-parallel effects, and the destabilizing nonlinear ones, so that both have to be taken into account. This is done here, where DNS is used to investigate the impulse response of the boundary layer encountered in a rotating disk configuration with radial throughflow.

Simulation of turbulent rotating cavities flows remains a challenge. Due to the skewing of the boundary layers [11], Reynolds stresses are not aligned with the mean flow vector that invalidates eddy-viscosity models. Besides, such models clearly fail to predict turbulent flow by delaying the transition along the rotor. Second moment closures provide a more appropriate level of modeling, but the Reynolds stress behavior is not fully satisfactory, particularly near the rotating disk (see a review in Launder *et al.* [4]). Moreover, the existence of persisting large-scale 3D precessing vortices at high Reynolds numbers requires strongly unsteady computations. As a consequence, LES seems to be right level of modeling. Nevertheless, the literature on the topic is rare. Wu and Squires [15] performed the first LES of the flow over a single rotating disk using dynamic and mixed dynamic Smagorinsky models to analyze the mechanisms promoting sweeps and injections. First computations treating the rotor-stator flow as unsteady were DNS of Lygren and Andersson [7] and Serre *et al.* [10]. Finally, Anderson and Lygren [1] carried out the first LES in a sector with periodic boundary conditions to reduce the costs. The use of spectrally accurate numerical schemes adds a difficulty due to the fact that spectral approximations are much less diffusive than low order ones, involving an accumulation of the energy on the high spatial frequencies which finally leads to the divergence of the computations. This is for all these reasons that we proposed in our group an original approach based on the spectral vanishing viscosity (SVV), first introduced by Tadmor [13] to stabilize resolution of hyperbolic equation. This alternative LES formulation is proposed and incorporated into the Navier-Stokes equations for controlling high-wavenumber oscillations [12]. Flow predictions at $Re = 10^6$ are provided in a closed rotor-stator cavity in order to avoid difficulties related to in-outflow conditions.

2 Modeling

We consider the flows within rotating cavities composed of two disks enclosing an annular domain of radii a and b with $b > a$. The domain can be open (rotating cavity) with a forced radial flow imposed at $r = a$ or eventually bounded by two co-axial cylinders of height $2h$ (rotor-stator cavity). The geometry is characterized by an aspect ratio $L = (b-a)/2h$ and a curvature parameter $R_m = (b+a)/(b-a)$. For the rotating cavity ($L = 10$, $R_m = 5$), both disks rotate at uniform angular velocity $\Omega = \Omega e_z$, e_z being the unit vector on the axis while in the rotor-stator cavity ($L = 5$, $R_m = 1.8$) one disk is at rest. These configurations are basic cavity elements of a turbine engine.

The incompressible fluid motion is governed by the three-dimensional Navier-Stokes equations in primitive variables. The equations are made dimensionless by considering h, Ω^{-1} and Ωb as length, time and velocity of reference. No-slip boundary conditions are applied at all walls so that all the near-wall regions were explicitly computed. At the inflow of the rotating cavity velocity profiles are prescribed while at the outflow a convective boundary condition is applied.

The temporal discretization adopted in this work is a projection scheme, based on backwards differencing in time (see the details in [12]).

The solutions are searched as truncated series of Chebyshev polynomials of degree at most equal to N_r and N_z in the non-homogeneous radial and axial directions (r,z) respectively and Fourier series in the 2π-periodic tangential direction of cut off frequency $N_\theta/2$. To perform LES, a well controlled diffusion has been incor-

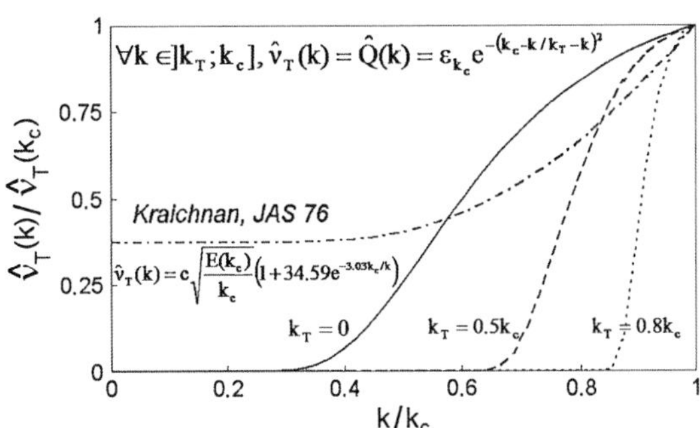

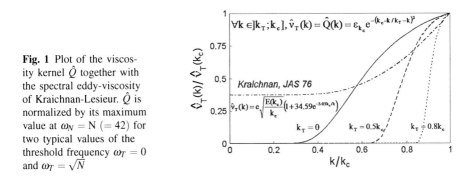

Fig. 1 Plot of the viscosity kernel $\hat{Q}$ together with the spectral eddy-viscosity of Kraichnan-Lesieur. $\hat{Q}$ is normalized by its maximum value at $\omega_N = \mathrm{N}\ (= 42)$ for two typical values of the threshold frequency $\omega_T = 0$ and $\omega_T = \sqrt{N}$

porated in the Navier-Stokes equations. A new diffusion operator Δ_{SVV} is simply implemented by combining the classical diffusion and the new SVV terms to obtain:

$$\nu \Delta_{SVV} \equiv \nu \Delta + \nabla.(\varepsilon_N Q_N \nabla) = \nu \nabla.S_N \nabla \tag{1}$$

where ν is the diffusive coefficient and where:

$$S_N = diag\left\{S^i_{N_i}\right\},\ S^i_{N_i} = 1 + \frac{\varepsilon^i_{N_i}}{\nu} Q^i_{N_i} \tag{2}$$

with $\varepsilon^i_{N_i}$ the maximum of viscosity and $Q^i_{N_i}$ a 1D viscosity operator acting in direction i, and defined in the spectral space by an exponential function: $\widehat{Q}^i_{N_i}(\omega) = 0$, if $0 \leq \omega \leq \omega^i_T$ and $\widehat{Q}^i_{N_i}(\omega) = \exp(-(\omega - \omega^i_T)^2/(\omega - \omega^i_T)^2)$ if $\omega^i_T \leq \omega \leq \omega^i_N$, where ω_T^i is the threshold after which the viscosity is applied and ω_N^i the highest frequency calculated in the direction i. An illustration of the viscosity kernel in the spectral space together with the classical spectral eddy-viscosity kernel of Kraichnan-Lesieur is presented in Figure 1. Such model is only active for the short length scales and it has been shown that it keeps the spectral convergence (details in [12]). It is noticeable that the SVV term is not scaled with Re that implied that for a fixed grid and given SVV parameters, the SVV term may become larger relatively to the classical diffusion term when increasing Reynolds number.

3 Results

3.1 Transition to turbulence

The breakdown to turbulence in a rotating disk boundary-layer is analyzed via direct numerical simulation (DNS) in a sector ($2\pi/4$ or $2\pi/8$ depending the mesh resolution) of an annular cavity with a forced inflow at the hub and free outflow at the rim. The largest size of the mesh in the 1/4 cavity is $649 \times 170 \times 65$ in radial, azimuthal and axial directions that corresponds to a grid resolution of approximately ten and six Kolmogorov length scales in the radial and axial directions, respectively. Our objective here was to obtain a scenario of transition based on a cascade of absolute instabilities as conjectured in Pier [9] from model equations.

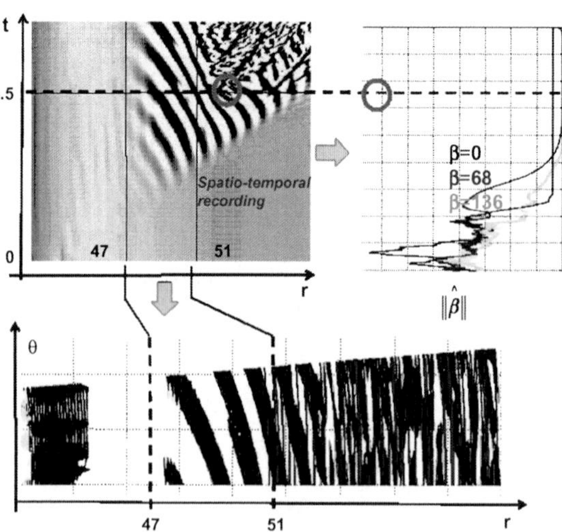

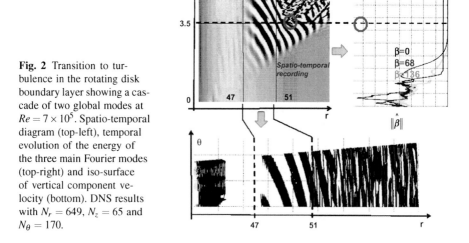

Fig. 2 Transition to turbulence in the rotating disk boundary layer showing a cascade of two global modes at $Re = 7 \times 10^5$. Spatio-temporal diagram (top-left), temporal evolution of the energy of the three main Fourier modes (top-right) and iso-surface of vertical component velocity (bottom). DNS results with $N_r = 649$, $N_z = 65$ and $N_\theta = 170$.

The global Reynolds number has been set to $Re = 7.10^5$, and the mass flow rate to $C_w(= Q/\nu b) = 1995$ which places the transition from convective to absolute instability in the first half of the cavity. Axisymmetric stationary base flow has been reached by damping convective axisymmetric modes of instability (see details in [14]), generated by the inflow condition at each time-step. The nonlinear dynamics of the flow has been analyzed by superimposing an initial spatially localized perturbation during a single time step at the beginning of the computations. The perturbation velocity field corresponds to a Stokes flow over a hemispherical roughness located at the wall near the hub with an amplitude of 0.1% of the velocity maxima. Figure 2 presents the nonlinear evolution of such an initial perturbation with a 68-fold symmetry. The spatio-temporal diagram (Figure 2 (top-left)) shows only the saturated front of the wave packet. Shortly after the saturation of the first global instability, characterized by a trailing edge moving upstream and rapidly stabilizing close to $r = 47$, the saturated wave downstream of the trailing front starts being perturbed with order-unity close to $r = 51$ at $t = 3.5$. Behind that secondary front the state is disordered in time and space indicating incipient turbulence. Calculations eventually blow up at about $t = 10$ due to aliasing. Indeed, the resolution of the mesh being approximately ten and six Kolmogorov length scales in the radial and axial directions respectively, the turbulent dissipation scale is not resolved. Nevertheless, SVV computations can be run as long as wanted and confirm the limited in time DNS results. The associated spatial structure is shown by a horizontal cut on Figure 2 (bottom) showing the clockwise spiral (rotating counterclockwise in time) that saturates after the primary front at $r = 47$, developing secondary instability shortly before $r = 51$. At this secondary front, base flow modifications become of order unity (Figure 2 (top-right)). Close to this secondary front the harmonic component $m = 136$ also become of the same amplitude compared to the primary rolls, and higher harmonics start having a physically relevant amplitude. Finally, the associated turbulent kinetic energy in a vertical cut in Figure 3 confirms that non zero values occur just downstream the radial location of the secondary front at $r = 51$.

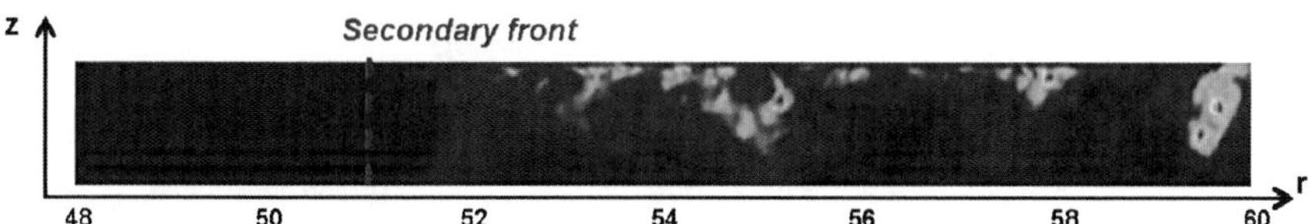

Fig. 3 Turbulent kinetic energy in the meridian plane showing the breakdown to turbulence downstream the primary front. Only the upper half-cavity is shown.

This scenario relies on low incoming noise upstream of the primary front, and a sufficiently strong impulsive perturbation as the first global bifurcation is known to be subcritical. For the first time it confirms the possibility of a direct transition to turbulence through an elephant cascade.

3.2 Turbulence

A detailed picture of the turbulent flow structure is provided by high-order LES in a closed rotor-stator cavity at Reynolds numbers $Re = 10^6$. The largest size of the mesh in the 1/2 cavity is $151 \times 240 \times 81$ in radial, azimuthal and axial directions that corresponds to a wall coordinate $z^+ \approx 1$ in both layers. SVV parameters have been set to $\varepsilon_N = (1/N, 1/N, 1/N)$ and $\omega_T = (2\sqrt{N}, 5\sqrt{N}, 4\sqrt{N})$ in (r, θ, z) directions respectively. The results published in [11] show that the mean flow is still composed of two boundary layers along the rotor and the stator separated by a geostrophic core like in the laminar regime. The agreement with LDV measurements is very satisfactory for the first and the second-order order statistics as well. The entrainment coefficient, $K = v/\Omega r$ at mid-radius, is equal to 0.38 compared to the experimental value $K = 0.42$. The prediction of this coefficient which is directly related to the pressure gradient (Taylor-Proudman theorem) is crucial to the design of thrust bearings. It emerges at this Reynolds number that the turbulence is mainly confined in the wall boundary layers including the layers along both cylinders closing the cavity, Figure 4. The stator boundary layer is fully turbulent. On the other hand, the rotor layer becomes progressively turbulent from the outer radial locations although the rotating hub is shown to destabilize the inner part of the boundary layers. The max of k is located at the jet impingement of the flow coming from the rotor. The turbulent flow is associated to quasi-axisymmetric coherent structures along the stator while in the transitional rotor-layer three-dimensional spiral arms can be still observed.

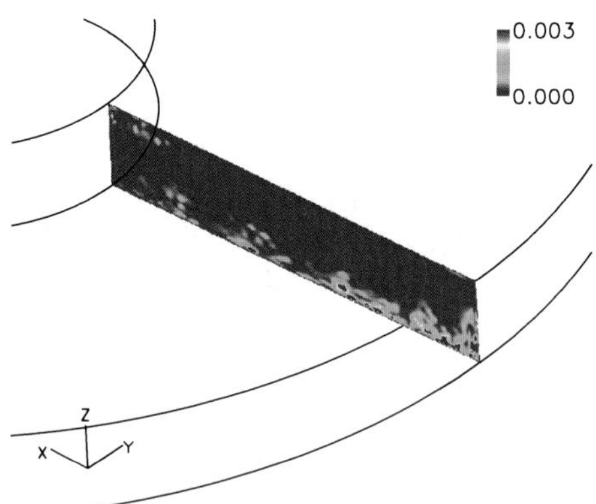

Fig. 4 Isolines map of the turbulent kinetic energy at $Re = 10^6$. LES-SVV results for a grid $N_r = 151$, $N_z = 81$, $N_\theta = 240$ and $\delta t = 10^-5$.

We have investigated the structural properties of the boundary layers along the disks. In particular, the structural parameter a_1 (the ratio of the magnitude of the shear stress vector to twice the turbulence kinetic energy) as well as the mean velocity angle ($\gamma_m = arctan(V_r/V_\theta)$) have been calculated. The truthfulness of the alignment assumption between the shear stress vector and the mean velocity gradient vector made in subgrid scale models using an eddy viscosity has also been checked. The mean velocity gradient angle ($\gamma_g = arctan((\partial_r V_r/\partial z)/(\partial_r V_\theta/\partial z))$) as well as the turbulence shear stress angle ($\gamma_\tau = arctan((\overline{V_r'V_z'}/\overline{V_\theta'V_z'}))$ along the radius of both disks have been evaluated. Results are summarized in the Table 1:

	γ_m (deg)	$\gamma_\tau - \gamma_g$ (deg)	max(a_1)
rotor	-0.5 / 16.5	0 / 94	0.058
stator	-34 / 0	0 / 151	0.062

Table 1 Structural properties of the boundary layers along the rotor and the stator. Mean velocity angle γ_m, mean velocity gradient angle γ_g and turbulence shear stress angle γ_τ as a function of the radius. a_1 is the Townsend parameter.

The strong depressed of a_1 (with respect ot the value 0.015 in 2D boundary layer) as well as the continuous change of γ_m with z show a skewing in the mean of the turbulent boundary layer resulting from the shear induced by rotation. Moreover, the low value of a_1 indicate that turbulence here is much less efficient than in 2D boundary layer to extract energy from the mean flow. Finally, results show that the shear stress vector is not aligned with the mean velocity gradient in both layers. As a consequence, turbulence models which rely on the assumption of an isotropic eddy viscosity should face problems because the difference $\gamma_\tau - \gamma_g$ is large.

4 Concluding remarks

In rotating cavities, the mean flow driving force results from the interaction between the rotation induced centrifugal forces and the viscosity dominated boundary layers that lead to numerical challenging investigations that might explain the limited literature. High-order DNS and LES approaches have been presented concentrating on preserving spectral accuracy. A scenario to turbulence through global secondary instability has been obtained for the first time using DNS that confirms the possibility of a direct route through absolute instability provided a low incoming noise upstream of the primary front. Our LES prediction confirms that rotation, curvature and confinement lead to a very anisotropic and inhomogeneous turbulence. Such results should encourage the LES community to increase its effort not only on the SGS modeling but also in the use of higher-order schemes.

Acknowledgements This work was granted access to the HPC resources of IDRIS under the allocation 2009-0242 made by GENCI (Grand Equipement National de Calcul Intensif). The work was supported by CNRS in the frame of the DFG-CNRS program "LES of complex flows". The author is indebted to all the following collaborators (cited by alphabetic order) for their active collaboration to this work: P. Bontoux, J.M. Chomaz, B.E. Launder, S. Poncet, E. Severac and B. Viaud.

References

1. Andersson, H.I. and Lygren, M. (2006). LES of open rotor-stator flow, *Int. J. Heat Fluid Flow*, 27, 551–557.
2. Crespo del Arco, E., Serre, E., Bontoux, P. and Launder, B.E. (2005). Stability, transition and turbulence in rotating cavities. In Advances in Fluid Mechanics (ed. M. RAHMAN), Dalhousie University Canada Series, 41:141–196. WIT press.
3. Davies, C. and Carpenter, P.W. (2003). Global behaviour corresponding to the absolute instability of the rotating-disk boundary layer. *J. Fluid Mech.*, 486:287–329.
4. Launder, B.E., Poncet, S. and Serre, E. (2010). Transition and turbulence in rotor-stator flows, *Ann. Rev. Fluid. Mech.*, 42:229–248.
5. Lingwood, R.J. (1997). Absolute instability of the Ekman layer and related rotating flows. *J. Fluid Mech.*, 331:405–428.
6. Littell, H.S. and Eaton, J.K. (1994). Turbulence characteristics of the boundary layer on a rotating disk. *J. Fluid Mech.*, 266:175–207.
7. Lygren, M. and Andersson, H. (2001). Turbulent flow between a rotating and a stationary disk, *J. Fluid Mech.*, 426, 297–326.
8. Owen, J.M. and Rogers, R.H. (1995). Heat transfer in rotating-disk system. Wiley.
9. Pier, B. (2003). Finite amplitude crossflow vortices, secondary instability and transition in the rotating-disk boundary layer. *J. Fluid Mech.*, 487:315–343.
10. Serre, E., Crespo del Arco, E. and Bontoux, P. (2001). Annular and spiral patterns between a rotating and a stationary disk, *J. Fluid Mech.*, 434,65–100.
11. Séverac, E., Poncet, S., Serre, E. and Chauve, M.P. (2007). Large eddy simulation and measurements of turbulent enclosed rotor-stator flows. *Phys. Fluids*, 19:085113.
12. Séverac, E. and Serre, E. (2007). A spectral vanishing viscosity LES model for the simulation of turbulent flows within rotating cavities. *J. Comp. Phys.*, 226(2):1234–1255.
13. Tadmor, E. (1989). Convergence of spectral methods for nonlinear conservation laws. *SIAM J. Numer. Anal.*, 26(1), 30.
14. Viaud, B., Serre, E. and Chomaz, J.M. (2008). Elephant mode sitting on a rotating disk in an annulus. *J. Fluid Mech.*, 598:451–464.
15. Wu, X. and Squires, K.D. (2000). Prediction and investigation of the turbulent flow over a rotating disk. *J. Fluid Mech.*, 418:231–264.

DNS of turbulent flow in a rotating rough channel

Vagesh D. Narasimhamurthy and Helge I. Andersson

1 Introduction

All solid surfaces in practice can be considered rough to a certain degree and this surface roughness is known to affect the fluid flow to a considerable extent. It is also known from the literature that system rotation affects both the mean fluid motion and the turbulence. Fluid flows involving both the surface roughness and the system rotation are therefore of major concern in industrial, geophysical and astrophysical applications.

The distinction between $d-$type and $k-$type roughness on wall turbulence is well established [1, 2] and related to the pitch-to-height ratio λ/k, where λ is the separation between roughness elements of height k. In the present study we considered a rod-roughened plane channel flow where both walls were roughened by square rods with height $k = 0.1h$ (h being the channel half-width) and pitch $\lambda/k = 8$. The rods were positioned in a non-staggered arrangement, as shown in Fig. 1. In order to explore the effect of system rotation direct numerical simulations (DNS) of pressure-driven flow in the rod-roughened channel have been performed. The driving pressure-gradient $-dP/dx$ was prescribed such that the Reynolds number based on the channel half-width h and the wall-friction velocity $u_\tau = (-\rho^{-1}hdP/dx)^{1/2}$ was equal to 400. This is essentially the same Reynolds number as the medium Re case reported by Moser et al. [3] for smooth channel flows and by Ashrafian et al. [4] for a rod-roughened channel flow. The computational domain comprised eight rods on each wall. In this study we considered two different grids, i.e. a coarse mesh of $256 \times 128 \times 128$ grid points and a fine mesh of $512 \times 320 \times 200$ grid points. Results

Vagesh D. Narasimhamurthy
Fluids Engineering Division, Dept. of Energy and Process Engineering, Norwegian University of Science and Technology (NTNU), 7491 Trondheim, Norway, e-mail: vagesh@ntnu.no

Helge I. Andersson
Fluids Engineering Division, Dept. of Energy and Process Engineering, Norwegian University of Science and Technology (NTNU), 7491 Trondheim, Norway, e-mail: helge.i.andersson@ntnu.no

H. Kuerten et al. (eds.), *Direct and Large-Eddy Simulation VIII*, ERCOFTAC Series 15, DOI 10.1007/978-94-007-2482-2_65,

from both grids will be presented and compared. In addition, the underlying physics will be highlighted.

2 Description of numerical method

The Navier-Stokes equations for an incompressible and isothermal flow in a constantly rotating frame of reference are considered. A Coriolis force term was implemented in the Navier-Stokes solver to account for system rotation, and the centrifugal effects are absorbed in the effective pressure term. The rotation number $Ro = 2\Omega h/u_\tau$, where Ω is the constant angular velocity, was set to 6 and data for a non-rotating case ($Ro = 0$) are included for comparative purposes.

The governing equations are solved in 3-D space and time using a parallel Finite Volume code called MGLET [5]. The code uses staggered Cartesian grid arrangements. Spatial discretization of the convective and diffusive fluxes are carried out using a 2nd order central differencing scheme. The momentum equations are advanced in time by a fractional time stepping using a 2nd order explicit Adams-Bashforth scheme. The Poisson equation for the pressure is solved by a full multi-grid method based on pointwise velocity-pressure iterations. The computational grid is divided into an arbitrary number of subgrids that are treated as dependent grid blocks in parallel processing. In the present study, the size of the computational domain in each coordinate direction $L_x \times L_y \times L_z$ is $2560 \times 800 \times 1280$ in wall units, i.e. practically the same as in the DNS case of Moser et al. [3] and Ashrafian et al. [4]. Uniform grid spacing is adopted in the streamwise and the spanwise directions, while a non-uniform mesh is used in the wall-normal direction. Periodic boundary conditions are employed in the streamwise and spanwise directions. No-slip and impermeability conditions are imposed on all the rigid surfaces.

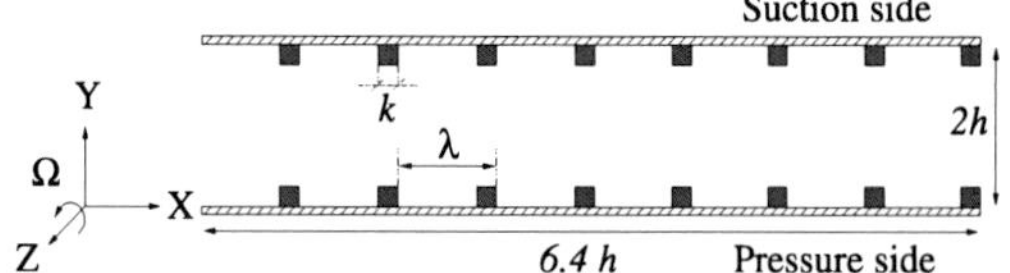

Fig. 1 Flow configuration (not to scale). The system is rotating with constant angular velocity Ω about the spanwise Z-axis.

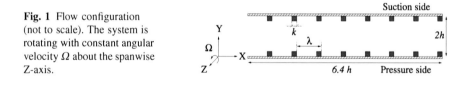

3 Results

The mean velocity field and the turbulence statistics have been deduced by first averaging in time and in the homogeneous spanwise direction. The time-averaging was performed over 400 statistical sample fields, each separated by $0.05h/u_\tau$. Advantage was then taken of the streamwise periodicity over the pitch length λ to further improve the quality of the sampling.

When the channel is rotated with a constant angular velocity Ω about a spanwise axis, cyclonic and anti-cyclonic behaviours are observed along the two channel walls, similarly as in the rotating smooth-walled channel studied by Kristoffersen & Andersson [6]. The mean velocity profiles in Fig. 2(a) show that $U(Y)$ exhibits a substantial region with a linearly increasing velocity. The slope is close to 2Ω, which implies that the absolute mean vorticity is driven to zero, just as in the smooth-walled case [6]. The peak of U^+ is shifted towards the suction side of the channel due to the rotation. Furthermore, if we compare the velocity profiles from both meshes it is evident that the coarse mesh in general overpredicts the velocity magnitude.

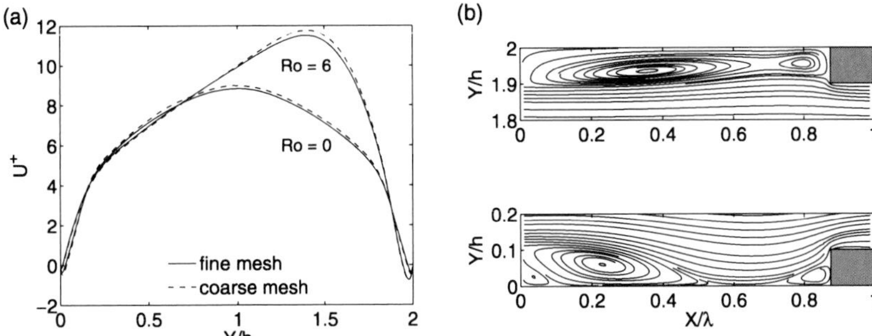

Fig. 2 (a) Streamwise mean velocity profiles $U(Y)$ non-dimensionalized by the wall friction velocity u_τ. Profiles from the present simulations are taken at $X/\lambda = 0.5$. (b) Streamlines illustrating the separated flow regions in the present rotating case Ro = 6.

The streamlines in Fig. 2(b) show that the near-wall behavior is rather different on the two sides. At the lower wall where the flow is anti-cyclonic, the reverse flow within the cavity between two subsequent rods is modest, whereas the reverse flow extends over the entire cavity near the top wall where the flow is cyclonic. The results of the present study show that a $k-$type roughness with $\lambda/k = 8$ is turned into $d-$type roughness by the action of the Coriolis force due to imposed system rotation. The explanation of this remarkable phenomenon may either be that cyclonic vortices are more stable than anti-cyclonic vortices, as convincingly argued by Cambon et al. [7], or that the turbulent fluctuations are enhanced near the pressure side and dampened near the suction side [6] and thus promote or reduce the spreading of the mixing-layer emanating from the downstream corner of a rod.

Iso-contours of the instantaneous vorticity field in Fig. 3 suggest that the turbulent vorticity has been enhanced along the pressure side as compared to the non-rotating case ($Ro = 0$). The Reynolds shear stress $< uv >$ profiles are shown in Fig. 4(a). The imposed system rotation has broken the conventional anti-symmetric variation of the turbulent shear stress since $< uv >$ is enhanced near the pressure side and reduced near the suction side. The turbulent kinetic energy in Fig. 4(b) is similarly damped along the suction wall as a result of the interaction between the instantaneous Coriolis force and the instantaneous velocity vector which nearly suppress

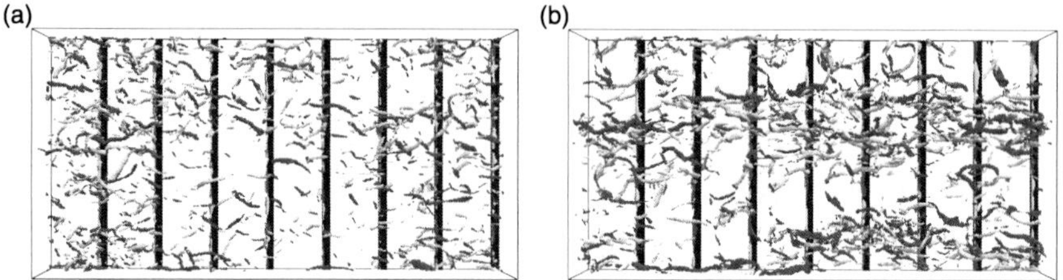

Fig. 3 Instantaneous vortical structures near the lower wall: (a) Ro = 0; (b) Ro = 6. Both snapshots are from the present fine mesh case.

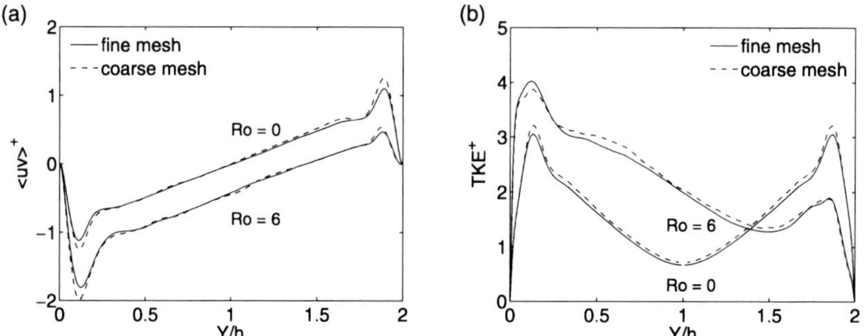

Fig. 4 (a) Reynolds shear stress $< uv >^+$; (b) Turbulence kinetic energy TKE^+. Profiles from the present simulations are taken at $X/\lambda = 0.5$.

the wall-normal velocity fluctuations. Again here, the coarse mesh overpredicts the turbulent shear stress and the TKE level.

The Reynolds stress transport equation can be written as:

$$\frac{D\overline{u_i u_j}}{Dt} = P_{ij} + C_{ij} + D_{ij} + G_{ij} + T_{ij} + \Pi_{ij} - \varepsilon_{ij} \tag{1}$$

where the right-hand-side terms, namely, production due to mean shear (P_{ij}), production due to rotation (C_{ij}), viscous diffusion (D_{ij}), pressure diffusion (G_{ij}), turbulent diffusion (T_{ij}), pressure-strain rate term (Π_{ij}) and the viscous dissipation term (ε_{ij}) are defined as:

$$P_{ij} = -\overline{u_i u_k}\frac{\partial U_j}{\partial x_k} - \overline{u_j u_k}\frac{\partial U_i}{\partial x_k}; \qquad C_{ij} = -2\Omega_k(\overline{u_j u_m}\,\varepsilon_{ikm} + \overline{u_i u_m}\,\varepsilon_{jkm}) \tag{2}$$

$$D_{ij} = \nu\left(\frac{\partial^2 \overline{u_i u_j}}{\partial x_k \partial x_k}\right); \quad G_{ij} = -\frac{1}{\rho}\frac{\partial}{\partial x_k}(\overline{pu_i}\delta_{jk} + \overline{pu_j}\delta_{ik}); \quad T_{ij} = -\frac{\partial}{\partial x_k}(\overline{u_i u_j u_k}) \tag{3}$$

$$\Pi_{ij} = \overline{\frac{p}{\rho}\left(\frac{\partial u_i}{\partial x_j} + \frac{\partial u_j}{\partial x_i}\right)}; \qquad \varepsilon_{ij} = 2\nu\left(\overline{\frac{\partial u_i}{\partial x_k}\frac{\partial u_j}{\partial x_k}}\right) \tag{4}$$

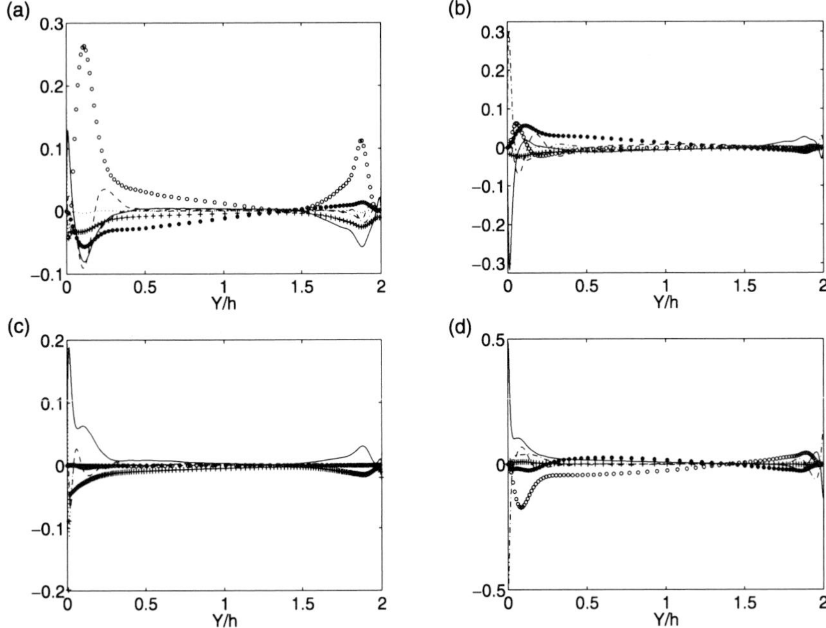

Fig. 5 Budgets of the individual Reynolds stress components for $Ro = 6$. All terms are scaled with u_τ^4/ν. $(a) < uu >; (b) < vv >; (c) < ww >; (d) < uv >$. Profiles are taken at $X/\lambda = 0.43$. $P_{ij}(\circ), C_{ij}(\bullet), D_{ij}(...), G_{ij}(-.-), T_{ij}(---), \Pi_{ij}(-), -\varepsilon_{ij}(+)$. Please note that the advection term $U_k \partial < u_i u_j > / \partial x_k$ (not shown here) is non-zero in the vicinity of ribs and therefore contribute to the balance.

The Reynolds stress budgets in Fig. 5 are distinctly asymmetric with the highest levels of the various contributions near the pressure (anti-cyclonic) side. The rotational production C_{ij}, which is identically zero in the direction of the axis of rotation, plays a major role in the two other coordinate directions. In the wide range $0.2 < Y/h < 1.4$ where the mean velocity U^+ varies linearly with Y, C_{xx} almost outweighs the mean shear production P_{xx}. In the wall-normal budget, on the other hand, P_{yy} is vanishingly small except in the vicinity of the cavities between two consecutive roughness elements and $C_{yy} = -C_{xx} = -4\Omega\overline{uv}$ becomes an essential source term. This Coriolis source enhances wall-normal velocity fluctuations along the pressure side and thereby indirectly increases $\overline{uv}$. Near the roughness elements, i.e. for $Y/h < 0.2$, almost all terms contribute to the Reynolds stress balances in Fig. 5. The substantial impact of turbulent diffusion T_{xx} due to wall-normal fluctuations is particularly noteworthy.

4 Concluding remarks

In the present study the focus has been on the combined effects of system rotation and wall roughness. The flow configuration closely resembled that of Ashrafian et al. [4]. The rotation number 6.0 was slightly lower than the highest rotation rate 7.6 considered in the plane channel flow by Kristoffersen & Andersson [6]. Results from a coarse grid simulation using 4.2 million grid points compared surprisingly well with data from a fully resolved DNS using 32.8 million points. In the presence of system rotation, the flow along the roughness elements at the stabilized suction side became of the skimming type; i.e. d-type. This striking phenomenon will be further explored over a wide range of rotation numbers.

Acknowledgements This work has received support from The Research Council of Norway (Programme for Supercomputing) through a grant of computing time and a research fellowship for the first author.

References

1. Perry, A.E., Schofield, W.H. & Joubert, P.N. (1969). Rough wall turbulent boundary layers. *J. Fluid Mech.* **37**, 383–413.
2. Jiménez, J. (2004). Turbulent flows over rough walls. *Annu. Rev. Fluid Mech.* **36**, 173–196.
3. Moser, R.D., Kim, J. & Mansour, N.N. (1999). Direct numerical simulation of turbulent channel flow up to $Re_\tau = 590$. *Phys. Fluids* **11**, 943–945.
4. Ashrafian, A., Andersson, H.I. & Manhart, M. (2004). DNS of turbulent flow in a rod-roughened channel. *Int. J. Heat Fluid Flow* **25**, 373–383.
5. Manhart, M. (2004). A zonal grid algorithm for DNS of turbulent boundary layers. *Computers & Fluids* **33**, 435–461.
6. Kristoffersen, R. & Andersson, H.I. (1993). Direct simulations of low-Reynolds-number turbulent flow in a rotating channel. *J. Fluid Mech.* **256**, 163–197.
7. Cambon, C., Benoit, J.-P., Shao, L. & Jacquin, L. (1994). Stability analysis and large-eddy simulation of rotating turbulence with organized eddies. *J. Fluid Mech.* **278**, 175–200.

LES of heated fuel bundle arranged into triangular array

S. Rolfo, J. C. Uribe and D. Laurence

1 Introduction

A rod bundle is a key constitutive element of a wide range of nuclear reactor designs. It is composed by a set of long thin rods containing the nuclear fuel and fluid-flow between them and generally parallel to the rods. As experiments in a such densely packed geometries are difficult, thermal-hydraulics simulations are valuable to study heat transfer, fluid-forces, homogeneity of temperatures and flow rates and their fluctuations for current or and future reactor designs.

The geometry is fully described using only the rod diameter (D) and pitch (P), i.e. the distance between the center of two adjacent rods. However the flow presents complicated features limiting the simple hydraulic diameter analogies. Experimental studies [5, 6] found that the distribution of the turbulent intensities is different from that in pipes and plane channels, and a higher level of mixing between sub-channels was also observed. In particular, in the gap region between two sub-channels, the maximum turbulent intensities are located far from the wall. Indeed the turbulent quantities are strongly dependent from the pitch-to-diameter ratio (P/D) of the configuration. When P/D is very tight, an energetic and almost periodic azimuthal flow pulsation is present in the gap region between two fuel elements. The phenomenon was already observed in [9] and confirmed in [3], then in [5] it was proposed as the cause for higher mixing between sub-channels. More recently in [2] a detailed analysis of these coherent large-scale structures was performed, but in a slightly different geometry than the one investigated herein.

Stefano Rolfo, Juan C. Uribe
School of MACE, The University of Manchester, PO Box 88, Manchester M60 1QD, UK, e-mail: rolfo.stefano@manchester.ac.uk, juan.uribe@manchester.ac.uk

Dominique Laurence
EDF R&D, 6 quai Watier, F-78400 Chatou
School of MACE, The University of Manchester, PO Box 88, Manchester M60 1QD, UK, e-mail: dominique.laurence@manchester.ac.uk

H. Kuerten et al. (eds.), *Direct and Large-Eddy Simulation VIII*, ERCOFTAC Series 15, DOI 10.1007/978-94-007-2482-2_66,

2 Case description

This work is a continuation of [8], where a reduced geometry was considered and a hybrid RANS/LES turbulence model was presented in order to analyse the high Reynolds number cases. Hereafter standard well resolved LES, using a Smagorinsky model [10], is considered and the analysis is carried out on an enlarged domain composed by two complete sub-channels. In the previous work some higher extra-frequencies were observed for the reduced geometry and one of the main objective herein is to verify their presence against a larger domain. Additionally, because of the very sparse availability of heat transfer data, calculations are carried out with several passive scalars in order to check the influence of the boundary conditions (Neumann and Dirichlet). The Prandtl number is kept constant and equal to 0.71. The geometry is characterized by a P/D ratio of 1.06 and two shear Reynolds numbers ($Re_\tau = u_\tau D_h/\nu$) of 600 and 800 are considered (bulk values of 6,000 and 13,000, respectively). As a consequence of flow pulsations the length in the stream-wise direction is equal to twelve times the hydraulic diameter. The near-wall mesh resolutions are: $0.7 \leq r^+ \leq 0.9$, $6 \leq r\Delta\theta^+ \leq 10$ and $11 \leq \Delta x^+ \leq 18$ for $Re_\tau = 400$ and $0.7 \leq r^+ \leq 1.0$, $7.5 \leq r\Delta\theta^+ \leq 15$, $17 \leq \Delta x^+ \leq 30$ for $Re_\tau = 800$.

The calculations are carried out using *Code_Saturne*, a finite volume, open source, incompressible Navier-Stokes solver. The collocated velocity-pressure coupling is ensured through a prediction/correction method based on a SIMPLEC algorithm and Rhie & Chow interpolation [7]. Second order central scheme is used both in time and space. For a general description of the code see [1] and [4] for LES applications.

3 Flow pulsations and Reynolds stresses

The phenomenon of the flow pulsations is visualized in Fig. 1, where velocity fluctuations in the stream-wise direction show very elongated structures located in the gap region. The velocity fluctuations in the span-wise direction (Fig. 1 middle) present an alternating negative/positive pattern, always limited to the gap, which confirms the enhanced inter-channel mixing. The final result is a meandering behaviour most visible on the temperature iso-surface (Fig. 1 right). The phenomenon is characterized in [5] by a constant Strouhal number defined as $St = \frac{fD}{\langle u \rangle_{GAP}} = 0.93$, where $\langle u \rangle_{GAP}$ is the bulk velocity in the gap region. In [8] we found three dominant frequencies, the first in accordance with the experimental value and the two higher ones corresponding to Strouhal numbers of 2 and 3 respectively (see Fig. 2 left). The reason was attributed to the small cross-section and the consequently large influence of the periodic boundary conditions between the sides. In the present work, where an extended cross-section is used, again three dominant frequencies at same Strouhal numbers are found, for both Reynolds numbers considered. An explanation of the two extra-frequencies may be found in the two-point correlation between inlet and middle of the domain. Velocity fluctuations in the stream-wise direction show a very low level of correlation (see Fig. 2 top-right), whereas the fluctuations in the span-wise are still highly correlated (see Fig. 2 bottom-right).

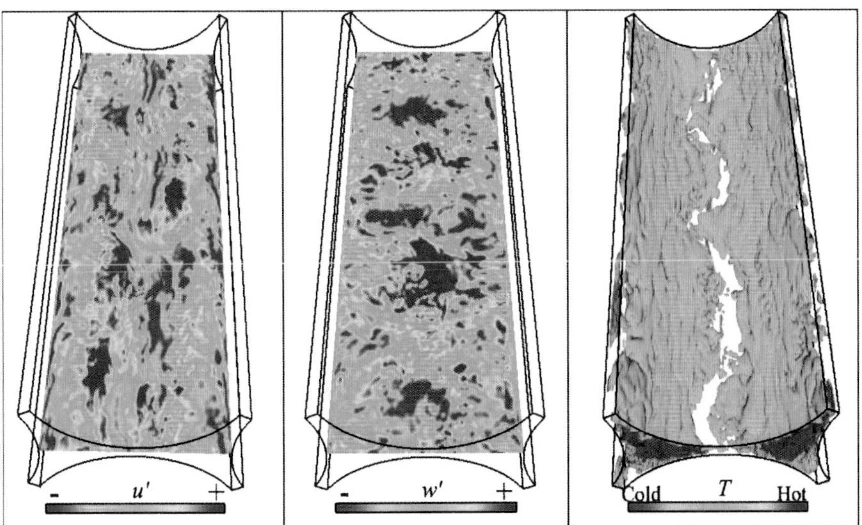

Fig. 1 Visualization of the flow pulsation with sub-channel mixing. Stream-wise velocity fluctuations (left), span-wise velocity fluctuations (middle), temperature iso-surfaces (right).

The effect of the coherent structures is clearly visible on some components of the Reynolds stresses reported in Fig. 3. Fluctuations in the stream-wise and wall-normal directions display profiles similar to the ones of pipes and plane channels, but fluctuations in the azimuthal direction show two peaks, of about the same intensity, one located in the open region close to the wall, where there is the maximum of transfer of energy from u'_{rms} to w'_{rms} by the pressure transfer terms. The second peak of w', in the middle of the gap region in contrast with the decrease of u', is clearly a result of the coherent structures discussed previously. $\langle u'w' \rangle$ and $\langle v'w' \rangle$ present maxima in the gap region then abruptly go to zero, as required by the symmetries of the geometry and Reynolds stress tensor. On the other hand $\langle u'w' \rangle$ corresponds to the main momentum transfer from axial flow to the tube walls (creating drag or head losses).

4 Heat transfer

Along with the dynamic field several passive scalars are used. Two different types of boundary conditions (BC) are employed: Neumann BC forcing constant wall heat flux ($54\,W/m^2$ for $Re = 6,000$ and $200\,W/m^2$ for $Re = 13,000$) and a Dirichlet

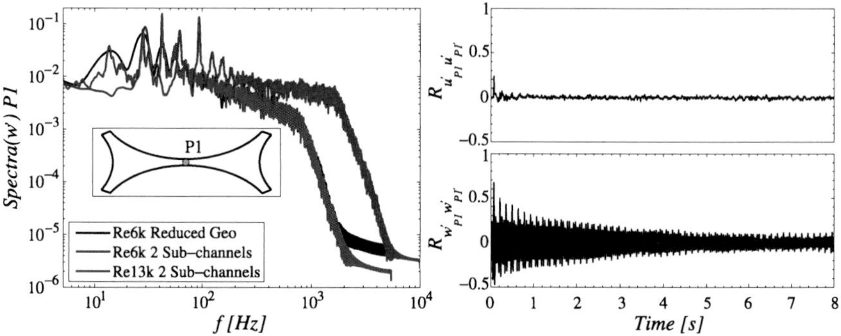

Fig. 2 Spectra of the span-wise velocity fluctuations in the middle of the gap region of the geometry (left). Two points correlation between points at the inlet and middle of the domain for the stream-wise direction (top right) and span-wise direction (bottom right).

constant wall temperature ($300\,K$ only for $Re = 6{,}000$). The dimensionless temperature profiles match the following log-law (see Fig. 4 left): $T^+ = 2.5\ln y^+ + 3.4$ with $T^+ = (T_w - T)/T_\tau$ and T_τ the local friction temperature[1]. Agreement with the log-law is is fairly good for Neumann BC, implying that wall functions might used for simulation at higher (reactor scale) Re numbers, whereas with a Dirichlet BC temperature profiles do not collapse so well, but this could still be a low Reynolds number effect. Temperature fluctuations have their maximum in the gap region and minimum in the center of the sub-channel (see Fig. 5). Heat fluxes $\langle u'\theta'\rangle^+$, $\langle v'\theta'\rangle^+$ and $\langle w'\theta'\rangle^+$ mirror the stresses u'_{rms}, $\langle u'v'\rangle^+$ and $\langle u'w'\rangle^+$, respectively. Temperature r.m.s. for the Dirichlet BC is very similar, in both trend and values, to the previous with the exception of the wall value where θ'_{rms} tends to zero in the Dirichlet BC case obviously. For the high Reynolds case temperature variance and heat fluxes presents similar features, with only a change in the apex values. The present database may be used in future for validation of second moment based URANS simulations.

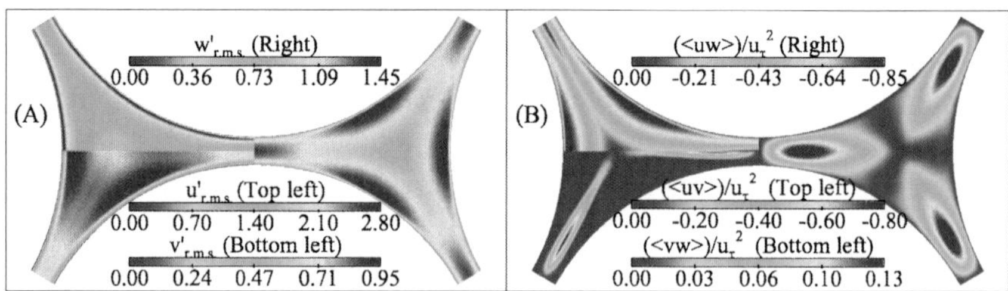

Fig. 3 Dimensionless Reynolds stresses for the rod bundle test case at $Re = 13000$. Side (A): u'_{rms}, v'_{rms} and w'_{rms}. Side (B): $\langle u'v'\rangle^+$, $\langle u'w'\rangle^+$ and $\langle v'w'\rangle^+$.

[1] In the case of Dirichlet BC the wall heat flux is varying as function of the circumferential direction and this is taken into account in the evaluation of T_τ.

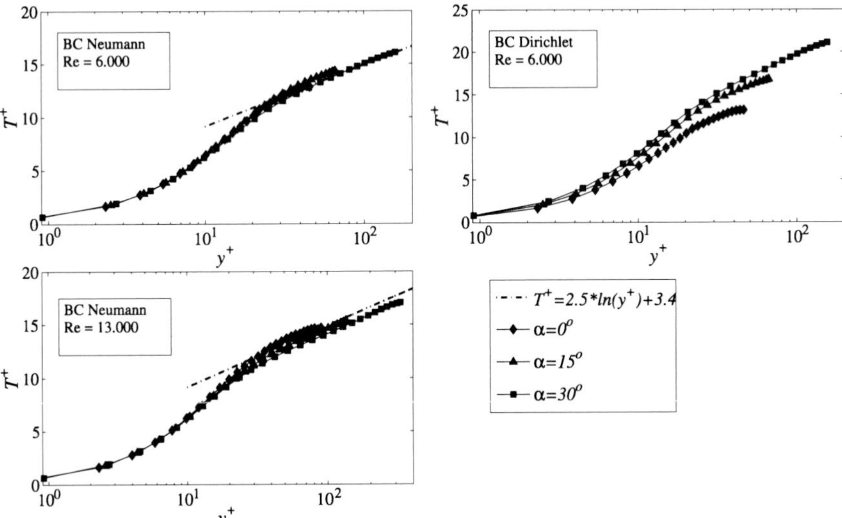

Fig. 4 Dimensionless temperature for different BC and Reynolds number. In all cases $Pr = 0.71$.

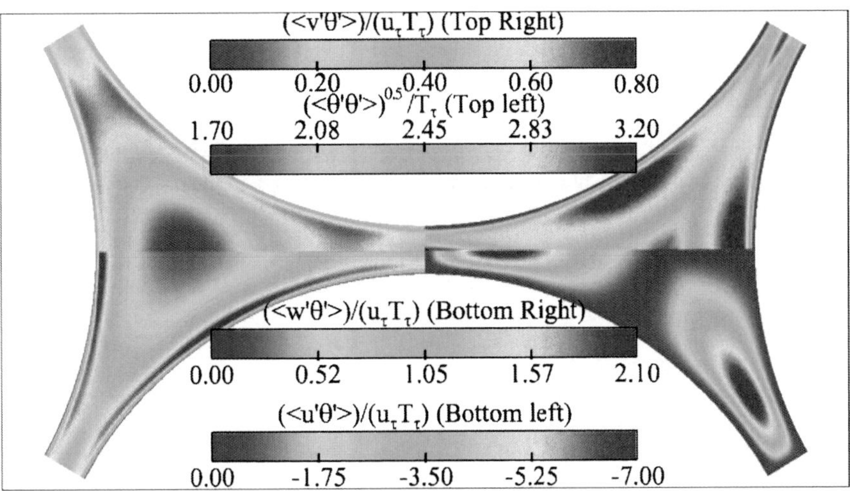

Fig. 5 Dimensionless temperature r.m.s. and heat fluxes at $Re = 6,000$.

5 Conclusions

Flow inside fuel rod bundle has been investigated using wall-resolved LES. The presence of flow pulsations in the gap region is reproduced and their influence on the mean quantities verified. However, frequency analysis shows the presence of two extra-frequencies, which are not reported in the experimental data. Their origin might be attributed to the fact that azimuthal velocity is still highly correlated between inlet and middle of the domain and verification against longer geometry is on going. Heat transfer analysis is also carried out finding that the variation of the thermal boundary condition from a constant wall heat flux to a constant wall temperature may be compatible with a wall function approach for high Re simulations

Acknowledgements This work was carried out as part of the TSEC programme KNOO and as such we are grateful to the EPSRC for funding under grant EP/C549465/1. The authors are also grateful to EDF R&D for the usage of their HPC facilities.

References

1. F. Archambeau, N. Mechitoua, and M. Sakiz. Code_Saturne: a finite volume code for the computation of turbulent incompressible flows – Industrial Applications. *Int. J. Finite Vol.*, 1(1), 2004.
2. F. Baratto, S. Bailey, and S. Tavoularis. Measurements of frequencies and spatial correlations of coherent structures in rod bundle flows. *Nuclear Engineering and Design*, 236(17):1830–1837, 2006.
3. J. Hooper and K. Rehme. Large-scale structural effects in developed turbulent flow through closely-spaced rod arrays. *Journal of Fluid Mechanics*, 145:305–337, 1984.
4. N. Jarrin, S. Benhamadouche, D. Laurence, and R. Prosser. A synthetic-eddy-method for generating inflow conditions for large-eddy simulations. *International Journal of Heat and Fluid Flow*, 27(4):585–593, 2006.
5. T. Krauss and L. Meyer. Experimental investigation of turbulent transport of momentum and energy in a heated rod bundle. *Nuclear Engineering and Design*, 180(3):185–206, 1998.
6. K. Rehme. The structure of turbulence in rod bundles and the implications on natural mixing between the subchannels. *International Journal of Heat and Mass Transfer*, 35(2):567–581, 1992.
7. C. Rhie and W. Chow. Numerical study of the turbulent flow past an airfoil with trailing edge separation. *AIAA Journal (ISSN 0001-1452)*, 21:1525–1532, 1983.
8. S. Rolfo, J. C. Uribe, and D. Laurence. LES and Hybrid RANS/LES of Turbulent Flow in Fuel Rod Bundle Arranged with a Triangular Array. *Direct and Large-Eddy Simulation VII*, pages 409–414, 2010.
9. D. S. Rowe, B. M. Johnson, and J. G. Knudsen. Implications concerning rod bundle crossflow mixing based on measurements of turbulent flow structure. *International Journal of Heat and Mass Transfer*, 17(3):407–419, 1974.
10. J. Smagorinsky. General circulation experiments with the primitive equations: I. The basic experiment. *Mon. Wea. Rev.*, 91:99–164, 1963.

Effect of wind-turbine surface loading on power resources in LES of large wind farms

Johan Meyers and Charles Meneveau

1 Introduction

As wind power grows as an important contributor to the worldwide overall energy portfolio, wind farms will cover increasingly larger surface areas. With the characteristic height of the atmospheric boundary layer (ABL) of about 1 km, wind farms with horizontal extents exceeding 10–20 km may therefore approach the asymptotic limit of 'infinite' wind farms, and the boundary layer flow may approach the fully developed regime. Envisioning such large-scale implementations calls for advancements in our understanding of the detailed interactions between wind turbines and the atmospheric surface layer. In the past, a number of studies have focussed on the effect wind-turbine arrays on the WTABL using elements of momentum theory, potential flow, and the superposition of wakes of individual turbines (cf. Lissaman 1979 [1], and Frandsen 1992 [2]). Several recent studies have focused on such dynamics of Wind Turbine Array Boundary Layers (WTABL) [3, 4, 5].

In this contribution, we present results of a WTABL LES study, and develop tools and concepts required to use these results to quantify and understand the energy extraction from a very large wind farm in the presence of an ABL which adapts its equilibrium based on the surface roughness induced by the wind farm. In particular, the contribution aims at determining optimal turbine spacing and surface loading for maximal power output at a given geostrophic wind speed. An important parameter appearing in this optimization, is the ratio of costs associated with use of land in use per turbine by cost of turbine installation and maintenance. For realistic values of this ratio, we find that the optimal average turbine spacing may be considerably

Johan Meyers
Department of Mechanical Engineering, Katholieke Universiteit Leuven, Celestijnenlaan 300A, B3001 Leuven, Belgium, e-mail: johan.meyers@mech.kuleuven.be

Charles Meneveau
Department of Mechanical Engineering, and Center for Environmental and Applied Fluid Mechanics, Johns Hopkins University, 3400 North Charles Street, Baltimore MD 21218, USA, e-mail: meneveau@jhu.edu

H. Kuerten et al. (eds.), *Direct and Large-Eddy Simulation VIII*,
ERCOFTAC Series 15, DOI 10.1007/978-94-007-2482-2_67,

higher ($\sim 15D$ to $25D$, with D the turbine-rotor diameter) then conventionally used in current wind-farm implementations ($\sim 7D$).

2 Description of numerical method

For our Large Eddy Simulation, we solve the filtered Navier-Stokes equation for an incompressible and neutral (non-bouyant) flow, i.e.,

$$\frac{\partial \widetilde{\mathbf{u}}}{\partial t} + \widetilde{\mathbf{u}} \cdot \nabla \widetilde{\mathbf{u}} = -\frac{1}{\rho} \nabla \widetilde{p} - \nabla \boldsymbol{\tau}_{sgs} + \widehat{\mathbf{f}}, \qquad \nabla \cdot \widetilde{\mathbf{u}} = 0; \tag{1}$$

$\widetilde{\mathbf{u}}$ is the velocity field (with components $\widetilde{u}_i$, $i = 1 \cdots 3$), $\widetilde{p}$ the pressure, $\boldsymbol{\tau}_{sgs}$ the subgrid-scale tensor. The term $\widehat{\mathbf{f}}$ represents additional forcing terms, which in the current paper correspond to the actuator-disk forces per unit mass effectuated by separate wind turbines on the boundary layer (see below). The effect of turbulence motions at scales smaller than the grid-size on the resolved field $\widetilde{\mathbf{u}}$ is modeled using the Smagorinksy subgrid-scale closure $\boldsymbol{\tau}_{sgs} = -2\ell^2 (2\mathbf{S} : \mathbf{S})^{1/2} \mathbf{S}$, with $\mathbf{S} = (\nabla \widetilde{\mathbf{u}} + (\nabla \widetilde{\mathbf{u}})^T)/2$ the strain-rate tensor, and ℓ is the characteristic length-scale for the eddy-viscosity. We use a non-dynamic version of the Smagorinsky model, which requires damping functions to account for effects near the ground, i.e. the length-scale is obtained [6] from $\ell^{-n} = [\kappa(z+z_0)]^{-n} + (C_s h)^{-n}$, with z_0 the roughness of the ground surface, κ the von Kármán constant (we use $\kappa = 0.4$), and $h = (h_1 h_2 h_3)^{1/3}$ the mesh spacing.

The forces that model the wind turbine are modeled according to [4]

$$f_x = -\frac{1}{2} C_T' \overline{V}^2, \qquad f_\theta = \frac{1}{2} C_P' \overline{V}^2 \frac{V}{\Omega r}. \tag{2}$$

Here, f_x, and f_θ are the axial and tangential forces (per unit surface and per unit density, and θ tangential to the turbine rotation) constituting the turbine forces $\mathbf{f}$, V is the axial velocity averaged over the rotor disk, and $\overline{V}$ the time-averaged value of V. Further C_T' and C_P' are respectively thrust and power coefficients. They are defined using the velocity at the disk, and hence differ from the usual thrust and power coefficients (C_T and C_P) commonly employed in the literature, which are based on the undisturbed upstream velocity V_∞ at hub height (e.g., $C_T' = 4/3$ for $C_T = 3/4$ [4]). In the current work, we will also employ the surface loading coefficient for analysis of results

$$c_{ft}' = C_T' A/S = \frac{C_T' \pi}{4s^2}, \tag{3}$$

with $A = \pi D^2/4$ the turbine rotor surface, S the land area available per turbine, and s the average turbine spacing normalized by D (in fact, it is the average turbine spacing s which will be optimized in §4).

To represent the turbine forces (2) defined in the turbine-rotor disk onto the LES grid, a Gaussian convolution filter $\mathscr{G}$ is used, such that $\widehat{\mathbf{f}} = \mathscr{G}\mathbf{f}$ [3, 4]. The equations (1) are discretized using a pseudo-spectral method in stream-wise and span-wise direction, with periodic boundary conditions in these directions. A fourth-order energy-conserving staggered finite-volume discretization is used in the wall-normal direction. At the top of the domain, a symmetry boundary condition is used. Time integration is performed by a classical four-stage fourth-order accurate Runge–Kutta time integration. Several simulations were performed with varying numbers of turbines, grid sizes and domain sizes. Details may be found in Ref. [3].

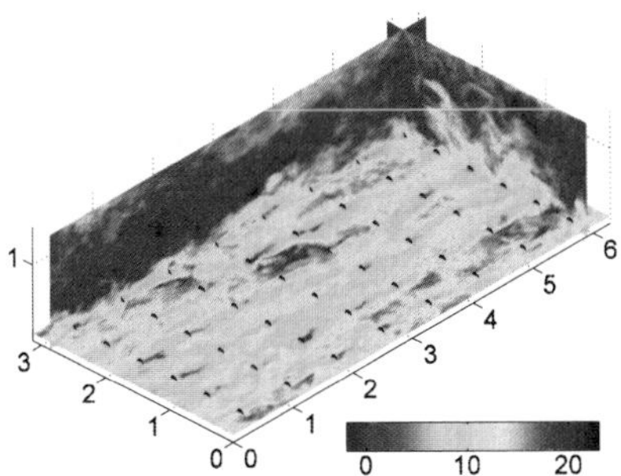

Fig. 1 Instantaneous stream velocity contours from LES of WTABL, in units of friction velocity

3 LES results

Figure 1 shows instantaneous velocity profiles on selected planes across the domain and thus gives a qualitative view of the simulations. In Fig. 2(a) time-averaged and plane-averaged velocity profiles are shown for different surface-loading coefficients c'_{ft}. The figure illustrates, that instead of one log-layer in the boundary layer, we now find a log-layer below the wind turbines, and another one on top of the wind turbines. As a consequence, the outer layer of the boundary layer experiences a surface roughness $z_{0,hi}$ which is higher then the surface roughness of the ground surface $z_{0,lo}$. This 'high'surface roughness, induced by the wind farm, may be linked to the upper log layer observed in Fig. 2(a). Another consequence is that the layer below the wind turbines experiences a friction velocity $u_{*lo} = \sqrt{\tau_w}$ which is lower then the friction velocity in the upper layer, since wind-turbines have taken out part of the momentum, with a momentum balance corresponding to

$$u_*{}^2 - u_{*lo}^2 = c'_{ft}\overline{V}^2/2. \tag{4}$$

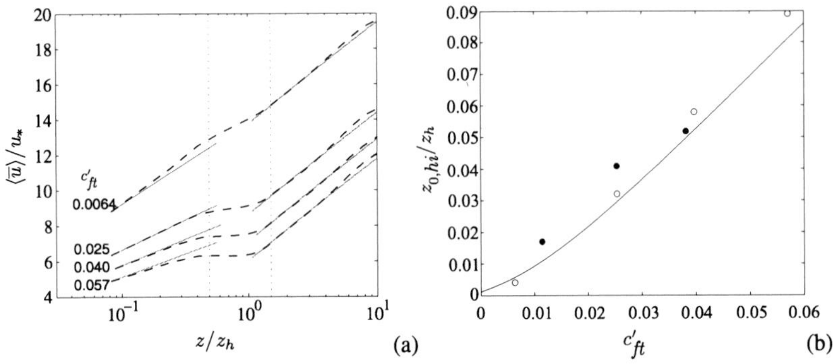

Fig. 2 (a) Plane averaged velocity profiles for WTABL cases with different surface loading c'_{ft}. (b) Effective roughness length determined from LES, as function of surface loading parameter c'_{ft} (symbols), and obtained from the surface roughness model Eq. (6) (full line). (Figures adapted from Ref. [3])

Based on the assumption of two log layers, and a connecting wake-mixing region, a model was derived in Ref. [3] for $z_{0,hi}$, and u_{*lo} as function of relevant WTABL parameters such as surface loading, turbine hub height, etc. The model corresponds to

$$\frac{u_{*lo}}{u_{*hi}} = \ln\left[\frac{z_h}{z_{0,hi}}\left(1+\frac{D}{2z_h}\right)^{\frac{\nu_w^*}{1+\nu_w^*}}\right] \Big/ \ln\left[\frac{z_h}{z_{0,lo}}\left(1-\frac{D}{2z_h}\right)^{\frac{\nu_w^*}{1+\nu_w^*}}\right], \tag{5}$$

where ν_w^* is a normalized "augmented wake eddy viscosity", estimated as $\nu_w^* \approx 28\sqrt{c_{ft}/2}$, with $c_{ft} = \pi C_T/4s^2 \approx 9/16 c'_{ft}$ [3]. Further,

$$\frac{z_{0,hi}}{z_h} = \left(1+\frac{D}{2z_h}\right)^{\frac{\nu_w^*}{1+\nu_w^*}} \exp\left(-\left[\frac{c_{ft}}{2\kappa^2} + \left(\ln\left[\frac{z_h}{z_{0,lo}}\left(1-\frac{D}{2z_h}\right)^{\frac{\nu_w^*}{1+\nu_w^*}}\right]\right)^{-2}\right]^{-1/2}\right). \tag{6}$$

In Fig. 2(b) the induced surface roughness $z_{0,hi}$ is shown as function of the surface loading c'_{ft}, either obtained directly from LES simulations, or by using the model Eq. (5,6). It is appreciated that the model provides a reasonable parametrization of the surface roughness. We will use this model next, to set up a model for wind-farm–ABL interaction with which we can optimize turbine spacing with respect to power output.

4 Optimization of turbine spacing

First of all, we define a relevant objective function for optimization based on the normalized farm power per unit cost and per unit geostrophic wind G (cubed), i.e.

$$P^*(s, C_T', \alpha) = \frac{P}{S\rho G^3/2} \frac{cost_L}{cost_T/S + cost_L} = \frac{C_T'}{4s^2/\pi + \alpha}\left(\frac{u_{*hi}}{G}\right)^3\left(\frac{\overline{V}}{u_{*hi}}\right)^3, \quad (7)$$

with $\alpha = cost_T/A/cost_L$. Further, $cost_L$ [\$/m^2] corresponds to wind-farm costs which are proportionally related to the area of land used, and $cost_T$ [\$] to costs which are proportional to the number of turbines employed. In the current work, we will not explore in detail values of $cost_L$, and $cost_T$, but instead explore a wide range of possible values for the non-dimensional ratio α.

In order to evaluate P^*, we need to have expressions for $\overline{V}/u_{*hi}$, and for u_{*hi}/G. A relation for the first ratio is readily obtained by combining Eqs. (4), (5), and (6). For the second ratio, we use standard ABL relations (see, e.g. [7]), i.e.

$$\frac{G}{u_*} = \sqrt{A^2 + \left[\frac{1}{\kappa}\ln\left(\frac{u_*}{G}Ro\right) - C\right]^2}, \quad (8)$$

with A, and C constants, and where the dimensionless group $Ro = G/(fz_0)$ has the form of a Rossby number, expressing a ratio between inertia and Coriolis forces, and f is the Coriolis parameter (e.g., at 40 degree latitude, $f = 9.34 \times 10^{-5}$ 1/s [7]). In the context of wind farms, we use $z_0 = z_{0,hi}$, and $u_* = u_{*hi}$. In the current work, we are mainly interested in the reaction of the ABL to changes in the surface roughness induced by wind turbines. Therefore, we introduce an alternative Rossby number, using the turbine hub-height as reference length scale, such that $Ro_h = G/(fz_{0,hi}) = Ro z_{0,hi}/z_h$, and we will evaluate the effect of variations in z_0/z_h, while keeping Ro_h constant. A representative reference value for Ro_h may, e.g., be estimated using $f = 9.34 \times 10^{-5}$ 1/s, $G = 20$m/s, and $z_h = 100$m, leading to $Ro_h \approx 2140$. Equation (8) provides us with an implicit relation for u_{*hi}/G, which we solve using MATLAB.

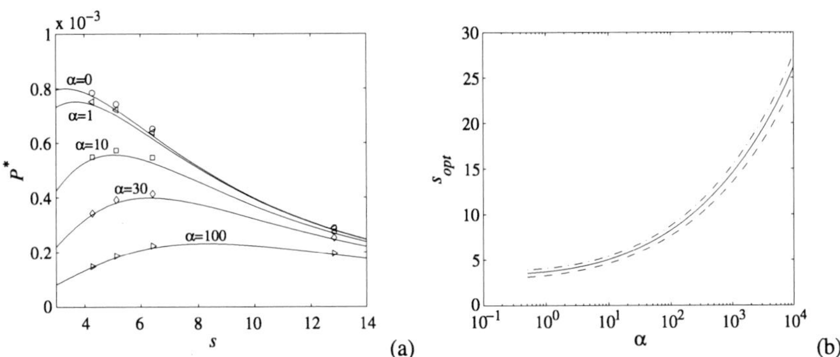

Fig. 3 (a) Normalized Power per unit cost P^* for different values of α as function of the average turbine spacing s (and $C_T' = 4/3$). Lines: obtained from model (Eqs. 7,5,6). Symbols: evaluating U_d/u_{*hi} directly from large-eddy simulations. (b) Optimal turbine spacing s_{opt} as function of α for (−−) $C_T' = 1.0$; (—) $C_T' = 4/3$; and (−·) $C_T' = 5/3$. More trends can be found in Ref. [8]

We now first evaluate P^* as function of s for different values of α, and $C_T' = 4/3$, in Figure 3(a). Results for $\alpha = 0$ show that the turbine spacing which achieves maximum power output per acre ($\alpha = 0$ presumes only land-related costs) is relatively small ($s_{opt} \approx 3$), and falls outside the range covered by the LES simulations for the development of the surface roughness model. However, at higher values for α, which is economically more relevant, the optimum shifts to higher values of s. In Figure 3(b) we further employ the model to optimize the turbine spacing, displaying the optimum spacing s_{opt} as function of α values, including three different values for C_T'. In practice, values for α may roughly be situated between 10^3 and 10^4, as it is appreciated that $cost_t/A \gg cost_L$. Looking at optimal turbine spacings in Figure 3(b) for this range, we find that $15 \lessapprox s_{opt} \lessapprox 25$, which is considerably larger than typical average spacings currently used in large wind farms both on and off shore (e.g., the well known Horns Rev wind farm off the coast of Denmark, has an average farm spacing of $s = 7$).

5 Conclusions

Based on results from a detailed LES study of the WTABL, a model for optimization of wind-farm power output as function of average turbine spacing has been formulated. It was found that the optimal average turbine spacing may be considerably higher then used in current wind-farm implementations.

Acknowledgements The research of JM is supported by OPTEC–K.U.Leuven. CM acknowledges the funding from the NSF's Energy for Sustainability Program (Project CBET 0730922).

References

1. Lissaman, P.B.S.: Energy effectiveness of arbitrary arrays of wind turbines. AIAA Paper 79-0114 (1979)
2. Frandsen, S.: On the wind speed reduction in the center of large clusters of wind turbines. J. Wind Eng. Indust. Aerodyn. **39**, 251–265 (1992)
3. Calaf, M., Meneveau, C., Meyers, J.: Large Eddy Simulation study of fully developed wind-turbine array boundary layers. Phys. Fluids **22**, 015110 (2010) doi: 10.1063/1.3291077
4. Meyers, J., Meneveau, C.: Large eddy simulations of large wind-turbine arrays in the atmospheric boundary layer. AIAA Paper 2010-827 (2010)
5. Cal, R.B., Lebrón, J., Kang, H.S., Castillo, L., Meneveau, C.: Experimental study of the horizontally averaged flow structure in a model wind-turbine array boundary layer. J. Renewable Sustainable Energy **2**, 013106 (2010) doi:10.1063/1.3289735
6. Mason, P.J., Thomson, D.J.: Stochastic backscatter in large-eddy simulations of boundary layers. J. Fluid Mech. **242**, 51 (1992)
7. Tennekes, H., Lumley, J.L.: A first Course in Turbulence. MIT Press, Massachusetts (1972)
8. Meyers, J., Meneveau, C.: Optimal turbine spacing in fully developed wind-farm boundary layers. Preprint submitted to Wind Energy (2010)

DNS of a turbulent channel flow with pin fins array: parametric study

B. Cruz Perez, J. Toro Medina, & S. Leonardi

1 Introduction

Gas turbines used in the aerospace industry are subject to extremely high temperatures from the combustor. Efficient cooling systems are required to avoid damage to the turbine blades and stators. There are two approaches mainly used in the cooling and protection of the turbine blades and stators, which are external and internal cooling. External cooling is achieved by jets in the exterior of the blade that creates a thin relative cold air film around the blade. The film prevents that the incoming hot air enters in direct contact with the turbine blade. Internal cooling is obtained by the use of internal channels with turbulators and pin fins exposed to a coolant fluid flow. At the trailing edge of the blade, pin fins are placed to increase the heat transfer while providing structural support to the blade itself (Metzger *et al.* 1984). Metzger *et al.* (1984) estimated that the pin heat transfer surface coefficients doubles the end-wall coefficients. They also showed how the orientation with respect to the mean flow affects the heat transfer and the pressure drop for pin fin arrays. On the other hand, Chyu *et al.* (1999) concluded that the heat transfer in the pin fins is 10 to 20 percent higher than the end-wall. Ames et al. (2004) studied the turbulence levels on pin fins arrays and its effect on the heat transfer, they also made a correlation that relates the turbulence levels, the heat transfer, and the Reynolds number (hereafter *Re*). Damerow *et al.* (1974) observed that ratio between the channel height (h) and the diameter (D) on staggered pin fin arrays has no effect on the friction factor, which contrasts with the data presented by Sparrow and Ramsey (1978). Cruz Perez *et al.* (2008) studied numerically and experimentally the length in a pin fin array to have a self similar flow. For more than five rows the friction factor only depends on the *Re* and not on the number of rows. The present study is focused on understanding the effect of streamwise and spanwise row spacing on the heat transfer and drag by performing Direct Numerical Simulations (DNS).

Dept. of Mech. Eng. University of Puerto Rico at Mayaguez, Puerto Rico US, e-mail: sleonardi@me.uprm.edu

H. Kuerten et al. (eds.), *Direct and Large-Eddy Simulation VIII*, ERCOFTAC Series 15, DOI 10.1007/978-94-007-2482-2_68,

2 Numerical Method

The non-dimensional Navier-Stokes, continuity and energy equations for incompressible flows are:

$$\frac{\partial U_i}{\partial t} + \frac{\partial U_i U_j}{\partial x_j} = -\frac{\partial P}{\partial x_i} + \frac{1}{Re}\frac{\partial^2 U_i}{\partial x_j^2} + \Pi \delta_{i1} \,, \; \nabla \cdot U = 0 \,, \tag{1}$$

$$\frac{\partial T}{\partial t} + \frac{\partial T U_j}{\partial x_j} = \frac{1}{Re\,Pr}\frac{\partial^2 T}{\partial x_j^2} \quad , \tag{2}$$

Π is the pressure gradient required to maintain a constant flow rate, δ_{ij} is the Kronecker delta, U_i is the component of the velocity vector in the i direction and P is the pressure, T is the temperature, α the thermal diffusivity and $Pr = \nu/\alpha = 0.71$ is the Prandtl number. The Reynolds number is $Re_b = U_b h/\nu = 4{,}000$ where, U_b is the bulk velocity and ν is the kinematic viscosity, h the channel half height.

The Navier-Stokes and energy equations have been discretized in an orthogonal coordinate system using the staggered central second-order finite-difference approximation. Here, only the main features are recalled since details of the numerical method can be found in Orlandi (2000). The pin fins are treated by the efficient immersed boundary technique described in detail by Orlandi & Leonardi (2006).

The computational box is $12h \times 2h \times 12h$ in x (streamwise), y (wall-normal) and z (spanwise direction) respectively. The grid has $512 \times 320 \times 512$ points in x, y, z respectively. The grid in the horizontal direction is uniform, while in the vertical direction the grid is non–uniform, the points being clustered near the walls (a hyperbolic tangent transformation is used; the first point is at $y = 2.6E-3$ from the wall).

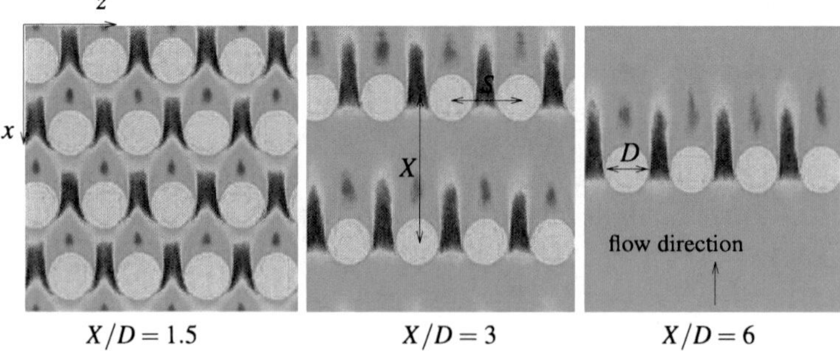

Fig. 1 Time averaged color contours of streamwise velocity at $0.1D$ from the hot wall, for $S/D = 1.5$: red, high; blue low.

Circular pin fins with $\frac{h}{D} = 1$ are considered (D is the diameter of the pin). The spanwise ($\frac{S}{D}$) and the streamwise distance ($\frac{X}{D}$) between the cylinder range from 1.5 to 6. The pin fins are in a staggered configuration, meaning that a row is shifted by a distance equivalent to $\frac{S}{2}$ with respect to the previous row (Figure 1). Periodic boundary conditions apply to the streamwise and the spanwise directions and no-slip condition to the top and bottom walls. The non-dimensional temperature on the lower wall is $T = 1$ while that on the upper wall is $T = -1$. Therefore, heat enters into the channel from the bottom wall and it is extracted at the upper wall.

3 Results

Time averaged streamwise velocity contours are shown in figure 1, at a distance $0.1D$ from the hot wall. The flow behavior around the pin fins resembles the flow around an infinite cylinder, but in these cases the length of the wake and the pressure gradient is affected by the near pin fins and the end wall interaction. When the flow reaches the pin fin a stagnation point occurs at the leading edge. As the fluid moves through the circumference the flow increases its velocity. Finally the boundary layer separates from the body, due to the adverse pressure gradient. Behind the pin fin a wake extends for a distance L which increases as the streamwise distance between the rows increases. For $X/D = 1.5$ the wake is limited by the next row and it is about $L/D = 0.5$. For $X/D = 6$, the flow impinging the fins is to a good approximation uniform and the cylinders act almost as isolated.

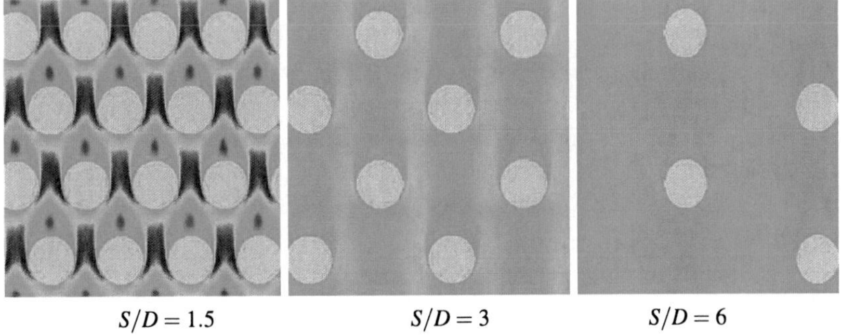

$S/D = 1.5$ $S/D = 3$ $S/D = 6$

Fig. 2 Time averaged color contours of streamwise velocity at $0.1D$ from the hot wall, for $X/D = 1.5$: red high; blue, low.

Figure 2 shows the time averaged streamwise velocity contours varying the spanwise distance between the cylinders (S/D). Since the flow rate is constant, by decreasing the distance between the pin fins, the velocity of the fluid increases. For large values of S/D the flow velocity is quite uniform, and streaks of low velocity are observed in proximity of the cylinders.

The heat flux ($q = 1/(RePr)d\langle T\rangle/dy$ at the wall) dependence on the spanwise spacing S/D and streamwise spacing X/D of the pins is shown in figure 3. The heat flux is normalized with the heat flux of a smooth channel with the same flow rate and Reynolds number (q_0). Pin fins arrays increase the heat flux with respect to a smooth

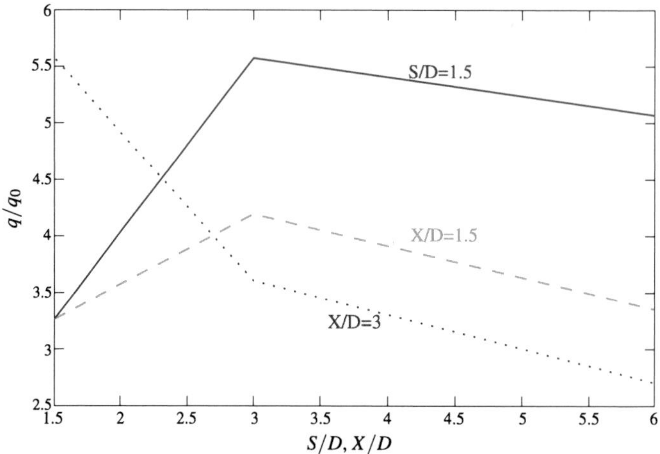

Fig. 3 Heat Flux (q) as a function of the spanwise spacing S/D and streamwise spacing X/D with respect to the heat flux over in a smooth channel (q_0): solid: $S/D = 1.5$, dashed: $X/D = 1.5$, dotted: $X/D = 3$.

channel. When the spanwise spacing is fixed to $S/D = 1.5$, the heat flux increases from $X/D = 1.5$ to $X/D = 3$ where it is maximum. For a larger streamwise spacing the heat flux decreases. In fact, in figures 1 and 2 we have observed that when the pin density is too large, the wake is reduced as well as the mixing promoted by the recirculation regions around and behind the cylinder. For a very large spacing, on the other hand, the flow impinging on the pins is almost uniform. Therefore, the case with $S/D = 1.5$ and $X/D = 3$ represents an optimum configuration for the heat transfer enhancement. A similar trend can be observed keeping the streamwise spacing constant $X/D = 1.5$. By increasing the spanwise spacing from 1.5 to 3, the heat flux increases. For a further increase of S/D the heat flux decreases. On the other hand, for $X/D = 3$ the heat flux decreases monotonically with S/D. In conclusion, the heat flux is maximized for $S/D = 1.5$ and $X/D = 3$ and $S/D = 3$ and $X/D = 1.5$ where q/q_0 is about 5.5 (the heat flux is 5.5 times larger than that in a smooth channel at the same flow rate and Reynolds number).

The pin fins arrays induce an heat transfer enhancement but at the same time an increase in the pressure drop. Over a streamwise wavelength S, the relative contributions of the frictional and form drag are $\overline{C_f} = \lambda^{-1} \int_0^{\lambda} \langle C_f \rangle \tau \cdot \mathbf{x} ds$ and $\overline{P_d} = \lambda^{-1} \int_0^{\lambda} \langle P \rangle \mathbf{n} \cdot \mathbf{x} ds$ respectively, where $\langle C_f \rangle = (\mu \partial \langle U \rangle / \partial y)$ is the the time average non-dimensional viscous shear stress at the wall, $\overline{P_d}$ is the form drag obtained by projecting $\langle P \rangle$ onto the x direction ($\mathbf{n}$ is the normal to the surface). The frictional and form drag decreases by decreasing S/D and X/D (figure 4). In fact, being the flow rate constant, the smaller the distance between the cylinders, the larger is the velocity. As a consequence, both the friction at the wall and the stagnation pressure on the cylinders increase. The friction and the form drag are more sensitive to S/D

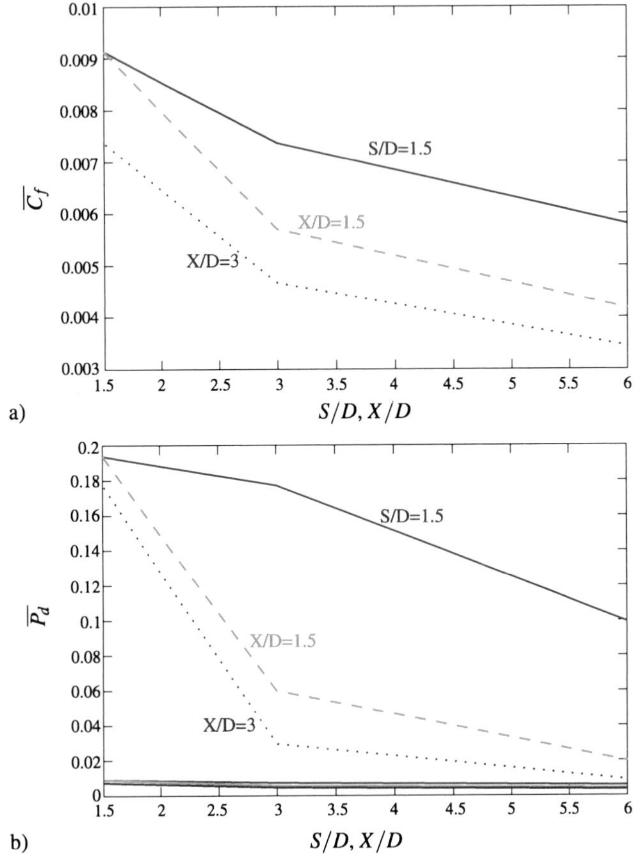

Fig. 4 Frictional (C_f) and Form Drag (P_d) for different values of S/D and X/D: legend as in Figure 3. In figure b) the curves of the frictional drag are included for reference.

than to the streamwise spacing. In fact, by doubling X/D from 1.5 to 3, the frictional drag decreases about 20%, while doubling S/D it decreases of about 40%. The form drag is even more sensitive to S/D. For $S/D = 1.5$, it is about 3 time the value for $S/D = 3$. In figure 4b both the frictional and form drag are shown. It can be observed that the frictional drag is a magnitude order smaller than the form drag. With respect to the channel, the total pressure drop in the channel is about 40-50 times larger, therefore the heat transfer enhancement is coupled with a large drag increase. An analysis of the thermodynamic cycle of the engine is necessary to determine whether the improved efficiency due to the heat transfer enhancement can balance the increased pressure drop.

4 Conclusions

Direct Numerical Simulations of the flow in a channel with pin fins array have been performed varying the spanwise and streamwise distance between consecutive rows. Heat transfer enhancement was observed with respect to a smooth (empty) channel. The layouts with $S/D = 1.5$ and $X/D = 3$ and $S/D = 3$ and $X/D = 1.5$ maximize the heat transfer. The pin fin arrays induce a drag increase in the channel, where the form (pressure) drag contribution is the largest. A functional analysis including the propulsion efficiency of the entire engine is necessary to understand whether the improved thermodynamic efficiency due to the increased heat transfer is larger than the increased pressure drop.

Acknowledgements This research was supported in part by the National Science Foundation through TeraGrid resources provided by TACC.

References

1. Ames, F. E., Dvorak, L. A., and Morrow, M. J. Turbulent augmentation of internal convection over pins in staggered pin fin arrays. *ASME Turbo Expo GT-2004-53889*, 2004.
2. Chyu, M. K., Hsing, Y. C., Shih, T. I.-P., and Nataran, V. Heat transfer contributions of pins and endwall in pin-fin arrays: Effects of thermal boundary condition modeling. *Journal of Turbomachinery*, 121:257–263, April 1999.
3. Cruz Perez, B., Toro Medina, J., Sundaram, N., Thole, K., and Leonardi, S. Direct numerical simulation and experimental results of a turbulent channel flow with pin fins array. *Direct Large Eddy Simulations 7*, 2008.
4. Damerow, W. P., Murtaugh, J. C., and Burgraf, F. Experimental and analytical investigation of the coolant flow characteristics in cooled turbine airfoils. *NASA CR-120883*, 1972.
5. Metzger, D. E., Fan, C. S., and Haley, S. W. Effects of pin shape and array orientation on heat transfer and pressure loss in pin fin arrays. *Journal of Engineering for Gas Turbines and Power*, 106:252–257, 1984.
6. Metzger, D. E., Fan, Z. X., and Shepard, W. B. Pressure loss and heat transfer through multiple rows of short pin fins. *Heat Transfer*, 3:137–142, 1999.
7. Orlandi, P. (2000) Fluid flow phenomena, a numerical toolkit. *Kluwer Academic Publishers*.
8. Orlandi, P. and Leonardi, S. DNS of turbulent channel flows with two- and three-dimensional roughness. *Journal of Turbulence*, 7, 2006.
9. Sparrow, E. M. and Ramsey, J. W. Heat transfer and pressure drop for a staggered wall-attached array of cylinders with tip clearance. *International Journal of Heat and Mass Transfer*, 21:1369–1377, 1978.

Drag Reduction on External Surfaces Induced by Wall Waves

H.C. de Lange and Luca Brandt

1 Introduction

Drag-reduction can be achieved by delaying of the onset of a turbulent flow as well as quenching turbulence itself. Due to the highly local nature of turbulent events and the rapid nature of breakdown a sensor-less (open-loop) strategy is highly preferable, since it prevents the necessity of large numbers of fast sensor/actuator combinations. Thus far, the success of the control strategies for boundary layer flows is limited and for bypass-transition none of the strategies has been successful. However, recent investigations indicate that sensorless (open-loop) control of transition to turbulence and drag reduction in turbulent flows is a feasible option.

In the publication by Min et al. [1] blowing/suction at the wall in the form of upstream traveling waves (UPTW) are applied in a turbulent channel flow. Their two- and three-dimensional numerical simulations show that upstream traveling waves in turbulent channel flow reduce the average friction coefficient to a (sub-)laminar level. Recent studies have explained this by the extra pumping provided by the wall-actuation [2], leading to negative power savings for the proposed strategy. Bewley [3] has theoretically shown that for any boundary control, the power exerted at the walls is always larger than the power saved by reducing to sub-laminar drag. The net power gain is, therefore, always negative if the uncontrolled flow is laminar. However, a positive gain can be achieved when the uncontrolled flow becomes turbulent but the controlled flow remains laminar. The conclusion is that the optimal control solution is to relaminarize the flow and transition control a viable approach.

H. C. de Lange
Department of Mechanical Engineering, Eindhoven University of Technology, The Netherlands, e-mail: h.c.d.lange@tue.nl

Luca Brandt
Linné Flow Centre, KTH Mechanics, SE-100 44 Stockholm, Sweden, e-mail: luca@mech.kth.se

H. Kuerten et al. (eds.), *Direct and Large-Eddy Simulation VIII*,
ERCOFTAC Series 15, DOI 10.1007/978-94-007-2482-2_69,

Inspired by the investigations of Min et al. [1], the influence on boundary-layer transition of traveling waves induced at the wall by blowing/suction over a finite length region is studied numerically. To this aim, control in the form of downstream traveling waves (DTW) is more promising as we will show here and as previously suggested by Lee [4]. These authors examine the linear stability of channel flow modulated by UPTW and DTW and show that DTW can have stabilizing effect on the flow, while UPTW are destabilizing.

A logical extension of the UPTW and DTW is the application of spanwise traveling, blowing-and-suction waves (SPTW). These waves sustain streaky structures which are not optimal for transition or to sustain turbulence and could thus reduce the drag. Du and Karniadakis [5] showed drag reduction for SPTW in turbulent channel flow using volume forcing. More recently, Quadrio and coworkers [6] examined the drag reduction in turbulent channel flow by wall actuation in the form of streamwise traveling waves of spanwise velocity perturbations. In these investigations a more feasible actuation is considered. However, to this point in time the effectiveness of this input has never been tested for boundary layer flows. The localization of the actuation in spatial flow is a new feature worth investigation here, therefore, the control acts only on a limited portion of the domain.

In this paper, a detailed numerical study of the mechanisms of drag reduction and transition delay in a boundary layer will be performed using three different forms of wall blowing/suction: upstream, spanwise and downstream traveling waves. The results will be used to quantify potential benefits of these sensor-less approaches when varying the different control parameter. The flow configuration investigated is a boundary layer subjected to a 4.7% main-stream turbulence level. This means that the transition process in the boundary layer will be bypass-transition. This is characterized by the appearance and growth of elongated streamwise velocity streaks followed by quick localized breakdown into turbulent patches induced by high-frequency secondary instability [7]. Thus, the present configuration can also be seen as a first step towards control of a fully turbulent flow.

2 Numerical method and flow parameter

The simulations are performed using a pseudo-spectral solver for the three-dimensional, time-dependent, incompressible Navier-Stokes equations [8]. The simulations are LES using the approximate deconvolution model relaxation term (ADM-RT) method [9]. This method has been proven robust and accurate for bypass-transition in boundary layers. The inflow is at $Re^*_\delta = 300$ while the computational domain has dimension $1000 \times 60 \times 50$ in units of inflow displacement thickness in the streamwise, wall-normal and spanwise directions respectively. Most of the simulations are performed using $256 \times 121 \times 36$ grid points. For all different configurations of the control-parameters separate runs have been performed. The large computational effort was mainly due to the number of runs necessary to find/determine the optimal control parameters. These are frequency and wavelength of the traveling wave, extension of the streamwise region where the control is active, amplitude of blowing and suction. The wall-normal velocity at the boundary is defined as

$$v(y=0,t) = Af(x)\cos(\alpha x + \beta z - ct)$$

where f is a smooth function rising from 0 to 1 in the control region, $\beta = \frac{2\pi}{\lambda_z} = 0$ for UPTW and DTW, $\alpha = \frac{2\pi}{\lambda_x} = 0$ for SPTW.

3 Results

First, the effect of UPTW has been studied. We have tried to identify possible successful configurations from a series of about 20 simulations. Although increasing the wave speed while decreasing the wave number and amplitude gives more stable solutions, transition is never delayed. In all simulated cases the introduction of UPTW proofs to enhance the transition to turbulence and, thus, increase the drag. Therefore, it has been concluded that UPTW are not suitable for boundary layer transition delay as suggested by previous linear analysis.

Similarly, simulations with SPTW showed that (independent of the spanwise wave-number, amplitude and phase-speed), the SPTW tend to promote transition. Only for wave-numbers close to those of the streaks naturally induced by the external turbulence, little transition delay has been observed. The optimal spanwise wavenumber has been found to be $\beta = 0.50$ (i.e. $\lambda_z = 12.5$). The delay is weakly dependent on the wave speed c for values between 1 and 3 and increases with the amplitude A for values lower than 0.15. This result seems to contradict the conclusions of Du and Karniadakis [5], a discrepancy which may be due to the difference in geometry (boundary layer vs. channel flow).

Simulations performed using DTW show that DTW are the most promising candidate for drag-reduction through transition delay. For high-enough wave speeds ($c > 2$) all waves reduce the drag by reducing the growth of streaks inside the boundary layer. For some wave-speeds ($c > 4$) and amplitudes ($A > 0.15$) we find that the streaks are damped to very low levels and completely disappear. This implies that, by correctly choosing the wave-number, amplitude and speed of the controlling waves, one can no longer observe the transition and turbulence within our computational domain. Our simulations indicate that the transition delay increases when increasing the streamwise length of the control region, decreasing to $\lambda_x = 50$ and $\lambda_x = 25$ and increasing the wave-speed to $c \in [4,5]$. The delay also increases by increasing the amplitude of blowing/suction, which was tested to $A = 0.25$. In this case, however, no net-power saving could be obtained owing to the cost of strong actuation.

An instantaneous flow visualization with this type of wall actuation is displayed in figure 1 and compared with the uncontrolled case. The uncontrolled bypass transition is displayed in the top panel: streaks are generated by free-stream vortices and

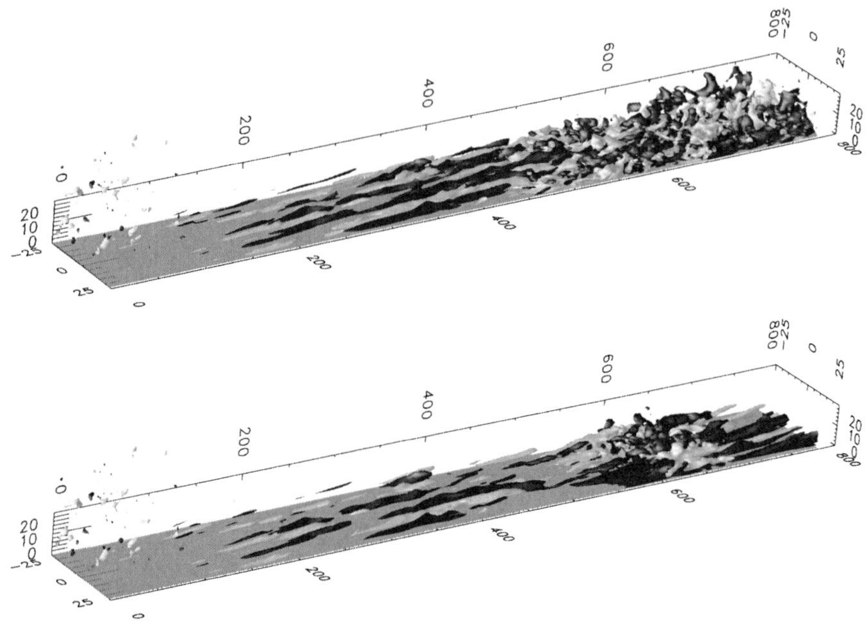

Fig. 1 Transition delay through the implementation of DTW given by contours the local stream-wise velocity minus the spanwise averaged stream-wise velocity (white is positive, black is negative). The top figure shows the uncontrolled case. Here, the bypass-transition regime leads to streak-generation which, subsequently breaks down quickly into turbulence (at about $x = 400$). In the lower figure a DTW-controlled case is shown for $A = 0.15$, $\lambda_x = 50$ and $c = 5$. Again, streaks are generated near the leading edge (before the control region), however, as soon as the streaks hit the control region, $x = 100$, the streaks start to decay and after some time they fully disappear. Only after the control region, $x > 500$, streak growth is recaptured and close to the end of the simulation domain (after $x = 700$) transitional flow is observed.

subsequently break down quickly into turbulence (at about $x = 400$). With blowing/suction in DTW, the streaks are stretched and decay, as soon as the streaks hit the control region at $x = 100$. Only downstream of the control region, $x > 500$, streak growth is again seen and (in the presented example) close to the end of the simulation domain (after $x = 700$) transitional flow is again observed. It must be noted that this depends on the depicted instant (transition is random in both time and place), as well as on the chosen control parameters.

The downstream evolution of the streamwise velocity fluctuations:

$$u_{rms}(x) = \sqrt{\frac{1}{\tau L_z}\int_\tau \int_{L_z} \left(u(x,2,z,t) - \frac{1}{L_z}\int_{L_z} u(x,2,z,t)dz \right)^2 dzdt}$$

is displayed in figure 2. UPTW waves enhance the amplification of the streaks leading to a steeper increase of u_{rms}, while the perturbations are not growing in the region when DTW are introduced.

The drag coefficient (C_f) (also shown in figure 2) is one of the primary properties to check when studying control of wall-bounded flows. The shape of the drag-profile clearly shows the decrease in drag due to the DTW. Behind the control region, C_f has a significantly lower value than the uncontrolled flow. For the UPTW-case, C_f sharply increases at the end of the control region, showing that the flow already becomes turbulent. The flow is fully turbulent at $x = 600$ for the controlled case.

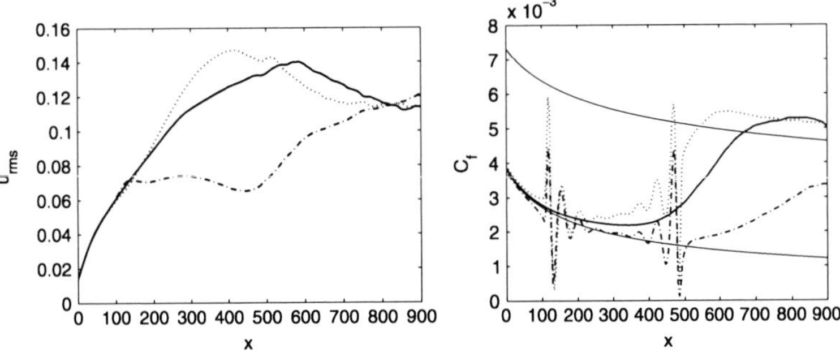

Fig. 2 Streamwise velocity fluctuations (u_{rms}) and drag coefficient (C_f) for the uncontrolled (—), DTW (–·) and UPTW (···) flows.

4 Conclusion

We performed numerical simulations of boundary layer transition in the presence of control at the wall in the form of traveling waves and found that for UPTW and SPTW there is no transition delay. In only a few cases of SPTW, one may observe a marginal drag reduction. On the other hand, DTW show substantial transition delay and drag reduction, leading to positive net power savings. The DTW are therefore a good candidate for sensor-less drag-reduction. The physical mechanisms for the observed transition delay are examined by looking at the flow statistics and production of turbulent kinetic energy. We observe that DTW induce a higher value of the shape factor in the laminar region, which has a stabilizing effect on the flow, whereas the opposite is true for UPTW. With DTW, the Reynolds stress becomes positive close to the wall in the upstream part of the control region and is significantly lower than in the uncontrolled case throughout the boundary layer. As a consequence, turbulence production is negative or reduced by about a factor 4 at downstream locations. The friction drag is reduced by the presence of DTW up to 55% at $x = 650$ and by 35% at $x = 900$, 400 units downstream of the control region. The efficiency of the process is important to verify whether energy is saved. Efficiency is computed as: required power for the uncontrolled and the controlled flow

$$\eta = \frac{P_{con} + P_{inp}}{P_{uncon}}$$

where P_{con} and P_{uncon} required power for the uncontrolled and the controlled flow, respectively and P_{inp} is the input power necessary for the actuation. The minimum value for η we obtained was 0.76 for $A = 0.15$, $\lambda_x = 50$, $c = 5$ and control active for $x \in [100, 500]$.

Finally, it is interesting to note that the present results can explain previous findings on drag reduction in fish swimming. Taneda [10] worked on fish like swimming motion and concluded that the appropriate phase speed of traveling waves may result in thrust amplification and significant drag reduction also owing to positive effects on flow separation. The combination of DTW and adverse pressure gradient may therefore deserve further investigations.

Acknowledgements We thank the DEISA Consortium (www.deisa.eu), co-funded through the EU FP6 project RI-031513 and the FP7 project RI-222919, for support within the DEISA Extreme Computing Initiative.

References

1. Min, T., Kang, S.M., Speyer, J.L. & Kim, J. (2006). Sustained sub-laminar drag in a fully developed channel flow. *J. Fluid Mech.* **558**, 309–318.
2. Hoepffner, J., & Fukagata, K. (2009). Pumping or drag reduction? *J. Fluid Mech.* **635**, 171–187.
3. Bewley, T.R. (2009). A fundamental limit on the balance of power in a transpiration-controlled channel flow *J. Fluid Mech.* **632**, 442–446.
4. Lee, C., Min, T. & Kim, J. (2008). Stability of channel flow subject to wall blowing and suction in the form of a traveling wave. *Phys. Fluids* **20**, 101513.
5. Du, Y. & Karniadakis, G.E. (2000). Suppressing wall turbulence by means of transverse traveling wave. *Science*, **288**.
6. Quadrio, M., Ricco, P. & Viotti, C. (2009). Streamwise-traveling waves of spanwise wall velocity in a turbulent channel flow. *J. Fluid Mech.* **627**, 161–178.
7. Schlatter, P., Brandt, L., Lange, H.C. de & Henningson, D.S. (2008). On streak breakdown in bypass transition. *Phys. Fluids* **20**, 101505.
8. Chevalier, M., Schlatter, P., Lundbladh, A. & Henningson, D.S. (2007). A Pseudo-Spectral Solver for Incompressible Boundary Layer Flows. *Tech. Rep. TRITA-MEK 2007:07. Royal Institute of Technology (KTH), Dept. of Mechanics, Stockholm.*
9. Schlatter, P., Stolz, S. & Kleiser, L. (2004). LES of transitional flows using the approximate deconvolution model. *Int. J. Heat Fluid Flow*, **25**, 549–558.
10. Taneda, S. (1977). Visual Study of unsteady separated flows around bodies, *Progr. Aerosp. Sci.* **17**, 287–348.

Impact of Secondary Vortices on Separation Dynamics in 3D Asymmetric Diffusers

Hayder Schneider, Dominic von Terzi, Hans-Jörg Bauer and Wolfgang Rodi

1 Introduction

The flow in two three-dimensional (3D) asymmetric diffusers with the same expansion but different aspect ratios was recently measured [1]. The results revealed complex 3D separation patterns with a severe sensitivity to the geometric variation. The setup served as a test case for two ERCOFTAC workshops [2] that aimed at assessing the predictive capabilities of various turbulence modeling approaches. Reynolds-Averaged Navier–Stokes (RANS) models based on the eddy-viscosity assumption yielded qualitatively wrong results. These models cannot reproduce secondary vortices (SV) in the inlet duct. Methods that account for SV or even resolve these structures fared better. In particular Large-Eddy Simulation (LES) was able to compute the flow in both diffuser geometries within measurement uncertainty [3]. The hypothesis that SV have a strong impact on the separation dynamics was further corroborated by recent experiments [4]. At the inlet of one of the diffusers, localized (steady and unsteady) perturbations were introduced. The authors conjectured that the forcing generated streamwise vortices and that these SV were responsible for the observed change in pressure recovery by up to 14%. In the present paper, the hypothesis is tested by controlled numerical experiments using LES and manipulation of (mean) SV in the inlet duct for both diffuser geometries.

2 Methodology

Figure 1 illustrates the part of the experimental setup of [1] simulated here. The computational domain consists of an inlet duct, the diffusers (D1 or D2) and an

Institut für Thermische Strömungsmaschinen, Karlsruher Institut für Technologie, Kaiserstr. 12, D-76131 Karlsruhe, e-mail: hayder.schneider@kit.edu, vonterzi@kit.edu, hans-joerg.bauer@kit.edu · Institut für Hydromechanik, Karlsruher Institut für Technologie, Kaiserstr. 12, D-76131 Karlsruhe, e-mail: rodi@kit.edu

H. Kuerten et al. (eds.), *Direct and Large-Eddy Simulation VIII*,
ERCOFTAC Series 15, DOI 10.1007/978-94-007-2482-2_70,

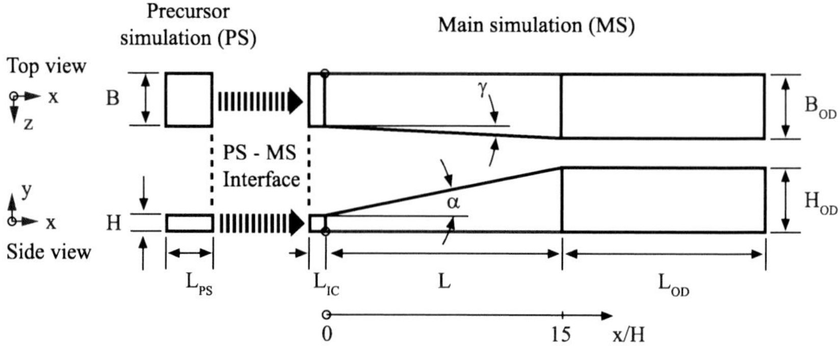

Fig. 1 Diffuser design with flow in positive x-direction (from left to right); origin of coordinate system indicated by circle.

	Dimension	D1	D2
Inlet channel	length (L_{PS})	3	
	length (L_{IC})	1	
	height (H)	1	
	width (B)	3.33	
Diffuser	length (L)	15	
	top angle of expansion (α)	11.3°	9°
	side angle of expansion (γ)	2.56°	4°
Outlet duct	length (L_{OD})	13	
	height (H_{OD})	4	3.37
	width (B_{OD})	4	4.51

Table 1 Dimensions of diffusers D1 and D2 (lengths are made dimensionless with H).

outlet duct, see table 1. All values reported here are made dimensionless using the bulk velocity U_b and the height H of the inlet duct. The origin of the Cartesian coordinate system was placed at the intersection of the two non-expanding walls and the beginning of the expansion. The flow in the experiments and the simulations is incompressible. The Reynolds number based on U_b and H is 10,000, and the Reynolds number based on the friction velocity u_τ in the inlet channel and H is 588.

All simulations were performed with the Finite Volume flow solver LESOCC2. The program solves the incompressible, filtered Navier–Stokes equations, see [5] for more details. The standard Smagorinsky model with $C_s = 0.065$ and van Driest wall-damping was employed as subgrid-scale model. Overall, the contribution of the subgrid-scale model was relatively small. The time-averaged fraction of turbulent to laminar viscosity ν_t/ν within the diffuser was of the order of $\mathcal{O}(10^{-2})$.

Instantaneous turbulent inflow data were generated by a precursor simulation (PS) running in parallel to the main simulation (MS). The data is generated by enforcing the experimental mass flux and periodicity within the PS–section. At every time-step, the data is fed into the MS through the PS–MS interface. At the outlet, a convective boundary condition (BC) was applied together with a buffer zone. No-slip BC were imposed at walls.

All simulations were carried out on the same computational grid covering the solution domain $x/H \in [-4;28]$ and containing 22×10^6 grid cells ($N_x \times N_y \times N_z = 896 \times 128 \times 192$). The grid was equidistantly spaced in the streamwise direction and refined towards the walls, such that $y^+_{\text{wall}} \approx z^+_{\text{wall}} \approx 1$. The time step was $4 \times 10^{-3} H/U_b$. In total, 600,000 time-steps were computed. Averaging started after $10L/U_b$, with a total averaging time T_{ave} of $150L/U_b$.

Simulation	Modification mode	Φ	Energy change of structures
BSL	baseline case (original state)	1	0
SV0	steady, weak	0	$\mathcal{O}(10^{-4}U_b^2)$
SVP2	steady, weak	2	$\mathcal{O}(10^{-4}U_b^2)$
SVP10	steady, strong	10	$\mathcal{O}(10^{-2}U_b^2)$
SVN10	steady, strong, opposite sense of rotation	-10	$\mathcal{O}(10^{-2}U_b^2)$
SVU10	unsteady, alternating sense of rotation	$10 \cdot \sin(\Omega \cdot t)$	$\mathcal{O}(10^{-4}U_b^2)$

Table 2 Overview of the simulation setup (for D1 and D2): In the case name BSL, SV, P, N and U denote baseline, secondary vortices, positive Φ, negative Φ and unsteady Φ, respectively; the number refers to the magnitude of Φ.

The manipulation of the SV occurs at the PS–MS interface. Decomposing the instantaneous velocity field $\mathbf{u}$ into the sum of its mean $\mathbf{U}$ and the fluctuating part $\mathbf{u}'$ yields $\mathbf{u} = \mathbf{U} + \mathbf{u}'$. Rewriting the mean yields $\mathbf{U} = (U,0,0)^T + (0,V,W)^T$, where the first term represents the streamwise component and the second term represents both the vertical and lateral components. U is important for determining both the Reynolds number and the mass flux, whereas V and W represent the (mean) SV in a cross-section. Hence, the procedure for the MS inlet BC with artificially manipulated SV reads:

$$\mathbf{u}|_{\text{MS}} = (U,0,0)^T|_{\text{PS}} + \Phi \cdot (0,V,W)^T|_{\text{PS}} + (u',v',w')^T|_{\text{PS}} \quad , \tag{1}$$

where Φ is an adjustable parameter. Note that by prescribing a BC that satisfies Eq. 1 it is possible to conduct all simulations at the same Reynolds number, mass flux, and with realistic turbulent fluctuations. Hence, the impact of SV can be separated from all other parameters. In total twelve simulations were conducted for D1 and D2. Table 2 summarizes the information. The frequency for the unsteady forcing was arbitrarily chosen as $\Omega = 0.471$. The Strouhal number based on inlet bulk velocity and duct height was hence $St = 0.075$.

3 Results

Figure 2 illustrates the impact of SV on flow separation for both diffusers. The SV are visualized by contours of mean streamwise vorticity ω_x in a cross-section at the diffuser inlet ($x/H = 0$). Their original state is shown in case BSL. According to

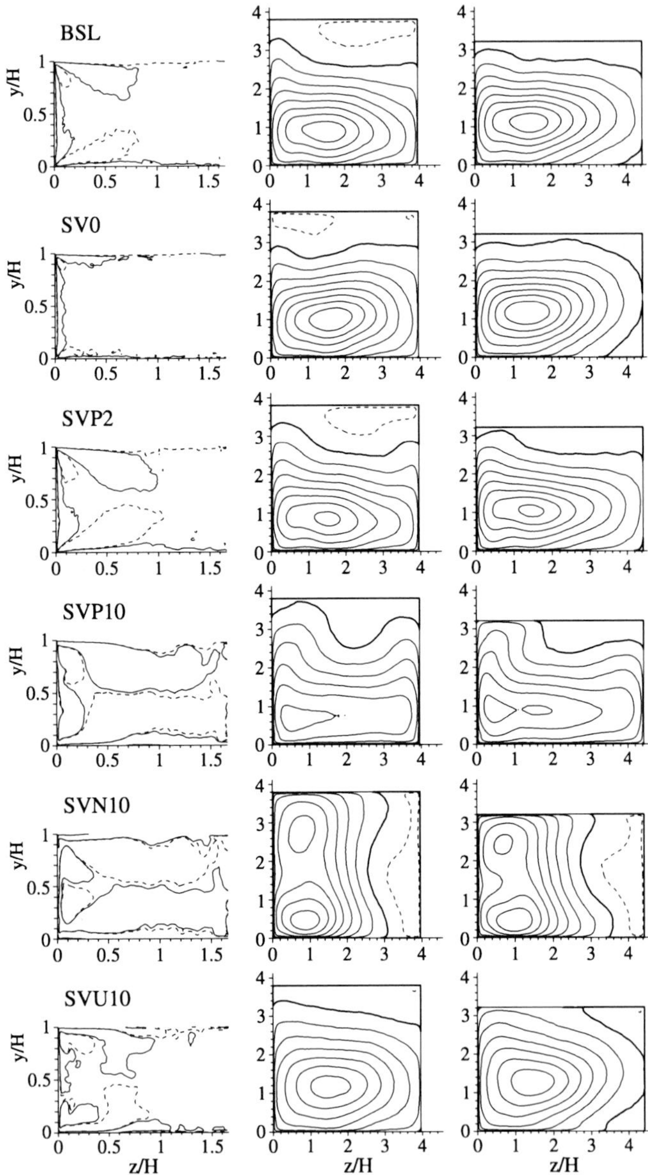

Fig. 2 Results for cases BSL, SV0, SVP2, SVP10, SVN10, and SVU10 (from top to bottom); left column shows mean streamwise vorticity contours with $\omega_x \pm 0.1$ in the inlet cross-section at $x/H = 0$; center and right columns show mean streamwise velocity contours U/U_b in a cross-section at $x/H = 14$ for diffusers D1 and D2, respectively; same U/U_b contours with interval 0.1 shown; thicker lines indicate zero-velocity and dashed lines negative levels.

the manipulation of the BC at the inlet of the computational domain ($x/H = -1$), the SV become weaker/smaller (SV0), stronger/larger (SVP2, SVP10), rotate in the opposite direction (SVN10), or are shaped differently (SVU10).

The qualitative consequences of the different SV on the separation dynamics are visualized by mean streamwise velocity contours U/U_b in a cross-section at $x/H = 14$, *i.e.* close to the diffuser exit. The region of separated flow is enclosed by the bold line. By comparing the BSL cases of D1 and D2, it can be discerned that the separation bubble develops differently according to the change in geometry. The separation in D1 is located only at the top wall, whereas in D2 the separation also extends to the right wall. This was already shown in [1, 3]. Attention is now turned to the flow field of the cases with manipulated SV. Clearly, manipulating the SV at the inlet has a strong impact on the separation bubble farther downstream! The state of the SV influences the shape, extent and also the location of both the separation bubble and the location of maximum forward velocity for both diffusers.

For SV0 and SVP2 the change in kinetic energy of the SV is of the order of $\mathcal{O}(10^{-4}U_b^2)$. Although the input of kinetic energy is relatively low, it is sufficient to cause a change in shape of the separation bubble for D1 and D2. For SV0 it seems that the contour levels are slightly rotated in counter-clockwise direction. The separation bubble is nearly horizontally aligned in D1, and extends further to the right wall in D2. For SVP2, the features of the separation bubble are more pronounced for both diffusers. The separation bubble is characterized by stronger curvature in the region of the top corners, and a higher extent towards the diffuser centerline.

By increasing the kinetic energy of the SV to the order of $\mathcal{O}(10^{-2}U_b^2)$, more drastic effects on the flow field become noticeable. Case SVP10 shows a strong deformation of the separation bubble with relatively little backward flow at the top corners, and a formation of a pronounced bump in the middle of the top wall. In SVN10 the sense of rotation of the SV is switched. This causes the separation bubble to be shifted to the right wall. Furthermore, two maxima of forward velocity are visible close to the left wall. The flow fields of D1 and D2 bear a striking resemblance. This is similar to case SVP10. Hence, it seems that stronger SV can suppress the sensitivity of the flow to the details of the diffuser geometry. For completeness, time-harmonic manipulation of the (mean!) SV (case SVU10) are shown as well.

With the manipulation, it was possible to either significantly increase or decrease the pressure recovery. Note that the manipulations performed here required only a fraction of the kinetic energy of the mean flow in the streamwise direction, but were applied in the whole cross-section. An increase was observed for enhancing the SV in their natural sense of rotation. For both diffusers an increase of the pressure recovery of up to 20% was achieved in the SVP10 case. In contrast, enhancing the SV against their natural sense of rotation yielded a decrease for the SVN10 case (20% for D2). On the other hand, prescribing a time-harmonic manipulation of the SV yielded again an increase in pressure recovery: up to 17% for D1 (all values are taken at the location $x/H = 15$ and compared to the BSL case). The consequences on the pressure recovery suggest that manipulating SV is an efficient means for flow control.

4 Conclusions

Controlled numerical experiments using well-resolved LES were conducted to investigate the impact of mean secondary vortices on the separated flow in 3D asymmetric diffusers. A methodology was presented that allowed for manipulating the SV at the diffuser inlet, while providing realistic time-dependant turbulent fluctuations. The SV were systematically manipulated to assess the sensitivity of the flow to these structures. Our results demonstrate that the SV impact strongly on the complex separation pattern in the diffuser. By removing, enhancing, switching the sense of rotation, or by combining these effects in time-harmonic forcing, it was possible to drastically alter the shape, location and extent of the separation zone farther downstream and to either increase or decrease the pressure recovery.

The good performance of the steady SV-manipulation implies that passive flow control devices may be successfully used in future diffuser designs. The results also provide an explanation why eddy-viscosity based RANS models, that inherently cannot account for mean SV, fail in predicting the location of the separated flow in such diffusers. Furthermore, the profound effect of the SV can explain why experimentalists and engineers in industry have repeatedly reported a sensitivity of the separated flow in diffusers. As demonstrated with case SVN10, for example, where the separation bubble relocated from the top to the side wall, our results show that a relatively small amount of energy input in the mean SV can have an effect of the order of unity in the diffuser.

Acknowledgements The work reported here was carried out within the "Research Group Turbo-DNS" at the Institut für Thermische Strömungsmaschinen. Its financial support by means of the German Excellence Initiative and Rolls–Royce Deutschland is gratefully acknowledged. The authors are grateful for the provision of computer time by the Steinbuch Centre for Computing.

References

1. Cherry EM, Elkins C, Eaton JK (2008) Geometric sensitivity of three-dimensional separated flows. Int J Heat Fluid Flow, 29(3):803–811
2. Jakirlić S, von Terzi D, Breuer M (2010) Lessons learned from the ERCOFTAC SIG 15 computational workshops on refined turbulence modelling: flow in a 3D diffuser as an example. In Proc. 5th ECCOMAS CFD 2010 (also to appear in ERCOFTAC Bulletin)
3. Schneider H, von Terzi DA, Bauer H-J, Rodi W (2010) Reliable and accurate prediction of three-dimensional separation in asymmetric diffusers using large-eddy simulation. J Fluids Eng, 132(3), doi: 10.1115/1.4001009
4. Grundmann S, Sayles EL, Eaton JK (2010) Sensitivity of an asymmetric 3D diffuser to plasma-actuator induced inlet condition perturbations. Exp Fluids, doi: 10.1007/s00348-010-0922-0
5. Breuer M, Rodi W (1996) Large eddy simulation of complex turbulent flows of practical interest. In Notes on Numerical Fluid Mech., Flow Simulations with High Performance Computers II. Ed.: Hirschel, EH Vieweg, Braunschweig, pp. 258–274

Fluid–Structure Interaction of a Flexible Structure in a Turbulent Flow using LES

M. Breuer, G. De Nayer, M. Münsch

1 Introduction

A structure placed in a fluid flow is always affected by the pressure and shear forces acting on the surface leading to structural deformations or deflections. Partially these can be neglected and such a rigid body assumption strongly reduces the complexity of a numerical simulation setup. However, in many circumstances this assumption does not hold and fluid–structure interaction (FSI) becomes of major interest. Technical applications are numerous such as artificial heart valves, lightweight roofage or tents. Therefore, a need for appropriate numerical simulation tools exists for such coupled problems and this is the objective of the present study. The developments are guided by the following ideas:

- Since it is well known that the RANS approach is not an appropriate choice for the prediction of instantaneous flow processes involving large–scale flow structures (separation, reattachment, vortex shedding) very often encountered in FSI problems, large–eddy simulation (LES) is the preferred technique.
- Since LES often requires small time steps, an explicit time–marching scheme relying on a predictor–corrector method is favored. The question of how to appropriately couple this scheme with the computation of the structure dynamics had to be solved.
- Since the usage of highly specialized codes for each subtask (fluid and structure) is assumed to be advantageous, a partitioned approach is chosen where the entire FSI problem is divided into a fluid and a structure domain. The coupling between both domains has to be done via an additional coupling interface. Here aspects

M. Breuer and G. De Nayer
Professur für Strömungsmechanik (PfS), Helmut–Schmidt–Universität Hamburg, Holstenhofweg 85, D–22043 Hamburg, Germany, e-mail: breuer/denayer@hsu-hh.de

M. Münsch
Lehrstuhl für Strömungsmechanik, Universität Erlangen–Nürnberg, Cauerstr. 4, D–91058 Erlangen, Germany, e-mail: mmuensch@lstm.uni-erlangen.de

H. Kuerten et al. (eds.), *Direct and Large-Eddy Simulation VIII*,
ERCOFTAC Series 15, DOI 10.1007/978-94-007-2482-2_71,

such as code-to-code communication or grid-to-grid interpolation, e.g., of loads and displacements, had to be considered.

These issues are discussed in the present contribution and then applied to a turbulent LES–FSI benchmark test case involving the flow around a flexible structure.

2 Computational Methodology

Within a FSI application the fluid forces lead to a displacement or deformation of the structure and thus the computational domain is no longer fixed but changes in time. In order to account for these variations, the most popular numerical technique is the so-called Arbitrary Lagrangian–Eulerian (ALE) formulation. Here the filtered Navier–Stokes equations for an incompressible fluid are re-formulated for a temporally varying domain on a temporally varying grid. Thus the size of the control volumes within the finite–volume scheme applied varies in time. The second–order accurate central discretization is done on a curvilinear, block-structured body–fitted grid with a collocated variable arrangement (code FASTEST–3D). To account for the subgrid scales, the classical Smagorinsky model as well as the dynamic model by Germano et al. are applied.

A challenge was to design a FSI coupling scheme, which on the one hand is appropriate for the explicit time–stepping scheme used and on the other hand avoids instabilities known from loose coupling schemes. In the scheme suggested [3, 2, 4, 7], each time step starts with an estimation of the structural displacement using a linear or quadratic extrapolation of the three preceding time steps. Within the predictor step the momentum equations are solved by an explicit time integration scheme, namely a low–storage Runge–Kutta scheme. Here the ALE formulation is taken into account, which ensures that the *space conservation law* is fulfilled. In the corrector step a Poisson equation for the pressure correction variable is solved, which guarantees a divergence–free velocity field and thus the fulfillment of the mass conservation. Whereas the predictor step is only done once per time step, the corrector step has to be repeated several times (5–10) until a predefined convergence criterion is reached. The coupling and thus the exchange of fluid forces in one direction and the resulting displacements in the other direction is conducted within the corrector step, which is repeated until a dynamic equilibrium between fluid and structure is achieved. Consequently, a strongly coupled but nevertheless still explicit time–stepping algorithm results.

Within the coupling procedure the resulting forces on the interface are computed by the finite–element CSD solver *Carat++* [5] especially developed by TU Munich for the prediction of shells or membranes. *Carat++* is based on several finite–element types and advanced solution strategies for form finding and nonlinear dynamic problems. For the dynamic analysis, different time–integration schemes such as the generalized–α method are used. In the modeling of thin-walled structures, membrane or shell elements are used for the discretization. Both, shell and membrane elements produce reduced structural models with a two–dimensional representation which can describe the three–dimensional physical properties by introducing mechanical assumptions for the thickness direction. In our test case, the plate is modeled with a seven-parameters shell element.

Due to different discretization techniques applied for the subtasks (finite volumes vs. finite elements) also different types of grids and different grid resolutions are used leading to non–matching surface meshes. Consequently, a grid–to–grid data interpolation and transfer becomes necessary. These interpolation steps including the transfer of the fluid loads to the CSD code and the structural displacements back to the CFD code are done via the coupling interface CoMA [6] also developed by TU Munich. A conservative interpolation is used for the transfer of the fluid loads from the cell vertices of the fluid domain to the grid nodes of the structure domain (CFD→CSD), whereas a bilinear interpolation of the displacements from the structure domain to the cell vertices of the fluid domain (CSD→CFD) is applied. The grid adaptation at the interface is based on an underrelaxation of the structural response by taking an underrelaxation factor (e.g. $\alpha = 0.5$) and the displacement of the previous sub-iteration loop into account. Based on the displacement at the boundaries, the grid adjustment is presently done by a transfinite interpolation technique, but more advanced techniques such as elliptic grid smoothing to better maintain the grid quality are under investigation. Subsequent to the grid adaptation solely the corrector step of the predictor–corrector scheme is performed again and a new velocity and pressure field is obtained. Afterwards new loads for the structure solver are generated leading to an update of the corresponding displacements. The dynamic equilibrium between fluid and structure is obtained if changes of the resulting displacements within the sub-iteration cycle reaches a convergence criterion. A convergence criterion based on the loads is also under consideration. The coupling interface is based on the Message–Passing–Interface (MPI) and thus runs in parallel to the fluid and structure solver. Since MPI is also used for the parallelization of the CFD code applied, efficient coupling with respect to high–performance computing is enabled.

A variety of different test cases considering either laminar flows and/or simplified structural models assuming elastically mounted cylinders or plates have been considered to validate and evaluate the present FSI algorithm, see e.g. [3, 2, 4, 8, 7].

3 Benchmark Test Case

The present configuration leans on the benchmark case FSI3 developed within the DFG research group FOR 493 on FSI [9] for laminar flows, but is extended to the turbulent flow regime. In a channel of length $L/D = 25$ and height $H/D = 4.1$ a cylinder of diameter D is mounted. The cylinder position is slightly off–centered, with the cylinder center located at $2D$ downstream of the inflow section and with a distance of $2D$ from the lower lateral channel wall. Based on D and the constant inflow velocity U_∞ the Reynolds number of the incompressible and Newtonian fluid ($\rho^f = 1000$ kg/m^3) is set to $Re = 10^4$. In the wake of the cylinder, a flexible plate is attached to the cylinder. The plate is of length $l = 3.5D$ and has a thickness of $h = 0.2D$. The cylinder is assumed to be fixed and rigid, whereas the material of the plate is of St. Venant–Kirchhoff type [1] characterized by a Poisson ratio of $\nu^s = 0.4$, a Young modulus of $E = 5.6 \cdot 10^6$ kg/(m s^2) and a density of $\rho^s = 1000$ kg/m^3. For the flow prediction a block–structured grid with about 17 million CVs is used. At the lateral channel boundaries a slip boundary condition is applied, whereas on the structure, no–slip boundary conditions are used. In spanwise direction periodic boundary conditions are chosen. At the channel outlet a convective outflow condition is specified. The elastic plate is resolved by the use of 10×10 four-nodes shell elements. All the nodes on the cylinder are fixed and the nodes at the free extremity are totally free. The z–displacement of the nodes on the sides are forced to be zero. Due to periodic boundary conditions set in the fluid solver there are always two nodes of the sides (one in a plane and its twin in the other plane) which have the same load. These two nodes must have the same displacement in x– and y–direction which is enforced in the CSD code.

4 Results and Discussion

For testing the entire coupling algorithm developed, the benchmark case FSI3 [9] mentioned above was considered first. With the exception of a parabolic inflow profile, the application of no–slip boundary conditions at the channel walls and a much lower Re–number of Re = 200, the setup is the same as for the LES case. For these flow predictions two different grids with either only about 27,000 or 90,000 CVs in a 2D plane were used. Exemplarily, Fig. 1(a) depicts a snapshot of the flow field by contours of the streamwise velocity in a x–y slice.

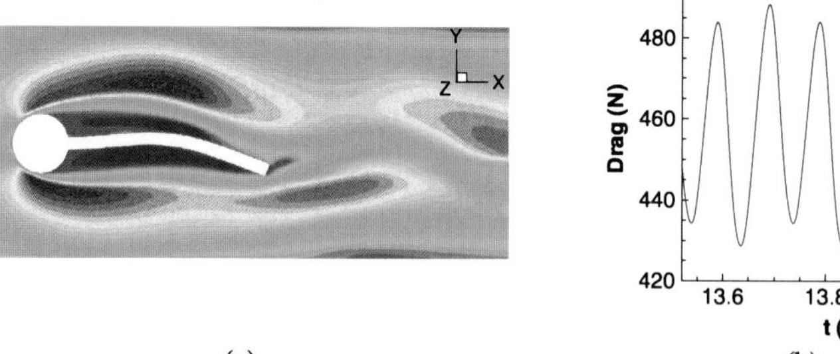

(a) (b)

Fig. 1 Laminar flow around the flexible structure, (a) Snapshot of contours of the streamwise velocity; (b) Time history of the drag on the structure.

Despite the splitter 'plate' behind the cylinder, vortex shedding occurs. The shed vortices travel downstream and start to interact with the flexible structure leading to an oscillating structure. Fig. 1(b) shows the time history of the drag force on the structure. After a transition phase (not shown) the amplitude of the oscillations reaches constant values. The same applies for the lift force and the displacements of point **A** located at the trailing edge on the center line of the flexible structure. The corresponding minimal and maximal values as well as the frequency of the lift force are given in Table 1 for both grids. For comparison the highly resolved finite–element simulation by Turek and Hron [9] is given. Despite some differences the overall agreement is satisfactory. Furthermore, an improvement of the results on the refined grid is visible with respect to the reference data. Partially the grid convergence seems to be small. In order to further investigate this issue, presently simulations on the next finer grid level are carried out. The coupling scheme was found to work very efficiently requiring only a few FSI iterations (5 to 7) to go below a convergence limit ensuring dynamic equilibrium, e.g. 10^{-4} for the L_2 norm of the displacements. Thus, by performing this test case the proper behavior of the whole partitioned FSI setup was proven.

Table 1 Results of the benchmark FSI3 (Re = 200) on two different grid levels.

	Resolution	Δy [$\times 10^{-3}$ m]		**Drag** [N]		**Lift** [N]		**Frequency**
		Min.	Max.	Min.	Max.	Min.	Max.	of Lift [Hz]
Coarse	27,040 CVs	-43.98	46.22	429.0	525.0	-216.0	230.0	4.99
Medium	89,952 CVs	-43.63	46.17	428.0	488.0	-166.0	210.0	5.04
Ref. [9]	highly resolved FE	-33.43	36.37	432.7	488.2	-151.4	156.4	5.46

For the turbulent case, the simulation was started with a rigid structure. The structure behind the cylinder acts like a splitter plate of length l/D = 3.5 attenuating the generation of a vortex street behind the cylinder. Fig. 2(a) depicts the pressure distribution in a x–y plane showing the shear layers with the Kelvin–Helmholtz instability leading to transition and two large vortices originating from the shedding process. The time history of the lift and drag coefficients confirms that the splitter plate does not suppress vortex shedding completely. Compared to a pure cylinder in free flow the Strouhal number decreases to about St = 0.172 which is the result of two opposing effects: splitter plate: St $\downarrow$, blockage: St $\uparrow$.

Then the plate was released and a fully coupled FSI–LES prediction was started. Owing to different loads on both sides the structure directly started to deflect in one direction, see Fig. 3(a). In this snapshot a region of low pressure is already visible at the lower side so that at a later instant the plate bent back to the initial position (Fig. 3(b)) and is then deflected in the other direction (Fig. 3(c)). To evaluate the entire deflection mode of the structure will be the task for the near future. Nevertheless, the study has shown that a reliable and efficient coupling scheme for the marriage of FSI and LES has been established.

Acknowledgements We gratefully acknowledge the cooperation with the Chair of Structural Analysis of TU Munich providing the codes Carat++ [5] and CoMA [6] including intensive support by Dipl.–Ing. Th. Gallinger, Dr.–Ing. R. Wüchner and Prof. Dr.–Ing. K.U. Bletzinger.

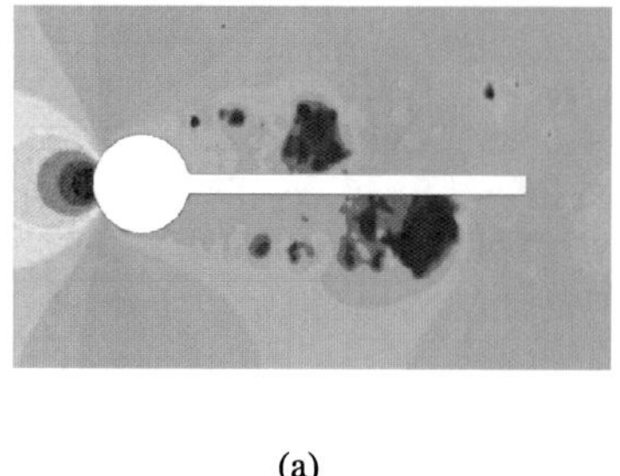

(a)

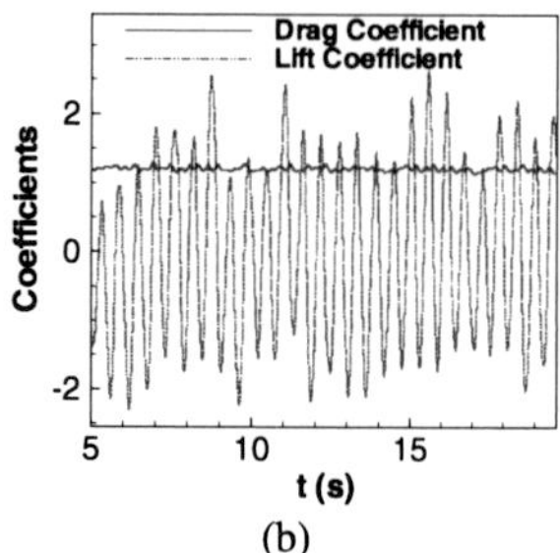

(b)

Fig. 2 Turbulent flow around the rigid structure, (a) Snapshot of contours of the pressure in one x–y plane; (b) Time history of the drag and lift coefficient on the structure.

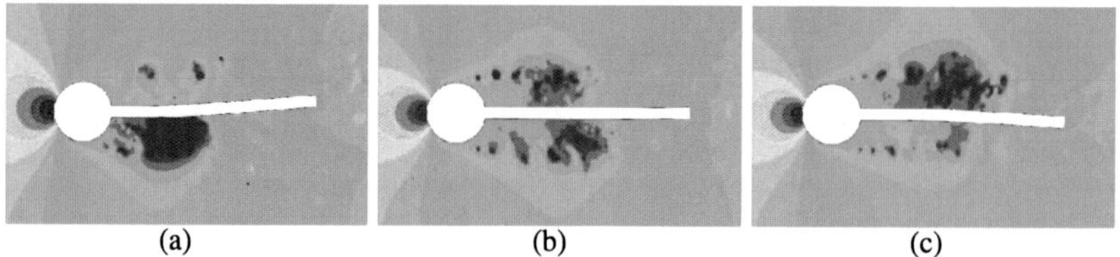

(a) (b) (c)

Fig. 3 Turbulent flow around the flexible structure, three different snapshots in time.

References

1. Belytschko, T., Liu, W.K., Moran, B.: Nonlinear Finite Elements for Continua and Structures, Wiley, New York (2000)
2. Breuer, M., Münsch, M.: LES Meets FSI – Important Numerical and Modeling Aspects, 7th Int. ERCOFTAC Workshop on DNS and LES: DLES-7, Trieste, Italy, Sept. 8–10, 2008, In: Armenio, V. et al. (eds.) DLES VII, ERCOFTAC Series, vol. **13**, pp. 245–251, Springer (2010)
3. Breuer, M., Münsch, M.: Fluid–Structure Interaction Using LES – A Partitioned Coupled Predictor–Corrector Scheme, PAMM, vol. **8**, pp. 10515–10516 (2008)
4. Breuer, M., Münsch, M.: FSI of the Turbulent Flow around a Swiveling Flat Plate Using Large-Eddy Simulation, In: Hartmann, S. et al. (eds.) Int. Workshop on Fluid-Structure Interaction (2008), 31–42, Kassel University Press, ISBN 978-3-89958-666-4, Kassel (2009)
5. Fischer, M., Firl, M., Masching, H., Bletzinger, K.-U.: Optimization of Nonlinear Structures based on Object–Oriented Parallel Programming, ECT2010: Seventh Int. Conf. Engineering Computational Technology, Valencia, Spain (2010)
6. Gallinger, T., Kupzok, A., Israel, U., Bletzinger, K.-U., Wüchner, R.: Computational Environment for Membrane–Wind Interaction. Int. Workshop on Fluid–Structure Interaction: Theory, Numerics and Applications, Herrsching, Germany, Sept. 29 – Oct. 1 (2008)
7. Münsch, M., Breuer, M.: Numerical Simulation of Fluid–Structure Interaction Using Eddy–Resolving Schemes, In: Bungartz, H.-J., Schäfer, M. (eds.) Fluid–Structure Interaction, Lecture Notes Comput. Sci. & Eng., LNCSE, **73**, pp. 221–253, Springer (2010)
8. Pereira Gomes, J., Münsch, M., Breuer, M., Lienhart, H.: Flow–induced Oscillation of a Flat Plate — A Fluid–Structure Interaction Study Using Experiment and LES, 16. DGLR–Fach–Symp. STAB, RWTH Aachen, Germany, Nov. 3–5, 2008, In: Notes on Numerical Fluid Mechanics and Multidisciplinary Design, vol. **112**, pp. 347–354, Springer (2010)
9. Turek, S., Hron, J.: Proposal for Numerical Benchmarking of Fluid–Structure Interaction between an Elastic Object and Laminar Incompressible Flow, In: Bungartz, H.-J., Schäfer, M. (eds.) Fluid–Structure Interaction, LNCSE, vol. **53**, pp. 371–385, Springer, Heidelberg (2006)